Image Recovery: Theory and Application

Image Recovery: Theory and Application

Edited by

Henry Stark

Department of Electrical, Computer, and Systems Engineering
Rensselaer Polytechnic Institute
Troy, New York

1987

ACADEMIC PRESS, INC.
Harcourt Brace Jovanovich, Publishers
Orlando San Diego New York Austin
Boston London Sydney Tokyo Toronto

ACADEMIC PRESS, INC.
Orlando, Florida 32887

United Kingdom Edition published by
ACADEMIC PRESS INC. (LONDON) LTD.
24–28 Oval Road, London NW1 7DX

Library of Congress Cataloging in Publication Data

Image recovery—theory and application.
Includes bibliographical references and index.
1. Image processing. 2. Signal processing.
I. Stark, Henry, Date .
TA1632.I4824 1986 621.36'7 86-10960
ISBN 0–12–663940–X (alk. paper)

PRINTED IN THE UNITED STATES OF AMERICA

87 88 89 90 9 8 7 6 5 4 3 2 1

Contents

Preface

Image recovery and, more generally, signal recovery refer to a class of problems that are among the most fundamental in the realm of science. They involve every known scale—from the cosmic to the atomic; from determining the structure of unresolved stars to determining the arrangement of the tiniest molecules. Stated in its most general form, the signal recovery problem is described like this: Given data g, determine the source f that produced g. Unfortunately, when stated at this level of generality, little more can be said. How is g related to f? Is g adequate for determining f? If not, can g be used to furnish an estimate $\hat{f}$ of f that, in some sense, is close to f? If g is corrupted by noise, does the noise play a big role in obscuring f? If yes, can we ameliorate the effects of the noise? Do small changes in g produce radical changes in $\hat{f}$? Even if g uniquely determines f, is there an algorithm for computing $\hat{f}$ from g? What about the role of *a priori* information? Can it be usefully incorporated in our attempts to recover f from g?

These (and others) are the kinds of questions that signal recovery concerns itself with. It is the purpose of this book to gather in one place many of the important results and applications of this field discovered in the past decade or so. As editor, naturally I hope I have been successful; this is for the reader to decide. But even as these words are being written, new results and novel applications are being discovered that do not or will not appear herein. Knowing this, I have tried to fill the pages of this book with theory that I thought was fundamental, with algorithms and applications that were clever and elegant.

It might be interesting to trace the recent history of signal recovery and conjecture why in the past decade or so it has attracted so much interest. First, superresolution and phase retrieval are problems of more than mere academic interest; indeed they are fundamental in many areas of the natural sciences. All physical instruments limit the amount of data we can collect, especially in the realm of the frequency domain. Often the fine details we seek are tantalizingly close to (but just beyond the reach of) the resolving power of the instrument no matter how large the instrument aperture is. Is this a permanent defect of humans—to want more knowledge than it is possible to get? Thus we spend much time and energy designing algorithms that will increase the resolving power of our telescopes and imaging systems to give us the superresolution of larger instruments that we do not have.

Astronomers and their mathematically oriented colleagues do not lack for ingenuity. To get nearly limitless resolution (submillionseconds of angle) they have devised *interferometric imaging* techniques. However, the theory behind interferometric imaging requires that we measure not just the modulus but also the phase of the available field (the Fourier transform of the object brightness distribution). The phase may be impossible to get. Hence the *phase retrieval problem*: what E. Leith calls "anti-holography," that is, to recover an image from a magnitude, while in holography we recover an image from the phase.

Astronomers are not the only ones devising schemes for recovering the phase. Materials scientists interested in the microstructure of matter use x-ray or electron diffraction patterns for this purpose. Unfortunately, only the *intensity* of the diffraction pattern is easily obtained. Without the phase, the complex three-dimensional structure of matter is lost. Recovering the phase is indispensable for getting the whole picture.

Thus signal recovery is useful in seeing the stars as well as the molecules. But these "old" problems are well known! What has happened in recent years to generate the current excitement? First and foremost, I believe, is the ready availability of large-scale computers that enable the realization of complex iterative algorithms involving thousands of cycles on million-pixel images. Second is the appearance (and rediscovery) of several key papers that proposed novel algorithms and new theory. Third is the appearance of new imaging modalities such as x-ray tomography, diffraction tomography, seismic imaging, and others. Finally, there is the increased mathematical sophistication of engineers, especially electrical engineers. Image recovery is a mathematically intensive activity; it requires more than basic calculus and Fourier transforms—the staple of engineers educated not many years ago. It is deeply rooted in functional analysis, linear algebra, and analytic function theory. The vocabulary of signal recovery includes such terms as Hilbert spaces, nonexpansive operators, Hadamard products, and irreducible polynomials in two variables. When I went to graduate school back in the sixties, very few engineers had any notions of what these terms meant.

It is always tricky (but not disagreeable) to make a list of the "best" or "worst" or "most important" items, be they movies, restaurants, or technical contributions. While the list of the most important recent contributions in the field may be debatable, still I believe few will argue with my choice of the following five works that, among others, had a galvanizing effect on signal recovery. They are the works by Gerchberg and Saxton [1] on iterative phase retrieval from two images; by Gerchberg [2] and Papoulis [3] on superresolution by iterative methods; and by Youla on the theory of alternating projections for both linear [4] and convex-nonlinear [5] image recovery. What I believe all these papers had in common was perhaps not so much their specific applications but rather the suggestion of new methods for tackling old (as well as new) problems.

Let me now briefly survey the contents of this book. In Chapter 1, "Signal Restoration, Functional Analysis, and Fredholm Integral Equations of the First Kind," Craig Rushforth gives a careful review of the fundamental properties of linear vector spaces and associated operators. Important image recovery concepts, such as the pseudoinverse operator, regularization, and the projection theorem, are presented. The discussions of iterative solutions to the Fredholm equations and the theory of alternating orthogonal projections immediately couple to Chapter 2, "Mathematical Theory of Image Restoration by the Method of Convex Projections," by Dan Youla, whose well-known paper on alternating orthogonal projections [4] generalized the Gerchberg–Papoulis algorithm to a broad class of linear, iterative algorithms and shed new light on the issues of uniqueness and stability. His chapter continues in this vein and extends his previous research to convex spaces and linear as well as convex nonlinear types of restoration. Among other things, Youla makes available to us the pioneering work of a group of Russian mathematicians on iteratively locating a point at the intersection of a specified number of convex sets. The method of convex projections is an extraordinarily useful and general algorithm that is already being applied to numerous practical problems. I hope its appearance in this book will broaden its impact even more.

While the first two chapters might be grouped under "Hilbert space techniques," the next three might be grouped under the heading "other methods." Ken Hanson, in Chapter 3, "Bayesian and Related Methods in Image Reconstruction from Incomplete Data," very convincingly demonstrates what can happen when inadequate data are used to reconstruct an image, such as in the limited angle problem in tomography. The author points out that obtaining a finite set of measurements on an arbitrary object function $f(x, y)$ will almost invariably produce blind spots about $f(x, y)$—the null space of the measurement. The result can be a reconstructed image containing severe artifacts. The alternative? Try using a statistical (Bayesian) approach. Thus given an unknown object, described in function space by a vector $\mathbf{f}$, and data described by a vector $\mathbf{g}$, we maximize $P(\mathbf{f}|\mathbf{g})$ i.e., the probability of the image $\mathbf{f}$, *given* the data $\mathbf{g}$. One of the excellent features of this approach is that it provides a method of supplementing the available measurements with information previously known about the shape or structure of the reconstructed object.

A different approach to image restoration is furnished by Rick Mammone in Chapter 4, "Image Restoration Using Linear Programming." The linear programming approach assumes that there exists an upper bound on the number of nonzero pixel values to be restored. Thus linear programming is particularly useful for sparse images, i.e., images with a large number of zero-value pixels, such as occur in astronomy and microscopy. An advantage of the linear programming approach over linear filtering, such as inverse or Wiener filters, is that the image degradation operator need not be shift-invariant or separable. Experimental results that speak well for this method of image recovery are furnished.

An altogether different approach to image recovery is presented by Zhuang, Østevold, and Haralick in Chapter 5, "The Principle of Maximum Entropy in Image Restoration." One version of the maximum entropy principle applied to image recovery is the following: When restoring an image of n pixels, assign a gray level distribution $\mathbf{f} = (f_1, \ldots, f_n)^T$ that maximizes the entropy $H(\mathbf{f}) = \Sigma_i f_i \log f_i$ consistent with the given data and subject to a normalizing constraint on $\Sigma_i f_i$. Does it work? It seems to. Zhuang and his coworkers furnish a review of the maximum entropy approaches advocated by various proponents (e.g., Frieden; Burch, Gull, Skilling) and then propose their own technique based on solving a system of ordinary differential equations with appropriate initial values. Their approach does not involve the usual optimization methods and relies on solving a one-dimensional search along some path. Convincing experimental results are furnished.

The next three chapters (Chapters 6–8) deal with magnitude and phase retrieval problems. These are, in my view, among the most interesting and challenging problems in the field. In Chapter 6, "The Unique Reconstruction of Multidimensional Sequences from Fourier Transform Magnitude or Phase," Monty Hayes asks under what conditions does the phase of the Fourier transform of a signal sequence uniquely specify the Fourier magnitude and *vice versa*? Since a signal and its Fourier transform are uniquely related, the above is equivalent to asking under what circumstances does the Fourier phase or magnitude uniquely specify the signal? It is interesting to note that except for a rather ethereal set of nondense, zero-measure functions essentially all two-dimensional signals are specified by their Fourier magnitude. However, this alone is no cause for rejoicing: one has to design algorithms to *recover* the phase or magnitude, one from the other. This Hayes addresses as well. All told, I recommend reading Chapter 6 not just for an understanding of the basic aspects of the magnitude and phase retrieval problems but as an introduction to the somewhat more specialized material in Chapters 7 and 8.

In Chapter 7, "Phase Retrieval and Image Reconstruction for Astronomy," Chris Dainty and Jim Fienup focus on the phase retrieval problem in astronomy. They carefully guide the reader through an extensive collection of phase retrieval algorithms and critically survey the strengths and weaknesses of the various techniques. Of particular interest is the iterative Fourier transform algorithm in the philosophy of the Gerchberg–Saxton algorithm. This algorithm requires four steps per cycle: (1) Fourier transform the *k*th estimate of the unknown signal; (2) apply the modulus constraint in the frequency plane; (3) inverse Fourier transform the modified estimate; and (4) apply the space domain constraint. An improved version, the so-called "input–output" algorithm, introduces a feedback parameter that can improve the rate of convergence. In the second half of the chapter, the authors critically survey object recovery from speckle interferometric data in a fashion that does not require one to be an expert either in astronomy or in interferometry to follow the discussion.

Another view of the phase (and modulus) retrieval problem is furnished by the theory proposed by Levi and Stark in Chapter 8, "Restoration from Phase and Magnitude by Generalized Projections." Adapting Youla's convex projection algorithm (Chapter 2) to problems in which either the data or *a priori* constraints do not form convex sets, the authors formulate an alternating "generalized" projections algorithm and define a criterion of convergence that measures how close (or how far) the signal estimate is from satisfying all of the data and *a priori* constraints. The occasional failure of phase retrieval algorithms to converge is explained in terms of "traps" and "tunnels." A technique for increasing the convergence rates of generalized projections algorithms by adjusting so-called "relaxation" parameters is furnished along with experimental results. The "input–output" algorithm of Chapter 7 and the generalized-projections phase-retrieval algorithm of Chapter 8 are in fact quite similar; the points of view, however, are somewhat different.

Applications of image recovery to the growing field of computer-aided tomography (CAT) are treated in Chapters 9–11. While Hanson in Chapter 3 is also interested in CAT, his approach there is statistical, while those in Chapters 9–11 are deterministic. In Chapter 9, "Image Reconstruction from Limited Data: Theory and Applications in Computerized Tomography," Barry Medoff introduces the reader to abstruse CAT-related phenomena such as the *bagel problem, shadows,* and *ghosts.* These rather unlikely terms are used to present a theory of image formation in tomography based on "generalized" inner products and infinite-dimensional linear spaces. The limited data problem is then treated in this inner product framework. An important observation is that the limited data problem in tomography can exhibit an inherent indeterminacy due to the presence of ghosts that are invisible to the data collector. Only additional *a priori* information can ameliorate this problem in practical implementations.

The very difficult image reconstruction and recovery problems in diffraction tomography are discussed by M. Kaveh and M. Soumekh in Chapter 10, "Computer-Assisted Diffraction Tomography." The mathematical model underlying diffraction tomography uses a wave equation with a spatially varying wave number. This in turn necessitates various approximations such as small perturbation theory and assumptions such as the validity of the Born or Rytov transformations in order to reconstruct an image. The authors derive the *backpropagation algorithm* and suggest that it contains the theoretical elegance of the backprojection algorithm of straight-line tomography except for being computationally more intensive. They then discuss the limited angle and incomplete data problems in diffraction tomography and alert us to the added difficulties in recovering the image when diffraction cannot be ignored.

The power and versatility of the convex projection algorithm is evidenced once more in Chapter 11, "Applications of Convex Projection Theory to Image Recovery in Tomography and Related Areas," by M. I. Sezan and H. Stark. Combining convex projections for image recovery with the direct Fourier method for image reconstruction from projections, the authors recover/reconstruct a simulated human thorax from limited-angle data. Numerous *a priori* known constraints are imposed and the results are compared with the traditional Gerchberg–Papoulis extrapolation algorithm that imposes only the classical space domain (finite size)/frequency domain (partially known spectrum) constraints. Needless to say, the systematic inclusion of additional information gives convex projections a significant edge over the Gerchberg–Papoulis algorithm. The latter, one should add, is also a projection type algorithm, although more modest in its use of *a priori* information.

The application of convex projections to radar and magnetic resonance imaging is also briefly discussed in Chapter 11, as is the use of convex projections in image synthesis from moment constraints.

Author Bahaa Saleh turns his attention to the image *discovery* problem in the penultimate Chapter 12, "Image Synthesis: Discovery Instead of Recovery." Saleh asks how would one design (i.e., "discover") an input image that, often being distorted by a given imaging system, yields the desired image at the output. Does such an input exist? Suppose the output image must be binary: does it exist? Suppose additionally that the input must be binary? These and other interesting questions are raised and answered, including the presentation of a smorgasbord of techniques for discovering the appropriate input image. Among these are solutions based on zero crossing theory, on linear programming, and on alternating orthogonal projections.

Last but not least, Mike Fiddy concludes with Chapter 13, "The Role of Analyticity in Image Recovery." The chapter could have been placed equally logically at the beginning in the "basic theory" group or along with the Hayes, Dainty, Fienup, et al. group on phase or magnitude retrieval, rather than here in the "general topics" group. Fiddy addresses the properties of band-limited functions and shows how the factorization of these functions is closely related to the uniqueness issues of image recovery from partial Fourier data such as reconstruction from Fourier magnitude or phase. Image recovery from zeros in one and higher dimensions, as well as the relationship between analyticity and phase unwrapping, is also discussed. Chapter 13 complements the discussions furnished in Chapters 6 and 7. By reminding us that image recovery is founded on the well-known mathematics of analytic function theory, Fiddy supplies the link that completes the chain between the old and the new and his chapter is an appropriate choice for closing the book.

Troy, New York HENRY STARK

REFERENCES

1. R. W. Gerchberg and W. O. Saxton, A practical algorithm for the determination of phase from image and diffraction plane pictures, *Optik* **35**, 237–246 (1972).
2. R. W. Gerchberg, Super resolution through error energy reduction, *Opt. Acta* **21**, 709–720 (1974).
3. A. Papoulis, A new algorithm in spectral analysis and band-limited extrapolation, *IEEE Trans. Circuits Syst.* **CAS-22**, 735–742 (1975).
4. D. C. Youla, Generalized image restoration by the method of alternating orthogonal projections, *IEEE Trans. Circuits Syst.* **CAS-25**, 694–702 (1978).
5. D. C. Youla and H. Webb, Image restoration by the method of convex projections: Part I, Theory, *IEEE Trans. Medical Imaging* **MI-1**, 81–94 (1982).

Acknowledgments

An editor's most important function is to choose the best possible group of contributing authors. In this respect I have been very fortunate; every colleague whom I felt could make an exciting contribution to this book agreed to participate. To be sure there are many excellent researchers who have made significant contributions to image recovery who are not direct contributors to this book. But, on balance, the authors of this book have furnished as broad a cross section of viewpoints, applications, and results as one could hope for. Thus I thank, first and foremost, my coauthors; they have truly demonstrated what a useful and theoretically challenging field image recovery is. I am also indebted to my present and former students from whom I have learned a great deal and who have generated such interesting results. Special thanks are due to Heywood Webb, formerly of the Rome Air Development Center (RADC), for introducing me to the subject of image recovery, for suggesting stimulating problems, and for helping to arrange for funding. The financial aid of RADC and the National Science Foundation was greatly appreciated. On behalf of all the authors I want to thank all the other funding agencies who, directly or indirectly, have made this book possible. Additional well-deserved thanks are due Priscilla Magilligan and MeeLi Chew Leith for their outstanding typing and technical help, the administration of the Rensselaer Polytechnic Institute for recognizing the scholarly aspect of this work, the staff at Academic Press for their support and encouragement, and finally, my wife, Alice, who read portions of the manuscript and gave helpful advice.

CHAPTER **1**

Signal Restoration, Functional Analysis, and Fredholm Integral Equations of the First Kind

C. K. Rushforth

Department of Electrical Engineering
University of Utah
Salt Lake City, Utah 84112

1.1 INTRODUCTION

Many problems in the fields of inverse scattering, tomography, image restoration, antenna design, and system identification involve the solution of a linear integral equation of the form

$$\int_a^b k(x, s)u(s)\, ds = g(x), \qquad c \leqslant x \leqslant d, \tag{1.1-1}$$

where $g(x)$ is an observed function, $k(x, s)$ is a known kernel, and $u(s)$ is the unknown function to be determined. In an image restoration problem, for example, $g(x)$ would represent the observed image, $k(x, s)$ would be the point spread function of the imaging system, and $u(s)$ would be the object distribution. In the classic and widely studied case of perfect lowpass filtering of spatial or temporal frequencies, the kernel in Eq. (1.1-1) is

$$k(x, s) = \frac{\sin 2\pi B(x - s)}{\pi(x - s)}, \tag{1.1-2}$$

where B is the filter bandwidth.

Equation (1.1-1) is known as a Fredholm integral equation of the first kind. It is well known that this equation is generally ill-posed in the sense that small perturbations in the observed signal g can give rise to unacceptably large changes in the solution. Great care must be taken to obtain approximate solutions that achieve the proper balance between accuracy and stability. The objective of this chapter is to elucidate the properties of Eq. (1.1-1) and to describe some representative solution techniques. Our treatment will be brief, and we refer the reader to [3, 8, 9, 16] for further tutorial discussion and more extensive bibliographies.

In order to provide a useful framework for considering Eq. (1.1-1), we first present a few basic definitions and results from functional analysis and linear operator theory in Section 1.2. In Section 1.3, we discuss the existence and uniqueness of solutions to compact linear operator equations, of which Eq. (1.1-1) is an important special case when certain conditions on $k(x, s)$ are met.

In practical problems, the observed function $g(x)$ is usually perturbed by noise or other sources of error. In those circumstances, it is appropriate to seek an approximate solution to Eq. (1.1-1). We discuss the operator pseudoinverse and its use in developing least-squares solutions to Eq. (1.1-1) in Section 1.4.

Techniques for mitigating the ill-posed nature of Eq. (1.1-1) are referred to as regularization techniques, a subject that we consider in Section 1.5. As we shall see, finding a useful approximate solution to Eq. (1.1-1) usually entails a trade-off between resolution and the regularity or stability of the solution. We attempt to provide some insight into this matter by using the singular value decomposition (SVD).

In Section 1.6, we examine the solution of Eq. (1.1-1) from a filtering point of view. In particular, we consider the truncated singular value decomposition. We consider the Landweber iteration in Section 1.7, alternating orthogonal projections in Section 1.8, and some regularized iterative algorithms in Section 1.9. The problem of discretizing Eq. (1.1-1) for solution on a digital computer is briefly addressed in Section 1.10, and Section 1.11 contains an assessment of the current status of the problem and some comments on future directions.

1.2 HILBERT SPACES AND LINEAR OPERATORS

In order to establish a framework for considering solutions of Eq. (1.1-1), we provide in this section some basic definitions and results pertaining to Hilbert spaces and to linear operators defined on these spaces. Our treatment will necessarily be very brief, the main purpose being to establish some notation and to summarize for reference purposes a few properties that play a central role in the remainder of the chapter. Further discussion of these topics can be found in [3, 4, 17].

A *linear vector space* V is a nonempty set of elements, together with an operator $+$ (called addition), defined on $V \times V$ and satisfying, for all $f, g, h \in V$,

(1) $f + g \in V$.
(2) $f + g = g + f$.
(3) $f + (g + h) = (f + g) + h$.
(4) There exists a unique element $0 \in V$ having the property that $f + 0 = f$ for all $f \in V$.
(5) For each $f \in V$, there exists an element $(-f) \in V$ with the property that $f + (-f) = 0$.
(6) $\alpha f \in V$ for any scalar α and any $f \in V$.
(7) $\alpha(f + g) = \alpha f + \alpha g$.
(8) $(\alpha + \beta)f = \alpha f + \beta f$.
(9) $(\alpha\beta)f = \alpha(\beta f)$.
(10) $1 \cdot f = f$.

A finite set of vectors $\{f_1, f_2, \ldots, f_n\}$ from V is said to be *linearly dependent* if there exist scalars $\alpha_1, \alpha_2, \ldots, \alpha_n$, not all zero, such that $\alpha_1 f_1 + \alpha_2 f_2 + \cdots + \alpha_n f_n = 0$. Otherwise $\{f_1 \cdots f_n\}$ is said to be *linearly independent.* An arbitrary set S of vectors is linearly independent if and only if every finite subset of S is linearly independent. If V contains n linearly independent vectors, while every set of $n + 1$ vectors is linearly dependent, V is said to be finite-dimensional with dimension n. If V is not finite-dimensional, it is said to be infinite-dimensional. A finite set S of vectors is said to be a *basis* for V if S is linearly independent and every vector in V can be written as a linear combination of elements of S. Infinite-dimensional bases will be discussed later in conjunction with Hilbert spaces.

A nonempty subset M of V is called a *linear subspace* if it is itself a linear vector space under the operations defined on V. The smallest linear subspace of any vector space consists of the single element 0. If S is any subset of V, the set of all finite linear combinations of elements of S is a linear subspace of V. This subspace is referred to as the *linear span* of S and is written $[S]$.

Now suppose that M and N are two linear subspaces of V and assume that neither is equal to $\{0\}$. If each $f \in V$ can be written in the form $f = g + h$ with $g \in M$ and $h \in N$, we write $V = N + M$ and say that V is the *vector sum* of M and N. If, further, this representation is unique, we write $V = N \oplus M$ and say that V is the *direct sum* of N and M. It is easy to show that $V = M + N \Rightarrow V = M \oplus N$ if and only if $M \cap N = \{0\}$.

Vector spaces as defined above have algebraic, but not analytic, properties. In order to do analysis in vector spaces, we introduce an additional feature called a norm.

Definition. Let V be a vector space. To each $f \in V$ we assign a real number denoted $\|f\|$ satisfying

(1) $\|f\| \geqslant 0$ with equality iff $f = 0$.
(2) $\|\alpha f\| = |\alpha| \|f\|$ for any scalar α and any $f \in V$.
(3) $\|f + g\| \leqslant \|f\| + \|g\|$ for all $f, g \in V$.

The quantity $\|f\|$ is called the *norm* of f, and $\{V, \|\cdot\|\}$ is called a *normed linear vector space.*

One of the most important concepts of analysis in normed vector spaces is that of convergence.

Definition. A sequence $\{f_n\}$ of vectors from V is said to *converge* iff there is a vector $f \in V$ such that

$$\lim_{n \to \infty} \|f_n - f\| = 0.$$

The vector f is called the *limit* of the sequence, and we write $f_n \to f$. It is easy to show that if the limit exists, it must be unique.

Let S be a subset of V. The *closure* of S, written $\bar{S}$, is simply S together with all of its limit points. S is said to be *closed* iff $S = \bar{S}$. A set $T \subset V$ is said to be *open* iff its complement is closed. A set $C \subset V$ is said to be *compact* if every open cover of C contains a finite subcover.

Now suppose the sequence $\{f_n\}$ of elements of V has the property that $\lim_{m,n \to \infty} \|f_n - f_m\| = 0$. Such a sequence is referred to as a *Cauchy sequence.* If every Cauchy sequence in V is convergent, we say that V is *complete.* A complete normed linear vector space is called a *Banach space.*

Much useful analysis can be carried out in Banach spaces. By adding another algebraic operation called an inner product, however, we provide additional structure that allows us to apply geometric intuition developed by thinking about problems in Euclidean vector spaces.

Definition. Let V be a vector space. An *inner product* is a complex-valued function defined on $V \times V$ such that, for all $f, g, h \in V$ and all complex scalars α,

(1) $(f, f) \geqslant 0$ with equality iff $f = 0$.
(2) $(f, g + h) = (f, g) + (f, h)$.
(3) $(f, g) = \overline{(g, f)}$ where the overbar denotes complex conjugate.
(4) $(\alpha f, g) = \alpha (f, g)$.

A vector space with an inner product is called an inner product space or a pre-Hilbert space. Such a space becomes a normed linear vector space upon defining a norm $\|f\| = (f, f)^{1/2}$. If a pre-Hilbert space is complete with respect to this norm, it is called a *Hilbert space*. All the spaces we deal with in this chapter are Hilbert spaces and are denoted $\mathscr{H}$, $\mathscr{H}_1$, $\mathscr{H}_2$, etc.

Two vectors f and g from a Hilbert space $\mathscr{H}$ are said to be *orthogonal* if $(f, g) = 0$. If S is a subset of $\mathscr{H}$, its *orthogonal complement* is

$$S^{\perp} = \{f : (f, s) = 0 \text{ for all } s \in S\}.$$

A Hilbert space $\mathscr{H}$ is said to be *separable* if it contains a countable dense set, i.e., if there exists a countable set $S = \{u_1, u_2, \ldots\}$ with the property that for any $\varepsilon > 0$ and any $u \in \mathscr{H}$, there exists a $u_n \in S$ such that $\|u - u_n\| < \varepsilon$. Applying the Gram–Schmidt orthogonalization procedure to S yields an orthonormal basis for $\mathscr{H}$. We assume that all our Hilbert spaces are separable and that they therefore posses orthonormal bases.

Now suppose that K is a linear operator from the Hilbert space $\mathscr{H}_1$ to the Hilbert space $\mathscr{H}_2$. The *domain* $D(K)$ of K is that subset of $\mathscr{H}_1$ over which K is defined. The *range* $R(K)$ of K is defined by

$$R(K) = \{g : g = Ku \text{ for some } u \in D(K)\},$$

and the *null space* $N(K)$ of K is defined by

$$N(K) = \{u : Ku = 0\}.$$

We define the norm of K to be[1]

$$\|K\| = \sup_{\substack{u \in D(K) \\ u \neq 0}} (\|Ku\| / \|u\|), \tag{1.2-1}$$

and we say that K is a bounded operator if it has finite norm.

An important property of linear operators is given in the following theorem.

Theorem 1.2-1. Let K be a linear operator from $\mathscr{H}_1$ to $\mathscr{H}_2$. Then K is bounded if and only if it is continuous.

If K is a bounded linear operator from $\mathscr{H}_1$ to $\mathscr{H}_2$, its *adjoint* K^* is the bounded linear operator from $\mathscr{H}_2$ to $\mathscr{H}_1$ satisfying

$$(Ku, v) = (u, K^*v)$$

for all $u \in \mathscr{H}_1$ and $v \in \mathscr{H}_2$. If $K^* = K$, K is said to be *self-adjoint*. The following theorem expresses some important relationships between K and K^*.

[1] sup is short for *supremum* (least upper bound).

Theorem 1.2-2. Let K be a continuous linear operator from $\mathcal{H}_1$ to $\mathcal{H}_2$. Then $[R(K)]^{\perp} = N(K^*)$ and $[N(K)]^{\perp} = \overline{R(K^*)}$. Furthermore, $\|K^*\| = \|K\|$.

Now let K be a linear operator mapping the Hilbert space $\mathcal{H}$ into itself. The *spectrum* of K is the set of complex numbers $\sigma(K)$ defined by

$$\sigma(K) = \{\lambda : K - \lambda I \text{ does not have a bounded inverse}\}. \tag{1.2-2}$$

The *spectral radius* of K is the real number

$$|\sigma(K)| = \sup\{|\lambda| : \lambda \in \sigma(K)\}. \tag{1.2-3}$$

If K is self-adjoint, $\sigma(K)$ is a nonempty set of real numbers with spectral radius $|\sigma(K)| = \|K\|$.

If $Ku = \lambda u$ for some nonzero $u \in \mathcal{H}$ and some complex number λ, we call λ an eigenvalue of K and u the associated eigenvector. If K is self-adjoint, the eigenvectors associated with distinct eigenvalues are orthogonal.

A linear operator K mapping $\mathcal{H}_1$ into $\mathcal{H}_2$ is called *compact* if $\overline{K(\mathcal{B})}$ is compact for every bounded subset $\mathcal{B}$ of $\mathcal{H}_1$. The following properties of compact operators will be useful in what follows [3].

Theorem 1.2-3. A compact operator is continuous.

Theorem 1.2-4. Consider the operator K defined by Eq. (1.1-1), and suppose that the kernel $k(\cdot, \cdot)$ is square-integrable over $[a, b] \times [c, d]$. Then K is a compact operator from $L_2[a, b]$ to $L_2[c, d]$.

Theorem 1.2-5. Let K be a compact self-adjoint operator from the Hilbert space $\mathcal{H}$ into itself. Then each nonzero element of the spectrum $\sigma(K)$ is an isolated point which is an eigenvalue of K. Furthermore, for each nonzero eigenvalue λ of K the set $N(K - \lambda I)$ is finite-dimensional. The eigenvalues $\lambda_1, \lambda_2, \ldots$ form a sequence that converges to zero unless their number is finite.

Theorem 1.2-6. Let K be a compact self-adjoint linear operator with eigenvalues $\lambda_1, \lambda_2, \ldots$ repeated according to the dimension of the associated spaces $N(K - \lambda I)$. Let $u_1, u_2, \ldots$ be the corresponding orthonormal eigenvectors. Then for any $u \in \mathcal{H}$, we have

$$Ku = \sum_{n=1}^{\infty} \lambda_n (u, u_n) u_n. \tag{1.2-4}$$

If K has only a finite number of eigenvalues, the sum in Eq. (1.2-4) will be finite, and we say that K has finite rank. In this case, $R(K)$ is a finite-dimensional set.

A useful decomposition of a Hilbert space is provided by the projection theorem.

Theorem 1.2-7 (the projection theorem). Let M be a closed linear subspace of the Hilbert space $\mathscr{H}$. Then $M^\perp$ is also a closed linear subspace, and every $u \in \mathscr{H}$ has a unique representation of the form $u = s + t$ with $s \in M$ and $t \in M^\perp$. Furthermore, s is the unique element in M closest to u. We define the projection operator P_M by the mapping $P_M u = s$, and we call s the projection of u onto M.

A final concept that plays an important role in the solution of Eq. (1.1-1) is that of a singular system. Let K be a compact linear operator from $\mathscr{H}_1$ to $\mathscr{H}_2$. Then K^*K is a compact self-adjoint linear operator mapping $\mathscr{H}_1$ into itself. The eigenvalues of K^*K are all nonnegative; listed in order of decreasing magnitude, they are $\lambda_1 \geqslant \lambda_2 \geqslant \cdots \geqslant 0$. Let $v_1, v_2, \ldots$ be a corresponding sequence of orthonormal eigenvectors of K^*K, and define

$$\mu_n = 1/\sqrt{\lambda_n} \tag{1.2-5}$$

and

$$u_n = \mu_n K v_n. \tag{1.2-6}$$

Then $\{u_n\}$ is an orthonormal sequence in $\mathscr{H}_2$ satisfying

$$\mu_n K^* u_n = v_n. \tag{1.2-7}$$

Theorem 1.2-8. $\{u_n\}$ is an orthonormal basis for $\overline{R(K)} = [N(K^*)]^\perp$ and $\{v_n\}$ is an orthonormal basis for $\overline{R(K^*)} = [N(K)]^\perp$. The sequence of triples $\{u_n, v_n; \mu_n\}$ is called a *singular system* for K. It follows from Eq. (1.2-5) and Theorem 1.2-5 that the singular values μ_n become arbitrarily large as n increases unless K is finite-dimensional. This property has important implications for the solution of Eq. (1.1-1).

1.3 EXISTENCE OF SOLUTIONS

The basic results on the existence of solutions of Eq. (1.1-1) are due to Picard [11] and are discussed extensively in [15]. These results are stated for the slightly more general case of compact linear operators [of which Eq. (1.1-1) is a special case when appropriate conditions on $k(x, s)$ are met] in the next theorem.

Theorem 1.3-1. Let K be a compact linear operator mapping $\mathcal{H}_1$ into $\mathcal{H}_2$, and let $\{u_n, v_n; \mu_n\}$ be a singular system for K. The equation $Ku = g$ has a solution if and only if the following two conditions hold:

(a) $g \in [N(K^*)]^{\perp}$.
(b) $\sum_{n=1}^{\infty} \mu_n^2 |(g, u_n)|^2 < \infty$.

When these conditions are met, the solution is given by

$$u = \sum_{n=1}^{\infty} \mu_n (g, u_n) v_n, \tag{1.3-1}$$

where equality is interpreted to mean that

$$\lim_{N\to\infty} \left\| u - \sum_{n=1}^{N} \mu_n (g, u_n) v_n \right\| = 0.$$

Condition (b) of Theorem 1.3-1 tells us that g must be sufficiently smooth in order that a solution exist. In particular, the sequence of inner products (g, u_n) must decrease rapidly enough to counter the increase in the singular values μ_n. As we have seen, $\mu_n \to \infty$ except in the degenerate case in which K is of finite rank.

If we use Eq. (1.3-1) to obtain a solution, the inner product (g, u_n) is weighted by the singular value μ_n. Since these singular values increase without bound in the nondegenerate case as n increases, small perturbations in the higher modes are multiplied by successively larger factors. This feature can cause small changes in g to manifest themselves as very large changes in the solution. This unfortunate property is not a result of using a particularly poor solution method but is inherent in the problem itself. Proper resolution of this difficulty requires a modification of the problem usually referred to as regularization. Before we discuss regularization procedures, however, we first develop a framework for obtaining and evaluating approximate solutions of Eq. (1.1-1).

1.4 LEAST-SQUARES SOLUTIONS AND THE OPERATOR PSEUDOINVERSE

It is convenient for many purposes, particularly when noisy data are involved, to consider approximate rather than exact solutions to Eq. (1.1-1). More generally, let K be a bounded linear operator from $\mathcal{H}_1$ to $\mathcal{H}_2$. We define a *least-squares solution* to the equation $Ku = g$ to be any $u \in \mathcal{H}_1$ that satisfies[2]

$$\|Ku - g\| = \inf_{x\in\mathcal{H}_1} \|Kx - g\|.$$

[2] inf is short for *infimum* (greatest lower bound).

A least-squares solution u thus has the property that Ku is as close as possible to the observed signal g. The following theorem [3] provides alternative characterizations of least-squares solutions.

Theorem 1.4-1. Let K be a bounded linear operator from $\mathscr{H}_1$ to $\mathscr{H}_2$. The following conditions are equivalent:

(a) $\|Ku - g\| = \inf_{x\in\mathscr{H}_1}\|Kx - g\|$.
(b) $K^*Ku = K^*g$.
(c) $Ku = Pg$, where P is the projection operator mapping $\mathscr{H}_2$ onto $\overline{R(K)}$.

It follows from (c) of Theorem 1.4-1 that Eq. (1.1-1) has a least-squares solution if and only if $Pg \in R(K)$ or, equivalently, if and only if $g \in R(K) + [R(K)]^\perp$. When this condition is satisfied, there is a unique least-squares solution with minimum norm, which we denote by $K^\dagger g$. The operator $K^\dagger$ defined in this way maps $R(K) + [R(K)]^\perp$ into $\mathscr{H}_1$, and is called the pseudoinverse of K. We have $N(K^\dagger) = [R(K)]^\perp$ and $R(K^\dagger) = [N(K)]^\perp$. $K^\dagger g$ is the unique least-squares solution of Eq. (1.1-1) lying in $R(K^\dagger) = [N(K)]^\perp$.

Even with this more general notion of solution, however, the ill-posed nature of the problem is still with us.

Theorem 1.4-2. Let K be a linear compact operator mapping $\mathscr{H}_1$ into $\mathscr{H}_2$. Then $K^\dagger$ is bounded if and only if K has finite rank.

Thus, except in the simple case of a degenerate kernel, Eq. (1.1-1) is still ill-posed even when least-squares solutions are sought. Small perturbations in the data signal g will generally lead to unacceptably large perturbations in $K^\dagger g$. The singular system of K provides the following useful characterization of $K^\dagger$ when K is compact.

Theorem 1.4-3. Let K be a compact linear operator from $\mathscr{H}_1$ to $\mathscr{H}_2$ with singular system $\{u_n, v_n; \mu_n\}$, and assume that $g \in R(K) + [R(K)]^\perp$. Then

$$K^\dagger g = \sum_{n=1}^{\infty} \mu_n (Pg, u_n) v_n = \sum_{n=1}^{\infty} \mu_n (g, u_n) v_n, \tag{1.4-1}$$

where P is the projection operator mapping $\mathscr{H}_2$ onto $\overline{R(K)}$.

The nature of the trouble with $K^\dagger g$ is clear from Eq. (1.4-1). When K is compact, $\mu_n \to \infty$, and perturbations in the higher modes can be greatly amplified. In order to obtain useful results, we seek alternative methods of obtaining approximate solutions that are less sensitive to small perturbations in the data.

1.5 REGULARIZATION

As we have seen, the equation $Ku=g$ is generally ill-posed when K is a linear compact operator. In particular, Eq. (1.1-1) is ill-posed when the kernel $k(\cdot,\cdot)$ is square-integrable and nondegenerate, a case of great practical importance. If we are to obtain useful approximate solutions to either of these equations, we must modify the problem in such a way that the solution becomes less sensitive to small perturbations in the data. At the same time, the solution to the modified problem must be close to the solution of the original problem. The process of modifying the original problem to achieve an acceptable compromise between these two conflicting goals is referred to as regularization. We briefly discuss the principles of regularization in this section.

The ill-posed nature of the equation $Ku = g$ is a consequence not only of the nature of the operator K but also of the domain $D(K)$ over which K is defined. Hence, we may be able to regularize the problem at hand either by modifying $D(K)$ or by modifying K itself. A classical result regarding $D(K)$ is provided by the following theorem [9].

Theorem 1.5-1. Let K be a bounded linear operator mapping $\mathscr{H}_1$ into $\mathscr{H}_2$, and let $\mathscr{C} \subset \mathscr{H}_1$ be a compact set. If K is restricted to $\mathscr{C}$, then K^{-1} is a bounded mapping of $K(\mathscr{C})$ into $\mathscr{H}_1$.

Thus, if we know *a priori* that the original object lies in a compact set, the problem of finding a solution becomes much less sensitive to perturbations in the data.

A second approach to regularization involves replacing the operator K by an operator $\tilde{K}$. Suppose further that the "true" g is perturbed, yielding an observed signal $\tilde{g}$ satisfying

$$\|g - \tilde{g}\| \leqslant \varepsilon. \tag{1.5-1}$$

We seek a least-squares solution to the modified equation

$$\tilde{K}u = \tilde{g}. \tag{1.5-2}$$

If the modified operator $\tilde{K}$ has a bounded pseudoinverse $\tilde{K}^\dagger$, the least-squares solution

$$\tilde{u} = \tilde{K}^\dagger \tilde{g} \tag{1.5-3}$$

will be stable and is a potential candidate as an approximate solution to the original equation. We seek modifications of the original operator K that yield approximate solutions of the form Eq. (1.5-3) having an appropriate balance of accuracy and stability.

We express the original least-squares problem in the form

$$Ku \simeq \tilde{g}. \tag{1.5-4}$$

As it stands, this least-squares problem still suffers from instability and must be regularized if a useful approximate solution is to be found. A regularizer for this problem is defined as follows [3].

Definition. A regularizer for Eq. (1.5-4) is a one-parameter family of linear operators $\{R_\alpha, \alpha \in A\}$, where A is a set of positive real numbers with $0 \in A$, satisfying

(a) For each $\alpha > 0$, R_α is bounded,

and

(b) $\lim_{\alpha \to 0} \|R_\alpha K u - u\| = 0$ for $u \in R(K^\dagger)$.

Thus, R_α is a bounded approximation to $K^\dagger$ and can therefore be used to obtain a stable approximate solution to Eq. (1.5-4) with an appropriate choice of the regularizing parameter α.

For a given value of α, let us take

$$u_\alpha = R_\alpha \tilde{g}$$

as the approximate solution to Eq. (1.5-4). We then have

$$\begin{aligned}\|u_\alpha - u\| &= \|R_\alpha \tilde{g} - K^\dagger g + R_\alpha g - R_\alpha g\| \\ &\leqslant \|R_\alpha \tilde{g} - R_\alpha g\| + \|R_\alpha g - K^\dagger g\| \\ &\leqslant \|R_\alpha\| \varepsilon + \|R_\alpha g - K^\dagger g\|.\end{aligned}$$

The second term is the error due to regularization (i.e., to the fact that R_α is only an approximation to $K^\dagger$), and it goes to zero as $\alpha \to 0$. The first term is an amplified version of the observation error. This term does not generally vanish; indeed, it almost always blows up, a fact that will be established momentarily. Thus, the choice of regularizing parameter α represents a compromise between fidelity (small regularization error) and stability (controlled noise amplification). Too large a value for α yields a stable solution of poor fidelity, while too small a value provides insufficient regularization to overcome the ill effects of noise amplification.

The fact that the amplified noise term generally blows up is established by the following theorem [3].

Theorem 1.5-2. If $\tilde{g} \notin D(K^\dagger) = R(K) + [R(K)]^\perp$, then $\lim_{\alpha\to 0} \|R_\alpha \tilde{g}\| = \infty$.

Since $\tilde{g}$ is a noisy version of g, it will amost certainly have a component outside $D(K^\dagger)$. This component will be amplified without bound as $\alpha \to 0$.

Any regularization procedure will entail a proper choice of the regularization parameter α. Many procedures for choosing α have been described, most of them *ad hoc* in nature. We adopt as a general guideline a concept referred to by Groetsch [3] as the discrepancy principle. We suppose as before that $\|\tilde{g} - g\| \leqslant \varepsilon$, and we consider the residual $(Ku_\alpha - \tilde{g})$, where $u_\alpha = R_\alpha \tilde{g}$ is the approximate solution obtained using the regularization paramater α. The discrepancy principle tells us to choose α such that the norm of the residual, $\|Ku_\alpha - \tilde{g}\|$, is equal to the discrepancy ε in the observed data. Groetsch [3] provides substantial theoretical justification for this choice, but it suffices for our purposes to note that reducing α to achieve $\|Ku_\alpha - \tilde{g}\| < \varepsilon$ generally increases the noise amplification and is probably unwarranted in view of the uncertainty in the data.

Before examining some particular approaches to regularization in greater detail, we make one more general point. Practical computational schemes for solving Eq. (1.5-4) are necessarily carried out in finite-dimensional spaces. There are two general approaches to obtaining a regularized finite-dimensional problem whose numerical solution provides a satisfactory approximate solution to Eq. (1.5-4). The first entails regularization of the infinite-dimensional problem followed by an appropriate discretization, while in the second these steps are carried out in the opposite order. A thorough discussion of this issue is found in [9].

Having introduced some basic principles of regularization in this section, we now turn in the next few sections to some representative procedures for obtaining useful approximate solutions to Eq. (1.1-1).

1.6 THE TRUNCATED SVD EXPANSION AND FILTERING

Let K be a compact linear operator from $\mathscr{H}_1$ to $\mathscr{H}_2$, and let $\{u_n, v_n; \mu_n\}$ be a singular system for K written in order of increasing μ_n. We have seen that the (unregularized) minimum-norm least-squares solution of the original equation $Ku = g$ can be expressed in the form

$$K^\dagger g = \sum_{n=1}^{\infty} \mu_n(g, u_n) v_n \tag{1.4-1}$$

provided that $g \in D(K^\dagger) = R(K) + [R(K)]^\perp$. When we are faced instead with the solution of Eq. (1.5-4), with $\tilde{g}$ almost certain not to lie in $D(K^\dagger)$, the solution provided by Eq. (1.4-1) with g replaced by $\tilde{g}$ will be unsatisfactory since the increasingly large singular values will amplify without bound the component of $\tilde{g}$ lying outside $D(K^\dagger)$. Perhaps the simplest conceptual approach to regularizing this problem is to truncate Eq. (1.4-1) to N terms and then to replace g by $\tilde{g}$. This yields the approximation

$$u^{(N)} = \sum_{n=1}^{N} \mu_n(\tilde{g}, u_n)v_n. \tag{1.6-1}$$

The regularization parameter in this case is N, the number of terms retained. A large value of N will yield a small regularization error, but at the same time will lead to high noise amplification because of the large value of μ_N. According to the discrepancy principle, a good choice for N would be the smallest integer such that $\|Ku^{(N)} - \tilde{g}\| \leqslant \varepsilon$.

The use of Eq. (1.6-1) to obtain an approximate least-squares solution is intuitively reasonable, but we can also provide a more formal justification for this approach [3]. Suppose we seek a finite-dimensional approximation to the minimum-norm least-squares solution of $Ku = g$. A reasonable approach to finding such an approximation is the following. First define the subspace V_N of $\mathscr{H}_1$ by

$$V_N = [v_1, v_2, \ldots, v_N], \tag{1.6-2}$$

where, as before, $\{u_n, v_n; \mu_n\}$ is a singular system for K and $[S]$ is the linear span of S. Then let K_N be the restriction of K to V_N, and denote by $K_N^\dagger g$ the minimum-norm least-squares solution to the modified equation $K_N u = g$. The resulting finite-dimensional approximation is precisely

$$K_N^\dagger g = \sum_{n=1}^{N} \mu_n(g, u_n)v_n, \tag{1.6-3}$$

and upon replacing g by $\tilde{g}$ we obtain Eq. (1.6-1). This lends further credence to $u^{(N)}$ as a reasonable approximation to $K^\dagger g$.

It is helpful to view Eq. (1.4-1) as a filter, the magnitude of whose response to the unit vector u_n is μ_n [14]. This filter can be interpreted as an inverse filter whose function is to counteract the effect of the operator K. To see this, suppose the original object is $u = v_m$. Then

$$g = Ku = Kv_m = \frac{1}{\mu_m} u_m$$

and

$$K^\dagger g = \sum_{n=1}^{\infty} \mu_n \left(\frac{1}{\mu_n} u_m, u_n\right) v_n = v_m = u.$$

Although it operates on elements of $D(K^\dagger)$ in precisely the correct way, the inverse filter produces catastrophic results when operating on a vector having a component outside $D(K^\dagger)$. One approach to mitigating this unfortunate property is to truncate the infinite series as discussed above. This process can be viewed as replacing the inverse filter with a filter whose response F_n to the unit vector u_n is

$$F_n = \begin{cases} \mu_n, & 1 \leqslant n \leqslant N, \\ 0, & n > N. \end{cases} \tag{1.6-4}$$

The response of this filter is exactly equal to that of the inverse filter for modes up to the Nth, and is identically zero thereafter. With an appropriate choice of N, this filter can achieve a reasonable balance between truncation error and excessive noise amplification caused by the larger singular values.

There are, of course, many other filter responses that can achieve such a compromise. Moreover, instead of truncating the SVD expansion abruptly, we may wish to effect a more gradual transition between the inverse filter response and a zero or constant response. An example of such a filter, obtained as the solution to a minimum mean-squared-error optimization problem when the statistics of the desired signal and the noise are known, is the Wiener filter [13]. Other filters are discussed in Section 1.9.

A tacit assumption made throughout this section is that the singular system $\{u_n, v_n; \mu_n\}$ for K is known. Although singular systems for certain simple operators are known (e.g., prolate spheroidal wave functions and their associated eigenvalues for perfect lowpass filtering of an object of finite support), they will generally have to be calculated numerically. This can be a difficult problem, and furthermore the computational burden associated with the calculation of Eq. (1.6-1) may be impractically large. Thus, it is of considerable interest to seek alternative methods of obtaining approximate least-squares solutions. Even when other procedures (e.g., iterative algorithms) are used, however, the filtering point of view discussed in this section can still be a valuable means of assessing performance [14].

1.7 THE ITERATIVE ALGORITHM OF LANDWEBER

Landweber's iterative algorithm [6] for solving a Fredholm equation of the first kind is motivated by considering the properties of the Fredholm equation of the *second* kind

$$Ku + \alpha u = g, \tag{1.7-1}$$

where K is a compact linear operator from $\mathscr{H}$ into itself and α is a small positive number. The following result is an immediate consequence of the well-known Fredholm alternative theorem [4].

Theorem 1.7-1. Suppose α is neither 0 nor the negative of any eigenvalue of K, and denote by I the identity operator. Then $(K+\alpha I)^{-1}$ exists and is bounded, and Eq. (1.7-1) has the unique solution

$$u = (K+\alpha I)^{-1} g. \tag{1.7-2}$$

In contrast to equations of the first kind, equations of the second kind are well-posed for values of α satisfying the conditions of Theorem 1.7-1. Furthermore, Eq. (1.7-1) can be solved by the iterative procedure of Liouville and Neumann:

$$u^{(n+1)} = \alpha^{-1}(g - Ku^{(n)}) \tag{1.7-3}$$

with initial condition $u^{(0)} = 0$. If the conditions of Theorem 1.7-1 are met, the iteration of Eq. (1.7-3) converges to the unique solution of Eq. (1.7-1).

Landweber's approach to the solution of the equation $Ku = g$ was to write this equation in a form which lends itself at least formally to the application of the Liouville–Neumann iteration. He first operated on both sides of the equation with the adjoint operator K^* to yield

$$K^*Ku = K^*g. \tag{1.7-4}$$

Adding u to both sides of this equation, rearranging, and converting to an iterative expression yields

$$u^{(n+1)} = K^*g + (I - K^*K)u^{(n)}. \tag{1.7-5}$$

If the operator K is self-adjoint, it is unnecessary to operate on the original equation with K^* and we can write

$$u^{(n+1)} = g + (I - K)u^{(n)}. \tag{1.7-6}$$

For simplicity, we investigate the properties of this last equation under the assumption that K is self-adjoint. Many physical problems give rise to operators having this property.

In order to analyze the iteration Eq. (1.7-6), we take $\{u_n, v_n; \mu_n\}$ to be a singular system for K. Since we are assuming that K is self-adjoint, we have $u_n = v_n$ for all n. We assume further that g satisfies the conditions of Theorem 1.3-1, from which it follows that the equation $Ku = g$ has a unique solution. Under these conditions, $u^{(n)}$ as defined by Eq. (1.7-6) converges to this solution.

Upon taking $u^{(0)} = 0$ and applying successive substitution, we can write Eq. (1.7-6) in the form

$$u^{(n+1)} = \sum_{m=0}^{n} (I - K)^m g. \tag{1.7-7}$$

Writing g in the form

$$g = \sum_{j=1}^{\infty} (g, u_j) u_j \tag{1.7-8}$$

and using the fact that

$$Ku_j = (1/\mu_j) u_j,$$

we obtain

$$\begin{aligned} u^{(n+1)} &= \sum_{j=1}^{\infty} (g, u_j) u_j \sum_{m=0}^{n} \left(1 - \frac{1}{\mu_j}\right)^m \\ &= \sum_{j=1}^{\infty} \mu_j \left[1 - \left(1 - \frac{1}{\mu_j}\right)^{n+1}\right] (g, u_j) u_j. \end{aligned} \tag{1.7-9}$$

Without loss of generality, the original equation $Ku = g$ can be scaled in such a way that no singular value of K is less than one. Assuming that this has been done, and assuming further that g satisfies the conditions of Theorem 1.3-1, the right-hand side of Eq. (1.7-9) approaches the right-hand side of Eq. (1.3-1) as $n \to \infty$, which is known to be the correct solution.

Unfortunately, the iterative procedure Eq. (1.7-6) suffers from the noise sensitivity inherent in the ill-posed nature of the original operator equation. As we have seen, some sort of regularization procedure is required to stabilize the solution. The simplest method of regularizing Eq. (1.7-6) is simply to terminate the procedure after an appropriate number of iterations. The number of iterations then plays the role of a regularizing parameter.

To illustrate this fact, we assume that Eq. (1.7-6) is terminated after n iterations; our approximate solution is then given by

$$u^{(n)} = \sum_{j=1}^{\infty} \mu_j \left[1 - \left(1 - \frac{1}{\mu_j}\right)^n\right] (g, u_j) u_j.$$

As in Section 1.6, we can view this result as the output of a linear restoration filter whose input is g and whose response to u_j is

$$F_j = \mu_j \left[1 - \left(1 - \frac{1}{\mu_j}\right)^n\right]. \tag{1.7-10}$$

This response is sketched as a function of j for fixed n in Fig. 1.7-1. Also shown for purposes of comparison are the responses of the inverse filter

$$F_j = \mu_j, \qquad \text{all } j, \tag{1.7-11}$$

and the truncated inverse filter

$$F_j = \begin{cases} \mu_j, & 1 \leqslant j \leqslant N, \\ 0, & j > N. \end{cases} \tag{1.7-12}$$

The extreme noise sensitivity of the inverse filter is caused by the ever-increasing values of F_j. Truncating the inverse filter as in Eq. (1.7-12) can substantially reduce the noise sensitivity; the number N of terms retained plays the role of regularizing parameter and represents a compromise between fidelity and stability.

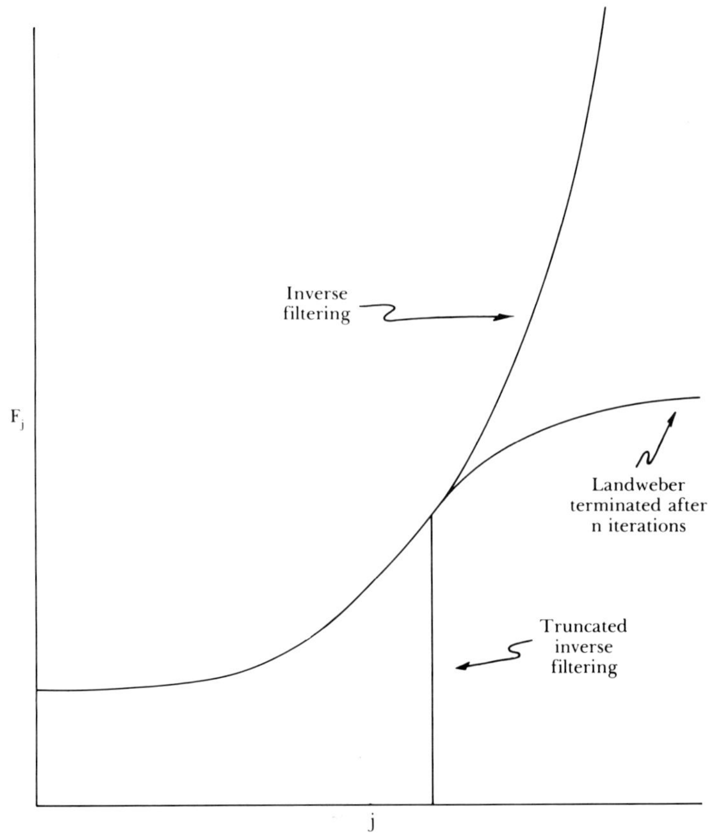

Fig. 1.7-1. The iterative algorithm of Landweber [Eq. (1.7-10)] versus the inverse filtering solution.

The iterative procedure of Landweber, whose filter response is given by Eq. (1.7–10) when the procedure is terminated after n iterations, falls between the extremes of the inverse filter and the truncated inverse filter. Its response follows that of the inverse filter for small values of j and then flattens out as j continues to increase. The asymptote is given by

$$\lim_{j\to\infty} \mu_j[1-(1-(1/\mu_j))^n] = n \tag{1.7-13}$$

since $\mu_j \to \infty$ as we noted previously.

It has been observed in practice that as the number of iterations of Eq. (1.7-6) increases, the quality of the resulting approximate solution first improves and then actually begins to deteriorate. This behavior reflects the trade-off between truncation error and noise amplification. A reasonable choice for the number of iterations would be the smallest value of n for which

$$\|g - Ku^{(n)}\| \leq \varepsilon, \tag{1.7-14}$$

where ε is the uncertainty in g.

The iterative procedure of Landweber was rediscovered and analyzed by Gerchberg [2] and Papoulis [10] some years later for the special case of a perfect lowpass filter. This algorithm is commonly referred to as the Gerchberg algorithm or the Gerchberg–Papoulis algorithm in the engineering and physics literature.

1.8 ALTERNATING ORTHOGONAL PROJECTIONS

An approach similar to that of Landweber, Gerchberg, and Papoulis has been investigated from a different point of view by Youla [18], who considers the case in which $\mathcal{H}_1 = \mathcal{H}_2 = \mathcal{H}$ and $K = P_A$, an orthogonal projection operator. He further assumes that u is known *a priori* to lie in a closed linear subspace M_B of $\mathcal{H}$, and hence that u satisfies

$$u = P_B u, \tag{1.8-1}$$

where P_B is the orthogonal projection operator from $\mathcal{H}$ onto M_B. Similarly, M_A is the range of P_A and is a closed linear subspace of $\mathcal{H}$. Defining $Q_A = I - P_A$ and noting that a projection operator is self-adjoint, we can combine Eq. (1.8-1) and Eq. (1.7-6) to yield the recursion

$$u^{(n+1)} = g + Q_A P_B u^{(n)}. \tag{1.8-2}$$

As before, we take $u^{(0)} = 0$. It is this equation that Youla analyzes. The interesting feature of his work is the novel geometric interpretation he provides. In Chapter 2, Youla discusses various extensions of the theory in this chapter to the more general class of closed convex sets.

In order to analyze Eq. (1.8-2), we further define $Q_B = I - P_B$, and we note that Q_A and Q_B are the orthogonal projection operators from $\mathscr{H}$ onto $M_A^\perp$ and $M_B^\perp$, the orthogonal complements of M_A and M_B, respectively. The iteration of Eq. (1.8-2) can now be described as follows: to obtain $u^{(n+1)}$ from $u^{(n)}$, we first project $u^{(n)}$ onto M_B and then project the result onto $M_A^\perp$. We then add g in order to restore the correct projection onto M_A. This process continues until a stopping criterion is satisfied. We wish to know the conditions under which the original equation has a unique solution, the conditions under which the problem is well-posed, and the conditions under which Eq. (1.8-2) converges to the correct solution.

Youla's analysis of this problem relies on the notion of the angle between two subspaces M_1 and M_2 of a parent Hilbert space $\mathscr{H}$. This notion is motivated by the Schwartz inequality, which states that

$$0 \leq |(f, g)|/\|f\|\,\|g\| \leq 1$$

for any vectors f and g in $\mathscr{H}$. By analogy with R^3, we define

$$\cos \psi(f, g) \triangleq |(f, g)|/\|f\|\,\|g\|, \tag{1.8-3}$$

where ψ is the angle between f and g, by convention taken to lie in the first quadrant. The angle $\psi(M_1, M_2)$ between the subspaces M_1 and M_2 is then defined to be

$$\psi(M_1, M_2) = \inf_{\substack{f \in M_1 \\ g \in M_2}} \psi(f, g).$$

Youla's central result can now be stated [18, p. 697, Theorem 1].

Theorem 1.8-1. Let M_A and M_B be closed linear subspaces of the Hilbert space $\mathscr{H}$, and let P_A, Q_A, P_B, and Q_B be the orthogonal projection operators from $\mathscr{H}$ onto M_A, $M_A^\perp$, M_B, and $M_B^\perp$, respectively. Suppose $u \in M_B$. Then

(a) u is uniquely determined by its projection $P_A u$ onto M_A iff

$$M_B \cap M_A^\perp = \{0\}.$$

(b) The problem of restoring u from $P_A u$ is well-posed iff

$$\psi(M_B, M_A^\perp) > 0,$$

which in turn is satisfied iff

$$\rho \triangleq \|Q_A P_B\| < 1.$$

(c) In both cases (a) and (b), the sequence $\{u^{(n)}\}$ generated by the recursion of Eq. (1.8-2) converges to u, i.e.,

$$\lim_{n\to\infty} \|u^{(n)} - u\| = 0.$$

The proof of this theorem is found in [18]. We confine ourselves here to a few remarks about its relationship to some results discussed earlier in this chapter.

The equation of interest in this section and in [18] is

$$P_A P_B u = g. \tag{1.8-4}$$

Since we have been primarily concerned in earlier sections with compact operators (e.g., Fredholm integral equations with square-integrable kernels), we take P_A to be compact. It follows that $P_A P_B$ is compact [4], and we can therefore analyze Eq. (1.8-4) by using methods described previously. In particular, it is of interest to consider the conditions under which Eq. (1.8-4) is ill-posed. Theorem 1.4-2 asserts that Eq. (1.8-4) is ill-posed iff $P_A P_B$ has infinite rank. Furthermore, the latter condition obtains iff the singular values of $P_A P_B$ become arbitrarily large as $n \to \infty$. Thus, we conclude that if the singular values of $P_A P_B$ become arbitrarily large, the angle $\psi(M_B, M_A^\perp)$ between M_B and $M_A^\perp$ will be zero and Eq. (1.8-4) will be ill-posed. This result coincides with [18, Theorem 2] when P_A is compact.

1.9 REGULARIZED ITERATIVE ALGORITHMS

The iterative procedures described in the two previous sections share the common drawback of all unregularized procedures for solving ill-posed problems, namely their sensitivity to noise. Although termination of the iteration after a finite number of steps serves as a regularization procedure of sorts, this method is inadequate for many purposes and we seek more effective approaches to the regularization of iterative techniques. Procedures of this type are discussed in [12]. Most of the results in [12] were obtained for compact linear operators that are also self-adjoint, and we shall confine our remarks to this class of operators.

The basic idea, here as elsewhere in this chapter, is to replace the original ill-posed problem

$$Ku = g \tag{1.9-1}$$

with a well-posed problem

$$\tilde{K}u = g \tag{1.9-2}$$

in which $\tilde{K}$ is an approximation to K. In order that Eq. (1.9-2) be well-posed, we demand further that $\tilde{K}$ have a bounded inverse (or pseudoinverse). One method of obtaining a suitable alternative equation is to incorporate additional *a priori* constraints on the solution u. For example, we may have some *a priori* information about the norm of u or its degree of smoothness.

We confine ourselves here to constraints of the form

$$\|Cu\| \leqslant E, \tag{1.9-3}$$

where C is a linear self-adjoint (but not compact) operator and E is a known upper bound. We seek an approximate solution $\tilde{u}$ that simultaneously satisfies Eq. (1.9-3) and the equation

$$\|Ku - g\| \leqslant \varepsilon. \tag{1.9-4}$$

That is, $\tilde{u}$ must satisfy the given constraint and must at the same time yield a small residual. Ross [12] takes as his approximate solution $\tilde{u}$ the vector that minimizes the functional

$$\|Ku - g\|^2 + (\varepsilon/E)^2\|Cu\|^2, \tag{1.9-5}$$

and he refers to $\tilde{u}$ as the *quasi-solution.* He further points out that the quasi-solution can be obtained as the solution to the equation

$$(K^2 + \alpha^2 C^2)u = Kg, \tag{1.9-6}$$

where $\alpha = \varepsilon/E$.

The operator C should be chosen to reflect some practical constraints on the solution, to provide a reasonable trade-off between resolution and stability, and to lend itself to reasonable implementation. In particular, it should be chosen in such a way that $(K^2 + \alpha^2 C^2)^{-1}K$ is bounded, thereby establishing a continuous relationship between the data g and the quasi-solution $\tilde{u}$.

We first consider the case

$$C = I. \tag{1.9-7}$$

In this case,

$$\|Cu\|^2 = \|u\|^2 = \text{signal energy}, \tag{1.9-8}$$

E^2 becomes an *a priori* bound on the signal energy, and Eq. (1.9-6) simplifies to

$$(K^2 = \alpha^2 I)u = Kg. \tag{1.9-9}$$

In order to develop an iterative algorithm for solving Eq. (1.9-9), we rewrite this equation in the form

$$u = Kg + [(1 - \alpha^2)I - K^2]u. \tag{1.9-10}$$

From this latter equation follows the iterative algorithm

$$u^{(n+1)} = Kg + [(1 - \alpha^2)I - K^2]u^{(n)}, \tag{1.9-11}$$

and as before we take $u^{(0)} = 0$. If $\alpha = 0$, Eq. (1.9-11) reduces to the Landweber iteration Eq. (1.7-5). More generally, α is a regularization parameter whose value must be determined from the properties of the noise and the signal energy.

Let $(u_n, v_n; \mu_n)$ be a singular system for K, and note that $u_n = v_n$ since K is self-adjoint. Writing

$$u = \sum_{i=1}^{\infty} (u, u_i)u_i \tag{1.9-12}$$

and

$$g = \sum_{i=1}^{\infty} (g, u_i)u_i \tag{1.9-13}$$

and substituting these expressions into Eq. (1.9-9), we obtain

$$(u, u_i) = F_i(g, u_i), \tag{1.9-14}$$

where

$$F_i = \mu_i/(1 + \mu_i^2\alpha^2). \tag{1.9-15}$$

Thus, we can view the solution to Eq. (1.9-9) as the output of a linear filter whose input is g and whose response to the singular function u_i is F_iu_i. A typical plot of the filter response F_i as a function of i is shown in Fig. 1.9-1. This plot should be compared with those shown in Fig. 1.7-1. This algorithm has been disucssed by Cesini *et al.* [1].

Although Eq. (1.9-11) is convenient to implement, it generally converges rather slowly. This drawback can be mitigated by using the alternative constraint operator

$$C^2 = K. \tag{1.9-16}$$

Equation (1.9-6) then becomes

$$(K + \alpha^2 I)u = g \tag{1.9-17}$$

with, as before, $\alpha = \varepsilon/E$. Once again we can write

$$(u, u_i) = F_i(g, u_i), \tag{1.9-14}$$

but now

$$F_i = \mu_i/(1 + \alpha^2\mu_i). \tag{1.9-18}$$

This quantity is also plotted in Fig. 1.9-1. Note that this filter response flattens out at a value $1/\alpha^2$ as $i \to \infty$, whereas the filter response described by Eq. (1.9-15) dies away to zero as $i \to \infty$.

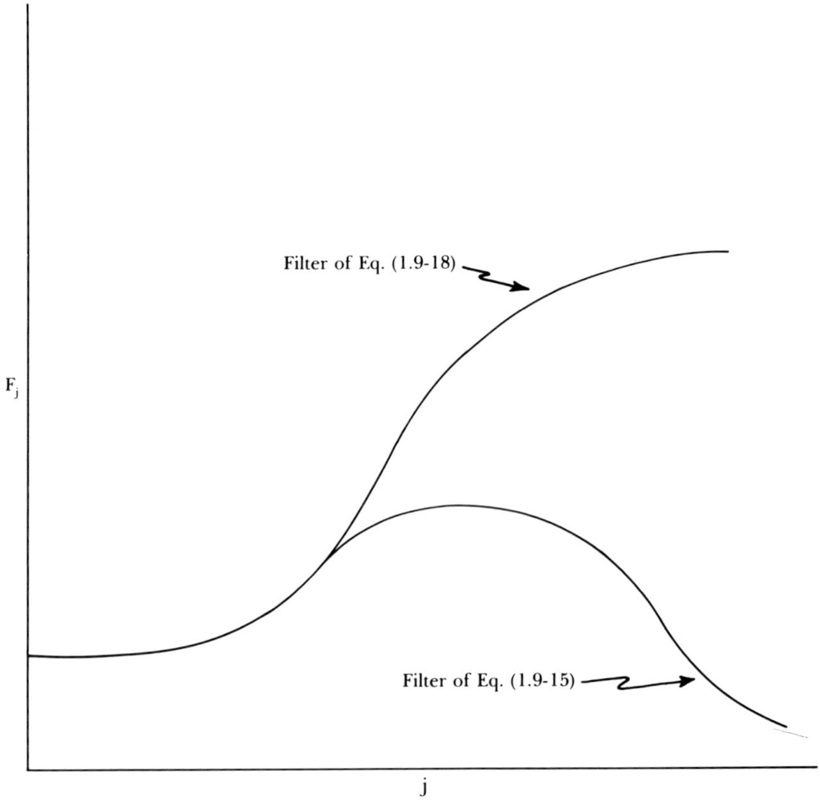

Fig. 1.9-1. The filter of Eq. (1.9-15) versus that of Eq. (1.9-18).

Rewriting Eq. (1.9-17) in the form

$$u = g + [(1 - \alpha^2)I - K]u \tag{1.9-19}$$

yields the iteration

$$u^{(n+1)} = g + [(1 - \alpha^2)I - K]u^{(n)}, \tag{1.9-20}$$

and, as before, we take $u^{(0)} = 0$. Ross [12] establishes the weak convergence [4] of this iterative algorithm, and he further demonstrates that convergence is faster than for the algorithm defined by Eq. (1.9-11). On the other hand, Eq. (1.9-20) cannot be applied if K has an eigenvalue close to $-\alpha^2$, whereas Eq. (1.9-11) can always be used. Moreover, the choice between the filters of Eqs. (1.9-15) and (1.9-18) will also depend on the signal and noise statistics of the particular problem being considered. It is likely that both algorithms will find application, as will other algorithms that arise out of alternative choices of the constraint operator C.

1.10 MOMENT DISCRETIZATION

If an approximate regularized solution to Eq. (1.1-1) is to be obtained numerically, a discretization must eventually take place at some stage of the solution process. As an example of such a discretization, we consider in this section an approach referred to as moment discretization [7].

For simplicity, we take $a = c = 0$ and $b = d = 1$, and we assume that $k(x, s)$ is a real continuous function on $[0, 1] \times [0, 1]$ and that $g(x)$ is a real continuous function on $[0, 1]$. We suppose that $g(x)$ has been measured at a finite number of points $0 \leq x_1 < x_2 < \cdots < x_n \leq 1$, and we consider the moment discretization

$$\int_0^1 k(x_i, s)u(s)\, ds = g(x_i), \qquad i = 1, 2, \ldots, n. \tag{1.10-1}$$

We define

$$k_i(s) \triangleq k(x_i, s) \tag{1.10-2}$$

and

$$g_i \triangleq g(x_i), \tag{1.10-2}$$

and we write Eq. (1.10-1) in the form

$$(u, k_i) = g_i, \tag{1.10-3}$$

where $(\cdot, \cdot)$ is the usual L_2 inner product.

We further define

$$T_n u \triangleq \begin{bmatrix} (u, k_1) \\ (u, k_2) \\ \vdots \\ (u, k_n) \end{bmatrix} \quad \text{and} \quad g = \begin{bmatrix} g_1 \\ \vdots \\ g_n \end{bmatrix}.$$

Equation (1.10-1) can then be written in the form

$$T_n u = g, \tag{1.10-4}$$

where T_n is an operator mapping $L_2[0, 1]$ into R^n. Since $R(T_n)$ is finite-dimensional, Eq. (1.10-4) has a least-squares solution of the form

$$\tilde{u} = T_n^\dagger g + z, \tag{1.10-5}$$

where $T_n^\dagger$ is the pseudoinverse of T_n and $z \in N(T_n)$. Equivalently, $\tilde{u}$ can be found as a solution to

$$T_n^* T_n u = T_n^* g, \tag{1.10-6}$$

where T_n^* is the adjoint of T_n.

Nashed [7] has shown that

$$T_n^* g = \sum_{i=1}^{n} g_i k_i(s), \tag{1.10-7}$$

and therefore that

$$R(T_n^* T) = R(T_n^*) = [k_1(s), \ldots, k_n(s)], \tag{1.10-8}$$

where [ ] denotes the linear span.

Equation (1.10-4) has a unique least-squares solution $\hat{u}_n \in R(T_n^*)$, which is also the least-squares solution of minimum norm:

$$\hat{u}_n(s) = T_n^\dagger g = \sum_{i=1}^{n} b_i k(x_i, s). \tag{1.10-9}$$

If $g(x) \in R(K)$, where K is the original infinite-dimensional integral operator, then $\hat{u}_n$ can be written explicitly in the form

$$\hat{u}_n(s) = g' Q_n^\dagger k(s),$$

where $k'(s) = (k_1(s), \ldots, k_n(s))$, the prime denotes transpose, and

$$Q_{ij} = \int_0^1 k(x_i, s) k(x_j, s)\, ds. \tag{1.10-11}$$

If $\{k_1(s), \ldots, k_n(s)\}$ are independent, then the least-squares solutions are in fact solutions, $Q_n^\dagger$ can be replaced by Q_n^{-1}, and the $\{b_i\}$ in Eq. (1.10-9) can be uniquely determined by solving the finite-dimensional system of equations

$$\sum_{j=1}^{n} (k_i, k_j) b_j = g_i, \qquad i = 1, 2, \ldots, n. \tag{1.10-11}$$

Once the inner products are computed by using a numerical quadrature procedure, the discretization is complete. Results for the general case are developed in [7].

The choice of the grid size in the discretization process can itself be part of the regularization process, as discussed in some detail in [5]. If the grid is very fine, the resulting algebraic equation will be ill-conditioned and will require additional regularization. The regularization techniques described in this chapter and in the references all have finite-dimensional counterparts in numerical linear algebra, which can then be applied.

Alternatively, Eq. (1.1-1) could first be regularized in the original infinite-dimensional space by modifying the operator. For example, the original equation of the first kind could be replaced by an equation of the second kind as in Section 1.9. This modified equation could then be discretized as outlined above. If done carefully, this alternative approach would yield an already regularized matrix equation whose solution would be an acceptable approximation to the solution of Eq. (1.1-1). A more complete discussion of the issues involved in regularization, discretization, and the commutativity of the various operations involved is contained in [9].

1.11 SUMMARY AND CONCLUSIONS

Many signal restoration problems can be formulated as Fredholm integral equations of the first kind, which are known to be generally ill-posed and therefore sensitive to small perturbations in the data. We have reviewed the salient properties of these equations and have discussed a few of the many techniques proposed for their solution.

Much is now known about these equations, but there nevertheless remain many unanswered questions. The choice of constraint operators and the associated regularizing parameters still needs more study. Nonlinear constraints (e.g., positivity) can have a powerful effect on the solution, and more analysis of such constraints is called for. For more information on this matter, see Youla [19]. More work is also needed on the problem of approximating equations defined in infinite-dimensional space by finite-dimensional equations that can be solved on a digital computer. Thus, the study of Fredholm equations of the first kind may be approaching maturity, but it is definitely not moribund.

ACKNOWLEDGMENT

The author would like to thank Professor Frank Stenger for his helpful comments.

REFERENCES

1. G. Cesini, G. Guattari, G. Lucarini and C. Palma (1978). An iterative method for restoring noisy images. *Optica Acta* **25** (6), 501–508.
2. R. W. Gerchberg (1974). Super-resolution through error energy reduction. *Optica Acta* **21** (9), 709–720.
3. C. W. Groetsch (1984). "The Theory of Tikhonov Regularization for Fredholm Equations of the First Kind," Pitman Publishing Limited, London.
4. V. Hutson and J. S. Pym (1980). "Applications of Functional Analysis and Operator Theory," Academic Press, New York.
5. L. S. Joyce and W. L. Root (1984). Precision bounds in super-resolution processing. *J. Opt. Soc. Am. A* **1** (2), 149–168.
6. L. Landweber (1951). An iteration formula for Fredholm integral equations of the first kind. *American Journal of Mathematics* **73**, 615–624.
7. M. Z. Nashed (1976). On moment-discretization and least-squares solutions of linear integral equations of the first kind. *Journal of Mathematical Analysis and Application* **53**, 359–366.

8. M. Z. Nashed (1976). "Generalized Inverses and Applications," Academic Press, New York.
9. M. Z. Nashed (1981). Operator-theoretic and computational approaches to ill-posed problems with applications to antenna theory. *IEEE Trans. on Antennas and Propagation,* **AP-29** (2), 220–231.
10. A. Papoulis (1975). A new algorithm in spectral analysis and bandlimited extrapolation. *IEEE Trans. on Circuits and Systems,* **CAS-22**, 735–742.
11. E. Picard (1910). Sur un theoreme generale relatif sux equations integrales de premier espece et sur quelques problemes de physique mathematique. *R. E. Mat. Palermo* **29**, 79–97.
12. G. Ross (1982). Iterative methods in information processing for object restoration. *Optica Acta* **29** (11) 1523–1542.
13. C. K. Rushforth and R. W. Harris (1968). Restoration, resolution, and noise. *J. Opt. Soc. Am.* **58** (4), 539–545.
14. C. K. Rushforth and R. L. Frost (1980). Comparison of some algorithms for reconstructing space-limited objects. *J. Opt. Soc. Am.* **70** (12), 1539–1544.
15. F. Smithies (1965). "Integral Equations," Cambridge University Press, Cambridge.
16. A. N. Tikhonov and V. Y. Arsenin (1977). "Solutions of Ill-Posed Problems," V. H. Winston and Sons, Washington, D.C.
17. A. Wouk (1979). "A Course of Applied Functional Analysis," Wiley, New York.
18. D. C. Youla (1978). Generalized Image Restoration by the Method of Alternating Orthogonal Projections. *IEEE Trans. on Circuits and Systems* **CAS-25** (9), 694–702.
19. D. C. Youla. This Volume, Chapter 2.

CHAPTER 2

Mathematical Theory of Image Restoration by the Method of Convex Projections

D. C. Youla

Department of Electrical Engineering and Computer Science
Polytechnic University
Farmingdale, New York 11735

2.1 INTRODUCTION

Some time ago we published a paper [1] in which it was suggested that many problems of image restoration are probably geometric in character and admit the following initial linear formulation: The original f is a vector known *a priori* to belong to a linear subspace $\mathscr{P}_b$ of a parent Hilbert space $\mathscr{H}$, but all that is available to the observer is its image $P_a f$, the projection of f onto a known linear subspace $\mathscr{P}_a$ (also in $\mathscr{H}$).

(1) Find necessary and sufficient conditions under which f is uniquely determined by $P_a f$ and (2) find necessary and sufficient conditions for the stable linear reconstruction of f from $P_a f$ in the face of noise. (In the latter case, the reconstruction problem is said to be well posed.) The answers turn out to be remarkably simple.

(a) f is uniquely determined by $P_a f$ iff $\mathscr{P}_b$ and the orthogonal complement of $\mathscr{P}_a$ have only the zero vector in common.

(b) The reconstruction problem is well posed iff the angle between $\mathscr{P}_b$ and the orthogonal complement of $\mathscr{P}_a$ is greater than zero. (All angles lie in the first quadrant.)

(c) In the absence of noise, there exists in both cases a and b an effective recursive algorithm for the recovery of f employing only the operations of projection onto $\mathscr{P}_b$ and projection onto the orthogonal complement of $\mathscr{P}_a$. These operations define the necessary instrumentation.

This point of view has proved to be very fruitful and can be made to encompass many important applications [2].

However, it appears that a linear formulation is only achievable by discarding information concerning the original f, and as a result the associated restoration problem is often ill posed [1, Theorem 2]. The entire process of smoothing can be conceptualized as a technique for reintroducing some of the missing information and therefore plays an essential role in combating the effects of noise.

As is shown in detail in Section 2.4, a *linear* image restoration problem transforms, under smoothing, into a *nonlinear* one of the following kind: The original f is *known a priori* to belong to the intersection $\mathscr{C}_0$ of m well-defined closed convex sets $\mathscr{C}_1, \mathscr{C}_2, \ldots, \mathscr{C}_m$, i.e.,

$$f \in \mathscr{C}_0 = \bigcap_{i=1}^{m} \mathscr{C}_i. \tag{2.1-1}$$

Given only the projection operators P_i onto the individual $\mathscr{C}_i$'s, $i = 1 \to m$, restore f, preferably by an iterative scheme. Thus, the P_i's define the necessary instrumentation in an arbitrary Hilbert space setting. We are now in a position to summarize the contents of Chapter 2.

Section 2.2 provides some basic tutorial material on convex sets and projection operators onto closed convex sets. Section 2.3 is devoted to an in-depth leisurely study of nonexpansive maps and their fixed points, with special emphasis on iterative methods. Because of the importance of these methods, we have supplied full proofs in order to make Chapter 2 self-contained. Section 2.4 applies the theorems in Section 2.3 to the problem of finding points belonging to the intersection of a finite number of closed convex sets. It is shown that the latter is a particular case of the problem of finding a fixed point of an asymptotically regular nonexpansive operator by iteration. In general, the convergence of this iteration is only weak and the question of strong convergence is examined in some detail. In this regard, Theorem 2.4-2 may be a new result and should prove useful for further study.

In Section 2.5, we develop algorithms for the realization of important operators that project onto closed convex sets. As we shall see in Chapter 3, some of these are directly applicable to image recovery in computer-aided tomography and elsewhere. We complete the chapter with a summary (Section 2.6) that also describes an extension of the theory to *unrestricted* iterations involving an *infinite* number of convex constraints. In our opinion, such an extension can provide fundamental insight into problems of adaptive control, robotics, econometrics, and decision making.

2.2 SOME PROPERTIES OF CONVEX SETS IN HILBERT SPACE

Definition 2.2-1. A subset $\mathscr{C}$ of $\mathscr{H}$ is said to be convex if together with any x_1 and x_2 it also contains $\mu x_1 + (1-\mu)x_2$ for all μ, $0 \leqslant \mu \leqslant 1$.

Theorem 2.2-1. Let $\mathscr{C}$ denote any closed convex subset of $\mathscr{H}$ and let f be any element of $\mathscr{H}$. Then there exists a unique $x_0 \in \mathscr{C}$ such that

$$\inf_{x \in \mathscr{C}} \|f - x\| = \|f - x_0\|. \tag{2.2-1}$$

(Clearly, x_0 is the element in $\mathscr{C}$ closest to f in norm.)

Proof. Let us first demonstrate uniqueness. Suppose that

$$\inf_{x \in \mathscr{C}} \|f - x\| = \delta = \|f - x_0\| = \|f - y_0\|, \tag{2.2-2}$$

where $x_0 \in \mathscr{C}$ and $y_0 \in \mathscr{C}$. Then, by replacing x and y in the identity

$$\left\|\frac{x+y}{2}\right\|^2 + \left\|\frac{x-y}{2}\right\|^2 = \frac{\|x\|^2 + \|y\|^2}{2} \tag{2.2-3}$$

by $f - x_0$ and $f - y_0$, respectively, we obtain

$$\left\|f - \frac{x_0 + y_0}{2}\right\|^2 = \delta^2 - \left\|\frac{x_0 - y_0}{2}\right\|^2 \leqslant \delta^2. \tag{2.2-4}$$

But, because $\mathscr{C}$ is convex, $(x_0 + y_0)/2 \in \mathscr{C}$ so that

$$\left\|f - \frac{x_0 + y_0}{2}\right\|^2 \geqslant \delta^2; \tag{2.2-5}$$

hence, $\|x_0 - y_0\| = 0$ and $x_0 = y_0$.

From the definition of infimum, there exists a sequence $\{x_n\}$ of elements contained in $\mathscr{C}$ such that

$$\lim \|f - x_n\| = \delta. \tag{2.2-6}$$

By replacing x and y in Eq. (2.2-3) by $f - x_n$ and $f - x_m$, respectively, we get

$$\|x_n - x_m\|^2 = 2(\|f - x_n\|^2 + \|f - x_m\|^2) - 4\left\|f - \frac{x_n + x_m}{2}\right\|^2. \tag{2.2-7}$$

Thus, since $(x_n + x_m)/2 \in \mathscr{C}$ for $n, m = 1 \to \infty$, (2.2-5) and (2.2-7) yield

$$0 \leqslant \lim_{n,m \to \infty} \|x_n - x_m\|^2 \leqslant 4\delta^2 - 4\delta^2 = 0, \tag{2.2-8}$$

which implies that

$$\lim_{n,m\to\infty} \|x_n - x_m\| = 0. \tag{2.2-9}$$

The sequence $\{x_n\}$ is therefore Cauchy and converges to a limit $x_0 \in \mathscr{C}$ because $\mathscr{C}$ is closed. From (2.2-6)

$$\|f - x_0\| = \delta, \tag{2.2-10}$$

Q.E.D.

Theorem 2.2-1 is fundamental and leads immediately to the notion of a projection operator: For any $x \in \mathscr{H}$ the projection $P_{\mathscr{C}}x$ of x onto $\mathscr{C}$ is the element in $\mathscr{C}$ closest to x. As we have seen, if $\mathscr{C}$ is closed and convex, $P_{\mathscr{C}}x$ exists and is uniquely determined by x and $\mathscr{C}$ from the minimality criterion

$$\|x - P_{\mathscr{C}}x\| = \min_{y\in\mathscr{C}}\|x - y\|. \tag{2.2-11}$$

This rule, which assigns to every $x \in \mathscr{H}$ its nearest neighbor in $\mathscr{C}$, defines the (in general) nonlinear projection operator $P_{\mathscr{C}}$: $\mathscr{H} \to \mathscr{C}$ without ambiguity. Later we shall develop several alternative important characterizations of $P_{\mathscr{C}}$.

Of course, any closed linear manifold (CLM) is automatically convex and closed and the corollary that follows is the classical orthogonal projection theorem.

Corollary 1. Let $\mathscr{G}$ denote any CLM in $\mathscr{H}$. Then, every $x \in \mathscr{H}$ possesses a unique decomposition of the form $x = x_1 + x_2$, where $x_1 = P_{\mathscr{G}}x_1 \in \mathscr{G}$ and $x_2 \in \perp\mathscr{G}$, the orthogonal complement[1] of $\mathscr{G}$.

Proof. Suppose that $x = x_1 + x_2 = x_3 + x_4$, where x_1 and x_3 belong to $\mathscr{G}$ and x_2 and x_4 belong to $\perp\mathscr{G}$. Then, $x_1 - x_3 = x_4 - x_2$ is a member of both $\mathscr{G}$ and $\perp\mathscr{G}$. Thus, from the previous footnote, $x_1 = x_3, x_2 = x_4$, and the decomposition, if it exists, must be unique.

Choose $x_1 = P_{\mathscr{G}}x$ and set $x_2 = x - P_{\mathscr{G}}x$. Then, $x = x_1 + x_2$ and it only remains to show that x_2 is orthogonal to $\mathscr{G}$. That is, $(x_2, y) = 0$ for all $y \in \mathscr{G}$. From the definition of $P_{\mathscr{G}}$ and the convexity of $\mathscr{G}$, for any $\mu, 0 < \mu < 1$, $\mu y + (1 - \mu)P_{\mathscr{G}}x \in \mathscr{G}$ for every $y \in \mathscr{G}$ so that

$$\|x - P_{\mathscr{G}}x\|^2 \leq \|x - \mu y - (1 - \mu)P_{\mathscr{G}}x\|^2. \tag{2.2-12}$$

Or, upon expansion of the right-hand side of (2.2-12),

$$\|x - P_{\mathscr{G}}x\|^2 \leq \|x - P_{\mathscr{G}}x\|^2 + \mu^2\|y - P_{\mathscr{G}}x\|^2 - 2\mu\,\mathrm{Re}(x - P_{\mathscr{G}}x, y - P_{\mathscr{G}}x). \tag{2.2-13}$$

[1] For any linear manifold $\mathscr{G}$, $y \in \perp\mathscr{G}$ iff$(x, y) = 0$ for all $x \in \mathscr{G}$. It is easily shown that $\perp\mathscr{G}$ is always a CLM and that if $\mathscr{G}$ is a CLM, then $\perp(\perp\mathscr{G}) = \mathscr{G}$. Evidently, $\mathscr{G} \cap (\perp\mathscr{G}) = \{\phi\}$ and $\mathscr{G} \subset \perp(\perp\mathscr{G})$.

Hence,

$$2\,\mathrm{Re}(x - P_{\mathscr{G}}x, y - P_{\mathscr{G}}x) \leqslant \mu \|y - P_{\mathscr{G}}x\|^2 \tag{2.2-14}$$

and by letting $\mu \to 0$ we find that

$$\mathrm{Re}(x - P_{\mathscr{G}}x, y - P_{\mathscr{G}}x) \leqslant 0 \tag{2.2-15}$$

for all $y \in \mathscr{G}$. Now, since $\mathscr{G}$ is also a linear manifold it contains λy for every real λ and (2.2-15) yields, for λ real,

$$\lambda\,\mathrm{Re}(x - P_{\mathscr{G}}x, y) \leqslant \mathrm{Re}(x - P_{\mathscr{G}}x, P_{\mathscr{G}}x). \tag{2.2-16}$$

This inequality can be true only if

$$\mathrm{Re}(x - P_{\mathscr{G}}x, y) = 0 \tag{2.2-17}$$

for all $y \in \mathscr{G}$. If now y in (2.2-17) is replaced by iy, we get

$$\mathrm{Im}(x - P_{\mathscr{G}}x, y) = 0 \tag{2.2-18}$$

so that finally

$$(x - P_{\mathscr{G}}x, y) = 0, \qquad y \in \mathscr{G}, \tag{2.2-19}$$

Q.E.D.

Corollary 2. Let $\mathscr{C}$ be a closed convex set. Then, for any $x \in \mathscr{H}$,

$$\mathrm{Re}(x - P_{\mathscr{C}}x, y - P_{\mathscr{C}}x) \leqslant 0 \tag{2.2-20}$$

for every $y \in \mathscr{C}$. Conversely, if some $z \in \mathscr{C}$ has the property

$$\mathrm{Re}(x - z, y - z) \leqslant 0 \tag{2.2-21}$$

for all $y \in \mathscr{C}$, then $z = P_{\mathscr{C}}x$.

Proof. A review of the proof of Corollary 1 reveals that (2.2-15) is valid for any closed convex set $\mathscr{C}$. Conversely, (2.2-21) implies that

$$\begin{aligned} \|x - y\|^2 = \|x - z + z - y\|^2 &= \|x - z\|^2 - 2\,\mathrm{Re}(x - z, y - z) + \|z - y\|^2 \\ &\geqslant \|x - z\|^2 \end{aligned} \tag{2.2-22}$$

for all $y \in \mathscr{C}$. Thus, by Theorem 2.2-1, $z = P_{\mathscr{C}}x$, Q.E.D.

Corollary 3. Any closed bounded convex set $\mathscr{C}$ is weakly compact.

Proof. What we have to show is that from every sequence $\{x_n\} \subset \mathscr{C}$ it is possible to extract a subsequence $\{x_{n'}\}$ that converges weakly[2] to some limit f, and that all such weak limits belong to $\mathscr{C}$. By hypothesis, $\mathscr{C}$ is bounded,

[2] This means that $\lim(g, x_{n'}) = (g, f)$ for every $g \in \mathscr{H}$.

as is therefore every sequence $\{x_n\}$ contained in $\mathscr{C}$. Hence, since Hilbert space $\mathscr{H}$ is weakly compact, there exists a subsequence $\{x_{n'}\}$ of $\{x_n\}$ such that $x_{n'} \rightharpoonup f, f \in \mathscr{H}$. Now according to (2.2-20)

$$\mathrm{Re}(f - P_{\mathscr{C}} f, x_{n'} - P_{\mathscr{C}} f) \leqslant 0 \qquad (2.2\text{-}23)$$

for all n'. Thus, $x_{n'} \rightharpoonup f$ implies that

$$0 \geqslant \mathrm{Re}(f - P_{\mathscr{C}} f, f - P_{\mathscr{C}} f) = \|f - P_{\mathscr{C}} f\|^2 \geqslant 0 \qquad (2.2\text{-}24)$$

or $\|f - P_{\mathscr{C}} f\| = 0$. Consequently, $f = P_{\mathscr{C}} f \in \mathscr{C}$, Q.E.D.

The proof of Corollary 3 reveals that a closed convex set $\mathscr{C}$, bounded or unbounded, contains all its weak limits and is therefore weakly closed. Now, since a strong limit is automatically weak, it is clear that weak closure of any set implies its strong closure but the converse is in general false. From this point of view, convex sets appear to be quite exceptional.

In a real Hilbert space, the inequality (2.2-20) for closed convex sets $\mathscr{C}$ reads

$$(x - P_{\mathscr{C}} x, y - P_{\mathscr{C}} x) \leqslant 0, \qquad \text{all } y \in \mathscr{C}. \qquad (2.2\text{-}25)$$

In this guise it can be interpreted to mean that the vector $x - P_{\mathscr{C}} x$ is supporting to $\mathscr{C}$ at the point $\mathscr{P}_{\mathscr{C}} x \in \mathscr{C}$. As Fig. 2.2-1 suggests, $x - P_{\mathscr{C}} x$ is "normal" to the "tangent plane" to $\mathscr{C}$ erected at the point $P_{\mathscr{C}} x$. This plane has $\mathscr{C}$ and x on opposite sides and therefore separates one from the other. Note also that the angle ψ between the vectors $x - P_{\mathscr{C}} x$ and $y - P_{\mathscr{C}} x$ is never less than 90°.

Theorem 2.2-2. Let $\mathscr{C}$ be any closed convex set. Then, for every pair of elements x and y in $\mathscr{H}$,

$$\|P_{\mathscr{C}} x - P_{\mathscr{C}} y\|^2 \leqslant \mathrm{Re}(x - y, P_{\mathscr{C}} x - P_{\mathscr{C}} y). \qquad (2.2\text{-}26)$$

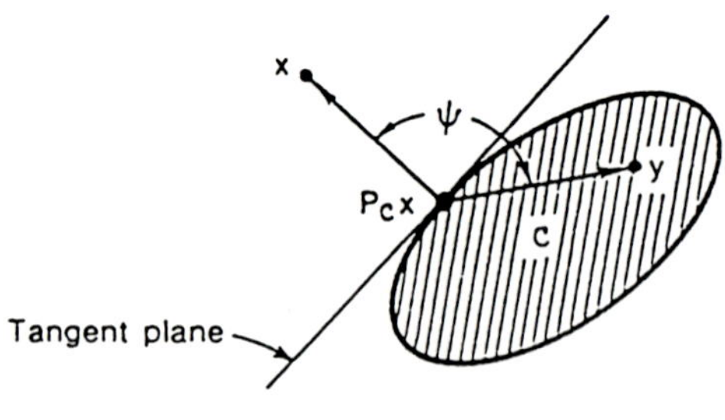

Fig. 2.1-2. $\mathscr{C}$ is a closed convex set.

Proof. Since $P_{\mathscr{C}}x$ and $P_{\mathscr{C}}y$ both belong to $\mathscr{C}$ it follows from (2.2-20) that

$$\operatorname{Re}(x - P_{\mathscr{C}}x, P_{\mathscr{C}}y - P_{\mathscr{C}}x) \leq 0 \tag{2.2-27}$$

and

$$\operatorname{Re}(y - P_{\mathscr{C}}y, P_{\mathscr{C}}x - P_{\mathscr{C}}y) \leq 0. \tag{2.2-28}$$

Thus, (2.2-26) is obtained immediately by addition, Q.E.D.

Corollary 1. Projection operators onto closed convex sets $\mathscr{C}$ are nonexpansive and therefore continuous.

Proof. Schwarz's inequality applied to (2.2-26) yields, for every x and y in $\mathscr{H}$,

$$\|P_{\mathscr{C}}x - P_{\mathscr{C}}y\| \leq \|x - y\|. \tag{2.2-29}$$

Therefore, under the operator $P_{\mathscr{C}}$, the distance between two images never exceeds the distance between the two originals, which is precisely the definition of nonexpansivity. *A fortiori*, $\|x - y\|$ "small" implies $\|P_{\mathscr{C}}x - P_{\mathscr{C}}y\|$ "small" so that $P_{\mathscr{C}}$ is continuous, Q.E.D.

It is possible to strengthen (2.2-29) in various directions by imposing one or more constraints on the "curvature" of $\mathscr{C}$. Admittedly, this strengthening appears more natural in a finite-dimensional setting and leads to several technical refinements not fully exploited in this chapter. Nevertheless, we present the results for possible future applications and because their proofs depend on Theorems 2.2-1 and 2.2-2 in an essential way.

Definition 2.2-2. A convex set $\mathscr{C}$ is said to be strictly convex if $x \in \mathscr{C}$, $y \in \mathscr{C}$, and $x \neq y$ imply that $(x + y)/2$, the midpoint of the "chord" connecting x and y, is an interior point of $\mathscr{C}$.

Definition 2.2-3. A convex set $\mathscr{C}$ is said to be uniformly convex if there exists a function $\delta(\tau)$, positive for $\tau > 0$ and zero only for $\tau = 0$, such that $x, y \in \mathscr{C}$ and

$$\|z - (x + y)/2\| \leq \delta(\|x - y\|) \tag{2.2-30}$$

imply $z \in \mathscr{C}$.

Definition 2.2-4. A uniformly convex set $\mathscr{C}$ for which it is possible to choose

$$\delta(\tau) = \mu\tau^2, \tag{2.2-31}$$

μ a positive constant, is said to be strongly convex. (Evidently, strong convexity implies uniform convexity, which implies strict convexity.)

Theorem 2.2-3 [3]. If $\mathscr{C}$ is strictly convex, $y \in \mathscr{C}$, $x \notin C$, $y \neq P_{\mathscr{C}}x$, we have

$$\operatorname{Re}(x - P_{\mathscr{C}}x, y - P_{\mathscr{C}}x) < 0. \tag{2.2-32}$$

Proof. By definition, if $y \in \mathscr{C}$ and $P_{\mathscr{C}}x \neq y$, there exists an $\varepsilon > 0$ (which depends in general on the choice of x and y) such that

$$w = z + \frac{P_{\mathscr{C}}x + y}{2} \in \mathscr{C} \tag{2.2-33}$$

for all z that satisfy $\|z\| \leqslant \varepsilon$. Hence, if $x \notin \mathscr{C}$ we can choose

$$z = \frac{\varepsilon(x - P_{\mathscr{C}}x)}{\|x - P_{\mathscr{C}}x\|} \tag{2.2-34}$$

and then replace y in (2.2-20) by w to obtain

$$\mathrm{Re}(x - P_{\mathscr{C}}x, y - P_{\mathscr{C}}x) \leqslant -2\varepsilon\|x - P_{\mathscr{C}}x\| < 0, \tag{2.2-35}$$

Q.E.D.

Corollary 3. If $\mathscr{C}$ is uniformly convex, $y \in \mathscr{C}$ and $x \notin \mathscr{C}$, then

$$\mathrm{Re}(x - P_{\mathscr{C}}x, y - P_{\mathscr{C}}x) \leqslant -2\delta(\|y - P_{\mathscr{C}}x\|) \cdot \|x - P_{\mathscr{C}}x\|. \tag{2.2-37}$$

If $\mathscr{C}$ is strongly convex,

$$\|P_{\mathscr{C}}x - y\| \leqslant \rho\|x - y\|, \tag{2.2-38}$$

where

$$\rho = (1 + 2\mu\|x - P_{\mathscr{C}}x\|)^{-1}. \tag{2.2-39}$$

Proof. Observe first that if $y = P_{\mathscr{C}}x$, (2.2-35) is correct with $\varepsilon = 0$. Consequently, in view of the properties of $\delta(\tau)$, (2.2-36) follows immediately by substituting $\delta(\|y - P_{\mathscr{C}}x\|)$ for ε in (2.2-35). To derive (2.2-37) we rewrite (2.2-35) in the form

$$\|P_{\mathscr{C}}x - y\|^2 \leqslant \mathrm{Re}(x - y, P_{\mathscr{C}}x - y) - 2\varepsilon\|x - P_{\mathscr{C}}x\|, \tag{2.2-39}$$

and then use Schwarz's inequality to obtain

$$\|P_{\mathscr{C}}x - y\|^2 - \|P_{\mathscr{C}}x - y\| \cdot \|x - y\| + 2\varepsilon\|x - P_{\mathscr{C}}x\| \leqslant 0. \tag{2.2-40}$$

Or, with ε replaced by $\mu\|P_{\mathscr{C}}x - y\|^2$,

$$\|P_{\mathscr{C}}x - y\|(\|P_{\mathscr{C}}x - y\| + 2\mu\|P_{\mathscr{C}}x - y\| \cdot \|x - P_{\mathscr{C}}x\| - \|x - y\|) \leqslant 0, \tag{2.2-41}$$

so that always,

$$\|P_{\mathscr{C}}x - y\| \leqslant \frac{\|x - y\|}{1 + 2\mu\|x - P_{\mathscr{C}}x\|}, \tag{2.2-42}$$

Q.E.D.

The closed half-circle $\mathscr{C}$ shown in Fig. 2.2-2a is convex but not strictly convex because the midpoint z is not an interior point. On the other hand, the full closed circle in Fig. 2.2-2b is the prototype of a strongly convex set.

Fig. 2.2-2. (a) Closed half-circle is convex but not strictly convex; (b) the closed full-circle is strongly convex (and therefore convex).

2.3 NONEXPANSIVE MAPS AND THEIR FIXED POINTS—BASIC THEOREMS

Definition 2.3-1. A mapping $T: \mathscr{C} \to \mathscr{H}$ is said to be a contraction if there exists a positive constant $\theta, 0 < \theta < 1$, such that

$$\|Tx - Ty\| \leqslant \theta\|x - y\| \tag{2.3-1}$$

for all $x, y \in \mathscr{C}$.

Any contraction has at most one fixed point. For if $Tx_1 = x_1$ and $Tx_2 = x_2$, then

$$\|x_1 - x_2\| = \|Tx_1 - Tx_2\| \leqslant \theta\|x_1 - x_2\| \tag{2.3-2}$$

which implies $\|x_1 - x_1\| = 0$ and hence $x_1 = x_2$.

Theorem 2.3-1 (the contraction principle). If $\mathscr{C}$ is a nonempty closed subset of $\mathscr{H}$, any contraction mapping T of $\mathscr{C}$ into itself possesses a unique fixed point x_∞. Moreover, starting from any element x_0 of $\mathscr{C}$, $T^n x_0 \to x_\infty$ as $n \to \infty$.

Proof. Since we already know that there is at most one fixed point, we need only show that a fixed point exists and is obtainable by iteration. Let x_0 be any member of $\mathscr{C}$ and set

$$x_n = Tx_{n-1}, \qquad n = 1 \to \infty. \tag{2.3-3}$$

Then,

$$\|x_{n+1} - x_n\| \leqslant \theta\|x_n - x_{n-1}\| \leqslant \cdots \leqslant \theta^n\|x_1 - x_0\| \tag{2.3-4}$$

so that for any $m > n$

$$\begin{aligned}\|x_m - x_n\| &\leqslant \|x_m - x_{m-1}\| + \|x_{m-1} - x_{m-2}\| + \cdots + \|x_{n+1} - x_n\| \\ &\leqslant \theta^n(1 + \theta + \cdots + \theta^{m-n-1})\|x_1 - x_0\| \\ &\leqslant \theta^n(1 - \theta)^{-1}\|x_1 - x_0\|. \end{aligned} \tag{2.3-5}$$

Since $0 < \theta < 1$, it follows that $\{x_n\}$ is a Cauchy sequence. Therefore, as $\mathscr{C}$ is closed, it converges to an element $x_\infty \in \mathscr{C}$. But clearly,

$$\begin{aligned}\|Tx_\infty - x_\infty\| &\leqslant \|Tx_\infty - x_{n+1}\| + \|x_{n+1} - x_\infty\| \\ &\leqslant \theta\|x_\infty - x_n\| + \|x_\infty - x_{n+1}\| \qquad (2.3\text{-}6)\end{aligned}$$

and the right side tends to zero as $n \to \infty$. Thus, $\|Tx_\infty - x_\infty\| = 0$, i.e., $Tx_\infty = x_\infty$. It should also be noted that (2.3-5) yields the estimate

$$\|x_\infty - x_n\| \leqslant \theta^n(1-\theta)^{-1}\|x_1 - x_0\|, \qquad (2.3\text{-}7)$$

Q.E.D.

Corollary. Let T^l, l an integer $\geqslant 1$, be a contraction of a nonempty closed set $\mathscr{C}$ into itself. Then T possesses a unique fixed point x_∞ obtainable as the limit of any sequence $\{T^n x_0\}$, $x_0 \in \mathscr{C}$.

Proof. According to Theorem 2.3-1, there exists a unique $x_\infty \varepsilon \mathscr{C}$ such that $T^l x_\infty = x_\infty$. Hence,

$$Tx_\infty = T(T^l x_\infty) = T^l(Tx_\infty) \qquad (2.3\text{-}8)$$

and $Tx_\infty \in \mathscr{C}$ is also a fixed point of T^l. By uniqueness, $Tx_\infty = x_\infty$ and x_∞ is therefore a fixed point of T. Consider the sequence $\{T^n x_0\}$ as $n \to \infty$ and write $n = ql + r$ where q is an integer and $0 \leqslant r < l$. Then, as $n \to \infty$, $q \to \infty$ and

$$\|T^n x_0 - x_\infty\| = \|T^r T^{ql} x_0 - T^r x_\infty\| \leqslant \|T^{ql} x_0 - x_\infty\| \to 0.$$

Consequently, $T^n x_0 \to x_\infty$ as $n \to \infty$, Q.E.D.

Contraction is a very strong requirement to impose on a mapping. A weaker notion is that of "shrinking."

Definition 2.3-2. A mapping $T\colon \mathscr{C} \to \mathscr{H}$ is said to be shrinking if

$$\|Tx - Ty\| < \|x - y\|, \qquad x \neq y, \qquad (2.3\text{-}9)$$

for all $x, y \in \mathscr{C}$.

Theorem 2.3-2. A shrinking map T of a compact[3] set $\mathscr{C}$ into itself always possesses a fixed point.

[3] A subset $\mathscr{C}$ of $\mathscr{H}$ is said to be compact if every sequence of elements in $\mathscr{C}$ has a subsequence which converges to a limit element in $\mathscr{C}$.

Proof. The numerical function $\|Tx - x\|$ is obviously continuous on $\mathscr{C}$. Hence, since $\mathscr{C}$ is compact there exists an element $x_\infty \in \mathscr{C}$ such that

$$\inf_{x \in \mathscr{C}} \|Tx - x\| = \|Tx_\infty - x_\infty\|. \tag{2.3-10}$$

But we must have $Tx_\infty = x_\infty$ because otherwise

$$\|T(Tx_\infty) - Tx_\infty\| < \|Tx_\infty - x_\infty\|, \tag{2.3-11}$$

a contradiction, Q.E.D.

In many applications even shrinking is difficult to achieve and our final weakening of the contraction idea is embodied in the concept of a nonexpansive operator (mapping).

Definition 2.3-3. A mapping $T: \mathscr{C} \to \mathscr{H}$ is said to be nonexpansive if

$$\|Tx - Ty\| \leqslant \|x - y\| \tag{2.3-12}$$

for all $x, y \in \mathscr{C}$.

Theorem 2.3-3. Let $T: \mathscr{C} \to \mathscr{C}$ be a nonexpansive map whose domain $\mathscr{C}$ is a nonempty closed *bounded convex* set. Then T has at least one fixed point.

Proof. Let y_0 be any preselected member of $\mathscr{C}$ and let the set $\mathscr{C}_0 = \{x : x = y - y_0, y \in \mathscr{C}\}$. The translate $\mathscr{C}_0$ is evidently also a closed bounded convex set which, moreover, contains the zero vector ϕ. Clearly, every $x \in \mathscr{C}_0$ possesses a unique decomposition $x = y - y_0, y \in \mathscr{C}$. Let $F: \mathscr{C}_0 \to \mathscr{C}_0$ be defined by

$$Fx = Ty - y_0. \tag{2.3-13}$$

This map F is nonexpansive because $x_1 = y_1 - y_0$ and $x_2 = y_2 - y_0$ imply that

$$\begin{aligned}\|Fx_1 - Fx_2\| &= \|Ty_1 - Ty_2\| \leqslant \|y_1 - y_2\| \\ &= \|(y_1 - y_0) - (y_2 - y_0)\| = \|x_1 - x_2\|.\end{aligned} \tag{2.3-14}$$

For any fixed k, $0 < k < 1$, the map $G = kF$ is a contraction of $\mathscr{C}_0$ into itself. Indeed, for any $x \in \mathscr{C}_0$, $kFx = k(Fx) + (1 - k)\phi \in \mathscr{C}_0$ and for all $x_1, x_2 \in \mathscr{C}_0$,

$$\|Gx_1 - Gx_2\| = k\|Fx_1 - Fx_2\| \leqslant k\|x_1 - x_2\|. \tag{2.3-15}$$

Hence, invoking Theorem 2.3-1, there exists for every constant k, $0 < k < 1$, a unique $x_k \in \mathscr{C}_0$ such that

$$x_k = kFx_k. \tag{2.3-16}$$

If it can now be shown that $x_k \to g$ as $k \to 1$ from below, then by the continuity of F and $g \in \mathscr{C}_0$, it is seen from (2.3-16) that $g = Fg$. Or, since $g = f - y_0, f \in \mathscr{C}$, we obtain $f = Tf$, so f is a fixed point of T. We shall actually prove that

$$\lim_{k\to 1;\, 0<k<1} x_k = g \tag{2.3-17}$$

where g is the (unique) fixed point of F in $\mathscr{C}_0$ of minimum norm.

Assume that $0 < k < l \leq 1$, $x_k = kFx_k$, $x_l = lFx_l$ and let $h = x_l - x_k$. Then, since $\|Fx_l - Fx_k\| \leq \|x_l - x_k\|$, we obtain

$$(l^{-1}(x_k + h) - k^{-1}x_k, l^{-1}(x_k + h) - k^{-1}x_k) \leq \|h\|^2 \tag{2.3-18}$$

or

$$(l^{-1} - k^{-1})^2\|x_k\|^2 + (l^{-2} - 1)\|h\|^2 \leq 2l(k^{-1} - l^{-1}) \operatorname{Re}(x_k, h). \tag{2.3-19}$$

Thus,

$$\operatorname{Re}(x_k, h) \geq 0 \tag{2.3-20}$$

which, together with the identity

$$\|x_l\|^2 = \|x_k + h\|^2 = \|x_k\|^2 + \|h\|^2 + 2\operatorname{Re}(x_k, h), \tag{2.3-21}$$

leads to the inequality

$$\|x_l\|^2 \geq \|x_k\|^2 + \|x_l - x_k\|^2. \tag{2.3-22}$$

To sum up, for any choice of sequence $0 < k_1 < k_2 < \cdots$ such that $k_i \to 1$, the sequence $\{\|x_{k_i}\|\}$ is monotone nondecreasing and bounded (because $\mathscr{C}_0$ is bounded). It therefore converges and, in particular, from (2.3-22)

$$\|x_l - x_k\|^2 \leq \|x_l\|^2 - \|x_k\|^2 \to 0 \tag{2.3-23}$$

as $l, k \to \infty$. By the completeness of $\mathscr{H}$, $x_{k_i} \to g \in \mathscr{C}_0$ because $\mathscr{C}_0$ is closed. (Of course, the limit g is independent of the particular selection of sequence $k_i \to 1$.)

Lastly, let e be any fixed point of F in $\mathscr{C}_0$. Then, $e = 1 \cdot Fe$ and we can apply (2.3-22) with $x_l = e$, $l = 1$, $x_k = x_{k_i}$, and $k = k_i$ for any $i = 1 \to \infty$. As $i \to \infty$, $x_{k_i} \to g$, so

$$\|e\|^2 \geq \|g\|^2 + \|e - g\|^2 \geq \|g\|^2. \tag{2.3-24}$$

Therefore, $\|g\| = \inf\|e\|$, as e ranges over the fixed points of F in $\mathscr{C}_0$, Q.E.D.

Theorem 2.3-3 is due to Browder [4] but the proof we have presented is, in all its essentials, that given by Halpern [5]. Unlike Theorem 2.3-2 (which it does not subsume), the compactness assumption on $\mathscr{C}$ is dispensed with completely and replaced instead by the much weaker constraints of convexity and boundedness. Nevertheless, in many signal-processing applications even this assumption of boundedness may be questionable because accurate *a priori* numerical bounds are not always available. However, if the existence of a fixed point is known in advance from, say, physical considerations, the boundedness requirement on $\mathscr{C}$ can be dropped. Our immediate objective therefore is to reach Theorem 2.3-4 (Opial), and to accomplish this we need three important preparatory lemmas, the last two of which we also owe to Opial [6].

Lemma 2.3-1. The set of fixed points $\mathscr{T}$ of a nonexpansive mapping T with closed convex domain $\mathscr{C}$ and range $\mathscr{H}$ is a closed convex set.

Proof. Let $x_i = Tx_i$, $i = 1 \to \infty$, and suppose that $x_i \to x$. Since $\{x_i\} \subset \mathscr{C}$, which is closed, $x \in \mathscr{C}$ and Tx is well defined. Thus, invoking nonexpansivity,

$$\|Tx - x\| = \|Tx - Tx_i + x_i - x\| \leqslant 2\|x - x_i\| \to 0, \qquad (2.3\text{-}25)$$

so $Tx = x$ and $\mathscr{T}$ is closed. To establish convexity we need a very useful inequality.

For any $x, y \in \mathscr{C}$ we have the easily verified identity

$$\|x - y\|^2 - \|Tx - Ty\|^2 = 4\,\mathrm{Re}(Px - Py, (1 - P)x - (1 - P)y), \quad (2.3\text{-}26)$$

where 1 is the identity operator and

$$P = (1 + T)/2. \qquad (2.3\text{-}27)$$

But T is nonexpansive, hence

$$\mathrm{Re}(Px - Py, (1 - P)x - (1 - P)y) \geqslant 0 \qquad (2.3\text{-}28)$$

for every $x, y \in \mathscr{C}$. Since P and T have precisely the same fixed points it suffices to show that the set of fixed points of P is convex.

Let y be any fixed point of P. Then (2.3-28) reduces to

$$\mathrm{Re}(Px - y, x - Px) \geqslant 0 \qquad (2.3\text{-}29)$$

for all $x \in \mathscr{C}$. Conversely, if some $y \in \mathscr{C}$ satisfies (2.3-29) for every $x \in \mathscr{C}$ it satisfies it for $x = y$, which implies that $\|y - Py\| \leqslant 0$ or $y = Py$. In short, the set of fixed points of T is the set of all $y \in \mathscr{C}$ that satisfy (2.3-29) for all $x \in \mathscr{C}$. But this set is obviously convex, Q.E.D.

Corollary. The map T with closed convex domain $\mathscr{C}$ is nonexpansive iff $P = (1 + T)/2$ satisfies (2.3-28) for all $x, y \in \mathscr{C}$.

Definition 2.3-4. A map $T: \mathscr{C} \to \mathscr{H}$ is said to be demiclosed if from

$$\{x_n\} \subset \mathscr{C}, \qquad x_n \rightharpoonup x_0, \qquad x_0 \in \mathscr{C}, \qquad Tx_n \to y_0 \tag{2.3-30}$$

follows $Tx_0 = y_0$.[4]

Definition 2.3-5. A map $T: \mathscr{C} \to \mathscr{C}$ is said to be asymptotically regular if for every $x \in \mathscr{C}$, $T^n x - T^{n+1} x \to \phi$.

Lemma 2.3-2. In a Hilbert space $\mathscr{H}$ let the sequence $\{x_n\}$ converge weakly to x_0. Then, for any $x \neq x_0$,

$$\liminf_{n\to\infty} \|x_n - x\| > \liminf_{n\to\infty} \|x_n - x_0\|. \tag{2.3-31}$$

Proof. Since a weakly convergent sequence is bounded, both limits in (2.3-31) are finite. Thus, to prove this inequality it suffices to note that in the estimate

$$\begin{aligned}\|x_n - x\|^2 &= \|x_n - x_0 + x_0 - x\|^2 \\ &= \|x_n - x_0\|^2 + \|x_0 - x\|^2 + 2\operatorname{Re}(x_n - x_0, x_0 - x)\end{aligned} \tag{2.3-32}$$

the last term tends to zero as $n \to \infty$, Q.E.D.

Lemma 2.3-3. Let T be any nonexpansive map with closed convex domain $\mathscr{C} \subset \mathscr{H}$. Then, $1 - T$ is demiclosed.

Proof. Let $\{x_n\} \subset \mathscr{C}$ converge weakly to x_0 and let $\{x_n - Tx_n\}$ converge strongly to y_0. Then, since T is nonexpansive,

$$\begin{aligned}\liminf_{n\to\infty}\|x_n - x_0\| &\geqslant \liminf_{n\to\infty}\|Tx_n - Tx_0\| = \liminf_{n\to\infty}\|Tx_n - x_n + x_n - Tx_0\| \\ &= \liminf_{n\to\infty}\|x_n - y_0 - Tx_0\| \geqslant \liminf_{n\to\infty}\|x_n - x_0\|\end{aligned} \tag{2.3-33}$$

by Lemma 2.3-2. Hence, again by Lemma 2.3-2, $x_0 = y_0 + Tx_0$ or $(1 - T)x_0 = y_0$ so that $1 - T$ is demiclosed, Q.E.D.

Theorem 2.3-4. Let $T: \mathscr{C} \to \mathscr{C}$ be an asymptotically regular nonexpansive map with closed convex domain $\mathscr{C} \subset \mathscr{H}$ whose set of fixed points $\mathscr{T} \subset \mathscr{C}$ is nonempty. Then, for any $x \in \mathscr{C}$, the sequence $\{T^n x\}$ is weakly convergent to an element of $\mathscr{T}$.

[4] In words, T is demiclosed if for any sequence $\{x_n\} \subset \mathscr{C}$ which converges weakly to $x_0 \in \mathscr{C}$, the strong convergence of the sequence $\{Tx_n\}$ to y_0 implies that $Tx_0 = y_0$.

Proof. For every $y \in \mathcal{T}$, the sequence $\{d_n\} = \{\|T^n x - y\|\}$ is nonincreasing because

$$d_{n+1} = \|T^{n+1}x - y\| = \|T(T^n x) - Ty\| \leqslant \|T^n x - y\| = d_n. \quad (2.3\text{-}34)$$

Thus, the nonnegative limit

$$d(y) = \lim_{n\to\infty} \|T^n x - y\| \quad (2.3\text{-}35)$$

exists as a finite number for every $y \in \mathcal{T}$.

According to Lemma 2.3-1, $\mathcal{T}$ is a closed convex subset of $\mathscr{C}$ and it follows that for any fixed $\delta \geqslant 0$, the set

$$\mathcal{T}_\delta = \{y \in \mathcal{T} : d(y) \leqslant \delta\} \quad (2.3\text{-}36)$$

is a closed bounded convex subset of $\mathcal{T}$ which is nonempty for δ large enough. Indeed, convexity and closure are obvious from (2.3-35), and boundedness is implied by the inequalities

$$\|y\| = \|y - T^n x + T^n x\| \leqslant \|T^n x - y\| + \|T^n x\| \quad (2.3\text{-}37)$$

and

$$\|T^n x\| = \|T^n x - y_0 + y_0\| \leqslant \|T^n x - y_0\| + \|y_0\|, \quad (2.3\text{-}38)$$

where y_0 is any preselected member of $\mathcal{T}$. Explicitly, for any $y \in \mathcal{T}_\delta$

$$\|y\| \leqslant \delta + d(y_0) + \|y_0\|. \quad (2.3\text{-}39)$$

Since bounded closed convex sets are weakly compact (Theorem 2.2-1, Corollary 3), the intersection of all such nonempty $\mathcal{T}_\delta$'s is a nonempty closed bounded convex set $\mathcal{T}_{\delta_0}$. Clearly, δ_0 is the smallest value of δ for which $\mathcal{T}_\delta$ is nonempty. The set $\mathcal{T}_{\delta_0}$ can contain only one element, say y_0, for if we suppose that it also contains $y_1 \neq y_0$, the identity

$$\left\|T^n x - \frac{y_0 + y_1}{2}\right\|^2 = \frac{1}{2}(\|T^n x - y_0\|^2 + \|T^n x - y_1\|^2) - \left\|\frac{y_0 - y_1}{2}\right\|^2 \quad (2.3\text{-}40)$$

yields $d[(y_0 + y_1)/2] < \delta_0$, which contradicts the meaning of δ_0.

The sequence $\{T^n x\}$ converges weakly to y_0. In fact, since this sequence is bounded, it suffices to prove that all possible weak limits of its various subsequences equal y_0.[5] Let $T^{n'}x \rightharpoonup y_1 \neq y_0$. Then, from the asymptotic regularity of T,

$$T^{n'}x - T^{n'+1}x = (1 - T)T^{n'}x \to \phi \quad (2.3\text{-}41)$$

[5] Assume that such is the case but $T^n x \not\rightharpoonup y_0$. Then, for some $f \in \mathscr{H}$ it is true that the sequence $\{(T^n x, f)\}$ fails to converge to (y_0, f). Hence, there exists a subsequence $\{T^{n'}x\}$ of $\{T^n x\}$ such that $\lim(T^{n'}x, f)$ exists and is unequal to (y_0, f). But by hypothesis the sequence $\{T^{n'}x\}$ itself contains a subsequence $\{T^{n''}x\}$ converging weakly to y_0; i.e., $(y_0, f) = \lim(T^{n''}x, f) = \lim(T^{n'}x, f) \neq (y_0, f)$, a contradiction.

and invoking the demiclosed character of $1 - T$ (Lemma 2.3-3), $(1 - T)y_1 = \phi$; i.e., y_1 is a fixed point of T. By Lemma 2.3-2,

$$\delta_0 = \lim\|T^{n'}x - y_0\| > \lim\|T^{n'}x - y_1\| = d(y_1), \tag{2.3-42}$$

which is incompatible with the meaning of δ_0. Thus, for every $x \in \mathscr{C}$, the sequence $\{T^n x\}$ is weakly convergent to a fixed point of T, Q.E.D.

Corollary. The sequence $\{T^n x\}$ converges strongly to y_0 iff at least one of its subsequences converges strongly.

Proof. We know that $T^n x \rightharpoonup y_0$, a fixed point of T, for every choice of $x \in \mathscr{C}$. Clearly, since the weak and strong limits of a sequence must coincide, the only possible strong limit of any subsequence of $\{T^n x\}$ is y_0. Now, from the equality

$$\|T^n x - y_0\|^2 = \|T^n x\|^2 - 2\,\mathrm{Re}(T^n x, y_0) + \|y_0\|^2 \tag{2.3-43}$$

we find that

$$\lim\|T^n x\|^2 = d^2(y_0) + \|y_0\|^2 \tag{2.3-44}$$

as $n \to \infty$. In particular, for any subsequence $\{T^{n'}x\}$ of $\{T^n x\}$,

$$\lim\|T^{n'}x\|^2 = d^2(y_0) + \|y_0\|^2. \tag{2.3-45}$$

But as we have already remarked, if any subsequence $\{T^{n'}x\}$ converges strongly, it converges strongly to y_0 and therefore $\lim\|T^{n'}x\|^2 = \|y_0\|^2$ so that necessarily, $d^2(y_0) = 0$ and

$$T^n x \to y_0, \tag{2.3-46}$$

Q.E.D.

This "all or nothing" aspect of strong convergence exhibited by the iterated sequence $\{T^n x\}$ appears to be the rule at the present level of generality. Unfortunately, the lack of strong convergence can present an obstacle from a numerical standpoint, and a more in-depth analysis is undertaken in the next section.

Definition 2.3-6. A mapping $T: \mathscr{C} \to \mathscr{C}$ is said to be a reasonable wanderer if for every $x \in \mathscr{C}$

$$\sum_{n=0}^{\infty} \|T^n x - T^{n+1}x\|^2 < \infty. \tag{2.3-47}$$

It should be evident that a reasonable wanderer is automatically asymptotically regular (Definition 2.3-5).

An example of a nonexpansive operator that is not asymptotically regular is easily constructed. On the real line choose $\mathscr{C} = [-1, 1]$ and define T by $Tx = -x$. Clearly, T is nonexpansive, maps $\mathscr{C}$ into $\mathscr{C}$, and its only fixed point is $x = 0$. However, $T^n x - T^{n+1} x = 2(-1)^n x \nrightarrow 0$ as $n \to \infty$ unless $x = 0$. Nevertheless, Browder and Petryshyn [7] have shown that the convex linear combination

$$T_\alpha = \alpha 1 + (1 - \alpha) T \tag{2.3-48}$$

of T and the identity operator 1 is a reasonable wanderer for all α in the interval $0 < \alpha < 1$ and any nonexpansive T.

Theorem 2.3-5 [7]. Let $T: \mathscr{C} \to \mathscr{C}$ be a nonexpansive map with closed convex domain $\mathscr{C}$ whose set of fixed points is nonempty. Then for any fixed $\alpha, 0 < \alpha < 1$, $T_\alpha: \mathscr{C} \to \mathscr{C}$ is a reasonable wanderer and the sequence $\{T_\alpha^n x\}$ converges weakly to a fixed point of T for every $x \in \mathscr{C}$.

***Proof*.** Clearly, T_α maps $\mathscr{C}$ into $\mathscr{C}$ so that the iterates $x_n = T_\alpha^n x$ are well defined for all $x \in \mathscr{C}$. In addition, for $\alpha \neq 1$ the fixed points of T and T_α coincide and it only remains to prove that T_α is a reasonable wanderer since it is obviously nonexpansive. Let $y \in \mathscr{C}$ be a fixed point of T. Then, $y = Ty = T_\alpha y$ and

$$\begin{aligned} \|x_{n+1} - y\|^2 &= \|\alpha x_n + (1 - \alpha) T x_n - y\|^2 \\ &= \|\alpha(x_n - y) + (1 - \alpha)(T x_n - y)\|^2 && \text{(2.3-49)} \\ &= \alpha^2 \|x_n - y\|^2 + 2\alpha(1 - \alpha) \operatorname{Re}(x_n - y, T x_n - y) \\ &\quad + (1 - \alpha)^2 \|T x_n - y\|^2. && \text{(2.3-50)} \end{aligned}$$

Similarly,

$$\begin{aligned} \|x_n - T x_n\|^2 &= \|x_n - y - (T x_n - y)\|^2 \\ &= \|x_n - y\|^2 - 2 \operatorname{Re}(x_n - y, T x_n - y) + \|T x_n - y\|^2, && \text{(2.3-51)} \end{aligned}$$

which, after multiplication by $\alpha(1 - \alpha)$ and addition to (2.3-51), yields

$$\begin{aligned} \|x_{n+1} - y\|^2 + \alpha(1 - \alpha)\|x_n - T x_n\|^2 &= \alpha \|x_n - y\|^2 \\ &\quad + (1 - \alpha)\|T x_n - y\|^2 \\ &\leqslant \|x_n - y\|^2. && \text{(2.3-52)} \end{aligned}$$

Hence,

$$\alpha(1 - \alpha)\|x_n - T x_n\|^2 \leqslant \|x_n - y\|^2 - \|x_{n+1} - y\|^2 \tag{2.3-53}$$

so that

$$\alpha(1-\alpha)\sum_{n=0}^{N}\|x_n - Tx_n\|^2 = \frac{\alpha}{1-\alpha}\sum_{n=0}^{N}\|T_\alpha^n x - T_\alpha^{n+1}x\|^2$$
$$\leqslant \|x-y\|^2 - \|x_{N+1}-y\|^2 \leqslant \|x-y\|^2. \tag{2.3-54}$$

Consequently, since $\alpha \neq 0$,

$$\sum_{n=0}^{\infty}\|T_\alpha^n x - T_\alpha^{n+1}x\|^2 < \infty, \tag{2.3-55}$$

Q.E.D.[6]

2.4 ITERATIVE TECHNIQUES FOR IMAGE RESTORATION IN A HILBERT SPACE SETTING

According to the discussion initiated in the introduction, there exists a class of image restoration problems in which the unknown f can be ascertained to lie in an intersection

$$\mathscr{C}_0 = \bigcap_{i=1}^{m}\mathscr{C}_i \tag{2.4-1}$$

of well-defined closed convex sets $\mathscr{C}_i$, $i = 1 \to m$. Evidently, $\mathscr{C}_0$ is also a closed nonempty convex set containing f, and we shall denote the respective projection operators projecting onto $\mathscr{C}_0$ and $\mathscr{C}_i$ by P_0 and P_i, $i = 1 \to m$. Clearly, f is a fixed point of P_0 and all the P_i's. More generally, f is a fixed point of P_0 and of every

$$T_i = 1 + \lambda_i(P_i - 1) \tag{2.4-2}$$

for any choice of relaxation constants $\lambda_1, \lambda_2, \ldots, \lambda_m$. Hence, under the same conditions, f is a fixed point of P_0 and the composition operator

$$T = T_m T_{m-1} \cdots T_1. \tag{2.4-3}$$

Conversely, can we say that any fixed point f of T lies in $\mathscr{C}_0$, and does the sequence $\{T^n x\}$ converge (weakly or strongly) to a point of $\mathscr{C}_0$ as n tends to infinity, irrespective of the initialization $x \in \mathscr{H}$?

Theorem 2.4-1. Let $\mathscr{C}_0$ be nonempty. Then, for every $x \in \mathscr{H}$ and every choice of constants $\lambda_1, \lambda_2, \ldots, \lambda_m$ in the interval $0 < \lambda < 2$, the sequence $\{T^n x\}$ converges weakly to a point of $\mathscr{C}_0$.

[6] For $\alpha = 1/2$, $T_\alpha = (1 + T)/2$, in which case Theorem 2.3-5 is easily deduced from the inequality (2.3-28).

Proof. For $\lambda_i \geqslant 0$, every

$$T_i = 1 + \lambda_i(P_i - 1) = (1 - \lambda_i)1 + \lambda_i P_i \tag{2.4-4}$$

is nonexpansive. The assertion is obviously correct for $0 \leqslant \lambda_i \leqslant 1$, but if $1 < \lambda_i$ we have $1 - \lambda_i < 0$ and it is necessary to reason differently. With the aid of Eqs. (2.2-27) and (2.2-30), it is found that

$$\begin{aligned}\|T_i x - T_i y\|^2 &= \|(1 - \lambda_i)(x - y) + \lambda_i(P_i x - P_i y)\|^2 &(2.4\text{-}5)\\ &= (1 - \lambda_i)^2\|x - y\|^2 + 2\lambda_i(1 - \lambda_i)\operatorname{Re}(x - y, P_i x - P_i y)\\ &\quad + \lambda_i^2\|P_i x - P_i y\|^2\\ &\leqslant (1 - \lambda_i)^2\|x - y\|^2 + (\lambda_i^2 + 2\lambda_i(1 - \lambda_i))\|P_i x - P_i y\|^2\\ &= (1 - \lambda_i)^2\|x - y\|^2 + \lambda_i(2 - \lambda_i)\|P_i x - P_i y\|^2\\ &\leqslant (\lambda_i(2 - \lambda_i) + (1 - \lambda_i)^2)\|x - y\|^2 = \|x - y\|^2 &(2.4\text{-}6)\end{aligned}$$

and nonexpansivity is established. We now show that T is a reasonable wanderer.

For $m = 1$, we have $T = T_1$, $\mathscr{C}_0 = \mathscr{C}_1$, and

$$\|x - Tx\|^2 = \lambda_1^2\|x - P_1 x\|^2. \tag{2.4-7}$$

Moreover, for any $y \in \mathscr{C}_0$, $Ty = P_1 y = y$ and

$$\begin{aligned}\|Tx - y\|^2 &= \|x - y + \lambda_1(P_1 x - x)\|^2 &(2.4\text{-}8)\\ &= \|x - y\|^2 + 2\lambda_1\operatorname{Re}(x - y, P_1 x - x) + \lambda_1^2\|x - P_1 x\|^2\\ &= \|x - y\|^2 - \lambda_1(2 - \lambda_1)\|x - P_1 x\|^2\\ &\quad + 2\lambda_1\operatorname{Re}(x - P_1 x, y - P_1 x) &(2.4\text{-}8a)\\ &\leqslant \|x - y\|^2 - \lambda_1(2 - \lambda_1)\|x - P_1 x\|^2 &(2.4\text{-}9)\end{aligned}$$

because the last term in Eq. (2.4-8a) is nonpositive. Thus, by combining Eqs. (2.4-7) and (2.4-9) it is found that

$$\|x - Tx\|^2 \leqslant \frac{\lambda_1}{2 - \lambda_1}(\|x - y\|^2 - \|Tx - y\|^2) \tag{2.4-10}$$

for $0 < \lambda_1 < 2$.

For arbitrary $m \geqslant 1$, a straightforward induction on m yields the inequality

$$\|x - Tx\|^2 \leqslant b_m 2^{m-1}(\|x - y\|^2 + \|Tx - y\|^2), \tag{2.4-11}$$

where

$$b_m = \sup_{1 \leqslant i \leqslant m}\left\{\frac{\lambda_i}{2 - \lambda_i}\right\}. \tag{2.4-12}$$

[Clearly, Eq. (2.4-10) subsumes the case $m = 1$.] In fact, let $T = T_m K$, where

$$K = T_{m-1}T_{m-2}\cdots T_1, \tag{2.4-13}$$

and observe that for $m \geq 2$,

$$\begin{aligned}\|x - Tx\|^2 &= \|x - Kx + Kx - Tx\|^2 \\ &\leq (\|x - Kx\| + \|Kx - Tx\|)^2 \\ &\leq 2(\|x - Kx\|^2 + \|Kx - Tx\|^2) \\ &\leq 2(\|x - Kx\|^2 + 2^{m-2}\|Kx - T_m Kx\|^2).\end{aligned} \tag{2.4-14}$$

Thus, by the induction hypothesis,[7]

$$\begin{aligned}\|x - Tx\|^2 &\leq b_m 2(2^{m-2}\|x - y\|^2 - 2^{m-2}\|Kx - y\|^2 \\ &\quad + 2^{m-2}\|Kx - y\|^2 - 2^{m-2}\|Tx - y\|^2) \\ &= b_m 2^{m-1}(\|x - y\|^2 - \|Tx - y\|^2),\end{aligned} \tag{2.4-15}$$

the desired inequality.

It now follows immediately from Eq. (2.4-16) that

$$\sum_{n=0}^{\infty} \|T^n x - T^{n+1}x\|^2 \leq b_m 2^{m-1}\|x - y\|^2 < \infty \tag{2.4-16}$$

and T is therefore a reasonable wanderer and, *a fortiori*, asymptotically regular. By Theorem 2.3-4, the sequence $\{T^n x\}$ converges weakly to a fixed point of T.

However, the fixed points of T coincide precisely with the points of $\mathscr{C}_0$, the intersection of the $\mathscr{C}_i$'s. Indeed, it is obvious that $x \in \mathscr{C}_0$ implies $x = Tx$ since $x \in \mathscr{C}_i$, $i = 1 \to m$. Conversely, if $x = Tx$ and $y \in \mathscr{C}_0$,

$$\|x - y\| = \|Tx - Ty\| \leq \|T_1 x - T_1 y\| = \|T_1 x - y\| \leq \|x - y\|; \tag{2.4-18}$$

hence $\|x - y\| = \|T_1 x - y\|$. In view of Eq. (2.4-10) this is possible only if $x = T_1 x$ so that $x = T_m T_{m-1} \cdots T_2 x$, and a repetition of the argument finally leads to $x \in \mathscr{C}_i$, $i = 1 \to m$; i.e., $x \in \mathscr{C}_0$ and the proof is complete, Q.E.D.[8]

We are now in a position to address the question of strong convergence. Let $T: \mathscr{C} \to \mathscr{C}$ denote a nonexpansive asymptotically regular operator with closed convex domain $\mathscr{C} \subset \mathscr{H}$ whose set $\mathscr{T}$ of fixed points is nonempty. From the corollary to Theorem 2.3-4, the sequence $\{T^n x\}$ converges strongly to a member of $\mathscr{T}$ iff at least one of its subsequences converges strongly. Clearly, such is assured if all iterates $T^n x$ lie ultimately in some compact or finite-dimensional subset of $\mathscr{H}$.

[7] $b_m \geq \lambda_m/(2 - \lambda_m)$ and $b_m \geq \sup_{1=i\leq m-1}\{\lambda_i/(2 - \lambda_i)\}$.

[8] That a composition of appropriately relaxed projection operators is automatically a reasonable wanderer is of course directly attributable to the special properties of projections onto closed convex sets.

The latter possibility exists even if $\mathscr{C}$, the domain of T, is itself not compact, as is seen in the case of the composition operator T of Eq. (2.4-3) with all λ_i's equal to one. Here, $\mathscr{C} = \mathscr{H}$, $T_i = P_i$, $i = 1 \to m$,

$$T = P_m P_{m-1} \cdots P_1 \tag{2.4-19}$$

and all iterates $T^n x$, $n \geq 1$, lie in $\mathscr{C}_m$. It suffices therefore that $\mathscr{C}_m$ be compact or finite-dimensional.[9] [Assumptions of this kind appear to be quite natural for those digital restoration problems falling within the scope of Theorem (2.4-1).]

It is interesting and important to note that if all $\mathscr{C}_i$'s in Theorem (2.4-1) are CLM's, the sequence of iterates $\{T^n x\}$ not only converges strongly but actually converges to the projection of x onto $\mathscr{C}_0$. This result is an extension of I. Halperin's neat generalization of von Neumann's celebrated alternating projection theorem [8] from $m = 2$ to arbitrary m [9].

Corollary 1. Under the constraints of Theorem 2.4-1, if every $\mathscr{C}_i$ is a CLM, T^n converges strongly to P_0, the orthogonal projection operator onto the CLM $\mathscr{C}_0$.

Proof.[10] Let us first remark that now all P_i's and P_0 are bounded linear self-adjoint operators. Consequently,

$$T_i^* = 1 + \lambda_i(P_i^* - 1) = 1 + \lambda_i(P_i - 1) = T_i \tag{2.4-20}$$

is self-adjoint and linear, $i = 1 \to m$, and T^n is also linear for all $n \geq 0$. Moreover, since T is nonexpansive and $T\phi = \phi$,

$$\|Tx\| = \|Tx - T\phi\| \leq \|x - \phi\| = \|x\| \tag{2.4-21}$$

for all $x \in \mathscr{H}$ so that $\|T\|$ and therefore $\|T^n\| \leq 1$, $n \geq 0$.

As has already been shown in Theorem 2.4-1, for every $x \in \mathscr{H}$,

$$T^n x - T^{n+1} x = T^n(1 - T)x \to \phi; \tag{2.4-22}$$

hence $T^n y \to \phi$ for all y in $R(1 - T)$, the range of $1 - T$. Now [10] the closure $[R(1 - T)]$ of the range of the bounded operator $1 - T$ is the orthogonal complement of the null space of

$$(1 - T)^* = 1 - (T_m T_{m-1} \cdots T_1)^* = 1 - T_1 T_2 \cdots T_m, \tag{2.4-23}$$

from which it is deduced that x belongs to this null space iff

$$x = T_1 T_2 \cdots T_m x. \tag{2.4-24}$$

[9] Actually, it is enough that at last one of the m $\mathscr{C}_i$'s be compact.
[10] Since $\phi \in \mathscr{C}_i$, $i = 1 \to m$, $\phi \in \mathscr{C}_0$, which is therefore never empty.

In Theorem 2.4-1 the update rule is given by $x_{n+1} = Tx_n$ with $x_0 = x$ an arbitrary element of $\mathscr{H}$. More generally, we can also work with the single-step cyclic scheme

$$\begin{aligned} x_1 &= T_1 x \\ x_2 &= T_2 x_1 \\ &\vdots \\ x_m &= T_m x_{m-1} = T_m T_{m-1} \cdots T_1 x = Tx \qquad (2.4\text{-}40) \\ x_{m+1} &= T_1 x_m \\ &\vdots \\ x_{2m} &= T_m x_{2m-1} = T_m T_{m-1} \cdots T_1 T x = T^2 x \\ x_{2m+1} &= T_1 x_{2m} \\ &\vdots \end{aligned}$$

Or, put more succinctly,

$$x_{n+1} = T_{\alpha(n)} x_n, \qquad (2.4\text{-}41)$$

where $x_0 = x$ and[13]

$$\alpha(n) = 1 + n \bmod m, \qquad n \geqslant 0. \qquad (2.4\text{-}42)$$

It is easily shown that the sequences $\{T^n x\}$ and $\{x_n\}$ converge weakly to the same limit $x^* \in \mathscr{C}_0$. Write $n = lm + r$, $0 \leqslant r \leqslant m - 1$. Then, for $n \geqslant 1$

$$x_n = T_r T_{r-1} \cdots T_1 T^l x \qquad (2.4\text{-}43)$$

and $l \to \infty$ as $n \to \infty$. Let $T^l x \rightharpoonup x^*$ and let the subsequence $\{x_{n'}\}$ of $\{x_n\}$ converge weakly to x'.

Then, $n' = ml' + r'$, $0 \leqslant r' \leqslant m - 1$, and $l' \to \infty$ as $n' \to \infty$. Moreover, making use of Eq. (2.4-11) with T replaced by $T_{r'} T_{r'-1} \cdots T_1$ and x by $T^{l'} x$, we obtain, for every $y \in \mathscr{C}_0$,

$$\|T^{l'} x - x_{n'}\|^2 \leqslant b_m 2^{m-1} (\|T^{l'} x - y\|^2 - \|x_{n'} - y\|^2) \qquad (2.4\text{-}44)$$

$$\leqslant b_m 2^{m-1} (\|T^{l'} x - y\|^2 - \|T^{l'+1} x - y\|^2). \qquad (2.4\text{-}45)$$

[13] $n \bmod m$ is the remainder obtained on division of n by m and is one of the integers $0, 1, \ldots, m-1$.

Hence, $\|T^{l'}x - x_{n'}\| \to 0$ as $n' \to \infty$. But $T^{l'}x - x_{n'} \rightharpoonup x^* - x'$, and since[14]

$$\|x^* - x'\| \leqslant \liminf_{n' \to \infty} \|T^{l'}x - x_{n'}\| = 0, \tag{2.4-46}$$

$x^* = x'$, as was to be shown. [To the best of our knowledge, the first proof of weak convergence of the iteration (2.4-41) to a point of $\mathscr{C}_0$ is due to Bregman [12]. However, as we have seen, the underlying theory is the same as that of the full composition operator $T = T_m T_{m-1} \cdots T_1$ and its associated fixed-point recursion $x_{n+1} = Tx_n$.]

The first lemma of this section is preparatory to a deeper study of the strong convergence of the sequence $\{T^n x\}$.

Lemma 2.4-1. Let T be any nonexpansive map whose set of fixed points includes a given closed convex set $\mathscr{T}$, and let $P_{\mathscr{T}}$ denote the projection operator onto $\mathscr{T}$. Then, for any $x \in$ domain T,

$$\operatorname{Re}(x - Tx, x - P_{\mathscr{T}}x) \geqslant 0 \tag{2.4-47}$$

and

$$\|x - Tx\| \leqslant 2\|x - P_{\mathscr{T}}x\|. \tag{2.4-48}$$

Proof. Clearly, for every x, y in domain T,

$$\|Tx - Ty\|^2 - \|x - y\|^2 \leqslant 0. \tag{2.4-49}$$

In particular, if $y = P_{\mathscr{T}}x$,

$$\|Tx - P_{\mathscr{T}}x\|^2 \leqslant \|x - P_{\mathscr{T}}x\|^2 \tag{2.4-50}$$

because $P_{\mathscr{T}}x \in \mathscr{T}$ and $TP_{\mathscr{T}}x = P_{\mathscr{T}}x$, by hypothesis. Thus,

$$\|x - Tx - (x - P_{\mathscr{T}}x)\|^2 \leqslant \|x - P_{\mathscr{T}}x\|^2 \tag{2.4-51}$$

and an easy simplification yields

$$\|x - Tx\|^2 \leqslant 2\operatorname{Re}(x - Tx, x - P_{\mathscr{T}}x), \tag{2.4-52}$$

$x \in$ domain T. From Eq. (2.4-52) and Schwarz's inequality we now obtain Eqs. (2.4-47) and (2.4-48) immediately, Q.E.D.

[14] Let $y_n \rightharpoonup f$. Then, for every $y \in \mathscr{H}$, $(f, y) = \lim(y_n, y)$ as $n \to \infty$. In particular, $|(f, y)| = \lim|(y_n, y)|$. But from Schwarz's inequality, $|(y_n, y)| \leqslant \|y\| \cdot \|y_n\|$ so that

$$|(f, y)| \leqslant \|y\| \cdot \liminf_{n \to \infty} \|y_n\|.$$

Thus, for $y = f$,

$$\|f\| \leqslant \liminf_{n \to \infty} \|y_n\|.$$

Inequality (2.4-47) simply states that Tx lies on the side of the support plane through x, parallel to that through $P_{\mathcal{T}}x$, containing $\mathcal{T}$ (Fig. 2.4-1).

Lemma 2.4-2. The iterated sequence $\{T^n x\}$ of Theorem 2.3.4 converges strongly iff

$$\|T^n x - P_{\mathcal{T}} T^n x\| \to 0 \tag{2.4-53}$$

as $n \to \infty$.[15]

Proof. As shown in Theorem 2.3-4, the sequence $\{T^n x\}$ converges weakly to some point $x^* \in \mathcal{T}$. Evidently, if the convergence is actually strong,

$$d(T^n x, \mathcal{T}) = \|T^n x - P_{\mathcal{T}} T^n x\| \leq \|T^n x - x^*\| \to 0 \tag{2.4-54}$$

as $n \to \infty$. Conversely, suppose that $d(T^n x, \mathcal{T}) \to 0$. Then, since T^k is nonexpansive for all integers $k \geq 0$ and includes $\mathcal{T}$ in its sets of fixed points, Eq. (2.4-48) yields, for all $x \varepsilon$ domain T,

$$\|x - T^k x\| \leq 2\|x - P_{\mathcal{T}} x\|, \qquad k \geq 0. \tag{2.4-55}$$

In particular, if x is replaced by $T^n x$ we obtain

$$\|T^n x - T^{n+k} x\| \leq 2\|T^n x - P_{\mathcal{T}} T^n x\| \to 0 \tag{2.4-56}$$

as $n \to \infty$. Thus, the sequence $\{T^n x\}$ is Cauchy and $T^n x \to x^*$, Q.E.D.

According to Lemma 2.4-2, convergence in Theorem 2.4-1 of the sequence $\{T^n x\}$ to a point of $\mathscr{C}_0$ is strong iff

$$d(T^n x, \mathscr{C}_0) = \|T^n x - P_0 T^n x\| \to 0 \tag{2.4-57}$$

as $n \to \infty$. (The set of fixed points of $T = T_m T_{m-1} \cdots T_1$ coincides with $\mathscr{C}_0$.) We can show without the imposition of any additional constraints that for $r = 1 \to m$,

$$d(T^n x, \mathscr{C}_r) = \|T^n x - P_r T^n x\| \to 0 \tag{2.4-58}$$

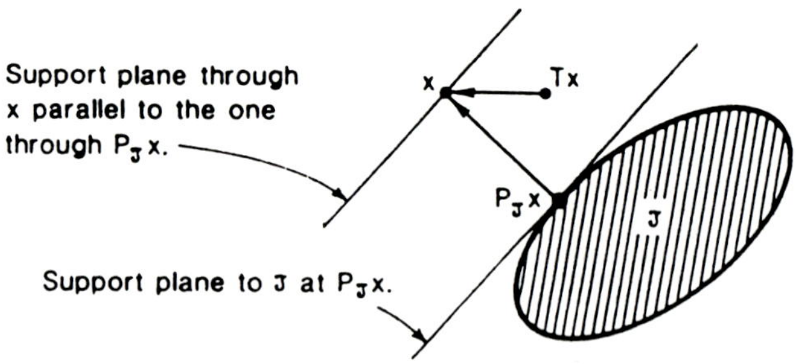

Fig. 2.4-1. Geometric interpretation of inequality (2.4-47).

[15] Clearly, $\|T^n x - P_{\mathcal{T}} T^n x\| = d(T^n x, \mathcal{T})$, the distance of $T^n x$ from $\mathcal{T}$.

as $n \to \infty$. (The distance of $T^n x$ from every $\mathscr{C}_r$ goes to zero as n goes to infinity.)

To see this, note that as a special case of Eq. (2.4-11),[16]

$$\|T^n x - P_r P_{r-1} \cdots P_1 T^n x\|^2 \leqslant 2^{r-1}(\|T^n x - y\|^2 - \|P_r P_{r-1} \cdots P_1 T^n x - y\|^2) \tag{2.4-59}$$

$$\leqslant 2^{r-1}(\|T^n x - y\|^2 - \|T^{n+1} x - y\|^2) \to 0 \tag{2.4-60}$$

for all $y \in \mathscr{C}_0$ and every $r = 1 \to m$ because

$$\|T^{n+1} x - y\| \leqslant \|P_r P_{r-1} \cdots P_1 T^n x - y\| \tag{2.4-61}$$

and

$$\lim_{n\to\infty} \|T^n x - y\| \tag{2.4-62}$$

exists (Theorem 2.3-4). Thus, since $P_r P_{r-1} \cdots P_1 T^n x \in \mathscr{C}_r$,

$$d(T^n x, \mathscr{C}_r) \leqslant \|T^n x - P_r P_{r-1} \cdots P_1 T^n x\| \to 0 \tag{2.4-63}$$

as $n \to \infty$ and Eq. (2.4-58) is established. Our problem therefore is to find some other restriction, preferably very weak, which together with Eq. (2.4-58) implies Eq. (2.4-57). The next theorem complements Theorem 2.4-1 and supplies two sufficient conditions for strong convergence, the second of which appeared for the first time in the excellent paper by Gubin *et al.* [3] already referred to in Theorem 2.2-3.

Theorem 2.4-2. Either of the two conditions (1) or (2) stated below suffices to guarantee the *strong* convergence of the sequence $\{T^n x\}$ of Theorem 2.4-1 to its weak limit x^*.

(1) At least one of the $\mathscr{C}_i$'s is uniformly convex and does not contain x^* in its interior.[17]

(2) For some α, $1 \leqslant \alpha \leqslant m$, the set intersection

$$\mathscr{C}_\alpha \cap \left(\bigcap_{i \neq \alpha}^{m} \mathscr{C}_i \right)^{\circ} \tag{2.4-64}$$

is nonempty (and therefore contains a point of $\mathscr{C}_0$).

In case (2) the strong convergence of $\{T^n x\}$ to x^* is at a *geometric* rate.

Proof. (1) To be specific, let us suppose that $\mathscr{C}_1$ is uniformly convex and contains x^* on its norm boundary. As we have already shown in Eq. (2.4-58)

$$d(T^n x, \mathscr{C}_1) = \|T^n x - P_1 T^n x\| \to 0 \tag{2.4-65}$$

[16] Identify T with $P_r P_{r-1} \cdots P_1$ and set $m = r$ and $\lambda_i = 1$, $i = 1 \to r$.

[17] The interior of any set $\mathscr{G}$ is denoted by $\mathscr{G}^{\circ}$.

as $n \to \infty$. Clearly,[18] every $y_n = P_1 T^n x$ lies in $\mathscr{C}_1$ and $y_n \rightharpoonup x^*$. Now, if the convergence of the sequence $\{y_n\}$ to x^* is not strong, there exists an $\varepsilon > 0$ and a subsequence $\{y_{n'}\}$ such that

$$\|y_{n'} - x^*\| \geqslant \varepsilon \tag{2.4-66}$$

for all n'.

Thus, since $x^* \in \mathscr{C}_0 \subset \mathscr{C}_1$ and the latter is uniformly convex (2.2-3), the point

$$z_{n'} = \frac{y_{n'} + x^*}{2} + h \tag{2.4-67}$$

is a member of $\mathscr{C}_1$ for all n' and all h such that

$$\|h\| \leqslant \delta(\varepsilon) > 0. \tag{2.4-68}$$

But for any such fixed choice of h, $z_{n'} \rightharpoonup x^* + h$, which belongs to $\mathscr{C}_1$ because the latter is weakly closed. Consequently, x^* is the center of a sphere of radius $\delta(\varepsilon) > 0$ contained in $\mathscr{C}_1$, and this contradicts the assumption that x^* is not an interior point of $\mathscr{C}_1$. Therefore the convergence of $\{y_n\}$ and hence of $\{T^n x\}$ to x^* must be strong. [Note that $\|T^n x - x^*\| \leqslant \|T^n x - y_n\| + \|y_n - x^*\| \to 0$ as $n \to \infty$.]

(2) Let z belong to the intersection (2.4-64). Then, $z \in \mathscr{C}_0$ and for some $\delta > 0$ it is true that $h \in \mathscr{C}_i$, $i \neq \alpha$, for all h such that

$$\|h - z\| \leqslant \delta. \tag{2.4-69}$$

Let $y = P_\alpha T^n x$. Then, for any $\varepsilon > 0$, no matter how small,

$$w = \frac{\varepsilon}{\varepsilon + \delta} z + \frac{\delta}{\varepsilon + \delta} \cdot y \tag{2.4-70}$$

is a member of $\mathscr{C}_0$ for all sufficiently large n.

In fact, since $z \in \mathscr{C}_\alpha$ and $y \in \mathscr{C}_\alpha$ and $\mathscr{C}_\alpha$ is convex, $w \in \mathscr{C}_\alpha$. Now observe that

$$\begin{aligned} \|y - P_i y\| = d(y, \mathscr{C}_i) &\leqslant \|y - P_i T^n x\| \\ &\leqslant \|T^n x - y\| + \|T^n x - P_i T^n x\| \qquad (2.4\text{-}71) \\ &= d(T^n x, \mathscr{C}_\alpha) + d(T^n x, \mathscr{C}_i) \leqslant \frac{\varepsilon}{2} + \frac{\varepsilon}{2} = \varepsilon \qquad (2.4\text{-}72) \end{aligned}$$

[18] For any $z \in \mathscr{H}$,

$$|(T^n x, z) - (y_n, z)| = |(T^n x - y_n, z)| \leqslant \|T^n x - y_n\| \, \|z\| \to 0$$

as $n \to \infty$. Hence, $\lim_{n\to\infty}(y_n, z) = \lim_{n\to\infty}(T^n x, z) = (x^*, z)$ so that $y_n \rightharpoonup x^*$.

for n large enough.[19] Thus, for all such n,

$$h = z + \frac{\delta}{\varepsilon}(y - P_i y) \tag{2.4-73}$$

belongs to $\mathscr{C}_i$, $i \neq \alpha$, because $\|h - z\| \leq \delta$. From the identity

$$w = \frac{\varepsilon}{\varepsilon + \delta}\left(z + \frac{\delta}{\varepsilon}(y - P_i y)\right) + \frac{\delta}{\varepsilon + \delta} P_i y = \frac{\varepsilon}{\varepsilon + \delta} h + \frac{\delta}{\varepsilon + \delta} P_i y \tag{2.4-74}$$

and the convexity of $\mathscr{C}_i$, we also conclude that $w \in \mathscr{C}_i$, $i \neq \alpha$; i.e., $w \in \mathscr{C}_i$, $i = 1 \to m$, and $w \in \mathscr{C}_0$.

Consequently,

$$d(T^n x, \mathscr{C}_0) \leq \|T^n x - w\| \leq \|T^n x - y\| + \|y - w\| \tag{2.4-75}$$

$$= d(T^n x, \mathscr{C}_\alpha) + \frac{\varepsilon}{\varepsilon + \delta}\|y - z\| \tag{2.4-76}$$

$$\leq \frac{\varepsilon}{2} + \frac{\varepsilon R}{\delta} = \left(1 + \frac{2R}{\delta}\right)\frac{\varepsilon}{2} = \frac{c\varepsilon}{2} \quad , \tag{2.4-77}$$

where

$$c = (1 + 2R/\delta) \tag{2.4-78}$$

and R is an upper bound on $\|y - z\|$.[20] Since ε is arbitrary, $d(T^n x, \mathscr{C}_0)$ converges to zero as $n \to \infty$ and $\{T^n x\}$ is therefore strongly convergent to x^*. It remains to establish that convergence is at a geometric rate.

Observe first that $\varepsilon/2$ in Eq. (2.4-77) is by construction, and for sufficiently large n, an upper bound on the m distances $d(T^n x, \mathscr{C}_i)$, $i = 1 \to m$:

$$d(T^n x, \mathscr{C}_i) = \|T^n x - P_i T^n x\| \leq \frac{\varepsilon}{2} \quad , \qquad i = 1 \to m. \tag{2.4-79}$$

Hence, for all large enough n,

$$d(T^n x, \mathscr{C}_0) \leq c \sup_{i=1 \to m} \|T^n x - P_i T^n x\|. \tag{2.4-80}$$

Choose the positive numbers ε_1 and ε_2 so that $\varepsilon_1 \varepsilon_2 / c^2 < 1$ and $0 < \varepsilon_1 \leq \lambda_i \leq 2 - \varepsilon_2$, $i = 1 \to m$. Then, from Eqs. (2.4-10) and (2.4-7) with T replaced

[19] Remember that according to Eq. (2.4-58) for $i = 1 \to m$, $\lim_{n\to\infty} d(T^n x, \mathscr{C}_i) = 0$.

[20] $\|y - z\| = \|P_\alpha T^n x - z\| \leq \|T^n x - z\|$ for all $z \in \mathscr{C}_0$. But the latter is a monotone non-increasing function of n and is therefore bounded by some finite number R.

by T_i, x by $T^n x$, and y by $P_0 T^n x$ we obtain, for every $i = 1 \to m$,

$$\begin{aligned} d^2(T^n x, \mathscr{C}_0) &\leqslant c^2 d^2(T^n x, \mathscr{C}_i) \\ &\leqslant \frac{c^2}{\lambda_i(2-\lambda_i)}(d^2(T^n x, \mathscr{C}_0) - \|T_k T^n x - P_0 T^n x\|^2) \qquad (2.4\text{-}81) \\ &\leqslant \frac{c^2}{\varepsilon_1\varepsilon_2}(d^2(T^n x, \mathscr{C}_0) - \|T^{n+1} x - P_0 T^n x\|^2) \qquad (2.4\text{-}82) \\ &\leqslant \frac{c^2}{\varepsilon_1\varepsilon_2}(d^2(T^n x, \mathscr{C}_0) - d^2(T^{n+1} x, \mathscr{C}_0)). \qquad (2.4\text{-}83) \end{aligned}$$

Therefore,

$$d(T^{n+1}x, \mathscr{C}_0) \leqslant \left(1 - \frac{\varepsilon_1\varepsilon_2}{c^2}\right)^{1/2} d(T^n x, \mathscr{C}_0), \qquad (2.4\text{-}84)$$

and by iteration,

$$\begin{aligned} d(T^n x, \mathscr{C}_0) &\leqslant \left(1 - \frac{\varepsilon_1\varepsilon_2}{c^2}\right)^{n/2} \cdot d(x, \mathscr{C}_0) \qquad (2.4\text{-}85) \\ &\leqslant R\left(1 - \frac{\varepsilon_1\varepsilon_2}{c^2}\right)^{n/2}. \qquad (2.4\text{-}86) \end{aligned}$$

The proof can now be brought to a quick conclusion. From Eq. (2.4-56), with $\mathscr{T} = \mathscr{C}_0$ and $P_{\mathscr{T}} = P_0$, Eq. (2.4-86) yields, for every integer $k \geqslant 1$,

$$\|T^n x - T^{n+k} x\| \leqslant 2R\left(1 - \frac{\varepsilon_1\varepsilon_2}{c^2}\right)^{n/2}. \qquad (2.4\text{-}87)$$

But for fixed n, $T^{n+k}x \to x^*$ as $k \to \infty$, so that a passage to the limit in Eq. (2.4-87) gives

$$\|T^n x - x^*\| \leqslant 2R\left(1 - \frac{\varepsilon_1\varepsilon_2}{c^2}\right)^{n/2}; \qquad (2.4\text{-}88)$$

i.e., $T^n x$ converges strongly to its limit x^* at least as fast as $2Ra^n$ where

$$a = \left(1 - \frac{\varepsilon_1\varepsilon_2}{c^2}\right)^{1/2} < 1, \qquad (2.4\text{-}89)$$

Q.E.D.[21]

[21] Except for some minor improvements, especially in the part dealing with geometric convergence, our proof of (2), Theorem 2.4-2, is in all respects the same as the original one given by Gubin *et al.* [3].

The idea behind the proof of (1), Theorem 2.4-2, is simple but the criterion itself is very effective. It differs fundamentally from (2) and, taken in the context of Theorem 2.4-1, appears to be new. Apart from the important uniform convexity requirement, (1) succeeds because the weak limit x^* is on the boundary of at least one $\mathscr{C}_i$, whereas the success of (2) depends entirely on the fact that at least one point of $\mathscr{C}_0$, not necessarily x^*, is interior to the set intersection of some $m - 1$ $\mathscr{C}_i$'s. In general, without the introduction of smoothing constraints, (2) is of limited applicability because for the class of image restoration problems we have in mind, the unknown f is usually a boundary point and convex sets with interiors are hard to come by.

For example, suppose that $\mathscr{H} = L_2$, the Hilbert space of all functions $f(t)$ of the real variable t square-integrable over $-\infty < t < \infty$. Let $F(\omega)$ denote its L_2-Fourier transform and let us denote the correspondence by $f(t) \leftrightarrow F(\omega)$. Then if $f_i(t) \leftrightarrow F_i(\omega)$, $i = 1, 2,$[22]

$$(f_1, f_2) \equiv \int_{-\infty}^{\infty} f_1(t)\bar{f}_2(t)\, dt$$

$$= \frac{1}{2\pi}\int_{-\infty}^{\infty} F_1(\omega)\bar{F}_2(\omega)\, d\omega. \tag{2.4-90}$$

For any prescribed $b > 0$ let $\mathscr{C}_b$ denote the subset of $\mathscr{H}$ composed of all f's band-limited to b; i.e., $f \in \mathscr{C}_b$ iff $F(\omega) = 0$, $|\omega| > b$. It is obvious that $\mathscr{C}_b$ is a CLM devoid of any interior points. Indeed, given any $\varepsilon > 0$ and any $f \in \mathscr{C}_b$ there exists a $g \notin \mathscr{C}_b$ such that

$$\|f - g\|^2 = \int_{-\infty}^{\infty} |f(t) - g(t)|^2\, dt < \varepsilon^2. \tag{2.4-91}$$

[Thus, as measured in terms of energy content, any size neighborhood of a band-limited signal contains signals that are not band-limited.]

As another illustration, for any prescribed $a > 0$, let $\mathscr{C}_a$ denote the set of all f's whose orthogonal projection $P_a f$ onto the closed interval $-a \leqslant t \leqslant a$ is a given function $g(t)$. [Of course, $g(t) = 0$, $|t| > a$.] It is easily seen that $\mathscr{C}_a$ is a linear variety whose interior is also empty. Thus $\mathscr{C}_a$ and $\mathscr{C}_b$ are two practical examples of *boundary* sets that occur repeatedly in many areas of signal processing.

Let $\mathscr{C}_1$ denote the closed unit sphere in $\mathscr{H}$ and let g be any fixed element of norm equal to one. Clearly, g lies on the boundary of $\mathscr{C}_1$, a closed

[22] For a scalar x, $\bar{x}$ denotes its complex conjugate, and as usual Re x and Im x are abbreviations for the real and imaginary parts of x, respectively.

uniformly convex set.[23] Let $\mathscr{C}_2$ denote a closed half-space whose boundary is tangent to $\mathscr{C}_1$ at g. Obviously, g is the only point common to the intersection $\mathscr{C}_1 \cap \mathscr{C}_2$. To find g we can generate the iterates $T^n x$, where $T = P_2 P_1$ and x is arbitrary. (Figure 2.4-2 should clarify the mechanics.)

Since $x^* = g$ lies on the boundary of $\mathscr{C}_1$, criterion (1) guarantees that $T^n x \to g$; however, (2) fails because $\mathscr{C}_2 \cap \mathscr{C}_1^\circ$ is empty. Nevertheless, geometric convergence is precluded by the tangency of the two boundaries at the point of contact x^*.

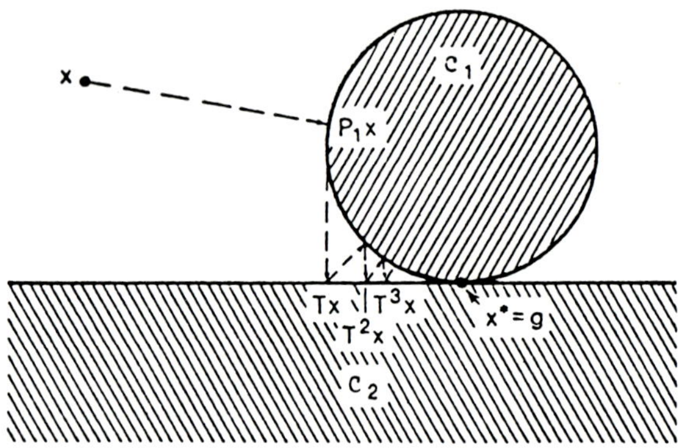

Fig. 2.4-2. Illustration of strong convergence guaranteed by Theorem 2.4-2, Condition (1).

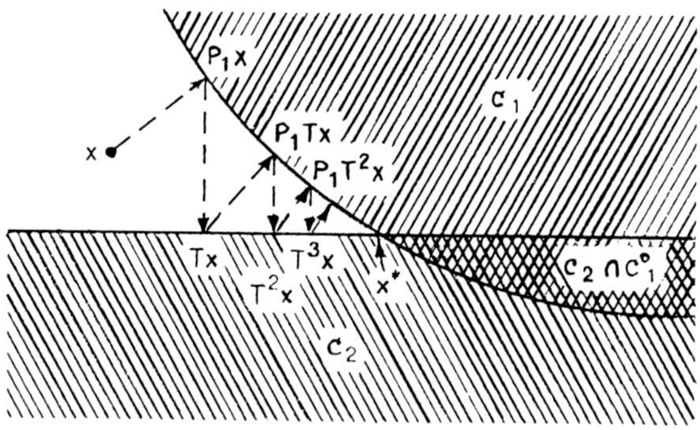

Fig. 2.4-3. Illustration of geometric convergence guaranteed by Theorem 2.4-2, Condition (2).

[23] By making use of the identity

$$\left\|\frac{x+y}{2}\right\|^2 + \left\|\frac{x-y}{2}\right\|^2 = \frac{\|x\|^2 + \|y\|^2}{2},$$

it is easily shown that the closed sphere of radius R is strongly convex (Definition 2.2-4) with $\delta(\tau) = \mu\tau^2$ and $\mu = 1/8R$.

In Fig. 2.4-3, (2) always succeeds and establishes that $T^n x \to x^*$ at a geometric rate. On the other hand, (1) predicts strong convergence only if $\mathscr{C}_1$ is uniformaly convex. Neither (1) nor (2) encompasses Corollary 2, Theorem 2.4-1, in which all $\mathscr{C}_i$'s are linear varieties.

2.5 USEFUL PROJECTIONS

Our examples are drawn from L_2, the space of all functions $f(t)$ of a single variable t square-integrable over $-\infty < t < \infty$, from $L_{2\times 2}$, the space of all functions $f(x, y)$ of two variables x, y square-integrable over $-\infty < x, y < \infty$, and from $\mathscr{H}$. Let $f_i(t) \Leftrightarrow F_1(\omega)$, $i = 1 \to 2$. Then, in L_2 an inner product is defined by[24]

$$(f_1, f_2) \equiv \int_{-\infty}^{\infty} f_1(t)\bar{f}_2(t)\, dt = \frac{1}{2\pi}\int_{-\infty}^{\infty} F_1(\omega)\bar{F}_2(\omega)\, d\omega. \tag{2.5-1a}$$

In this metric, $f_n \to f$ is equivalent to

$$\|f_n - f\|^2 = \int_{-\infty}^{\infty} |f_n(t) - f(t)|^2\, dt = \frac{1}{2\pi}\int_{-\infty}^{\infty} |F_n(\omega) - F(\omega)|^2\, d\omega \to 0. \tag{2.5-1b}$$

Similarly, in $L_{2\times 2}$,

$$(f_1, f_2) \equiv \iint_{-\infty}^{\infty} f_1(x, y)\bar{f}_2(x, y)\, dx\, dy$$

$$= \frac{1}{4\pi^2}\iint_{-\infty}^{\infty} F_1(\omega_1, \omega_2)\bar{F}_2(\omega_1, \omega_2)\, d\omega_1\, d\omega_2 \tag{2.5-1c}$$

and $f_n \to f$ translates into

$$\iint_{-\infty}^{\infty} |f_n(x, y) - f(x, y)|^2\, dx\, dy = \frac{1}{4\pi^2}\iint_{-\infty}^{\infty} |F_n(\omega_1, \omega_2) - F(\omega_1, \omega_2)|^2\, d\omega_1\, d\omega_2 \to 0. \tag{2.5-1d}$$

[24] For a scalar x, $\bar{x}$ denotes its complex conjugate.

(1) For any prescribed $b > 0$ let $\mathscr{B}_b$ denote the subset of L_2 composed of all f's band-limited to b rad/s; i.e., $f \in \mathscr{B}_b$ iff $F(\omega) = 0$ almost everywhere (a.e.) in $|\omega| > b$. It is obvious that $\mathscr{B}_b$ is a CLM devoid of interior points. Indeed, given any $\varepsilon > 0$ and any $f \in \mathscr{B}_b$, there exists an $h \notin \mathscr{B}_b$ such that

$$\|f - h\|^2 = \int_{-\infty}^{\infty} |f(t) - h(t)|^2 \, dt \leqslant \varepsilon^2. \tag{2.5-2}$$

Thus, as stated previously, any size L_2 neighborhood of a band-limited signal contains signals that are not band-limited.

(2) For any prescribed $a > 0$, let $\mathscr{T}_a$ be composed of all f's in L_2 that vanish a.e. in $|t| > a$. Clearly, $\mathscr{T}_a$ is also a CLM without interior.

(3) For any prescribed $g(t) \in L_2$, let $\mathscr{C}_a(g)$ denote the set of all f's whose projections onto $\mathscr{T}_a$ equal $g(t)$. From the projection theorem,

$$\mathscr{C}_a(g) = g \dotplus (\perp \mathscr{T}_a), \tag{2.5-3}$$

a linear variety obtained by displacing the CLM $\perp \mathscr{T}_a$ a distance g from the origin ϕ.[25]

(4) Let $\mathscr{B}_b^+$ denote the subset of $\mathscr{B}_b$ composed of those functions whose Fourier transforms are a.e. nonnegative. It is readily shown that $\mathscr{B}_b^+$ is a closed convex cone (with vertex ϕ).[26] For if $f(t)$, $f_1(t)$, and $f_2(t)$ belong to $\mathscr{B}_b^+$ and μ is any nonnegative number, $\mu f(t) \Leftrightarrow \mu F(\omega)$ is clearly in $\mathscr{B}_b^+$. Similarly, if $0 \leqslant \mu \leqslant 1$, then $\mu f_1(t) + (1 - \mu) f_2(t) \Leftrightarrow \mu F_1(\omega) + (1 - \mu) F_2(\omega) \geqslant 0$ a.e. and $\mathscr{B}_b^+$ is at least a convex cone. To demonstrate closure we must prove that $\{f_n\} \subset \mathscr{B}_b^+$ and $f_n \to f$ imply $f \in \mathscr{B}_b^+$. Let $f_n(t) \Leftrightarrow F_n(\omega)$, $n = 1 \to \infty$, and let $f(t) \Leftrightarrow F(\omega) \triangleq F_1(\omega) + jF_2(\omega)$, where $F_1(\omega)$ and $F_2(\omega)$ are real. Then, since all F_n's are real, Parseval's formula yields

$$\int_{-\infty}^{\infty} |F_n(\omega) - F_1(\omega)|^2 \, d\omega + \int_{-\infty}^{\infty} |F_2(\omega)|^2 \, d\omega \to 0, \tag{2.5-4}$$

as $n \to \infty$. Thus, $F_2(\omega) = 0$ a.e. and $F(\omega) = F_1(\omega)$ is real. It now follows from (2.5-4) and the vanishing of all F_n's for $|\omega| > b$ that

$$\int_{|\omega| \leqslant b} |F_n(\omega) - F(\omega)|^2 \, d\omega + \int_{|\omega| > b} |F(\omega)|^2 \, d\omega \to 0. \tag{2.5-5}$$

Consequently, $F(\omega) = 0$ a.e. in $|\omega| > b$, $f(t)$ is bandlimited to b, and

$$\int_{|\omega| \leqslant b} |F_n(\omega) - F(\omega)|^2 \, d\omega \to 0. \tag{2.5-6}$$

[25] In L_2 and $L_{2\times 2}$, ϕ is the zero function.
[26] A convex set $\mathscr{C}$ is a cone with vertex ϕ, if $f \in \mathscr{C}$ implies $\mu f \in \mathscr{C}$ for all $\mu \geqslant 0$.

Finally, since all F_n's are nonnegative, the assumption that $F(\omega) < 0$ on some set Δ of positive measure entails the inequality

$$\int_{|\omega|\leqslant b} |F_n(\omega) - F(\omega)|^2 \, d\omega \geqslant \int_\Delta |F(\omega)|^2 \, d\omega > 0, \qquad (2.5\text{-}7)$$

all n, which contradicts Eq. (2.5-6). Therefore, $f \in \mathscr{B}_b^+$.

(5) Let $\mathscr{B}_b^+(\rho)$ denote the subset of $\mathscr{B}_b^+$ composed of all f's that satisfy

$$\frac{1}{2\pi} \int_{-b}^{b} F(\omega) \, d\omega = \rho. \qquad (2.5\text{-}8)$$

Of course, $f(t) \Leftrightarrow F(\omega) \geqslant 0$ and $\rho \geqslant 0$ is prescribed. That $\mathscr{B}_b^+(\rho)$ is convex is evident, and as regards closure, suppose again that $\{f_n\} \in \mathscr{B}_b^+(\rho)$ and $f_n \to f$. Invoking (4), $f \in \mathscr{B}_b^+$; hence, if $f(t) \Leftrightarrow F(\omega)$, Schwarz's inequality yields

$$\begin{aligned} |\rho - \frac{1}{2\pi} \int_{-b}^{b} F(\omega) \, d\omega|^2 &= \frac{1}{4\pi^2} |\int_{-b}^{b} (F_n(\omega) - F(\omega)) \, d\omega|^2 \\ &\leqslant \frac{1}{4\pi^2} \left(\int_{-b}^{b} |F_n(\omega) - F(\omega)| \, d\omega \right)^2 \\ &\leqslant \frac{b}{2\pi^2} \int_{-b}^{b} |F_n(\omega) - F(\omega)|^2 \, d\omega \to 0; \qquad (2.5\text{-}9) \end{aligned}$$

i.e.,

$$\frac{1}{2\pi} \int_{-b}^{b} F(\omega) \, d\omega = \rho \qquad (2.5\text{-}10)$$

and $\mathscr{B}_b^+(\rho)$ is closed.

(6) Let $\mathscr{B}_b^+(\rho^2)$ denote the subset of $\mathscr{B}_b^+$ composed of all f's that satisfy the energy constraint

$$\int_{-\infty}^{\infty} |f(t)|^2 \, dt = \frac{1}{2\pi} \int_{-b}^{b} |F(\omega)|^2 \, d\omega \leqslant \rho^2. \qquad (2.5\text{-}11)$$

Since $\|\mu f_1 + (1-\mu) f_2\| \leqslant \mu \|f_1\| + (1-\mu)\|f_2\| \leqslant \mu\rho + (1-\mu)\rho = \rho$ for any f_1 and f_2 in $\mathscr{B}_b^+(\rho^2)$ and any μ, $0 \leqslant \mu \leqslant 1$, we conclude that $\mathscr{B}_b^+(\rho^2)$ is convex. In addition, $f_n \to f$ implies that $\rho \geqslant \|f_n\| \to \|f\|$ and closure is also obvious. It is worth mentioning that $\mathscr{B}_b^+$, $\mathscr{B}_b^+(\rho)$, and $\mathscr{B}_b^+(\rho^2)$, as subsets of $\mathscr{B}_b$, are all without interiors.

(7) For a *prescribed* real function $\varphi(\omega_1, \omega_2)$ defined on a prescribed set Δ in the $\omega_1 - \omega_2$ plane, let $\mathscr{C}_\varphi(\Delta)$ denote the collection of all functions $f(x, y) \in L_{2\times 2}$ whose Fourier transforms are of the form

$$F(\omega_1, \omega_2) = A(\omega_1, \omega_2) \, e^{j\varphi(\omega_1, \omega_2)} \qquad (2.5\text{-}12)$$

a.e. in Δ, where $A(\omega_1, \omega_2) = |F(\omega_1, \omega_2)|$. Clearly, $\mathscr{C}_\varphi(\Delta)$ is a convex cone. To establish closure, note that in the transform domain, $f_n \to f$ goes into

$$A_n(\omega_1, \omega_2)\, e^{j\theta_n(\omega_1,\omega_2)} \to A(\omega_1, \omega_2)\, e^{j\theta(\omega_1,\omega_2)}; \qquad (2.5\text{-}13)$$

or, letting Δ' denote the set complement of Δ,

$$\iint_{\Delta} |A_n e^{j\varphi} - A\, e^{j\theta}|^2\, d\omega_1\, d\omega_2 + \iint_{\Delta'} |F_n - F|^2\, d\omega_1\, d\omega_2 \to 0. \qquad (2.5\text{-}14)$$

Since $A_n \to A$, Eq. (2.5-14) remains valid if $\theta(\omega_1, \omega_2)$ is replaced by $\varphi(\omega_1, \omega_2)$ a.e. in Δ and left unchanged for $(\omega_1, \omega_2) \in \Delta'$. Thus, if this modified exponential is denoted by $e^{j\beta(\omega_1,\omega_2)}$, then $F_n \to A\, e^{j\beta}$. But a mean-square limit is essentially unique, whence, a.e., $e^{j\theta(\omega_1,\omega_2)} = e^{j\beta(\omega_1,\omega_2)}$, the desired result.

(8) For a *prescribed* nonnegative function $M(\omega_1, \omega_2)$ square-integrable over a prescribed set Δ in the $\omega_1 - \omega_2$ plane, let $\mathscr{C}_M(\Delta)$ denote the set of all f's in $L_{2\times 2}$ whose Fourier transforms have magnitudes subordinate to $M(\omega_1, \omega_2)$ over Δ; i.e., if $f(x, y) \in \mathscr{C}_M(\Delta)$, then $f(x, y) \Leftrightarrow F(\omega_1, \omega_2)$, where $|F(\omega_1, \omega_2)| \leqslant M(\omega_1, \omega_2)$ a.e. in Δ. Let $F_i(\omega_1, \omega_2) \Leftrightarrow f_i(x, y) \in \mathscr{C}_M(\Delta)$, $i = 1 \to 2$. Since

$$|\mu F_1 + (1-\mu)F_2| \leqslant \mu|F_1| + (1-\mu)|F_2| \leqslant \mu M + (1-\mu)M = M \qquad (2.5\text{-}15)$$

for $0 \leqslant \mu \leqslant 1$ and almost all $(\omega_1, \omega_2) \in \Delta$, $\mathscr{C}_M(\Delta)$ is convex. To verify closure, let $\{f_n\} \subset \mathscr{C}_M(\Delta)$ and suppose that $f_n \to f$. Then, $F_n \to F$. Now, as is well known [10], there exists a subsequence $\{F_{n'}\}$ of $\{F_n\}$ that converges pointwise to $F(\omega)$ in the ordinary sense. This means that

$$|F(\omega_1, \omega_2)| = \underset{n' \to \infty}{\text{ordinary limit}}\, |F_{n'}(\omega_1, \omega_2)| \leqslant M(\omega_1, \omega_2) \qquad (2.5\text{-}16)$$

a.e. in Δ; hence, $\mathscr{C}_M(\Delta)$ is closed.

(9) Let $\mathscr{T}^+(\Delta)$ denote the subset of L_2 composed of all functions $f(t)$ a.e. nonnegative on the prescribed set Δ. Clearly, $\mathscr{T}^+(\Delta)$ is a closed convex set with vertex ϕ.

(10) Let d_1 and d_2 be the prescribed width and height of a pixel. Given N^2 pixels in the x–y plane, N a positive integer, let $\mathscr{D}_N(d_1, d_2)$ denote the subset of $L_{2\times 2}$ composed of all $f(x, y)$ which equal zero off the pixels, take on the value 1 a.e. on pixel $(0, 0)$ and assume *nonnegative* constant values a.e. on the remaining pixels.[27] Convexity of $\mathscr{D}_N(d_1, d_2)$ is evident and closure

[27] The elements of $\mathscr{D}_N(d_1, d_2)$ are the nontrivial pictures of rectangular bounded support which have been digitized spatially by N^2 pixels and normalized to unity brightness level in pixel $(0, 0)$.

follows from two observations: first, if $f_n \to f$, we must obviously have $f(x, y) = 0$ off the pixels; second, there exists a subsequence $\{f_{n'}\}$ that converges a.e. to f in the ordinary sense. Thus, since every $f_{n'}(x, y)$ takes on nonnegative constant values on every pixel and the value 1 on pixel $(0, 0)$, the same is true for the limit function $f(x, y)$. Note that $\mathscr{D}_N(d_1, d_2)$ is contained in a linear manifold of dimension N^2 and is therefore finite-dimensional.

(11) In $\mathscr{H}$, the sphere $\mathscr{S}_R(h)$ of radius R and center h contains all $f \in \mathscr{H}$ such that

$$\|f - h\| \leqslant R. \tag{2.5-17}$$

This set is closed and uniformly convex.

Theorem 2.5-1. Let the projection operators onto the 11 closed convex sets defined in (1)–(11) be denoted by P_i, $i = 1 \to 11$. In both L_2 and $L_{2\times 2}$, let an arbitrary element be denoted by $f = f_1 + jf_2 \Leftrightarrow F_1 + jF_2 = A\,e^{j\theta}$, where f_1, f_2, F_1, F_2, A, and θ are all real functions of the respective number of variables in their respective domains. Also, let f designate an arbitrary element of $\mathscr{H}$ for case (11). The 11 P_i's are synthesized by means of the following rules.

(1) Onto $\mathscr{B}_b$:[28]

$$P_1 f \Leftrightarrow p_b(\omega) F(\omega). \tag{2.5-18}$$

(2) Onto $\mathscr{T}_a$:

$$P_2 f = p_a(t) f(t). \tag{2.5-19}$$

(3) Onto $\mathscr{C}_a(g)$:[29]

$$P_3 f = g(t) + [1 - p_a(t)] f(t). \tag{2.5-20}$$

(4) Onto $\mathscr{B}_b^+$:[30]

$$P_4 f \Leftrightarrow p_b(\omega) F_1^+(\omega). \tag{2.5-21}$$

(5) Onto $\mathscr{B}_b^+(\rho)$:

$$P_5 f \Leftrightarrow p_b(\omega)(F_1(\omega) + c)^+, \tag{2.5-21}$$

[28] $p_b(\omega) = \begin{cases} 1, & |\omega| \leqslant b, \\ 0, & |\omega| > b, \text{ etc.} \end{cases}$

[29] $g(t) = 0 \quad \text{for } |t| > a.$

[30] $F_1^+(\omega) = \begin{cases} F_1(\omega), & \text{for } F_1(\omega) \geqslant 0, \\ 0, & \text{otherwise.} \end{cases}$

where the real constant c is chosen so that

$$\frac{1}{2\pi}\int_{-b}^{b}(F_1(\omega)+c)^+\,d\omega=\rho. \tag{2.5-23}$$

(6) Onto $\mathscr{B}_b^+(\rho^2)$:

$$P_6 f \Leftrightarrow c p_b(\omega) F_1^+(\omega), \tag{2.5-24}$$

where

$$c=\begin{cases}1 & \text{if } \rho^2 \geqslant \dfrac{1}{2\pi}\displaystyle\int_{-b}^{b}(F_1^+(\omega))^2\,d\omega,\\ \rho \div \sqrt[+]{\dfrac{1}{2\pi}\displaystyle\int_{-b}^{b}(F_1^+(\omega))^2\,d\omega}, & \text{otherwise.}\end{cases} \tag{2.5-25}$$

(7) Onto $\mathscr{C}_\varphi(\Delta)$:

$$P_7 f \Leftrightarrow \begin{cases} e^{j\varphi(\omega_1,\omega_2)}A(\omega_1,\omega_2)\cos^+[\theta(\omega_1,\omega_2)-\varphi(\omega_1,\omega_2)], & (\omega_1,\omega_2)\in\Delta,\\ F(\omega_1,\omega_2), & (\omega_1,\omega_2)\notin\Delta.\end{cases} \tag{2.5-26}$$

(8) Onto $\mathscr{C}_M(\Delta)$:

$$P_8 f \Leftrightarrow A_8(\omega_1,\omega_2)\,e^{j\theta(\omega_1,\omega_2)}, \tag{2.5-27}$$

where

$$A_8(\omega_1,\omega_2)=\begin{cases}\min\{A(\omega_1,\omega_2), M(\omega_1,\omega_2)\}, & (\omega_1,\omega_2)\in\Delta,\\ A(\omega_1,\omega_2), & (\omega_1,\omega_2)\notin\Delta.\end{cases} \tag{2.5-28}$$

(9) Onto $\mathscr{T}^+(\Delta)$:

$$P_9 f=\begin{cases} f_1^+(t), & t\in\Delta\\ f(t), & t\in\Delta'.\end{cases} \tag{2.5-29}$$

(10) Onto $\mathscr{D}_N(d_1, d_2)$: let R_{mn} denote the region of the x–y plane occupied by pixel (m, n) and let

$$E_{mn}(f_1)\equiv\frac{1}{d_1 d_2}\int_{R_{mn}} f_1(x,y)\,dx\,dy. \tag{2.5-30}$$

Then,

$$P_{10} f=\begin{cases}1 & \text{on pixel } (0,0),\\ E_{mn}^+(f_1) & \text{on pixel } (m,n)\neq(0,0),\\ 0 & \text{off a pixel.}\end{cases} \tag{2.5-31}$$

(11) Onto $\mathscr{S}_R(h)$:

$$P_{11} f=\begin{cases} h+R\dfrac{f-h}{\|f-h\|}, & \|f-h\|\geqslant R,\\ f, & \|f-h\|\leqslant R.\end{cases} \tag{2.5-32}$$

These 11 projection operators $P_1 \to P_{11}$ are of frequent occurrence in a wide variety of signal-processing applications. As a typical illustration of how they can be used, let us consider the problem of recovering a spatially digitized picture from partial knowledge of the phase angle $\theta(\omega_1, \omega_2)$ of its Fourier transform $F(\omega_1, \omega_2)$.[31] Suppose therefore that $\theta(\omega_1, \omega_2)$ is specified as equal to the function $\varphi(\omega_1, \omega_2)$ over some finite rectangle in the $\omega_1 - \omega_2$ plane and that the unknown picture brightness is nonzero over pixel $(0, 0)$. Then, without loss of generality, we may assume that $f \in \mathscr{D}_N(d_1, d_2)$. Hence, $f \in \mathscr{C}_0 = \mathscr{D}_N(d_1, d_2) \cap \mathscr{C}_\varphi(\Delta)$ and the appropriate projection operators are P_{10} and P_7, respectively. Let

$$T_{10} = 1 + \lambda_{10}(P_{10} - 1), \qquad T_7 = 1 + \lambda_7(P_7 - 1) \tag{2.5-33}$$

and $T = T_{10}T_7$, where λ_{10} and λ_7 are any two positive numbers less than 2. Then, according to Theorem 2.4-1 and the fact that $\mathscr{D}_N$ is finite-dimensional, iteration $x_{n+1} = Tx_n$ converges *strongly* to a point $x^*(x, y)$ of $\mathscr{C}_0$, irrespective of the choice of initialization function $x_0(x, y) \in L_{2\times 2}$.

In general, an element of $\mathscr{D}_N(d_1, d_2)$ is not uniquely determined by the phase of its Fourier transform and it is possible for $\mathscr{C}_0$ to contain more than one point. This means that the limit function $x^*(x, y)$ can depend on the particular $x_0(x, y)$ used to initialize the recursion and it becomes necessary to impose additional conditions on $\theta(\omega_1, \omega_2)$ [13–15].

Similarly, since a prescription of $|F(\omega_1, \omega_2)| = M(\omega_1, \omega_2)$ a.e. in some set Δ places $f(x, y)$ in a known closed convex set $\mathscr{C}_M(\Delta)$, P_8 is the corresponding projector. Hence by Theorem 2.4-1 and the condition of finite-dimensionality, the iteration $x_{n+1} = T_{10}T_8x_n$ converges strongly to a point $x^* \in \mathscr{C}_0 = \mathscr{D}_N(d_1, d_2) \cap \mathscr{C}_M(\Delta)$. Unfortunately, all that can be said about the picture $x^*(x, y)$ is that the magnitude of its Fourier transform is subordinate to $M(\omega_1, \omega_2)$ over Δ but not necessarily equal to it. Nevertheless, we believe that this difficulty can be resolved.

***Proof of Theorem* 2.5-1.** We shall only establish the more difficult rules for P_5, P_6, and P_{10}. In the discussion of P_5 and P_6 it suffices, for our purposes, to assume that the Fourier transform $F(\omega)$ of $f(t)$ is a.e. continuous in $|\omega| \leq b$. Let $P_5 f \Leftrightarrow G(\omega)$. Then, since $P_5 f \in \mathscr{B}_b^+(\rho)$, $G(\omega)$ is nonnegative and vanishes for $|\omega| > b$. Thus, bearing in mind that $f(t) \Leftrightarrow F(\omega) = F_1(\omega) + jF_2(\omega)$,

$$2\pi\|f - P_5 f\|^2 = \int_{-b}^{b} (F_1 - G)^2\, d\omega + \int_{-b}^{b} F_2^2\, d\omega + \int_{|\omega|>b} |F(\omega)|^2\, d\omega. \tag{2.5-34}$$

[31] References 13–15 contain good general treatments of the problem of digital signal reconstruction from phase and magnitude. However, a certain modification is needed to incorporate the restriction that pixel brightness is nonnegative.

Consequently, $G(\omega)$ minimizes the integral

$$I = \int_{-b}^{b} (F_1 - G)^2 \, d\omega \tag{2.5-35}$$

subject to the side restriction

$$\frac{1}{2\pi} \int_{-b}^{b} G(\omega) \, d\omega = \rho. \tag{2.5-36}$$

Suppose therefore that $G(\omega)$ minimizes I and satisfies Eq. (2.5-36). Let ω_0 and $\omega_1 \neq \omega_0$ be any two continuity points of $G(\omega)$ in $|\omega| \leqslant b$ such that $G(\omega_0) > 0$.

Let $\delta_a(\omega; \omega_l, h)$ denote a rectangular pulse centered on $\omega = \omega_l$, of width $a > 0$ and height $h > 0$. Clearly, for any preassigned positive value of h and all sufficiently small positive a and μ,

$$\tilde{G}(\omega) = G(\omega) - \mu\delta_a(\omega; \omega_0, h) + \mu\delta_a(\omega; \omega_1, h) \tag{2.5-37}$$

is nonnegative for $|\omega| \leqslant b$ and also satisfies Eq. (2.5-36). Moreover, the value $\tilde{I}$ obtained by substituting $\tilde{G}(\omega)$ for $G(\omega)$ in Eq. (2.5-35) is given by

$$\tilde{I} = I + \frac{\mu}{\pi} \int_{-b}^{b} (F_1 - G)[\delta_a(\omega; \omega_0, h) - \delta_a(\omega; \omega_1, h)] \, d\omega$$
$$+ \frac{\mu^2}{2\pi} \int_{-b}^{b} [\delta_a(\omega; \omega_0, h) - \delta_a(\omega; \omega_1, h)]^2 \, d\omega. \tag{2.5-38}$$

Since $I \leqslant \tilde{I}$ and $\mu > 0$,

$$\lim_{\mu \to 0} \frac{\tilde{I} - I}{\mu} = \frac{1}{\pi} \int_{-b}^{b} (F_1 - G)[\delta_a(\omega; \omega_0, h) - \delta_a(\omega; \omega_1, h)] \, d\omega \geqslant 0. \tag{2.5-39}$$

Clearly, as $a \to 0$ and $h \to \infty$ so that $ah \to 1$, the two δ_a's tend to delta functions centered on ω_0 and ω_1 and Eq. (2.5-39) yields

$$F_1(\omega_0) - G(\omega_0) \geqslant F_1(\omega_1) - G_1(\omega_1). \tag{2.5-40}$$

But if $G(\omega_1)$ is also positive, the argument is reversible, whence

$$F_1(\omega_1) - G(\omega_1) \geqslant F_1(\omega_0) - G(\omega_0) \tag{2.5-41}$$

and so

$$F_1(\omega_0) - G(\omega_0) = F_1(\omega_1) - G(\omega_1). \tag{2.5-42}$$

In short, for this set of ω's, there exists a constant c such that

$$G(\omega) = F_1(\omega) + c. \tag{2.5-43}$$

The relationship (2.5-43) holds only in the interval $|\omega| \leq b$ and only if $F_1(\omega) + c > 0$. Consequently, at all continuity points of $F_1(\omega)$,

$$G(\omega) = p_b(\omega)(F_1(\omega) + c)^+, \tag{2.5-44}$$

where the constant c is chosen to satisfy the requirement

$$\rho = \frac{1}{2\pi} \int_{-b}^{b} (F_1(\omega) + c)^+ \, d\omega. \tag{2.5-45}$$

[The value of c can be found by merely shifting the curve $F_1(\omega)$ vertically a fixed amount c until the area under the positive loops of $F_1(\omega) + c$ included in the interval $|\omega| \leq b$ equals $2\pi\rho$.]

Let $G(\omega)$ denote the Fourier transform of $P_6 f \in \mathscr{B}_b^+(\rho^2)$. Again $G(\omega)$ is nonnegative, vanishes for $|\omega| > b$, and minimizes the integral I defined in Eq. (2.5-35), subject, however, to the new constraint

$$\frac{1}{2\pi} \int_{-b}^{b} G^2(\omega) \, d\omega \leq \rho^2. \tag{2.5-46}$$

As before, suppose that ω_0 and $\omega_1 \neq \omega_0$ are any two continuity points of $G(\omega)$ at which $G(\omega_0)$ and $G_1(\omega_0)$ are both positive.

Define $\tilde{G}(\omega)$ by

$$\tilde{G}^2(\omega) = G^2(\omega) - \mu\delta_a(\omega; \omega_0, h) + \mu\delta_a(\omega; \omega_1, h). \tag{2.5-47}$$

Clearly, for fixed arbitrary $h > 0$, the positive square root of $\tilde{G}^2(\omega)$ yields a function $\tilde{G}(\omega) \geq 0$ over $-b \leq \omega \leq b$, provided that $\mu > 0$ and $a > 0$ are sufficiently small. Furthermore,

$$\frac{1}{2\pi} \int_{-b}^{b} \tilde{G}^2(\omega) \, d\omega = \frac{1}{2\pi} \int_{-b}^{b} G^2(\omega) \, d\omega \leq \rho^2 \tag{2.5-48}$$

and the side condition (2.5-46) is also met.

By straightforward expansion,

$$\tilde{I} = \frac{1}{2\pi} \int_{-b}^{b} (F_1 - \tilde{G})^2 \, d\omega = \frac{1}{2\pi} \int_{-b}^{b} (F_1^2 - 2F_1\tilde{G} + \tilde{G}^2) \, d\omega \tag{2.5-49}$$

$$= \frac{1}{2\pi} \int_{-b}^{b} (F_1 - G)^2 \, d\omega + \frac{1}{\pi} \int_{-b}^{b} F_1(G - \tilde{G}) \, d\omega \tag{2.5-50}$$

because

$$ah = \int_{-b}^{b} \delta_a(\omega; \omega_0, h) \, d\omega = \int_{-b}^{b} \delta_a(\omega; \omega_1, h) \, d\omega. \tag{2.5-51}$$

But $G(\omega)$ minimizes I, whence

$$\int_{-b}^{b} F_1(G - \tilde{G})\, d\omega \geqslant 0 \tag{2.5-52}$$

for all sufficiently small $\mu > 0$ and $a > 0$.

From Eq. (2.5-47),

$$(G - \tilde{G})(G + \tilde{G}) = \mu[\delta_a(\omega; \omega_0, h) - \delta_a(\omega; \omega_1, h)] \tag{2.5-53}$$

and since $\tilde{G}(\omega) \to G(\omega)$ as $\mu \to 0$, we find that

$$\lim_{\mu \to 0} \frac{G(\omega) - \tilde{G}(\omega)}{\mu} = \frac{\delta_a(\omega; \omega_0, h) - \delta_a(\omega; \omega_1, h)}{2G(\omega)}. \tag{2.5-54}$$

Division of Eq. (2.5-52) by $\mu > 0$ yields, with the aid of Eq. (2.5-54) and an obvious limiting process,

$$\int_{-b}^{b} \frac{F_1(\omega)}{G(\omega)} \delta_a(\omega; \omega_0, h)\, d\omega \geqslant \int_{-b}^{b} \frac{F_1(\omega)}{G(\omega)} \delta_a(\omega; \omega_1, h)\, d\omega. \tag{2.5-55}$$

Reasoning as in Eq. (2.5-39), we obtain, in succession,

$$\frac{F_1(\omega_0)}{G(\omega_0)} \geqslant \frac{F_1(\omega_1)}{G(\omega_1)} \tag{2.5-56}$$

and

$$\frac{F_1(\omega_0)}{G(\omega_0)} = \frac{F_1(\omega_1)}{G(\omega_1)}. \tag{2.5-57}$$

Thus, for this set of ω's, there exists a constant c such that

$$\frac{F_1(\omega)}{G(\omega)} = c. \tag{2.5-58}$$

If $F_1(\omega) = 0$ a.e. in $|\omega| \leqslant b$, we must obviously choose $G(\omega) = 0$ a.e. in $|\omega| \leqslant b$ but otherwise $c \neq 0$. However, *a priori*, c can have either sign, and it follows from $G(\omega) \geqslant 0$ that[32]

$$G(\omega) = cp_b(\omega)F_1^+(\omega), \qquad c > 0, \tag{2.5-59}$$

or

$$G(\omega) = cp_b(\omega)F_1^-(\omega), \qquad c < 0. \tag{2.5-60}$$

[32] $F_1^-(\omega) = \begin{cases} F_1(\omega), & F_1(\omega) < 0, \\ 0, & F_1(\omega) \geqslant 0, \end{cases}$ *etc.*

With the aid of the identities $F_1(\omega) = F_1^+(\omega) + F_1^-(\omega)$ and $F_1^+(\omega)F_1^-(\omega) \equiv 0$, it is easily verified that the values of I paired with the choices Eqs. (2.5-59) and (2.5-60) are given by

$$I^+ = \frac{(1-c)^2}{2\pi}\int_{-b}^{b}(F_1^+(\omega))^2\,d\omega + \frac{1}{2\pi}\int_{-b}^{b}(F_1^-(\omega))^2\,d\omega \quad (2.5\text{-}61)$$

and

$$I^- = \frac{(1-c)^2}{2\pi}\int_{-b}^{b}(F_1^-(\omega))^2\,d\omega + \frac{1}{2\pi}\int_{-b}^{b}(F_1^+(\omega))^2\,d\omega, \quad (2.5\text{-}62)$$

respectively.

Since

$$\inf_{c<0} I^- = \frac{1}{2\pi}\int_{-b}^{b}|F_1(\omega)|^2\,d\omega \quad (2.5\text{-}63)$$

is never smaller than the value

$$\frac{1}{2\pi}\int_{-b}^{b}(F_1^-(\omega))^2\,d\omega \quad (2.5\text{-}64)$$

assumed by I^+ for $c = 1$, Eq. (2.5-59) for $G(\omega)$ is always the correct one. Moreover, Eq. (2.5-59) transforms the constraint (2.5-46) into

$$c^2 \leqslant \frac{\rho^2}{(1/2\pi)\int_{-b}^{b}(F_1^+(\omega))^2\,d\omega} \quad (2.5\text{-}65)$$

and it follows from Eq. (2.5-61) that the rule (2.5-22) for c is also valid.

Lastly, consider P_{10}. Evidently, the problem is one of determining a set of nonnegative constants k_{mn} such that

$$\iint_{R_{mn}} |f(x,y) - k_{mn}|^2\,dx\,dy = \text{minimum} \quad (2.5\text{-}66)$$

for $(m, n) \neq (0, 0)$. Thus, for any such pixel, a real constant c must be found that minimizes the expression

$$I(c) = \iint_{R_{mn}} |f_1 - c^2 + jf_2|^2\,dx\,dy = \iint_{R_{mn}} (f_1 - c^2)^2\,dx\,dy + \iint_{R_{mn}} f_2^2\,dx\,dy. \quad (2.5\text{-}67)$$

Since $dI(c)/dc = 0$ yields

$$c \iint_{R_{mn}} (f_1 - c^2)\, dx\, dy = 0, \tag{2.5-68}$$

either $c = 0$ or

$$c^2 = \frac{1}{d_1 d_2} \iint_{R_{mn}} f_1(x, y)\, dx\, dy = E_{mn}(f_1). \tag{2.5-69}$$

Clearly, if $E_{mn}(f_1) \geqslant 0$ we use Eq. (2.5-69), but otherwise, $c = 0$; hence,

$$c^2 = \begin{cases} 1, & \text{on pixel } (0, 0), \\ E^+_{mn}(f_1), & \text{on pixel } (m, n) \neq (0, 0), \end{cases} \tag{2.5-70}$$

Q.E.D.

2.6 SUMMARY AND NEW DEVELOPMENTS

In this chapter we have discussed the fundamental mathematical theory underlying a broad class of image restoration problems. Specifically, it has been argued that the reconstruction of an image f which is subject to a finite number of known compatible convex constraints is equivalent to the determination of a point belonging to the nonempty intersection of a finite number of well-defined closed convex sets. This latter problem admits an iterative solution involving a successive composition of relaxed projection operators. Typical examples of convex constraints are positivity, space-limitedness, band-limitedness, energy-boundedness, and consistency with both spatial and Fourier domain measurements.

If we turn our attention away from image restoration problems and consider certain system applications, such as adaptive control and robotics, it becomes necessary to study the convergence properties of iterations of nonexpansive operators drawn in an *unrestricted* fashion from an *infinite* family $\{T_t\}$.

More precisely, let $\{T_t: \mathscr{C} \to \mathscr{C}\}$, $t \in I$, denote an index pool of nonexpansive operators which have a common fixed point in some common closed convex set $\mathscr{C} \subset \mathscr{H}$. Any map $r(\cdot)$ from the nonnegative integers N into I defines an iteration

$$f_{k+1} = T_{r(k)} f_k, \qquad k = 0 \to \infty, \tag{2.6-1}$$

for any choice of initial $f_0 \in \mathscr{C}$. We say that the iteration is *unrestricted* if no constraints are imposed on the choice of $r(\cdot)$. Without some assumptions regarding the dependence of T_t on t and the nature of the map $r(\cdot)$, it is easily shown that (2.6-1) can diverge, even if $\mathscr{C}$ is compact and the pool is finite.[33] The following is a simple 2-dimensional counterexample.

In Fig. 2.6-1 below, the operators in the family are orthogonal projection operators onto straight lines through the origin parametrized by their angle t with the positive x-axis.

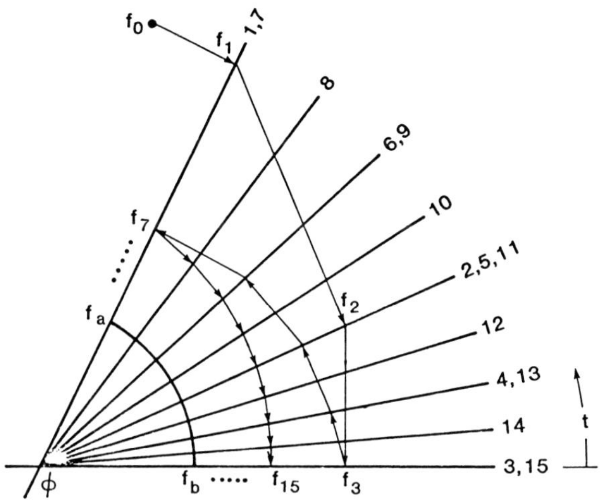

Fig. 2.6-1

The initial vector f_0 is projected onto CLM 1 followed by projections onto 2 and 3. From f_3 we project onto 4, 5, 6 and 7 to arrive at f_7. Note that line 4 bisects the angle between 2 and 3 while 6 bisects 1 and 2. From f_7 we project successively onto 8 which bisects 7 and 9, onto 9, onto 10, which bisects 9 and 11, etc., until we reach f_{15}. This bisection and projection process is repeated endlessly and eventually the iterates move along the heavy-lined limiting circular arc $f_a f_b$ of radius > 0. Clearly, all points on this arc are limit points of the sequence $\{f_k\}$ and have equal nonzero norm. Consequently, $\{f_k\}$ does not converge. [Note that every t between 0 and t_1 is a limit point of the sequence $\{r(k)\}$.]

Definition 2.6.1 [11, 16]: A *linear* nonexpansive operator $T: \mathscr{H} \to \mathscr{H}$ is said to have property (W) if $\|x\| = \|Tx\|$ implies $x = Tx$.

[33] In Theorem 2.4-1, the family $\{T_t\} = \{T_1, \ldots T_m\}$ is finite and $r(\cdot)$ is periodic. In fact, the iteration is exactly the single-step cyclic scheme shown in Eq. (2.4-40).

The following elegant result obtained by Amemiya and Ando in 1964 is perhaps the only convergent unrestricted iteration of the form (2.6-1) known to date!

Theorem 2.6.1 [11]. Let $\{T_1, \ldots, T_n\}$ denote a finite pool of linear nonexpansive operators with property (W). Let $r(\cdot)$ be any map from the set of nonnegative integers N into $\{1, \ldots, n\}$. Then, the iteration

$$f_{k+1} = T_{r(k)} f_k, \qquad k = 0 \to \infty, \tag{2.6-2}$$

converges weakly for any initial $f_0 \in \mathscr{H}$.[34]

Corollary 1 [16]. If at least one T_i which is repeated infinitely often has a range which is relatively compact for bounded sequences,[35] the convergence is strong.

Corollary 2 [16]. Let $\{T_1, \ldots, T_n\}$ denote a finite set of relaxed projection operators onto respective linear varieties $\mathscr{C}_1, \ldots, \mathscr{C}_n$. For $k \geqslant j$ let every $T_{r(k)}$ be repeated infinitely often and let the intersection $\mathscr{C}$ of the corresponding linear varieties be nonempty. Then, for any initial $f_0 \in \mathscr{H}$ and for any such map $r(\cdot)$, the iteration

$$f_{k+1} = T_{r(k)} f_k, \qquad k = 0 \to \infty, \tag{2.6-3}$$

converges weakly to $P_{\mathscr{C}} f_j$, the projection of f_j onto $\mathscr{C}$.

As we have already seen by counterexample, Theorem 2.6.1 is false if the family $\{T_t\}$ is infinite, even in a finite-dimensional setting, and additional constraints must be imposed.

Definition 2.6.2 [16]. Let the parameter t belong to a compact subset I of a complete metric space with distance function $d(\cdot, \cdot)$. The family $\{T_t\}$ is said to be Lipschitz if there exists a positive constant M such that

$$\|T_{t_1} - T_{t_2}\| \leqslant M d(t_1, t_2) \tag{2.6-4}$$

for all t_1, t_2 in I.

Definition 2.6.3 [16]. A map $r(\cdot)$ from N into I is said to be normal if the sequence $r(0), r(1), \ldots,$ has only a finite number of limit points. Denote

[34] Actually, Theorem 2.6.1 is a corollary of a more basic result given as Theorem 2 in Ref. 16.

[35] A set $\mathscr{S} \subset \mathscr{H}$ is relatively compact for bounded sequences, if every bounded sequence contained in $\mathscr{S}$ has a strongly convergent subsequence (whose limit is not necessarily in $\mathscr{S}$).

these limit points by $t_1, \ldots, t_m$.[36] Clearly, every $r(k)$ can be assigned to one of m nonoverlapping convergent subsequences $r^{(l)}(k)$, i.e.,

$$\lim_{k\to\infty} r^{(l)}(k) = t_l, \qquad l = 1 \to m. \tag{2.6-5}$$

Let $r'(k) = t_l$ if $r(k)$ belongs to the subsequence converging to t_l. For a prescribed normal map $r(\cdot)$, $r'(\cdot)$ is a well-defined function from N onto the set $\{t_1, \ldots, t_m\}$ and

$$\lim_{k\to\infty} d(r(k), r'(k)) = 0. \tag{2.6-6}$$

A normal map $r(\cdot)$ is said to be strongly normal if

$$\sum_{k=0}^{\infty} d(r(k), r'(k)) < \infty. \tag{2.6-7}$$

Of course, (2.6-7) implies (2.6-6) but not conversely.

Theorem 2.6.2 [16]. Let $\{T_t\}$, $t \in I$, be a Lipschitz family of linear nonexpansive operators with property (W) and Lipschitz constant M. Let $r(\cdot)$ be any strongly normal map from N into I. Then, the iteration

$$f_{k+1} = T_{r(k)} f_k, \qquad k = 0 \to \infty, \tag{2.6-8}$$

converges weakly for every initial $f_0 \in \mathscr{H}$.

Corollary 1. Let T_t denote a relaxed projection operator onto a linear variety $\mathscr{C}_t = g_t + \mathscr{S}_t$. Let $\lambda_t \in (0, 2)$ be the corresponding relaxation parameter and let P_t be the orthogonal projection operator onto the CLM $\mathscr{S}_t$, $t \in I$. Let the intersection

$$\mathscr{C} = \bigcap_{t\in I} \mathscr{C}_t \tag{2.6-9}$$

be nonempty and suppose that the families $\{P_t\}$ and $\{\lambda_t\}$, $t \in I$, are both Lipschitz. Then, the iteration

$$f_{k+1} = T_{r(k)} f_k, \qquad k = 0 \to \infty, \tag{2.6-10}$$

converges weakly for any initial $f_0 \in \mathscr{H}$ and any strongly normal map $r(\cdot)$.

Although a convenient counterexample is not at hand, it appears fairly certain that normality alone does not suffice. However, as the next theorem shows, it is sometimes possible to replace the inequality in (2.6-7) by the much weaker "tail-squared" constraint,

$$\sum_{k=0}^{\infty} d^2(r(k), r'(k)) < \infty. \tag{2.6-11}$$

A normal map $r(\cdot)$ which satisfies (2.6-11) is said to be weakly normal.

[36] Since I is compact, $m \geqslant 1$.

Theorem 2.6.3 [16]. Let $\{P_t\}$, $t \in I$, denote a Lipschitz family of linear orthogonal projection operators onto respective CLM's $\mathscr{S}_t$. Let $r(\cdot)$ denote any weakly normal map from N into I with only *one* limit point, say t_1. Then, the iteration

$$f_{k+1} = P_{r(k)} f_k, \qquad k = 0 \to \infty, \tag{2.6-12}$$

converges weakly for every $f_0 \in \mathscr{H}$ to a point in $\mathscr{S}_{t_1}$.

We conjecture that theorem 2.6.1 remains valid for *any* finite pool of *nonlinear* projection operators onto closed convex sets whose intersection is nonempty. Unfortunately, this conjecture is open but our final theorem comes very close to the mark.

Theorem 2.6.4 (Unpublished). Let $\{T_1 \ldots, T_n\}$ denote a finite pool of arbitrary relaxed projection operators onto respective closed convex sets $\mathscr{C}_1, \ldots, \mathscr{C}_n$. Let $r(\cdot)$ denote any map from N into the set $\{1, \ldots, n\}$ and let $\mathscr{C}$ denote the intersection of those $\mathscr{C}_i$'s associated with T_i's that are repeated infinitely often. Let us also assume that at least one of the latter "distinguished" T_i's has a range which is relatively compact for bounded sequences. Then, the iteration

$$f_{k+1} = T_{r(k)} f_k, \qquad k = 0 \to \infty, \tag{2.6-13}$$

converges strongly to a point in $\mathscr{C}$ for every initial $f_0 \in \mathscr{H}$.[37]

Any nontrivial extension of Theorem 2.6.4 to an infinite family would be immensely useful in the design of nonlinear adaptive controllers for deterministic plants with unknown parameters.

REFERENCES

1. Dante C. Youla, Generalized image restoration by the method of alternating projections. *IEEE Transactions on Circuits and Systems* **CAS-25** (9) (1978).
2. H. Stark, D. Cahana, and H. Webb, Restoration of arbitrary finite-energy optical objects from limited spatial and spectral information. *J. Optical Society of America* **71** (6) (1981).
3. L. G. Gubin, B. T. Polyak, and E. V. Raik, The method of projections for finding the common point of convex sets. *U.S.S.R. Computational Mathematics and Mathematical Physics* **7** (6) 1–24 (1967).
4. F. E. Browder, Fixed-point theorems for noncompact mappings in Hilbert space. *Proc. Nat. Acad. Sci., U.S.A.* **53**, 1272–1276 (1965).
5. B. Halpern, Fixed points of nonexpanding maps. *Bull of the Am. Math. Soc.* **73**, 957–961 (1967).
6. Z. Opial, Weak convergence of the sequence of successive approximations for nonexpansive mappings. *Bull. Am. Math. Soc.* **73**, 591–597 (1967).

[37] Naturally, Theorem 2.6-4 guarantees convergence in a finite-dimensional setting since the sequence $\{f_k\}$ is bounded.

7. F. Browder and W. Petryshyn, The solution by iteration of nonlinear functional equations in Banach space. *Bull. Am. Math. Soc.*, **72**, 571–575 (1966).
8. John von Neumann, "Functional Operators, Vol. II" (Annals of Mathematics Studies, No. 22, Theorem 13.7, p. 55), Princeton, N.J., 1950.
9. Israel Halperin, The product of projection operators. *Acta Sci. Math.* **23**, 96–99 (1962).
10. A. C. Zaanen, "Linear Analysis," New York: Interscience, 1953.
11. I. Amemiya and T. Ando, Convergence of random products of contractions in hilbert space. *Acta Sci. Math.* (*Szeged*) **26**, MR 32 (4570), 239–244 (1965).
12. L. M. Bregman, Finding the common point of convex sets by the method of successive projection. *Dokl. Akad. Nauk SSSR*, **162** (3), 487–490 (1965).
13. M. H. Hayes, J. S. Lim, and A. V. Oppenheim, Signal reconstruction from phase or magnitude. *IEEE Trans. Acoust., Speech, Signal Processing*, **ASSP-28** (1980).
14. A. V. Oppenheim, M. H. Hayes, and J. S. Lim, Iterative procedures for signal reconstruction from phase. *Proc. SPIE Conf.*, Washington, D.C., 10–11 (1980).
15. M. H. Hayes, The reconstruction of a multidimensional sequence from the phase or magnitude of its Fourier transform. *IEEE Trans. Acoust., Speech, Signal Processing* **ASSP-30** (1982).
16. D. C. Youla and V. J. Velasco, "Unrestricted Iterations of Nonexpansive Operators in Hilbert Space: Part I–Theory" (in preparation).

CHAPTER 3

Bayesian and Related Methods in Image Reconstruction from Incomplete Data

Kenneth M. Hanson

Los Alamos National Laboratory
Los Alamos, New Mexico 87545

3.1 INTRODUCTION

The problem of obtaining an artifact-free computed tomographic (CT) reconstruction from projection data that are limited in number and possibly angular coverage is a difficult one to solve in general. Similarly, restorations of blurred images are almost always marred by artifacts that appear to be related to the blurring function. These difficulties arise from a fundamental limitation inherent in incomplete data sets. This limitation may be viewed as resulting from an essential lack of information in the measurements about the unknown source function, which is codified in the concept of the null space of functions associated with the measurement geometry. The Bayesian approach allows one to incorporate *a priori* information about the source function based on the properties of the ensemble of source functions realizable in the specified imaging situation. If the prior knowledge is restrictive enough, reasonable estimates of the null-space components of the source function can be obtained, thereby reducing the artifacts in the reconstruction. The use of prior knowledge about the shape or structure of the source function will be emphasized here. The results of the maximum *a posteriori* (MAP) method will be compared with the fit and iterative reconstruction (FAIR) technique, in which the previously known shape of the object may be explicitly incorporated by means of a model of the object.

Finally, a more comprehensive and flexible Bayesian approach is suggested in which the ensemble mean and covariance specifications are estimated with the help of the measurements.

The present work deals with discretely sampled, linear imaging systems. Thus, the concept of null space and the Bayesian methods proposed for overcoming its limitations are relevant to a wide variety of digital image restoration problems. The term reconstruction will be used here to refer to image recovery in a broad sense, encompassing, for example, tomographic reconstruction, restoration of blurred images, and decoding of encoded images. Owing to the background of the author and the basis of this chapter on an earlier article related to tomography [1], the discussion is heavily weighted toward tomographic reconstruction.

3.2 MEASUREMENT SPACE—NULL SPACE

The CT reconstruction problem may be stated as follows: given a finite set of projections of a function of two dimensions $f(x, y)$ with compact support, obtain the best estimate of that function. Each projection may generally be written as a weighted two-dimensional (2-D) integral of $f(x, y)$

$$g_i = \iint_{-\infty}^{\infty} h_i(x, y) f(x, y)\, dx\, dy, \tag{3.2-1}$$

where the h_i are the weighting functions and $i = 1, 2, \ldots, N$ for N individual measurements. We will refer to the h_i as response functions. In the CT problem, the h_i typically have large values within a narrow strip and small or zero values outside the strip. If the h_i are unity within a strip of finite width and are zero outside, Eq. (3.2-1) becomes a strip integral. For zero strip width, it becomes a line integral. These are recognized as idealizations of the usual physical situation. The generality of Eq (3.2-1) allows it to closely represent actual physical measurements because it can take into account response functions that vary with position. For example, a slowly increasing strip width is implied by the divergence of the sampling beam inherent in proton computed tomography [2, 3], in diffraction tomography with ultrasound [4], which is discussed by Kaveh and Somekh in Chapter 10, or in well-to-well seismic tomography [5].

Equation (3.2-1) is clearly applicable to any discretely sampled, linear imaging system. What will be developed regarding the CT problem will therefore be applicable to the *image recovery problem* in general. For example, Eq. (3.2-1) applies to the standard reconstruction problem in image processing of the restoration of an image from its blurred rendition. In this case, each g_i is a sample of the blurred image and the response function $h_i(x, y)$ is the blur function, which defines how much the intensity of the original image f at each point (x, y) contributes to g_i. The CT problem may be thought of as a special case of the general image recovery problem in which the blur function is a thin strip that transverses the region of support.

It is often convenient to use vector notation to express relationships in linear imaging systems [6]. Assume the unknown function $f(x, y)$ and the response functions $h_i(x, y)$ belong to a Hilbert vector space (Chapter 1). Then Eq. (3.2-1) is identified as an inner product between the vectors $\mathbf{h}_i$ and $\mathbf{f}$. This may be expressed in vector notation as

$$g_i = (\mathbf{h}_i, \mathbf{f}) = \mathbf{h}_i^T\mathbf{f}, \tag{3.2-2}$$

where $(\mathbf{a}, \mathbf{b})$ denotes the inner product between vectors $\mathbf{a}$ and $\mathbf{b}$ and the superscript T denotes the transpose operation, which converts a column vector into a row vector. All N equations implied by Eq. (3.2-1) may be summarized as

$$\mathbf{g} = \mathbf{Hf}. \tag{3.2-3}$$

In the present context, $\mathbf{H}$ is interpreted as a linear operator corresponding to the measurement process, which maps the function $f(x, y)$ defined on the 2-D continuous domain of (x, y) onto the discrete set of measurements enumerated by the index i.

Returning to Eq. (3.2-2), each measurement g_i may be thought of as a projection of the unknown vector $\mathbf{f}$ onto the response vector $\mathbf{h}_i$. Only those components of $\mathbf{f}$ that lie in the subspace spanned by the set of all $\mathbf{h}_i$ contribute to the measurements. This subspace will be called the *measurement space.* The components of $\mathbf{f}$ in the remaining orthogonal subspace, the *null space*, do not contribute to the measurements. Any vector lying wholly in the null space yields zero for all measurements. As a consequence, the null-space contribution to $\mathbf{f}$ cannot be determined from the measurements alone. The fact that $f(x, y)$ is a function of the continuous variables x and y implies that $\mathbf{f}$ must belong to a vector space of infinite dimensions. A finite number of measurements can only provide information about a finite-dimensional subspace of the infinite-dimensional space of $\mathbf{f}$. In such circumstances, the null space must have infinite dimension! Therefore, given a finite number of measurements, it is never possible to completely determine an arbitrary $f(x, y)$. Hence, the seemingly frivolous statement "a finite set of radiographs tells nothing at all" [7] about a 3-D object. Virtually every digital image recovery problem possesses a null space.

The ambiguities associated with the existence of a null space arise because of the many-to-one (infinite-to-one) transformation from the original object to the finite number of measurements, which is inherently noninvertable. When a null space exists, the data are said to be incomplete. In other terminology, the inability to determine a correct solution to the measurement equations, Eq. (3.2-1), is called an ill-posed problem. Further reading on the null-space/measurement-space concept may be found in papers by Twomey [8–10], Tanabe [11], Jackson [12], Katz [13], Llacer [14], and Sanz and Huang [15].

Figure 3.2-1 illustrates that the null-space component of an image is a tangible entity, which is easily calculable. It is assumed that measurements are available along 11 projection directions, which span only 90° in view angle. Figure 3.2-1c is obtained by starting the algebraic reconstruction technique (ART) algorithm [16] with the source distribution of Fig. 3.2-1a and using zero for all the assumed projection measurements. On convergence ART produces an image whose projections match the input values, that is, zero, along each of the 11 assumed projection directions. As explained in the next section, ART affects only the measurement-space component of the image, leaving just the null-space component in this case. Figure 3.2-1c is just one of many possible null-space images that will not contribute anything to the available measurements in the assumed projection directions. The difference between Figs. 3.2-1a and 3.2-1c is the measurement-space component of the source, Fig. 3.2-1b, which is unambiguously determined by noiseless projection measurements. As explained below, this should be the result obtained by using standard, deterministic reconstruction algorithms whose solutions satisfy the measurement Eqs. (3.2-1).

Other authors have alluded to the problem of the null space in CT [17, 18] but have not explicitly considered its visual effect on reconstructions from limited projection data. The existence of the ambiguities associated with the null space is known well by applied mathematicians. For example, the range of the transpose of the projection-measurement matrix A, referred to by Lakshminarayan and Lent [18] in stating the restrictions to their Theorem 1, which concerns the convergence of Richardson's algorithm, is the measurement space in their square-pixel representation.

Specific discussion of the null space associated with the CT problem has appeared in many articles. Smith *et al.* [7] explored the null space of a finite number of projections to determine the convergence rate of the ART algorithm. This work was extended by Hamaker and Solmon [19], who went so far as to calculate the "angles" between the null spaces corresponding to such projections. Katz [13] made extensive use of the null-space concept to determine the conditions for uniqueness of a reconstruction, since the original function $f(x, y)$ can be unambiguously determined if and only if the null space associated with the measurements is empty (his Proposition VI.1). Louis [20] developed an explicit expression for functions

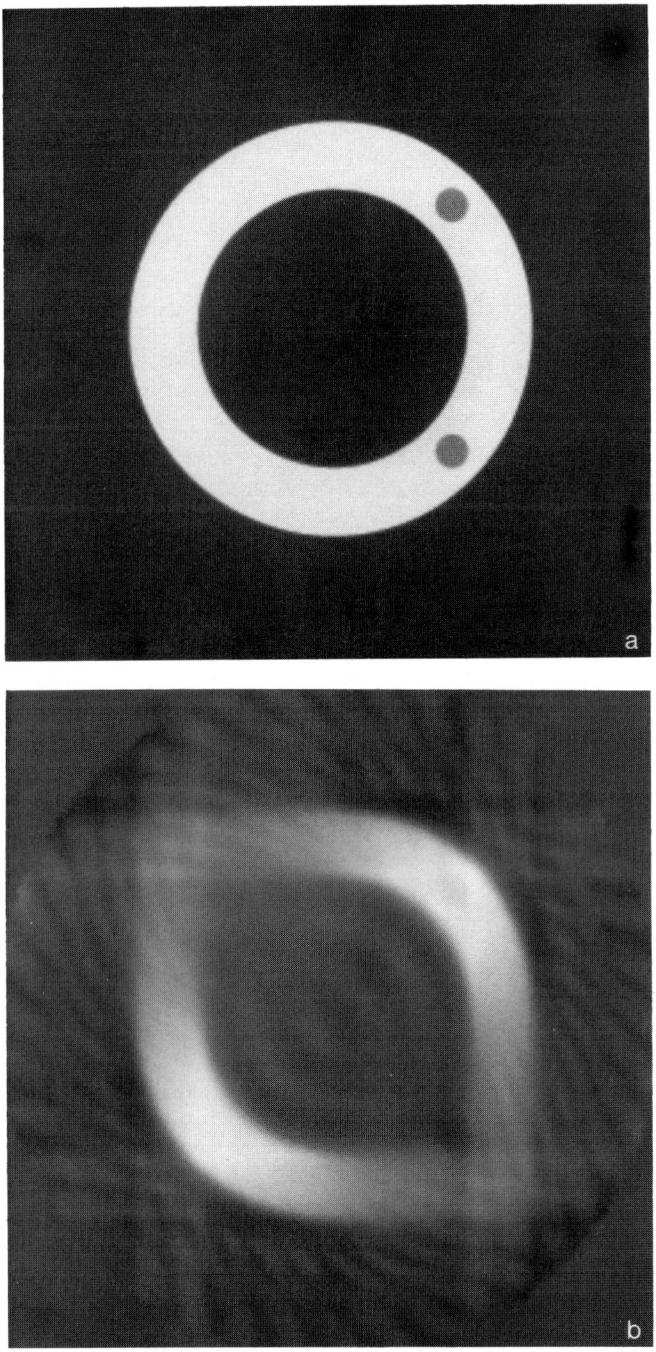
a
b

(continued)

belonging to the null space corresponding to a finite number of evenly spaced projections and showed that ghosts from the null space could appear as lesions. Louis [21] also considered the null space associated with projection imaging in higher dimensions and developed suitable series expansions. Medoff *et al.* [22, 23] recognized the consequences of the null space associated with limited data and introduced a method to reduce null-space ghosts through the application of known constraints on the reconstructed image. Further references on the limited-angle CT problem may be found in [24]. Andrews and Hunt [6] offer numerous cogent discussions of the difficulties introduced by the null space in image restoration without mentioning it by name. The importance of the null-space problem in other image inversion problems will be discussed in the next section.

Fig. 3.2-1. Decomposition of an object consisting of an annulus with two half-density holes (a) into its measurement-space (b) and null-space (c) contributions corresponding to the assumption that there are 11 measured projections available covering 90°. The projections of (c) along each of the 11 measurement directions are zero! This illustrates that for any given measurement scheme, every function is a sum of a component that is measured and a component that is not.

Fourier analysis is often used to visualize the CT problem. Its usefulness stems from the *projection-slice theorem*, which states that the 1-D Fourier transform of a projection taken with parallel strip or line integrals is the same as the 2-D Fourier transform of $f(x, y)$ evaluated along a line through the origin of the 2-D spatial-frequency domain. See Chapters 9, 10, and 11. Then the knowledge of a set of parallel projections that span a limited range of angles is tantamount to knowing the Fourier amplitudes of the object inside the corresponding sector of the 2-D frequency domain. The limited-angle reconstruction problem amounts to determining the Fourier amplitudes in the "missing" sector.

3.3 DETERMINISTIC SOLUTIONS

Deterministic solutions to the measurement Eq. (3.2-1) are defined as solutions that follow directly from the measurements and not from any other information. They typically do not take into account the effect of random noise, which is always present in actual measurements. As argued in the previous section, deterministic solutions should be restricted to the measurement space. This restriction will now be explored for many of the standard methods used to solve the image recovery problem.

It is natural to expand the estimate of $f(x, y)$ in terms of the response functions [25]

$$\hat{f}(x, y) = \sum_{i=1}^{N} a_i h_i(x, y). \tag{3.3-1}$$

Since the measurement subspace of $f(x, y)$ is spanned by the response functions, this expansion provides a complete basis for the measurement space. Because the $h_i(x, y)$ are orthogonal to the null space, the null-space components of $\hat{f}(x, y)$ are zero. This leads to artifacts in $\hat{f}(x, y)$ because it does not possess those components of the original function $f(x, y)$ that lie in the null space [24]. In vector notation Eq. (3.3-1) may be expressed as

$$\hat{\mathbf{f}} = \mathbf{H}^T \mathbf{a}, \tag{3.3-2}$$

where $\mathbf{a}$ is the vector of length N composed of the coefficients a_i and $\mathbf{H}^T$ is the transpose of $\mathbf{H}$.

The response function expansion, Eq. (3.3-1), is formally identical to the familiar *backprojection* process [26] where the value a_i is added to the image along the strip function $h_i(x, y)$. The transverse profile of the response function corresponds to the interpolation function used by the backprojection algorithm. In vector notation the backprojection operation is denoted by $\mathbf{H}^T$. Since the response functions are orthogonal to the null space, any backprojection process that includes just the available measurements can only affect the measurement-space components of the reconstruction.

One way often used to overcome the nonuniqueness difficulty associated with the existence of a null space is to seek that solution of Eq. (3.2-1) with minimum norm. This amounts to minimizing

$$\|\hat{f}(x, y)\|^2 = \iint_{-\infty}^{\infty} \hat{f}^2(x, y)\, dx\, dy \tag{3.3-3}$$

subject to satisfying all N constraints implied by Eq. (3.2-1). When this minimization problem is approached by the usual method of Lagrange multipliers, it is found that the solution must have the *form* given by Eq. (3.3-1). This is understandable since a necessary condition for its minimum-norm solution is that its null-space components are zero. In the desire to force a unique solution to the inversion problem, the fundamental ambiguities associated with the null space have been glossed over.

In most of the well-known CT reconstruction algorithms, the reconstruction is a linear combination of backprojections. These include filtered backprojection [27], algebraic reconstruction technique (ART) [16, 28], simultaneous iterative reconstruction technique (SIRT) [29], SIRT-like algorithms (least squares [30] and other variants [31]), and the "natural-pixel" matrix formulation by Buonocore *et al.* [25], based explicitly on the expansion given by Eq. (3.3-1). Such algorithms can only alter the measurement-space part of the initial estimate. When the initial estimate lies solely in the measurement space, as is normally the case, so will the final estimate. Then the null-space components of the solution will be zero, leading to artifacts.

Although it is feasible to consider the original function $\mathbf{f}$ to belong to a (Hilbert) vector space of infinite dimensions, it is more typical to limit the dimensionality of the estimate $\hat{\mathbf{f}}$ by writing it as a linear combination of a finite number of basis functions. A common choice for basis functions is that of square pixels placed on an equally spaced grid [6]. In this case, the continuous coordinates x and y in Eq. (3.2-1) are replaced by a single index j designating the jth pixel, the integration becomes a summation, and $\mathbf{H}$ is a matrix. Then the image is represented by the sequence of coefficients that describe each pixel's contribution to the image. This vector of coefficients is often referred to as the image and is denoted by $\hat{\mathbf{f}}$, thereby leaving the vector equations that describe the linear imaging system unchanged. This finite vector belongs to a Euclidean vector space. Other types of local basis functions have considerable merit [32]. In the image-deblurring problem, the sampled data typically consist of a blurred image of the same size as the desired deblurred image. Then $\mathbf{H}$ is a square matrix. In the CT problem, the projection measurements lie in an obviously different domain than the original image. The matrix **H** is very seldom square. The incongruity of the measurement samples and the image that is peculiar to the CT problem make it pedagogically useful for exploring new approaches to image recovery.

The difficulty of inverting the matrix Eq. (3.2-3) for image recovery has been appreciated for some time. The straightforward solution may be written in terms of a matrix inverse

$$\mathbf{f} = \mathbf{H}^{-1}\mathbf{g}. \tag{3.3-4}$$

The inverse exists, however, only if **H** is nonsingular, that is, possesses no null space. An illuminating way to approach the inversion of Eq. (3.2-3) is to consider the singular-value decomposition (SVD) of the measurement matrix **H** (Chapter 1). The number of nonzero singular values gives the rank of **H**, which corresponds to the number of linearly independent measurements of **f**. This has been called the number of degrees of freedom contained in the measurements [6] and is the same as the dimension of the measurement space. The singular vectors of the SVD with zero singular values span the null space. For the purpose of providing an inverse that involves only the measurement space, a *pseudoinverse* can be defined as an expansion of the inverse in terms of singular vectors with nonzero singular values [33]. As an example of the usefulness of this approach, McCaughey and Andrews [34] have calculated the eigenvalue map, which is the sequential ordering of the eigenvalues, of $\mathbf{HH}^T$ for equally spaced projection measurements. Their result for 32 samples per view, Fig. 3.3-1, shows that the number of degrees of freedom asymptotically approaches an upper limit of about 700 as the number of projection angles continually increases. This indicates that the dimension of the measurement space reaches an upper limit and that more projections than roughly 40 do not add much information (in the absence of noise considerations).

It is known that the Fourier analysis of a blurred image with a stationary point spread function (PSF) is closely related to the abovementioned eigenanalysis [6]. In fact, in the circulant approximation of the square, block-Toeplitz matrix **H**, the Fourier transform coefficients of the PSF, called the optical transfer function (OTF), are the eigenvalues of **H**. The zeros in the OTF are identified as being associated with the eigenvectors that span the null space of the PSF. It is this null space that is responsible for artifacts in deterministic reconstructions, which are seen clearly in restorations of blurred text [35]. When several blurred images of the same text are available, each subjected to a different PSF, the reconstruction can be nearly artifact-free [36] because the zeros of the OTSs have little overlap, leaving an almost empty null space.

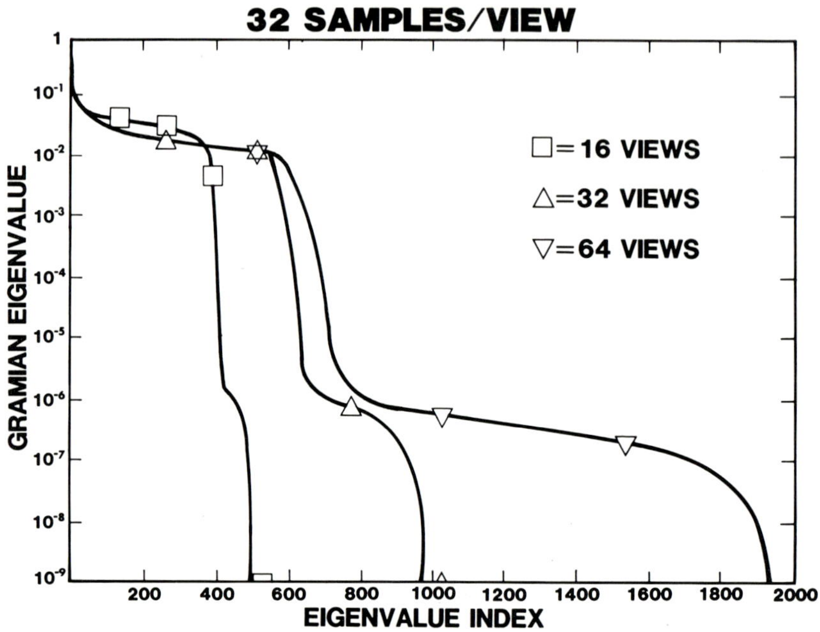

Fig. 3.3-1. Eigenvalue map corresponding to various numbers of equally spaced projections, each consisting of 32 samples. (From McCaughey and Andrews [34]. Copyright 1979 IEEE.)

Many standard deterministic algorithms applied to image recovery have no effect on the null-space component of the solution. For example, when the blur function is symmetric and stationary in the standard problem of the restoration of a blurred image, the response-function expansion, Eq. (3.3-1) or (3.3-2) or the $\mathbf{H}^T$ operation, amounts to a convolution with the blur function itself. Although restoration algorithms are not ordinarily expressed explicitly in terms of such a convolution, all deterministic restoration algorithms can be. In its complete form, the *Wiener filter* [6] is based on the Bayesian approach (next section). However, as it is often applied, the image and noise covariance matrices are assumed to be proportional to the identity matrix, reflecting a lack of prior knowledge. In this case the Wiener filter leaves zeros in the OTF unchanged. Thus the null-space component of $\hat{\mathbf{f}}$ is zero and it must lie in the measurement space. As such, it can be written as an expansion in terms of the response functions, that is, as Eq. (3.2-2). The same can be said of the homomorphic filter. Pseudoinverse methods, when expressed in terms of the singular vectors of $\mathbf{H}$, avoid the use of singular vectors with zero singular values, again leaving the null-space components zero. In the constrained least-squares approach, the norm of some linear operator times $\mathbf{f}$ is minimized subject to the condition that χ^2 (see next section) be equal to a prescribed value. It can be shown that when the operator is diagonal, the reconstruction is proportional to $\mathbf{H}^T$. When the operator is not diagonal, the null space can be impacted, but this nondiagonality implies the use of specific prior knowledge and the algorithm cannot be viewed as deterministic. It is concluded that deterministic restoration algorithms do not affect the null-space component of the unblurred image.

The restriction of deterministic solutions to the measurement space should not be viewed as a negative conclusion. Rather, it is simply a statement of what is possible for a given set of measurements in the absence of further information. It allows one formally to reinterpret the goal for obtaining an improved reconstruction from limited data as that of estimating the null-space contribution through the use of further information available about the function to be reconstructed.

3.4 THE BAYESIAN APPROACH

The essence of the Bayesian approach is the assumption that the image to be reconstructed is a random selection from an identifiable ensemble of similar images. In the context of medical imaging, an example of such an ensemble is the collection of all hearts imaged by the same kind of procedure. By using this prior information about the type of image that is expected, it is anticipated that a meaningful estimate of the null-space component of the reconstruction will be provided, thereby reducing artifacts. It is essential that the knowledge about the ensemble extend beyond what can be ascertained from an imaging modality that suffers from a limited-data geometry. Otherwise, no information about the null space can be added to the measurements themselves. The Bayesian approach permits prior information about the structure of the reconstructed object to be employed to estimate the null-space components of the solution. Of course, other types of prior information can also be incorporated in the Bayesian method of reconstruction.

Deterministic solutions to the image recovery problem tacitly assume the measurements, Eq. (3.2-3), can be made with infinite accuracy. In reality, all measurements of continuous quantities are subject to random fluctuations called noise. In general, the measurements should be written as

$$\mathbf{g} = \mathbf{H}\mathbf{f} + \mathbf{n}, \tag{3.4-1}$$

where $\mathbf{n}$ is the noise vector. It must be emphasized that the noise vector is a random variable. It is different for each set of measurements and its exact value cannot be correctly guessed. Each realization of the vector $\mathbf{n}$ may be regarded as a random selection from an infinitely large ensemble, or collection, of noise vectors. In general, the noise fluctuations may possess an arbitrary probability density distribution. Frequently, however, the assumption is made that the noise has a multivariate Gaussian distribution with a zero mean

$$P(\mathbf{n}) \sim \exp\{-\tfrac{1}{2}\textstyle\sum n_i[\mathbf{R}_n^{-1}]_{ij}n_j\},$$
$$\sim \exp\{-\tfrac{1}{2}\mathbf{n}^T\mathbf{R}_n^{-1}\mathbf{n}\}, \tag{3.4-2}$$

where the second expression is in the vector notation of the previous section and $\mathbf{R}_n$ is the noise covariance matrix, the *ij* element of which is

$$[\mathbf{R}_n]_{ij} = \langle n_i n_j \rangle. \tag{3.4-3}$$

The brackets $\langle\,\rangle$ indicate an average over all members of the ensemble of noise vectors. The above expression is general enough to characterize fully noise fluctuations that depend on the strength of the signal being measured or on the position of the measurement, or depend on each other, that is, are correlated. By its definition, $\mathbf{R}_n$ is a positive-definite matrix, and its inverse, needed in Eq. (3.4-2), is assured.

Under a wide range of reasonable conditions [37], when averaged over the full ensembles of noise and images, the best estimate for the reconstruction is that particular image **f** that maximizes the *a posteriori* conditional probability density of **f** given the measurements **g**. This probability is given by Bayes's formula

$$P(\mathbf{f}|\mathbf{g}) = \frac{P(\mathbf{g}|\mathbf{f})P(\mathbf{f})}{P(\mathbf{g})} \tag{3.4-4}$$

in terms of the conditional probability of **g** given **f**, $P(\mathbf{g}|\mathbf{f})$, and the *a priori* probability distributions of **f** and **g** separately, $P(\mathbf{f})$ and $P(\mathbf{g})$.

Herman and Lent [31] and Hunt [38] proposed using the Bayesian approach for image reconstruction. It is assumed that $P(\mathbf{g}|\mathbf{f})$ is given by $P(\mathbf{n})$, in which **g** is Gaussian distributed about the mean **Hf**, as in Eq. (3.4-2),

$$P(\mathbf{g}|\mathbf{f}) \sim \exp\{-\tfrac{1}{2}(\mathbf{g} - \mathbf{Hf})^T\mathbf{R}_n^{-1}(\mathbf{g} - \mathbf{Hf})\}. \tag{3.4-5}$$

This may be referred to as the measurement probability density distribution, since it follows solely from the distribution of the error fluctuations in the measurements. This probability is also often called the likelihood function [37]. The *a priori* probability density function for the ensemble of images $P(\mathbf{f})$ is assumed to be a multivariate Gaussian distribution with a mean value $\bar{\mathbf{f}}$ and with a covariance matrix $\mathbf{R}_f$

$$P(\mathbf{f}) \sim \exp\{-\tfrac{1}{2}(\mathbf{f} - \bar{\mathbf{f}})^T\mathbf{R}_f^{-1}(\mathbf{f} - \bar{\mathbf{f}})\}. \tag{3.4-6}$$

Under these assumptions, the maximum *a posteriori* (MAP) solution is easily shown to satisfy [38] the *MAP equation*

$$\mathbf{R}_f^{-1}(\bar{\mathbf{f}} - \mathbf{f}) + \mathbf{H}^T\mathbf{R}_n^{-1}(\mathbf{g} - \mathbf{Hf}) = 0, \tag{3.4-7}$$

where $\mathbf{H}$ is the linear operator (matrix) corresponding to the projection process described by the integral in Eq. (3.2-1). As shown in the previous section, the transpose of $\mathbf{H}$ is the familiar backprojection operation. The first term comes from the derivative of $P(\mathbf{f})$ given by Eq. (3.4-6) and the second from $P(\mathbf{g}|\mathbf{f})$, Eq. (3.4-5). It can be seen that the MAP solution strikes a balance between its deviation from the ensemble mean $\bar{\mathbf{f}}$ and the solution to the measurement equation ($\mathbf{g} = \mathbf{Hf}$). This balance is determined by the covariance matrices $\mathbf{R}_f$ and $\mathbf{R}_n$, which specify the confidence with which each deviation is weighted, as well as possible correlations between the deviations. Because the last operation in the second term is $\mathbf{H}^T$, it is only the measurement-space part of this term that has any relation to the first term. It is noted that in Eq. (3.4-7) the linearity in the unknown image $\mathbf{f}$ follows from the assumption of normal distributions for the *a priori* and measurement-error probability densities.

The MAP reconstruction method presented here is equivalent to the *minimum-variance linear estimator* with nonstationary mean and covariance ensemble characterizations [37]. It is also called the *minimum mean-square-error method* [6]. When the blur function, noise, and ensemble image properties are stationary (do not depend on position), then $\mathbf{H}$, $\mathbf{R}_f$, and $\mathbf{R}_n$ are Toeplitz matrices, and, in the circulant approximation, this is the same as the well-known *Wiener filter* [6].

When there is an absence of prior information, as is often the case, the *a priori* probability distribution $P(\mathbf{f})$ does not play a role in Eq. (3.4-4). The MAP approach then reduces to maximizing the conditional probability Eq. (3.4-5) or, what is equivalent, minimizing (-2) times the logarithm of that probability, which is called chi-squared

$$\chi^2 = (\mathbf{g} - \mathbf{Hf})^T\mathbf{R}_n^{-1}(\mathbf{g} - \mathbf{Hf}). \tag{3.4-8}$$

The *minimum chi-squared method*, or the *least-squares method* as it is often called, is a deterministic reconstruction procedure since no reference is made to the stochastic nature of $\mathbf{f}$. When $\mathbf{H}$ is a linear matrix operator, as assumed here, the solution to the χ^2 minimization problem is given in terms of the inverse (or pseudoinverse [33]), Eq. (3.3-4). If the matrix $\mathbf{H}$ is too large to invert directly, an iterative method might be employed [15, 30, 39]. When $\mathbf{H}$ is a nonlinear operator, the $\mathbf{f}$ that minimizes χ^2 can usually be found by an iterative procedure [40]. The minimum χ^2 solution is also often referred to as the *maximum-likelihood solution*, with the probability density distribution given by Eq. (3.4-5) termed the likelihood function [37]. This

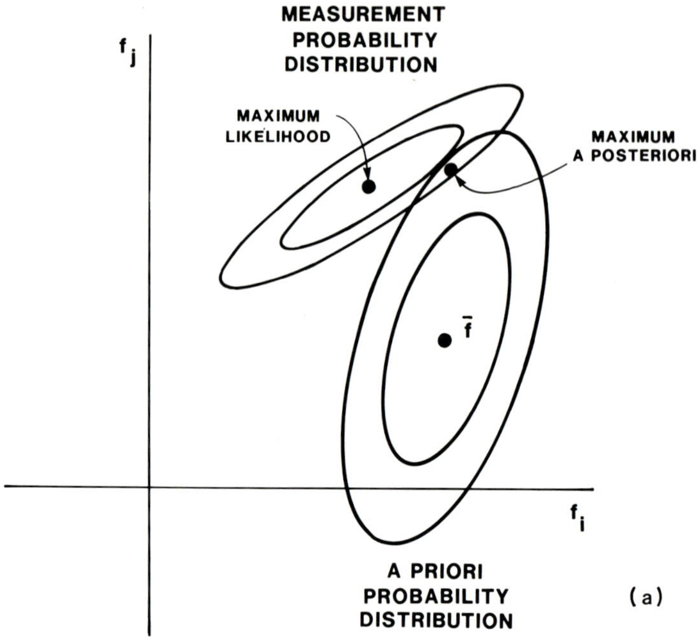
MEASUREMENT
PROBABILITY
DISTRIBUTION
f_j
MAXIMUM
LIKELIHOOD
MAXIMUM
A POSTERIORI
$\bar{f}$
f_i
A PRIORI
PROBABILITY
DISTRIBUTION
(a)

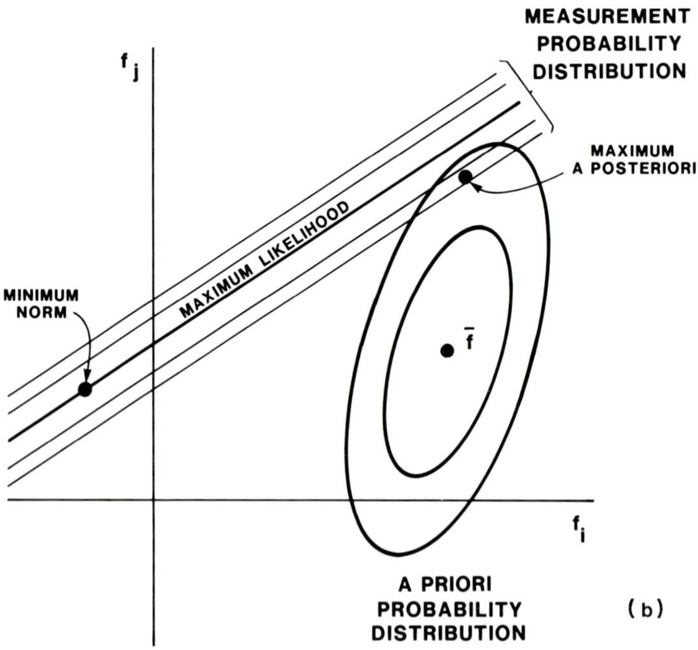
MEASUREMENT
PROBABILITY
DISTRIBUTION
f_j
MAXIMUM
A POSTERIORI
MAXIMUM LIKELIHOOD
MINIMUM
NORM
$\bar{f}$
f_i
A PRIORI
PROBABILITY
DISTRIBUTION
(b)

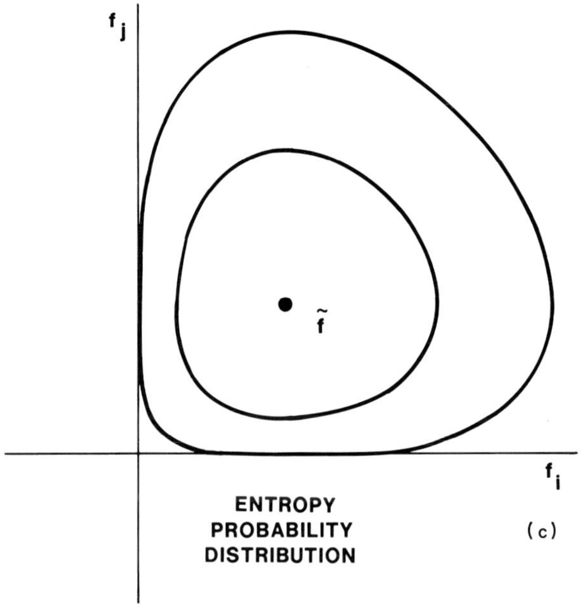

(*continued*)

usage is not uniformly employed, however [41], as the probability given by Eq. (3.4-4) is sometimes referred to as the likelihood. Reference to maximum likelihood is also legitimate under assumptions about the noise other than it is normally distributed, Eq. (3.4-2). For example, Shepp and Vardi [42] have developed a maximum-likelihood reconstruction algorithm appropriate to Poisson-distributed noise.

The above concepts may be visualized by considering the equal-probability contours in a plane defined by two of the components of the image vector **f**. Figure 3.4-1 shows these contours for a number of situations. Under the assumption of Gaussian probability distributions, the contours

---

Fig. 3.4-1. Contour plots of the probability distributions associated with prior information and the available measurements displayed as a function of two components of the image vector, f_i and f_j, for (a) Gaussian distributions, (b) same with the null space intersecting the f_i-f_j plane, and (c) distribution corresponding to the maximum-entropy principle. In the Bayesian approach, the information provided by the measurements is supplemented by prior knowledge cast in terms of an ensemble mean $\bar{\mathbf{f}}$ and its covariance matrix $\mathbf{R}_f$. The maximum *a posteriori* solution maximizes the product of the *a priori* probability and the measurement probability or likelihood function. Maximum-entropy algorithms may be interpreted in terms of the MAP method with the *a priori* probability distribution replaced by a non-Gaussian distribution specified only by its maximum $\tilde{\mathbf{f}}$. This probability function is nonzero only for positive image values.

are ellipses. The contours for the *a priori* probability distribution are centered on the ensemble mean vector $\bar{\mathbf{f}}$ and their shape and extent is controlled by the *ii*, *jj*, and *ij* elements of $\mathbf{R}_f$. The contours for the *a posteriori* distribution are centered on the deterministic maximum-likelihood (least-squares) solution. Their shape is affected by the noise covariance matrix $\mathbf{R}_n$ and the measurement matrix $\mathbf{H}$. If the f_i–f_j plane includes a component of the null space, there is no unique solution for f_i and f_j that satisfies the measurements Eq. (3.2-3). Figure 3.4-1b represents such a situation. Every point on the solution line of Fig. 3.4-1b yields exactly the same measurement vector. Deterministic algorithms based on the minimum-norm criterion select the point on the solution line that is closest to the origin. The MAP approach leads to a solution that is closer to the ensemble mean value in this plane, ignoring off-diagonal correlations in $\mathbf{R}_f$. The relationship between MAP and the maximum-entropy approach will be considered in the next section.

When the solution to the above MAP equation is averaged over the complete ensemble of noise and images, and the correct values are used for the *a priori* entities, the resulting $\langle \mathbf{f} \rangle$ can be shown to be *unbiased*; that is, it is the same as $\bar{\mathbf{f}}$ [37]. However, if the wrong $\bar{\mathbf{f}}$ is used and $\mathbf{R}_f^{-1}$ is nonzero, then $\langle \mathbf{f} \rangle$ will be biased. Of more importance is the fact that for subensembles of images whose mean characteristics are different from those of the ensemble, the MAP estimates are biased toward the ensemble mean. This is an important consideration when the objective of the imaging and reconstruction task is to identify abnormal occurrences, as in looking for diseased hearts. If the ensemble contains a preponderance of normal individuals, the abnormals will tend to be made to appear as normal by the MAP algorithm. This sort of consideration may require us to return to the fundamental statement of MAP and change the cost function to reflect the fact that a misidentified abnormal may be less acceptable than a misidentified normal (as in heart diagnosis). Alternatively, one might allow for greater deviations from $\bar{\mathbf{f}}$ than the true ensemble value of $\mathbf{R}_f$ indicates.

3.5 USE OF OTHER KINDS OF PRIOR KNOWLEDGE

One of the main features of the MAP approach presented in the previous section is that it provides a method of supplementing the available measurements with information previously known about the shape or structure of the reconstructed object. Various other extensions to deterministic algorithms have been proposed to improve the reconstructions from limited data through the use of *a priori* knowledge about the object. These include consistency, analytic continuation, region of support, upper and lower limits, and maximum entropy, which will be discussed in turn. The Bayesian approach may be argued to encompass all of these in one way or another.

Inouye [43] used the consistency between the measurements that is implied by the statement that all are projections of a single 2-D object. Inouye expressed this consistency through the allowed functional forms of the moments of the projections. Of course, consistency is inherent in any reconstruction algorithm that satisfies the measurement equations Eq. (3.2-1). Inouye [44] also advanced a method of calculating the missing views that is based on the Tschebyscheff polynomial expansion originally used by the CT vanguard, Allan Cormack [45]. This method is closely related to analytic continuation. Hanson [24] concluded that these proposed methods of coping with missing data are ineffective in overcoming the measurement-space restrictions presented above for the solution of the general problem.

Tam *et al.* [46] introduced a method to use the *a priori* known region of support of the source distribution. This method is the 2-D counterpart of the celebrated *Gerchberg–Papoulis algorithm* [47, 48] for obtaining "*super-resolution.*" It is an iterative technique in which the known properties of the image in the spatial and Fourier domains are alternatively invoked. The objective is to use the known spatial extent of the source to extend the 2-D Fourier transform of the reconstruction from the "known" sector into the "missing" sector (Section 3.2 and Chapter 9). This method has been studied extensively [49–52] and has been shown to have some merit when either the region of support is very restrictive or the angular region of the missing projections is fairly narrow. Tam and Perez-Mendez analyzed this problem by evaluating the eigenfunction map of the matrix that is the product of the constraints of a circular region of support and band-limitedness.

It is known to be possible to completely determine an integrable function with a finite domain of support from projections that are continuous in angle over some finite range and have infinite spatial resolution [7, 44]. Thus, the null space must be empty and the eigenvalues in the analysis of Tam and Perez-Mendez must not actually go to zero in the limit of an infinite number of angles. The function to be reconstructed must be representable by the response-function expansion. This seems at odds with the interpretation of the limited-angle CT problem as one of a missing sector in the 2-D Fourier domain. However, the projection-slice theorem holds only for projections over the infinite plane. When the object is restricted to a finite domain, the Fourier transform of each parallel projection set, although being concentrated along the slice, actually has infinite extent perpendicular to the slice. Thus, the projections of an object with known finite region of support do sample the missing sector. The above theorem may then be interpreted as saying that when there is an infinite number of samples from that sector, the Fourier amplitudes in the sector may be recovered because the totality of response functions provides a complete basis for the full 2-D frequency domain. The missing sector is not missing at all! Incidentally, the problem of superresolution may be similarly approached from the standpoint of a measurement-space expansion. The Gerchberg–Papoulis algorithm has been shown to be a special case of a general SIRT-like iterative procedure due to Bialy [15].

The region-of-support constraint may be incorporated more directly into many reconstruction algorithms. In virtually every iterative algorithm it is possible to invoke constraints on the reconstructed function by restricting the function as part of the updating step. Such iterative algorithms yield a solution that satisfies the region-of-support constraint, and Tam's procedure is not required. This constraint may be enforced through a redefinition of the response functions $h_i(x, y)$ in Eq. (3.2-1) to make them zero outside the region of support. Then the backprojection operation, Eq. (3.3-1), only affects the reconstruction within the region of support. With this redefinition of the $h_i(x, y)$, the measurement space includes only functions that fulfill the region-of-support constraint. From this standpoint, Tam's iterative procedure does not affect the null space associated with the available measurements. The natural-pixel formulation of Buonocore *et al.* [25, 53] may be revised in a similar manner, but that would probably ruin the properties of the measurement matrix they exploited to arrive at an efficient matrix calculation.

Often it is known that the physical quantity to be reconstructed cannot be negative, as, for example, light intensity, density, pressure, and linear attenuation coefficients. Thus, it has been suggested by numerous authors that reconstructions of such quantities should be constrained to be nonnegative—Gordon *et al.* [16], Huang *et al.* [54], Herman and Lent [39], Schafer *et al.* [55], Sezan and Stark [56], and Medoff *et al.* [22], to name just a few. A simple extension of this is the constraint of known upper and lower limits on the reconstruction function. Imposition of such constraints can be very effective when a major portion of the image is close to the limits. Likely candidates are renditions of text and photographs of the heavens. For this kind of image, the limits are likely to be violated by artifacts and noise. Thus, by invoking bounds on the reconstructed function, these transgressions would be reduced and the final image is likely to be more pleasing. The value of the lower-bound constraint in overcoming the difficulties associated with the null space can be inferred from the excellent maximum-entropy reconstruction of text shown in Burch *et al.* [57] and was explicitly demonstrated by Trussell and Civanlar [58]. Hanson [59] has found that in favorable circumstances, the nonnegativity constraint can improve the visualization of objects when it is hindered by either artifacts produced by the limited nature of the data or noise in the measurements. When most of the image values are far removed from legitimate limits, as in CT scans of the brain, the use of such limits cannot be expected to change the results.

Figure 3.4-1 brings to mind the concepts presented in Chapters 2, 8, and 11 having to do with *projections onto convex sets* (POCS) [56, 60]. This correspondence is valid if one restricts oneself to the set of solutions possessing prior or measurement probabilities greater than a fixed value. For the measurement probability distribution, this amounts to placing a bound on χ^2. Then the sets have well-defined boundaries. Furthermore, for multivariate Gaussian probability distributions, the sets are obviously convex. Of course, one of the great advantages of the POCS method is that it permits the use of numerous constraints that are beneficial in image restoration [56, 58] but cannot be easily incorporated in conventional algorithms. One disadvantage of the POCS method is that it yields only one solution out of a possibly infinite set of feasible ones if the sets overlap substantially. Furthermore, the POCS solution in this case depends on the starting point and on the ordering of the projection operations. Trussell and Civanlar [61] have used the POCS approach to solve the MAP problem with several side constraints by varying the probability level that defines the boundaries of the sets until the MAP equation is satisfied. In its full generality, however, the Bayesian approach is not restricted to this kind of arbitrary threshold in probability. Because the probability values are often continuous, the sets involved in the Bayesian analysis are fuzzy.

Many authors have proposed and used the *maximum-entropy* condition to ameliorate artifacts in reconstructions. For examples, see Chapter 5 as well as Frieden [41], Wernecke and d'Addario [62], Gull and Daniell [63], and Minerbo [64]. In the straightforward approach, it is desired to find the solution to the measurement equations, Eq. (3.2-1), that maximizes the entropy

$$S = -\sum p_i \ln(p_i), \tag{3.5-1}$$

where p_i is the portion of the total intensity that lies in the ith pixel

$$p_i = f_i / \sum f_i. \tag{3.5-2}$$

It is assumed that $p_i \geqslant 0$. Andrews and Hunt [6] and Minerbo [64] showed by means of the method of Lagrange multipliers that, when the sum of f_i is considered to be a constant, the maximum-entropy condition leads to a solution of the form

$$\ln\{\hat{f}(x, y)\} = C + \sum_{i=1}^{N} \lambda_i h_i(x, y), \tag{3.5-3}$$

where C is an additive constant and the λ_i are the Lagrange multipliers, which must be adjusted to satisfy the measurement equations. The maximum-entropy principle implies a solution whose logarithm is proportional to an expansion in terms of response functions, in much the same way as the minimum-norm condition yields the linear, response-function expansion form, Eq. (3.3-1). Equation (3.5-3) enforces the nonnegativity constraint inherent in maximum entropy.

Recognizing the fundamental difficulty of this approach when the data are inconsistent, that is, when no solution to Eq. (3.2-3) exists, Gull and Daniell [63] merged maximum entropy with the probabilistic concepts of random noise by proposing that one find the solution that maximizes entropy subject to the constraint that χ^2 be equal to a predetermined value x_0^2. Andrews and Hunt [6] suggested this as an alternative form of the constrained least-squares approach but did not show any results. The resulting expression for the quantity to be minimized appears very similar to the quadratic form of the logarithm of the *a posteriori* probability function based on Gaussian probability distributions, Eqs. (3.4-5) and (3.4-6), that led to the standard MAP equation (3.4-7). The similarity prompts one to interpret the maximum-entropy technique in terms of a Bayesian approach [65] in which the *a priori* probability density is given by

$$[(\mathbf{f}) \sim \exp\left\{-N \sum_i f_i \ln\left(\frac{f_i}{e\tilde{f}_i}\right)\right\}. \tag{3.5-4}$$

The constant e is required for normalization purposes to make the maximum of $P(\mathbf{f})$ coincide with the vector $\tilde{\mathbf{f}}$, which takes the place of the ensemble mean $\bar{\mathbf{f}}$ in Eq. (3.4-6) [57]. The constant N is the number of something—photons in the image, or pixels, or balls thrown by monkeys, or whatnot. Its interpretation is still being studied [66]. In any event, the value of N is computationally considered to be a Lagrange multiplier that is varied to bring χ^2 to the required value [57]. Independent of whether this Bayesian interpretation is correct or not, the performance of the maximum-entropy algorithm can be understood and interpreted in terms of what one expects for MAP under the assumption of a prior probability distribution given by Eq. (3.5-4). The contours of this probability distribution given by Eq. (3.5-4). The contours of this probability distribution are shown in Fig. 3.4-1c with $\tilde{\mathbf{f}}_i = \tilde{\mathbf{f}}_j = 0.0001$ and $N = 10{,}000$. It is observed that this entropy probability function is quite broad. Near its maximum $\tilde{\mathbf{f}}_i$, its behavior is similar to that of a Gaussian with an rms width of about $\tilde{\mathbf{f}}_i$. Thus, the conventional MAP approach, augmented to include a constraint of nonnegativity, might be expected to yield results quite similar to those obtained with a maximum-entropy algorithm. This has been our experience, as shown in Fig. 3.10-1, referred to in Section 3.10. As N gets smaller, the above probability distribution becomes even flatter.

The above interpretation of the maximum-entropy algorithms in terms of the Bayesian probabilistic philosophy is contrary to the axiomatic approach to maximum entropy espoused by many [66, 67]. Some would prefer to view MAP as a special case of the maximum-entropy principle [68]. However, others find it difficult to accept the interpretation of image intensities as probabilities [69]. To add to the confusion, the proper choice for the entropy expression has been a constant source of debate [70, 71]. One advantage of the Bayesian interpretation is that it permits one to ask whether the entropy probability distribution Eq. (3.5-4) represents an appropriate *a priori* probability distribution for the particular problem one is addressing.

3.6 MAP SOLUTIONS

We have adopted an iterative approach to the solution of Eq. (3.4-7) based on the scheme proposed by Herman and Lent [31]. The iteration scheme is given by

$$\mathbf{f}^0 = \bar{\mathbf{f}}, \tag{3.6-1a}$$

$$\mathbf{f}^{k+1} = \mathbf{f}^k + c^k \mathbf{r}^k, \tag{3.6-1b}$$

$$\mathbf{r}^k = \bar{\mathbf{f}} - \mathbf{f}^k + \mathbf{R}_f \mathbf{H}^T \mathbf{R}_n^{-1}(\mathbf{g} - \mathbf{H}\mathbf{f}^k), \tag{3.6-1c}$$

$$c^k = \mathbf{r}^{kT}\mathbf{s}^k / \mathbf{s}^{kT}\mathbf{s}^k, \tag{3.6-1d}$$

$$\mathbf{s}^k = (\mathbf{I} + \mathbf{R}_f \mathbf{H}^T \mathbf{R}_n^{-1} \mathbf{H})\mathbf{r}^k, \tag{3.6-1e}$$

where vector $\mathbf{r}^k$ is the residual of Eq. (3.4-7) (multiplied by $\mathbf{R}_f$), the scalar c^k is chosen to minimize the norm of $\mathbf{r}^{k+1}$, and $\mathbf{I}$ is the identity matrix. When the residual goes to zero, the corresponding $\mathbf{f}^k$ is clearly a solution of the MAP Eq. (3.4-7). This iterative scheme is very similar to the one proposed by Hunt [38] for MAP image restoration in the presence of a nonlinear scalar transfer function. His update scheme consisted in incrementing $\mathbf{f}^k$ by Eq. (3.6-1c) multiplied by $\mathbf{R}_f^{-1}$. This might introduce computational difficulty if $\mathbf{R}_f$ is nontrivial. Trussell and Hunt [72] refined Hunt's iteration procedure to improve the rate of convergence. We have found that Eq. (3.6-1) works well, although convergence typically requires 10 to 20 iterations. An advantage of any iterative reconstruction scheme is that constraints may readily be placed on the reconstructed function $\mathbf{f}^{k+1}$ after each update. Such constraints include upper and lower limits to the reconstruction value, known region of support, etc. The reconstruction algorithm can thus become quite nonlinear.

which consisted of a mosaic of irregularly shaped regions of nearly constant intensity. Certain kinds of constraints, such as region of support and nonnegativity, can also be incorporated in this algorithm in the form of hard constraints [84]. Thus simulated annealing offers an appealing alternative way to incorporate certain kinds of prior knowledge about the structure of the reconstructed object that may be difficult to implement in and may depart drastically from the simplifying assumptions of the standard MAP method. This method would appear to be most easily computed when the energy functions involve a small number of pixels.

3.7 FAIR—FIT AND ITERATIVE RECONSTRUCTION

The fit and iterative reconstruction technique introduced by Hanson [85] offers an alternative method for incorporating *a priori* information about the source function is used to construct a parametric model of the unknown function. What is known about the structure of the object is employed to restrict the structure of the model. The first step in this algorithm is to determine the parameters in the model from the available projection data by a least-squares [or minimum chi-squared, Eq. (3.4-8)] fitting procedure. In the second step of FAIR, an iterative reconstruction procedure is performed, using the fitted parametric model as the initial estimate. The iterative reconstruction procedure forces the result to agree with the measurements through alteration of its measurement-space contribution. The null-space contribution to the FAIR reconstruction arises solely from the parametric model fitted in the first step and hence from the *a priori* information used in specifying the model. In a certain sense, this parameter estimation problem may be viewed as a selection of that point in Hilbert space which satisfies the constraints placed on the function through specification of the parametric model and minimizes chi-squared. Although the ART algorithm [16] has been used in the second, iterative step of FAIR, other iterative algorithms such as the MAP algorithm above can be used advantageously (see Section 3.6). It might be noted that ART, as usually employed, forces the reconstruction to agree with each measurement in turn and is unable to accommodate inconsistent data (contaminated by noise, for example). However, when employed with a vanishingly small damping factor, or relaxation parameter, ART can cope with inconsistent data and will converge to a least-squares solution [86].

An important advantage of using the fitting procedure in the FAIR technique is its flexibility. Additional parameters may be introduced in the model to permit the position, orientation, and size of the object to be adjusted to match the data. Constraints may be placed on the extra parameters to avoid unacceptable objects. For example, in the annulus problems below, it is possible to allow each of the 2-D Gaussian functions to be centered at a variable radius instead of a fixed, predetermined radius. This would permit the size and the shape of the contour of the annulus to be determined from the measurements. The anular structure of the modeled object could still be enforced by adding a penalty function to chi-squared based on the quadratic differences between the radii of adjacent Gaussians. This MAP approach to parameter estimation would tend to make the radius a smooth function of polar angle.

3.8 COMPARISON OF MAP AND FAIR RESULTS

The results of application of various reconstruction methods to a specific two-dimensional, limited-angle reconstruction problem will be compared. Algorithms that are useful for handling incomplete data through the use of *a priori* information must have the following important characteristics. They must (1) significantly reduce artifacts that result from inappropriate null-space contributions, (2) gracefully respond to inconsistencies between the actual source function and the assumptions about it, and (3) tolerate noise in the projection data. It will be demonstrated that the MAP and FAIR algorithms conform to these requirements.

These reconstruction techniques have been applied to a source function consisting of a fuzzy annulus with variable amplitude, Fig. 3.8-1a, which roughly emulates the nuclear isotope distribution in the cross section of a heart [87]. The peak value of the distribution is 1.24. The available projection data consist of 11 views covering 90° in projection angle. At first no noise was added to the projections. Each projection contained 128 samples evenly spaced over the full width of the image. All reconstructions contain 128×128 pixels. The mean diameter of the annulus is 64 pixels. The measurement-space reconstruction obtained with ART [16], Fig. 3.8-1b, shows severe artifacts that tend to obscure much of the source distribution.

Figure 3.8-1c shows the reconstruction obtained with the maximum-entropy algorithm MENT provided to us by Minerbo [64]. This algorithm provides a modest improvement over ART, particularly in regard to the detection of the dip in the annulus at 50°. However, MENT does not have much effect on the splaying of the reconstruction along the central axis of the available views. In our experience the principal advantage of the maximum-entropy constraint appears to be its implicit constraint of non-negativity. ART reconstructions that are constrained to be nonnegative are

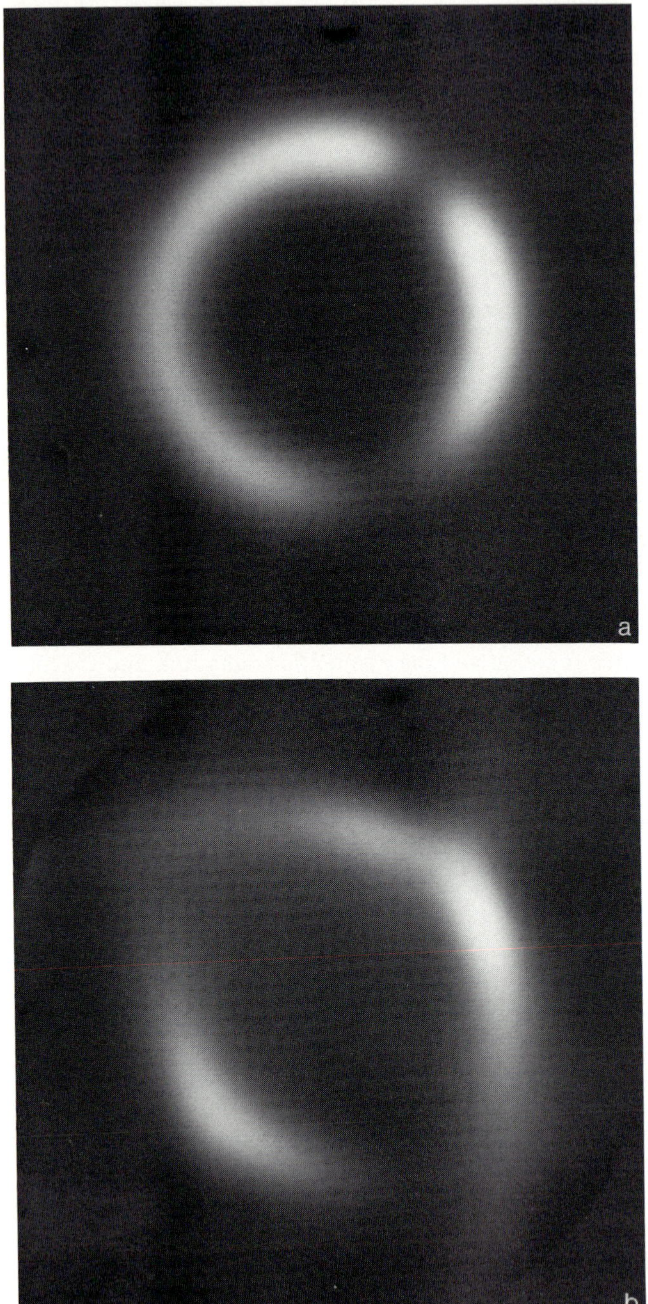
a
b

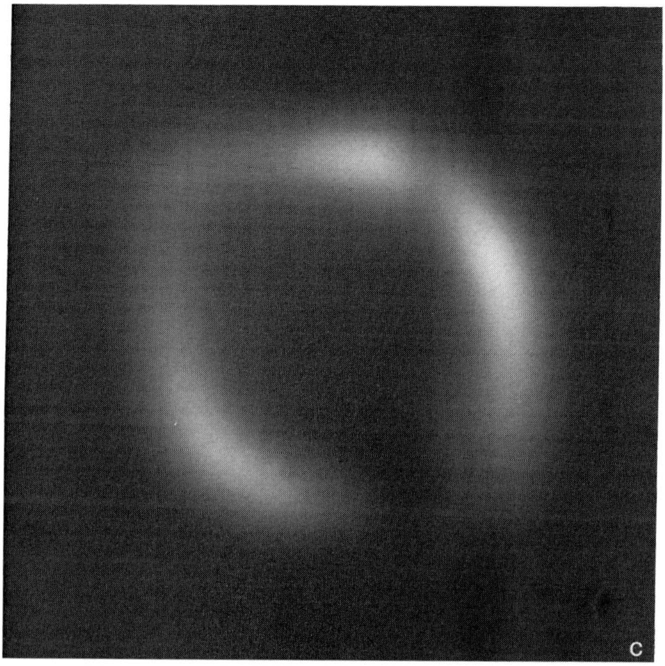

(continued)

very similar to the MENT results [85]. The nonnegativity constraint amounts to the incorporation of *a priori* knowledge about the source function. This constraint is generally applicable and is very effective in the reconstruction of certain types of source distributions, such as pointlike objects on a zero background. However, there are many source distributions and data collection geometries for which nonnegativity provides little benefit, such as the present test case. We will not apply the nonnegativity constraint in the remainder of these comparisons to emphasize the value of using *a priori* knowledge about the shape of the object.

It was assumed that the *a priori* information consisted of the knowledge that the source function had an annular structure with known position, radius, and width. Thus, in the MAP approach, $\bar{\mathbf{f}}$ was chosen to be an annulus with constant amplitude and Gaussian cross section. The mean radius and width of the annulus were chosen to be the same as the unknown source function. The covariance matrix $\mathbf{R}_f$ was assumed to be diagonal and thus could be represented as an image proportional to the ensemble variance about the mean $\bar{\mathbf{f}}$. The covariance image $\mathbf{R}_f$ was large (1.0) at the peak of

Fig. 3.8-1. (a) Source distribution used for the first example. (b) ART and (c) MENT reconstructions obtained using 11 views covering 90° in projection angle. Unconstrained ART was used, while MENT has an implicit nonnegativity constraint.

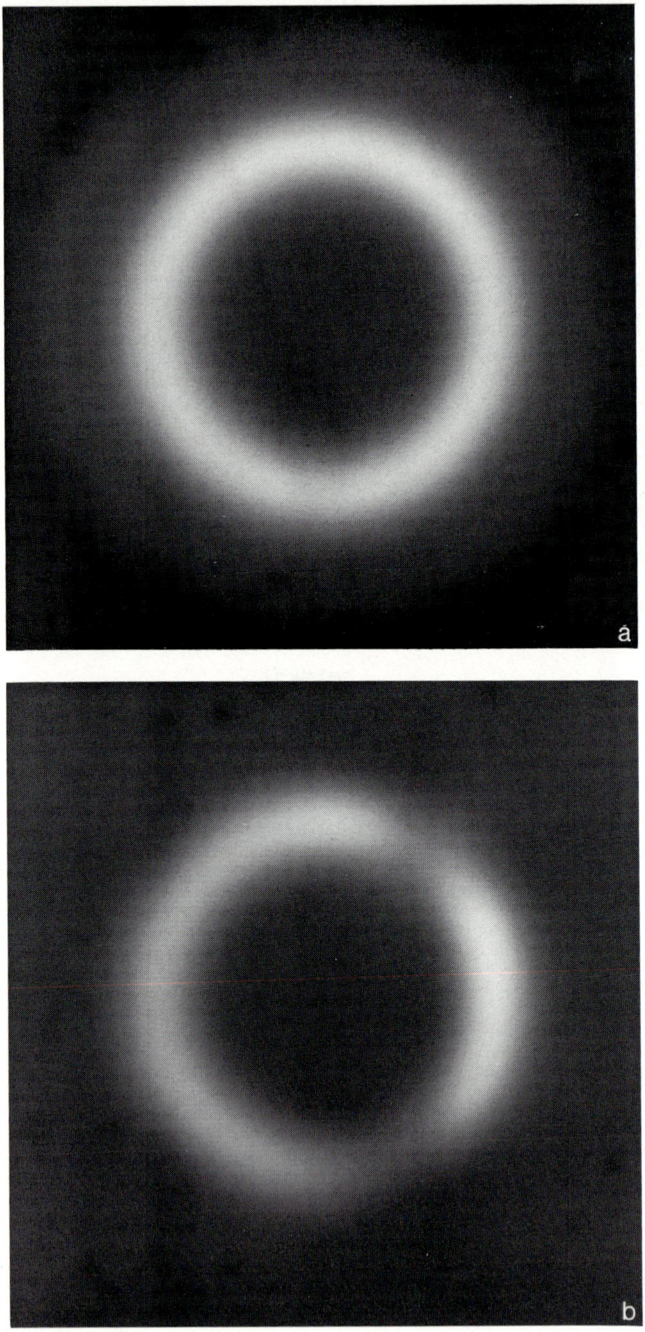
a
b

(*continued*)

the annulus and small (0.2) inside and outside, Fig. 3.8-2a. Because noiseless projections were used, the measurement noise was assumed to be uncorrelated, constant, and low in value. The resulting MAP reconstruction, Fig. 3.8-2b, is vastly superior to the ART and MENT results, eliminating essentially all the artifacts present in these deterministic solutions. The parametric model chosen for the FAIR method consisted of 18 two-dimensional Gaussian functions evenly distributed on a circle. The radius of the circle and the width of the Gaussians were chosen to be the same as those of the source function. The fitting procedure determined the amplitude of each of the Gaussian functions. The resulting fitted function was used as the initial estimate in ART to obtain the final result, Fig. 3.8-2c. This FAIR reconstruction is comparable to the MAP result.

Fig. 3.8-2. Reconstructions obtained by using the *a priori* information that the unknown source function is a fuzzy annulus with known radius and width. The MAP reconstruction (b) was obtained with a flat annulus for the ensemble mean $\bar{\mathbf{f}}$, and the variance image (a) as the diagonal entries of the ensemble covariance matrix $\mathbf{R}_f$ (nondiagonal entries assumed to be zero). The FAIR result (c) was based on a model of the image consisting of 18 Gaussian functions distributed around the circle. The use of *a priori* knowledge significantly reduces the artifacts present in the deterministic reconstruction in Fig. 3.8-1.

For a quantitative comparison, Fig. 3.8-3 shows the maximum reconstruction value obtained along radii as a function of angle for the various reconstruction methods presented in Figs. 3.8-1 and 3.8-2. The FAIR reconstruction is seen to follow the original source dependence most closely, with the MAP result a close second. The ART reconstruction has many quantitatively serious defects. The computation times on a CDC-7600 for the algorithms presented here are ART (10 iterations), 17 s; MAP (10 iterations), 73 s; FAIR (3 iterations), 25 s; MENT (6 iterations), 105 s. The corresponding execution time for filtered backprojection is 5 s.

A slightly different source function, Fig. 3.8-4a, was used to test the ability of the algorithms to deal with inconsistencies between assumptions about the source function and its actual distribution. This source function is the same as Fig. 3.8-1a with a narrow, 0.6 amplitude, 2-D Gaussian added outside the annulus at 330° and a broad, 0.1 amplitude, Gaussian added

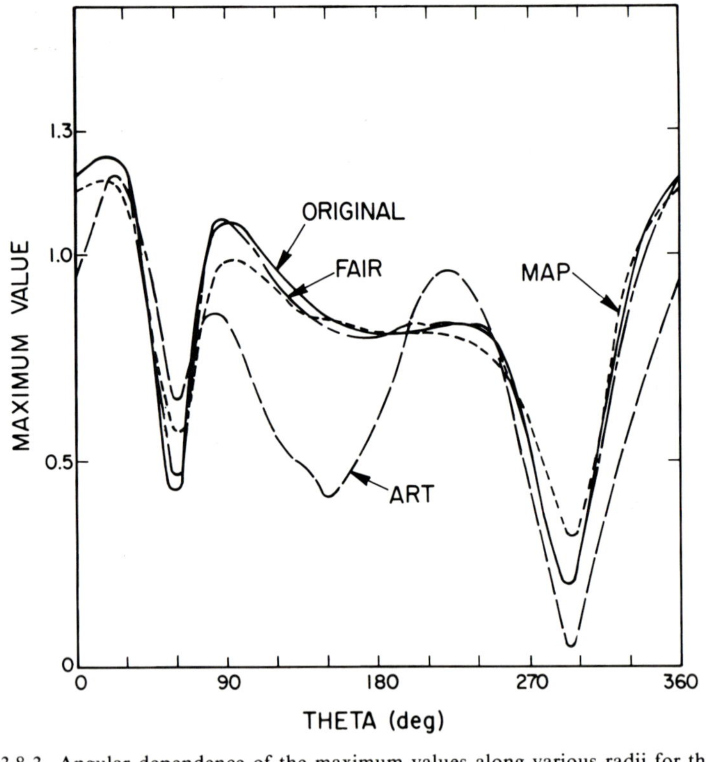

Fig. 3.8-3. Angular dependence of the maximum values along various radii for the ART, MAP, and FAIR reconstructions in Figs. 3.8-1 and 3.8-2 compared with that for the original function, Fig. 3.8-1a, quantitatively demonstrating the improvement afforded by MAP and FAIR.

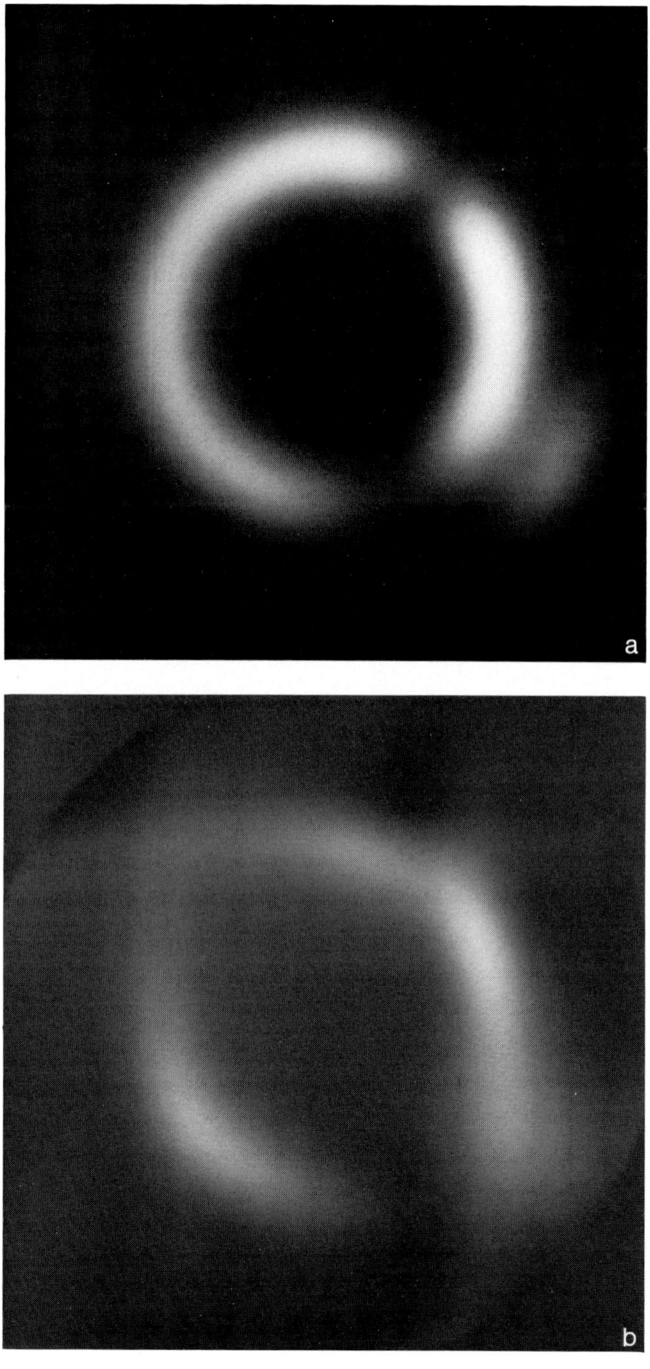
a
b

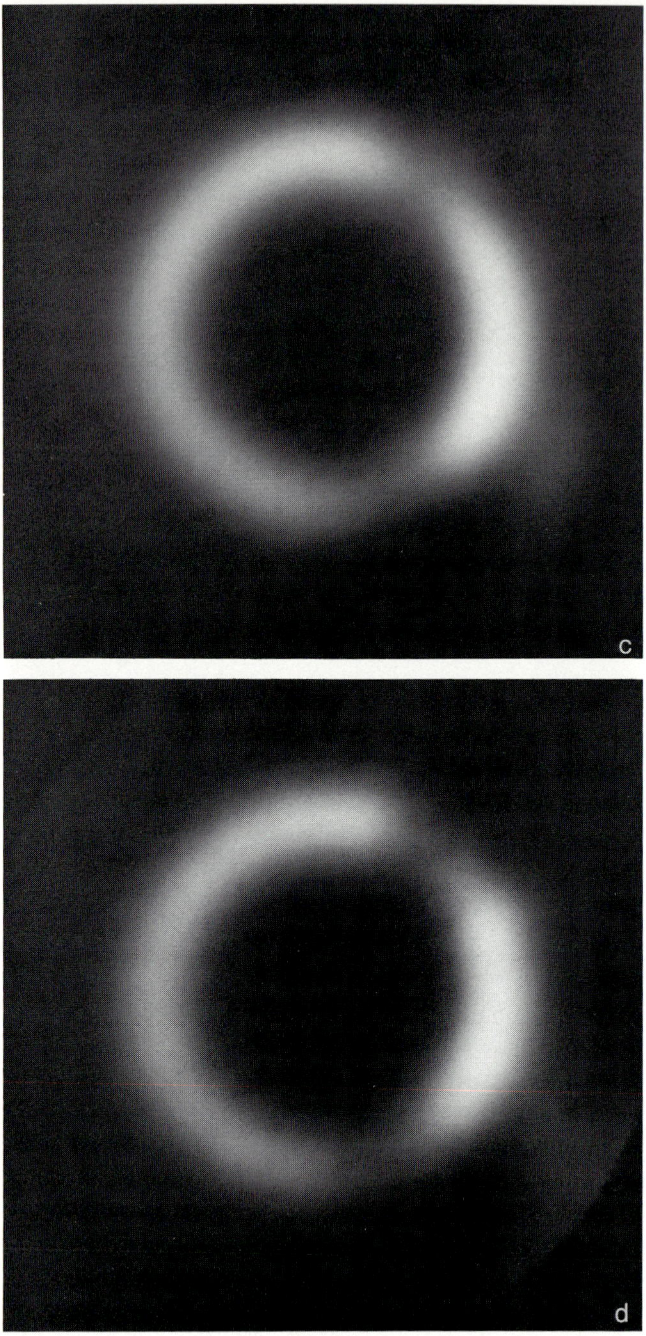

Fig. 3.8-4. Reconstructions of a source distribution (a) that does not conform to the annular assumption obtained from 11 views subtending 90° with the (b) ART, (c) MAP, and (d) FAIR algorithms. Both MAP and FAIR tend to move the additional source placed outside the annulus onto the annulus. However, they provide indications in the reconstructions that there is some exterior activity.

underneath the annulus at 162°. The reconstructions obtained using the same assumptions as above are shown in Figs. 3.8-4b to 3.8-4d. Both MAP and FAIR handle the inconsistencies similarly. The angular dependence of the maximum reconstruction value, Fig. 3.8-5, shows that both algorithms produce an excess near 330° since they tend to shift the discrepant exterior source to the annulus, consistent with the *a priori* assumptions. However, both methods do have a significant response in the region of the exterior source and, therefore, provide some information about the discrepancy. This would not be the case for the MAP algorithm were $\mathbf{R}_f$ chosen to be zero outside the annulus. This illustrates the need for conservatism in placing restrictions on the reconstructions that may be violated by the actual source distribution. The second, iterative reconstruction step in the FAIR method is needed for the same reason as it allows corrections to be made to the fitted model, if indicated by the available projections.

The final example is the reconstruction of Fig. 3.8-1a from noisy data. The same 11 projections were used as before but with random noise added,

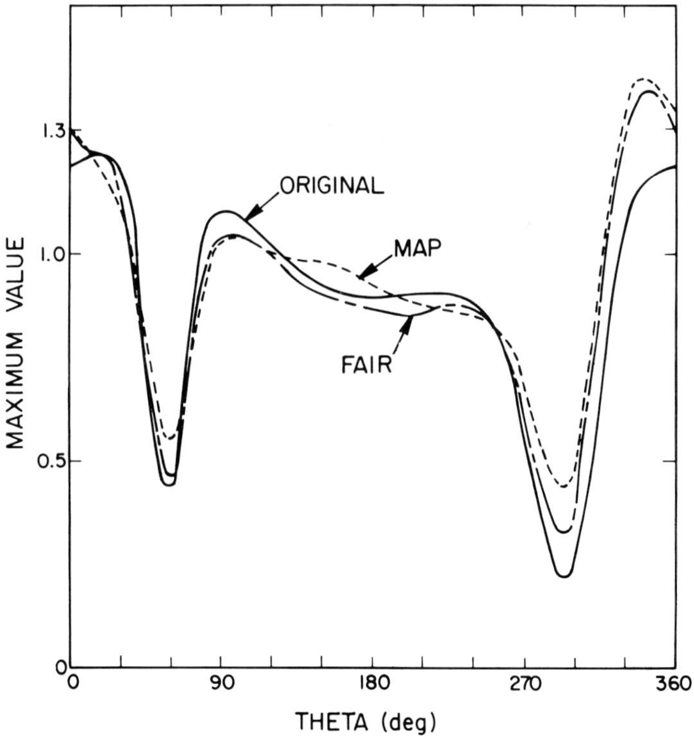

Fig. 3.8-5. Angular dependence of the maximum values in the MAP and FAIR reconstructions of Fig. 3.8-4.

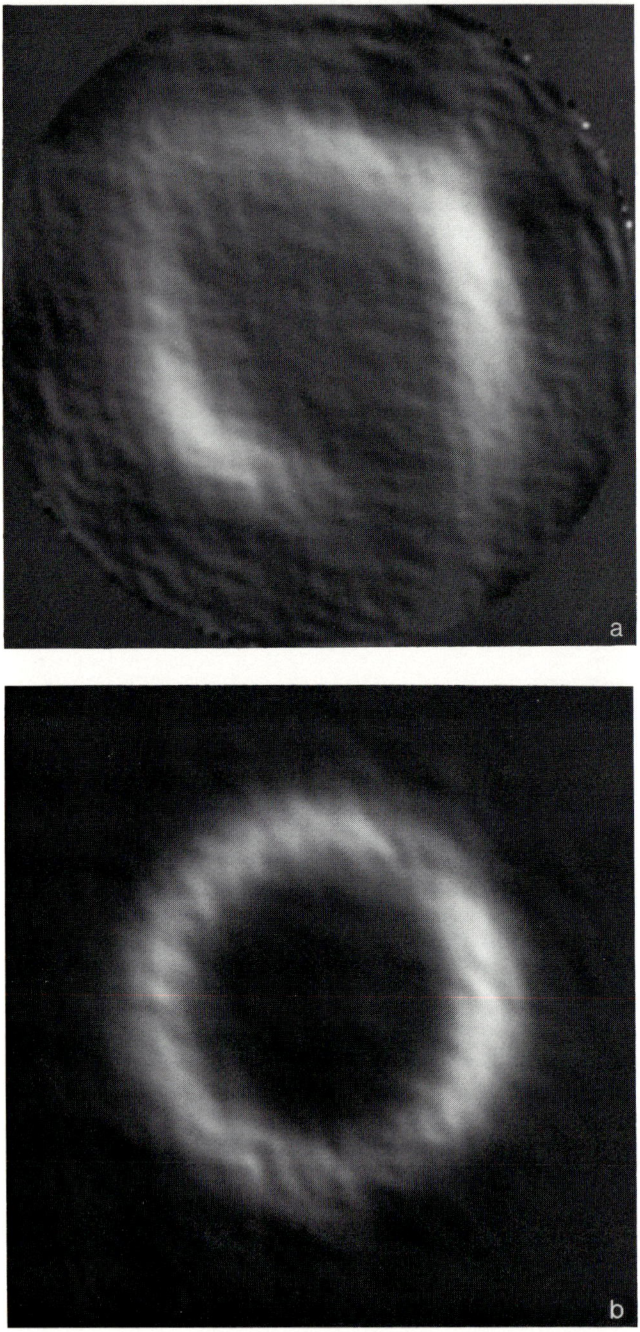
a
b

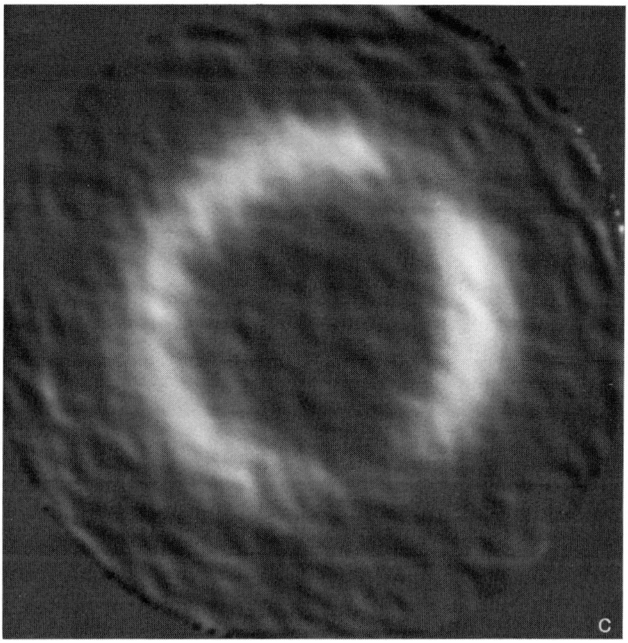

(*continued*)

which has an rms deviation of 10% relative to the maximum projection value. The reconstructions in Fig. 3.8-6 demonstrate that both MAP and FAIR simply yield noisy versions of those obtained from noiseless projections. There is no disastrous degradation as would be expected for algorithms based on analytic continuation [43, 44]. Although the FAIR result appears to be much noisier than the MAP reconstruction, careful observation reveals that both have nearly identical noise in the annular region, which is the only region of interest for diagnosis. The rms difference between the projection measurements and the ray sums of the MAP and FAIR reconstructions, respectively, are roughtly 0.8 and 0.5 times the actual rms deviation of the noise in the projections. This indicates that both algorithms have attempted to solve the measurement equations beyond what is reasonable. The MAP algorithm does balance the rms error in the projections against the deviation from $\bar{\mathbf{f}}$. However, when used with unity damping factor, ART simply attempts to reduce the rms projection error to zero, satisfying each projection

---

Fig. 3.8-6. Reconstructions of the source in Fig. 3.8-1a from 11 noisy projections with the (a) ART, (b) MAP, and (c) FAIR algorithms, showing that the latter two algorithms are tolerant of noise.

in turn. It may never converge to a solution if the data are inconsistent. It is possible to temper the adverse reaction of ART to inconsistent data by using small damping factors [86].

3.9 A GENERALIZED BAYESIAN METHOD

Both MAP and FAIR, as presented above, have drawbacks. The incorporation of *a priori* knowledge in the MAP algorithm presented above is quite restrictive. It does not readily accommodate prior source distributions that vary in size, shape, or location. However, the fitting procedure used in the first step of FAIR can easily handle such variations by including them as variables to be determined from the data. In the spirit of the Bayesian approach, constraints on these variables may be introduced to guide the fitting procedure toward a "reasonable" result. The use of ART in the second, iterative portion of FAIR has the disadvantage that ART tries to reduce the discrepancy in the measurement equations to zero without regard for the estimated uncertainties in the data. Thus, the FAIR result shown in Fig. 3.8-6c is quite noisy and is substantially farther from the actual source distribution (rms deviation = 0.154) than the intermediate fitted result (rms deviation = 0.031).

In a more *global Bayesian approach* to the problem, the fitting procedure in FAIR may be used to estimate suitable ensemble properties $\bar{\mathbf{f}}$ and $\mathbf{R}_f$ for input to a MAP algorithm. The fitting procedure may be viewed as defining a subensemble appropriate to the available data. This is similar to a radiologist's approach to interpreting a radiograph. The radiograph typically contains enough identifying features that the radiologist can determine which part of the body is being imaged with which imaging modality. This allows him to select the appropriate set of diagnostic criteria. For the present example, the fitting procedure used the same model as described above for the first step of FAIR and the fitted result was used for $\bar{\mathbf{f}}$. $\mathbf{R}_f$ was assumed to be 0.1 times the identity matrix. This is much smaller than the value used in the preceding MAP calculation to reflect the supposition that the fit is much closer to the desired result than the annulus of constant amplitude previously used for $\bar{\mathbf{f}}$. Figure 3.9-1 shows the MAP reconstruction using the 10% rms noise data, and the results are superior to those in Fig. 3.8-6. The rms deviation of this reconstruction relative to the source function is 0.035, whereas that for the earlier MAP result, Fig. 3.8-6, is 0.060. When this generalized MAP method is applied to the projections of the inconsistent source function shown in Fig. 3.8-4a, the result is very similar to that obtained with FAIR, Fig. 3.8-4d. These examples only hint at the power of

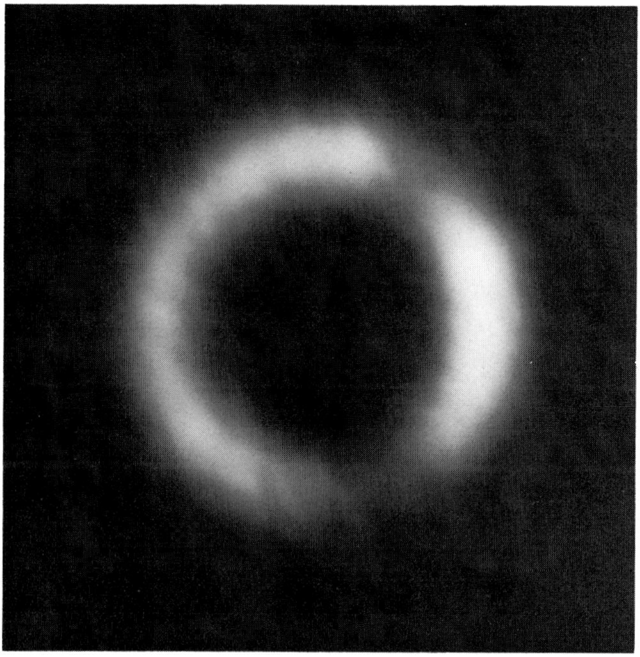

Fig. 3.9-1. Reconstruction from the same data used in Fig. 3.8-6 obtained by employing MAP as the second step in the FAIR procedure. This global Bayesian approach yields the best estimate of the original function and provides flexibility in the use of *a priori* information.

this global Bayesian approach where the MAP algorithm is used for the second, iterative step of FAIR. The true flexibility of this global approach awaits demonstration with more demanding problems.

3.10 DISCUSSION

In past comparisons of MAP results to those of more standard techniques in the areas of CT [31, 73] and in image restoration [38, 72, 88, 89], MAP yielded few or no benefits. The reasons for the success of MAP in the above limited-angle CT example problem are: (1) the solution is severely underdetermined because of the limited nature of the data set, and (2) the *a priori* assumptions about $\bar{\mathbf{f}}$ and $\mathbf{R}_f$ can be made quite restrictive because of the nature of the stated problem. It is expected that the Bayesian analysis will be most useful in situations where these two conditions hold because only then will there be both a significant contribution from the null space, which

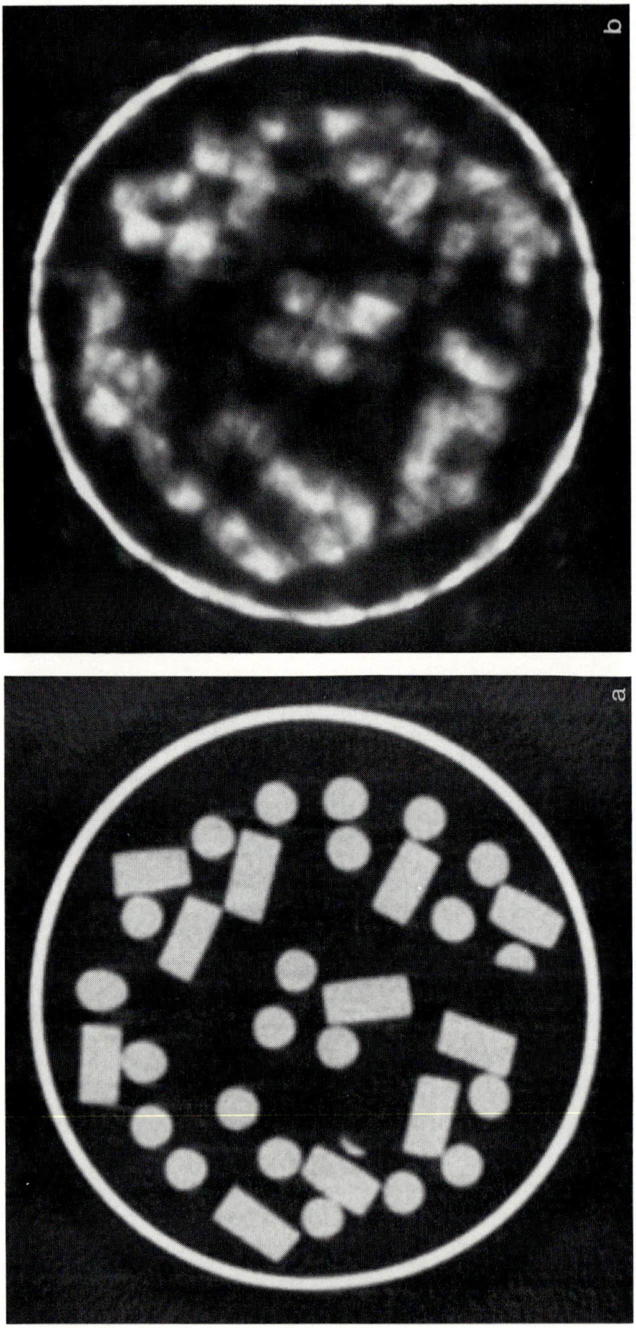
b
a

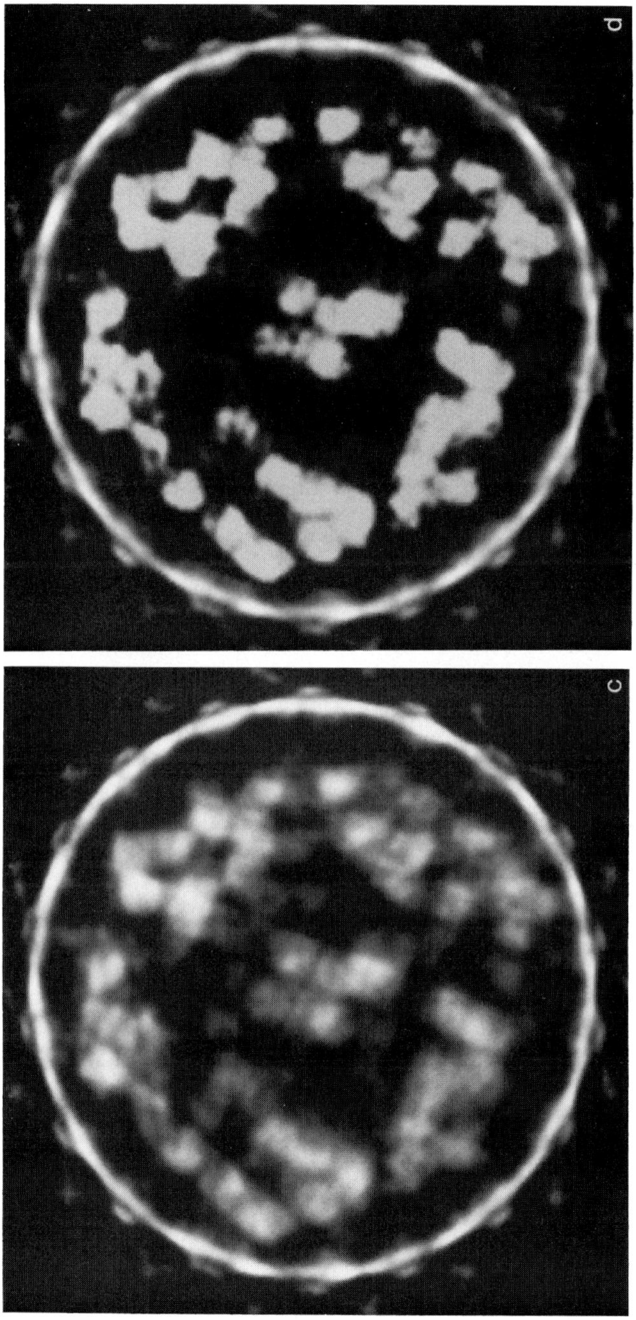

Fig. 3.10-1. Reconstructions from x-ray transmission data of a mock-up of propellant grains inside a fiberglass cannon bore. See Zoltani *et al.* [90, 91] for details. The true object is given by (a), the filtered backprojection reconstruction from 180 views. The maximum-entropy reconstructions from nine views (b) and five views (e) show artifacts similar to those in their counterparts (c) and (f) obtained by using ART with a nonnegativity constraint. The soft enforcement of binary reconstruction values in the interior of the cylinder, (d) and (g), respectively, results in ART reconstructions that are much more "reasonable". However, the reconstruction with only five views (g) is plainly wrong, even though it looks like what is expected. Incidentally, prior knowledge of the fiberglass cylinder was incorporated in (g).

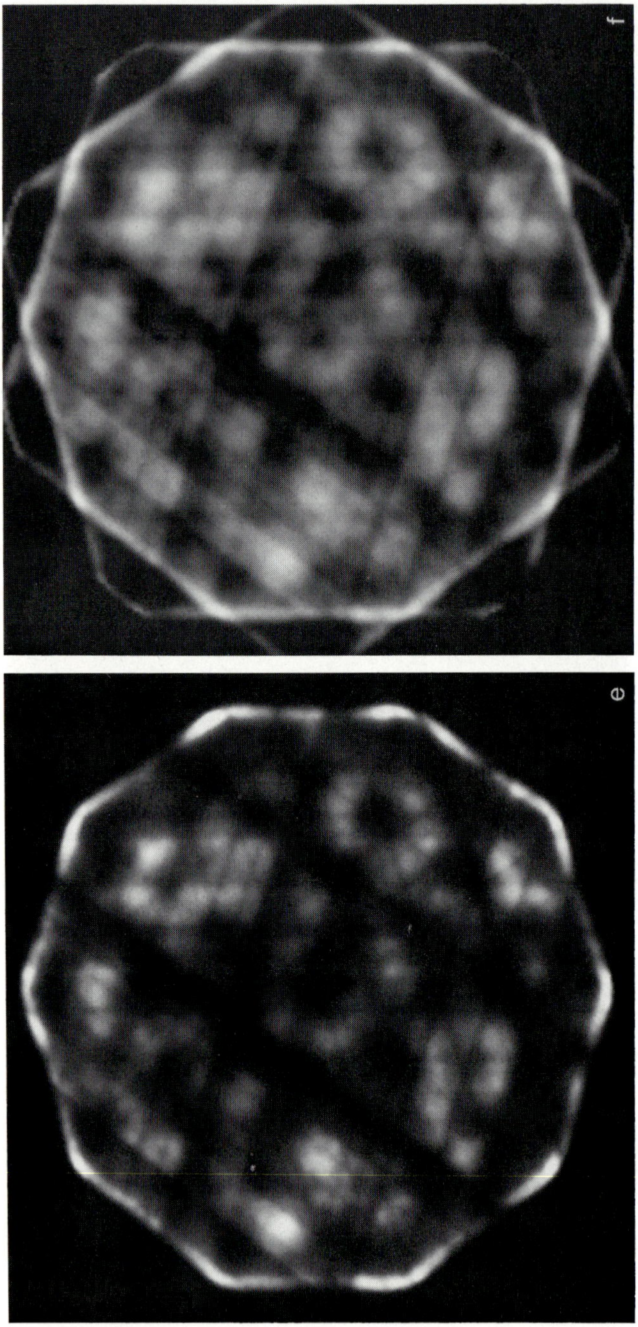
f
e

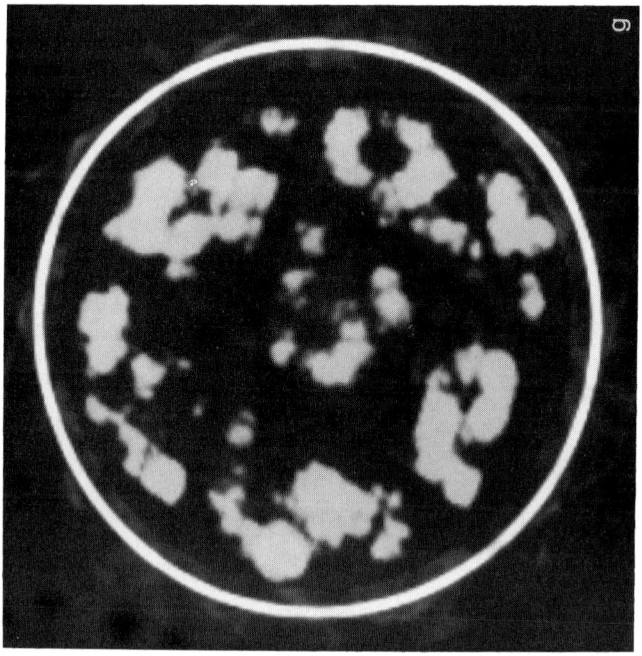

Fig. 3.10-1 (*continued*)

must be properly estimated to avoid artifacts, and the necessary information to accomplish that.

As indicated several times in this chapter, great care must be exercised in specifying the ensemble mean and covariance matrix when using the Bayesian approach. The same must be said of other supplementary forms of constraint. First, it is necessary to avoid assumptions that are inconsistent with the actual object as well as with the data. Otherwise the abundance of constraints may cause these inconsistencies to show up as artifacts that appear to be part of what was expected. The second consideration is that the MAP method necessarily produces results that are weighted toward $\bar{\mathbf{f}}$, as discussed in Section 3.4. This may not be appropriate if the sought-for entities are anomalous. The magnitude of the elements of $\mathbf{R}_f$ can be increased from their true ensemble values to allow for this. Thus one may view the selection of the ensemble characteristics as a method of "tuning" the reconstruction algorithm. This tuning must be done with the final purpose of the images kept in mind. It should be realized that this is a trade-off situation. The constraints must be restrictive in order to affect the null space and hence to reduce reconstruction artifacts. But overly restrictive constraints can obscure deviations from what one expects, which might be precisely what one is looking for. To make matters worse, in such a situation the resulting reconstructions may look very "reasonable," giving the observer a false sense of security. Figure 3.10-1shows what happens when excessively restrictive constraints (binary reconstruction values) are placed on data that possess insufficient information. This exercise suggests that the definition of an optimal reconstruction should include the ability to estimate the reliability of the interpretation from the reconstruction itself. It is this ability, among others, that distinguishes the human observer [76].

The judgment of the success of the Bayesian analysis is not a trivial process. Because the Bayesian approach is based on minimization of an ensemble variance, it can be accurately evaluated only by using a relatively large number of trials. The assessment of the technique cannot be made on the basis of a single image or even several images, as the reader of the open literature in general or of this chapter in particular is asked to do. Before a Bayesian method can be fully tested in any new imaging situation, the observers must be thoroughly trained. This requires a large set of training images for which the "truth" is known. The observer should be fully aware of the kinds of assumptions that have been made in the reconstruction process in order not to be misled by the results. Judgment of any new imaging technique is too often made on the basis of experience with past techniques without retraining the observers. Of course, such retraining is costly and demands real commitment to the new method.

3.11 SUMMARY

The deficiency in limited data is a consequence of a null space of reconstruction functions about which the available measurements say nothing. Artifacts in deterministic reconstructions can be reduced if the null-space components of the original image can be properly estimated through the use of prior information. The Bayesian approach permits the incorporation of information about the general shape or structure of the object. An iterative reconstruction algorithm is presented that finds the solution with the maximum *a posteriori* (MAP) probability. The incorporation of other kinds of prior knowledge, including maximum entropy, may be considered as special cases of the Bayesian approach. An alternative Bayesian scheme is the fit and iterative reconstruction (FAIR) algorithm that is based on a parametric model of the object, which is restricted according to prior knowledge about the object. Both MAP and FAIR are shown to work well in ameliorating reconstruction artifacts, even in the presence of noise and other sources of inconsistencies with the prior assumptions. A global Bayesian approach, which combines the flexibility of the fitting procedure in the FAIR method and the tolerance of the MAP method, promises to be even better. It is stressed that the prior constraints placed on any reconstruction must be carefully chosen so as not to automatically produce an acceptable looking result, irrespective of the data. Because the Bayesian method allows information that is not inherent in the measurement data to affect the resulting image, it is wise to take a cautious approach in interpreting its results. Thus, much personal experience with the Bayesian method is required before one can rely on it.

ACKNOWLEDGMENTS

This work was supported by the U.S. Department of Energy under contract W-7405-ENG-36. I wish to acknowledge helpful and interesting discussions with Gerald Minerbo, Michael Buonocore, Barry Medoff, Jorge Llacer, George W. Wecksung, and Henry Stark. George Wecksung is responsible for the computer codes used to generate the MAP examples shown here and for introducing me to the measurement-space/null-space concept. I thank Richard P. Kruger and Csaba Zoltani for the data used to obtain the reconstructions shown in Fig. 3.10-1.

REFERENCES

1. K. M. Hanson and G. W. Wecksung (1983). Bayesian approach to limited-angle reconstruction in computed tomography. *J. Opt. Soc. Amer.* **73**, 1501–1509.
2. K. M. Hanson, J. N. Bradbury, T. M. Cannon, R. L. Hutson, D. B. Laubacher, R. J. Macek, M. A. Paciotti and C. A. Taylor (1981). Computed tomography using proton energy loss. *Phys. Med. Biol.* **26**, 965–983.

3. K. M. Hanson, J. N. Bradbury R. A. Koeppe, R. J. Macek, D. R. Machen, R. Morgado, M. A. Paciotti, S. A. Sanford and V. W. Steward (1982). Proton computed tomography of human specimens. *Phys. Med. Biol.* **27**, 25-36.
4. A. J. Devaney (1983). A computer simulation study of diffraction tomography. *IEEE Trans. Biomed. Eng.* **BME-30**, 377-386.
5. A. J. Devaney (1984). Geophysical diffraction tomography. *IEEE Trans. Geosci. Remote Sens.* **GE-22**, 3-13.
6. H. C. Andrews and B. R. Hunt (1977). "Digital Image Restoration", Prentice-Hall, Englewood Cliffs, New Jersey.
7. K. T. Smith, D. C. Solmon and S. L. Wagner (1977). Practical and mathematical aspects of the problem of reconstructing objects from radiographs. *Bull. Amer. Math. Soc.* **83**, 1227-1270.
8. S. Twomey (1965). The application of numerical filtering to the solution of integral equations encountered in indirect sensing measurement. *J. Frank Inst.* **279**, 95-109.
9. S. Twomey and H. B. Howell (1967). Some aspects of the optical estimation of microstructure in fog and cloud. *Appl. Opt.* **6**, 2125-2131.
10. S. Twomey (1974). Information content in remote sensing. *Appl. Opt.* **13**, 942-945.
11. K. Tanabe (1971). Projection method for solving a singular system of linear equations and its applications. *Numer. Math.* **17**, 203-214.
12. D. D. Jackson (1972). Interpretation of inaccurate, insufficient and inconsistent data. *Geophys. J. R. Astr. Soc.* **28**, 97-109.
13. M. B. Katz (1979). "Questions of Uniqueness and Resolution in Reconstruction from Projections," Lecture Notes in Biomathematics, (S. Levin, ed.) Springer, Berlin.
14. J. Llacer (1979). Theory of imaging with a very limited number of projections. *IEEE Trans. Nucl. Sci.* **NS-26**, 596-602.
15. J. L. C. Sanz and T. S. Huang (1983). Unified Hilbert space approach to iterative least-squares linear signal restoration. *J. Opt. Soc. Am.* **73**, 1455-1465.
16. R. Gordon, R. Bender and G. T. Herman (1970). Algebraic reconstruction techniques for three-dimensional electron microscopy and x-ray photography. *J. Theor. Biol.* **29**, 471-481.
17. B. F. Logan and L. A. Shepp (1975). Optimal reconstruction of a function from its projections. *Duke Math. Jour.* **42**, 645-659.
18. A. V. Lakshminarayan and A. Lent (1976). The simultaneous iterative reconstruction technique as a least-squares method. *Proc. SPIE* **96**, 108-116.
19. C. Hamaker and D. C. Solmon (1978). The angles between the null spaces of x rays. *J. Math. Anal. Appl.* **62**, 1-23.
20. A. K. Louis (1981). Ghosts in tomography—The null space of the Radon transform. *Math. Meth. Appl. Sci.* **3**, 1-10.
21. A. K. Louis (1984). Orthogonal function series expansions and the null space of the Radon transform. *SIAM J. Math. Anal.* **15**, 621-633.
22. B. P. Medoff, W. R. Brody and A. Macovski (1982). Image reconstruction from limited data. *Proc. Int. Workshop on Physics and Engineering in Medical Imaging*, Pacific Grove, 1982, IEEE Computer Society, Silver Spring, 188-192.
23. B. P. Medoff, W. R. Brody, M. Nassi and A. Macovski (1983). Iterative convolution backprojection algorithms for image reconstruction from limited data. *J. Opt. Soc. Am.* **73**, 1493-1500.
24. K. M. Hanson (1982). CT reconstruction from limited projection angles. *Proc. SPIE* **347**, 166-173.
25. Buonocore, M. H., Brody, W. R., and Macovski, A. (1981). Natural pixel decomposition for two-dimensional image reconstruction. *IEEE Trans. Biomed. Eng.* **BME-28**, 69-78.
26. G. T. Herman (1980), "Image Reconstruction from Projections," Academic Press, New York.

27. Shepp, L. A., and Logan, B. F. (1974). The Fourier reconstruction of a head section. *IEEE Trans. Nucl. Sci.* **NS-21**, 21–43.
28. R. Gordon (1974). A tutorial on ART. *IEEE Trans. Nucl. Sci.* **NS-21**, 78–93.
29. P. Gilbert (1972). Iterative methods for the three-dimensional reconstruction of an object from projections. *J. Theor. Biol.* **36**, 105–117.
30. M. Goitein (1972). Three-dimensional density reconstruction from a series of two-dimensional projections. *Nucl. Instrum. Meth.* **101**, 509–518.
31. G. T. Herman and A Lent (1976). A computer implementation of a Bayesian analysis of image reconstruction. *Inform. Contr.* **31**, 364–384.
32. K. M. Hanson and G. W. Wecksung (1985). Local basis-function approach to computed tomography. *Appl. Opt.* **24**, 4028–4039.
33. A. Ben Israel and T. N. E. Greville (1974). "Generalized Inverses: Theory and Applications," John Wiley and Sons, New York, pp. 244–245.
34. D. G. McCaughey and H. C. Andrews (1977). Degrees of freedom for projection imaging. *IEEE Trans. Acoust., Speech, Signal Processing* **ASSP-25**, 63–73.
35. T. M. Cannon (1976). Blind deconvolution of spatially invariant image blurs with phase. *IEEE Trans. Accoust., Speech, Signal Processing* **ASSP-24**, 58–63.
36. D. C. Ghiglia (1984). Space-invariant deblurring given N independent blurred images of a common object. *J. Opt. Soc. Am.* **A1**, 398–402.
37. A. P. Sage and J. L. Melsa (1979). "Estimation Theory with Applications to Communications and Control," Robert E. Krieger, Huntington, p. 175.
38. B. R. Hunt (1977). Bayesian methods in nonlinear digital image restoration. *IEEE Trans. Comput.* **C-26**, 219–229.
39. G. T. Herman and A. Lent (1976). Iterative reconstruction algorithms. *Comput. Biol. Med.* **6**, 273–294.
40. P. R. Bevington (1969). "Data Reduction and Error Analysis for the Physical Sciences," McGraw-Hill, New York.
41. B. R. Frieden (1972). Restoring with maximum likelihood and maximum entropy. *J. Opt. Soc. Am.* **62**, 511–518.
42. L. A. Shepp and Y. Vardi (1982). Maximum likelihood reconstruction for emission tomography. *IEEE Trans. Med. Imaging* **MI-1**, 113–122.
43. T. Inouye (1979). Image reconstruction with limited-angle projection data. *IEEE Trans. Nucl. Sci.* **NS-26**, 2666–2669.
44. T. Inouye (1982). Image reconstruction with limited view angle projections. *Proc. Int. Workshop on Physics and Engineering in Medical Imaging*, Pacific Grove, 1982, IEEE Computer Society, Silver Spring, Maryland, 165–168.
45. A. M. Cormack (1963). Representation of a function by its line integrals, with some radiological applications. *J. Appl. Phys.* **34**, 2722–2727, and *J. Appl. Phys.* **35**, 2908–2913.
46. K. C. Tam, V. Perez-Mendez and B. Macdonald (1979). 3-D object reconstruction in emission and transmission tomography with limited-angular input. *IEEE Trans. Nucl. Sci.* **NS-26**, 2797–2805.
47. R. W. Gerchberg (1974). Super-resolution through error energy reduction. *Opt. Acta* **21**, 709–720.
48. A. Papoulis (1975). A new algorithm in spectral analysis and band-limited extrapolation. *IEEE Trans. Circuits Syst.* **CAS-22**, 735–742.
49. K. C. Tam and V. Perez-Mendez (1981). Tomographic imaging with limited-angle input. *J. Opt. Soc. Am.* **71**, 582–592.
50. K. C. Tam and V. Perez-Mendez (1981). Limits to image reconstruction from restricted angular input. *IEEE Trans. Nucl. Sci.* **NS-28**, 179–183.
51. F. A. Grünbaum (1980). A study of Fourier space methods for limited-angle image reconstruction. *Numer. Funct. Anal. and Optimiz.* **2**, 31–42.

CHAPTER 4

Image Restoration Using Linear Programming

Richard Mammone

Department of Electrical and Computer Engineering
Rutgers University
Piscataway, New Jersey 08854

The problem of restoring an image which has been degraded by a linear operator can be formulated as a discrete optimization problem. The solution of this problem can be found efficiently by linear programming methods. There are generally a number of optimal solutions to an ill-posed signal recovery problem. A unique solution can be defined if sufficient *a priori* information is available. Numerous methods of image restoration have been developed for various applications. Most methods have been developed to restore an image within the passband of the imaging systems. These methods include inverse filtering [1], Wiener filtering [2], least-squares solutions with consistent equality constraints or regularization [3], and the pseudoinverse solution [4]. There has also been much interest recently in methods which extrapolate the image spectrum outside the passband by employing *a priori* information; see [5–12]. This book provides a good representative survey of the modern work in this new area.

The discrete formulation of the image restoration problem requires the solution of a system of linear equations. A subset of these equations are linearly independent. The number of linearly independent equations is called the rank. The system of linear equations is called rank-deficient if the rank is less than the number of unknowns or the number of equations.

The method discussed in this chapter uses a linear programming algorithm. The linear programming approach makes the tacit assumption that the image to be restored has an upper bound on the number of nonzero pixel values. This upper bound is equal to the rank of the linear operator. Thus, the linear programming approach is particularly useful for sparse images, that is, images with a large number of pixels with zero value. Such images are frequently found in astronomy, microscopy, medical imaging, and electron microscopy. The restriction that the image to be restored must be sparse is no longer necessary when the degradation operator is not rank-deficient. In addition to the implicit constraint that the image be sparse, the linear programming approach allows for additional constraints such as upper and lower bounds. The linear programming method is general in the sense that the degradation operator does not have to be shift-invariant or separable.

In this chapter we describe the effects of rank deficiency. The problem of restoration is viewed as a combinatorial problem, which is solved by discrete optimization techniques. The discrete optimization problem is formulated so that various measures of the error are minimized. The choice of the error metric is shown to be related to the statistical properties of the error. The least absolute value or L_1 norm estimate is shown to be well suited for image restoration using linear programming. A method of minimizing the L_1 norm of the error in a computationally efficient manner is presented. The superresolving capability of the L_1 method is demonstrated by computer simulations.

4.1 IMAGE RESTORATION

The difficulties involved in image recovery can be demonstrated by an example. Consider an ideal one-dimensional coherent imaging system with an aperture of length A. The only source of degradation is the finite aperture size. The relationship between the exact image $f(x)$ of extent L and the degraded image $g(x)$ is given by

$$g(x) = \int_{-L/2}^{L/2} f(\varepsilon) \frac{\sin A(x - \varepsilon)}{(x - \varepsilon)} \, d\varepsilon. \tag{4.1-1}$$

Thus the observed image $g(x)$ can be considered to be the output of an ideal lowpass filter with cutoff frequency A, where the input is the exact image $f(x)$ of length L. An imaging system which is limited only in that its aperture is finite is called diffraction-limited. A practical imaging system will have addition aberrations which will yield a kernel other than the $\sin(Ax)/x$ kernel in the integral given by Eq. (4.1-1). However, this will not affect the linear programming method of image restoration because it does not depend on the specific kernel used.

Equation (4.1-1) can be solved for the exact image $f(x)$, given the observed image $g(x)$. The exact solution is obtained by using the prolate spheroidal wave functions [13]. This can be seen by reviewing the properties of these functions. The prolate spheroidal wave functions $\psi_n(x)$ are the eigenfunctions of the integral equation given by Eq. (4.1-1), i.e.,

$$\int_{-L/2}^{L/2} \psi_n(\varepsilon)\frac{\sin A(x-\varepsilon)}{(x-\varepsilon)}\,d\varepsilon = \lambda_n\psi_n(x), \qquad n = 0, 1, 2, \ldots, \tag{4.1-2}$$

where λ_n are the eigenvalues of the integral equation given by Eq. (4.1-1).

The prolate spheroidal functions are orthogonal in the intervals $(-\infty, \infty)$ and $(-L/2, L/2)$. Thus,

$$\int_{-\infty}^{\infty} \psi_n(x)\psi_m(x)\,dx = \begin{cases} 1, & n = m, \\ 0, & n \neq m, \end{cases} \tag{4.1-3}$$

and

$$\int_{-L/2}^{L/2} \psi_n(x)\psi_m(x)\,dx = \begin{cases} \lambda_n, & n = m, \\ 0, & n \neq m. \end{cases} \tag{4.1-4}$$

The functions are complete in both intervals. Thus the band-limited function $g(x)$ can be written as the series expansion

$$g(x) = \sum_{n=0}^{\infty} c_n\psi_n(x), \tag{4.1-5}$$

and the function $f(x)$ of finite support can be written as

$$f(x) = \sum_{n=0}^{\infty} d_n\psi_{Ln}(x), \tag{4.1-6}$$

where $\psi_{Ln(x)}$ are the prolate spheroidal wave functions truncated to the interval $(-L/2, L/2)$. The coefficients c_n and d_n can be obtained by forming the projections onto the orthonormal functions, i.e.,

$$c_n = \int_{-\infty}^{\infty} g(x)\psi_n(x)\,dx \tag{4.1-7}$$

and

$$d_n = \frac{1}{\lambda_n}\int_{-L/2}^{L/2} f(x)\psi_n(x)\,dx. \tag{4.1-8}$$

Given the integral equation

$$g(x) = \int_{-L/2}^{L/2} f(\varepsilon)\frac{\sin A(x-\varepsilon)}{(x-\varepsilon)}\,d\varepsilon, \tag{4.1-9}$$

we can replace $g(x)$ and $f(\varepsilon)$ in Eq. (4.1-9) with the series expansions given by Eqs. (4.1-5) and (4.1-6) to obtain

$$\sum_{n=0}^{\infty} c_n \psi_n(x) = \int_{-L/2}^{L/2} \sum_{n=0}^{\infty} d_n \psi_{Ln}(x) \frac{\sin A(x-\varepsilon)}{(x-\varepsilon)} \, d\varepsilon$$

$$= \sum_{n=0}^{\infty} \lambda_n d_n \psi_n(x), \tag{4.1-10}$$

where the last expression is obtained from the fact that the prolate spheroidal wave functions are the eigenfunctions of this integral equation. The left-hand-side series expansion must be equivalent to the right-hand-side series expansion term by term. Thus,

$$c_n = \lambda_n d_n \tag{4.1-11}$$

or

$$d_n = c_n / \lambda_n. \tag{4.1-12}$$

The exact solution thus involves finding the projections c_n of the measurement $g(x)$ onto the set of prolate spheroidal wave functions ψ_n. Each of these coefficients is then divided by the known eigenvalue λ_n and the exact solution to Eq. (4.1-1) can be written as

$$f(x) = \sum_{n=0}^{\infty} \frac{c_n}{\lambda_n} \psi_{Ln}(x). \tag{4.1-13}$$

Thus, the exact solution to Eq. (4.1-1) is given by Eq. (4.1-13). However, Eq. (4.1-1) does not fully represent a practical imaging system. There are invariably random errors (noise) which add to the measured image values of $g(x)$. When noise is considered, the solution to Eq. (4.1-1) given by Eq. (4.1-13) is found to be unstable.

The instability arises in that the eigenvalues λ_n exhibit a unit step type of behavior. The first $2AL$ eigenvalues are approximately equal to one and quickly fall to zero thereafter. The division by the small eigenvalues indicated in Eq. (4.1-13) makes this expression extremely sensitive to measurement errors. A frequently proposed scheme is to truncate the series given by Eq. (4.1-13) to about $n = 2AL$ [14]. This eliminates the high-order prolate spheroidal wave functions, which contain most of the high-frequency information. However, the high-frequency components are usually the quantities of greatest interest. Alternative methods of tapering the eigenvalues have also been proposed [14] to provide some noise immunity. But since the taper must also go to zero rapidly, this approach offers little gain in resolution.

The preceding analysis using prolate spheroidal wave functions is important more for the insight we gain into the image restoration than as a method of solution. The discrete formulation of the problem can be analyzed in a similar manner. In this case a set of orthonormal vectors is used in place of the orthonormal prolate spheroidal wave functions. The discrete formulation of the image restoration problem can be written in matrix notation as

$$\mathbf{g} = \mathbf{H}\mathbf{f}, \tag{4.1-14}$$

where $\mathbf{g}$ and $\mathbf{f}$ are vectors composed of the sampled values of the measured and actual images, and $\mathbf{H}$ is a matrix representing the integral operator by using an appropriate numerical integration rule. The integral equation given by Eq. (4.1-1) can be discretized by using any one of several numerical integration techniques. If the rectangle rule of integration is used, the resulting system of equations behaves very similarly to the continuous case [15]. The eigenvectors are the discrete prolate spheroidal wave sequences. The eigenvalues also exhibit a unit step function dependence on the order. The $\mathbf{H}$ matrix will be different if another integral equation is used or a different numerical integration technique is employed, but in general the $\mathbf{H}$ matrix will be effectively rank-deficient. A rank-deficient matrix of rank r has only r nonzero eigenvalues [16]. The rank of the $\mathbf{H}$ matrix, using the rectangular rule for a diffraction-limited imaging system, is therefore $2AL$ [15]. That is, any imaging system is effectively limited to some maximum spatial frequency A and the objects of interest have a maximum length L. The measured image $\mathbf{g}$ is therefore completely determined by $2AL$ samples. This is exactly the number of samples obtained if the measured image is sampled uniformly at the Nyquist spacing of $1/2A$ for the length of the desired image L. A larger number of samples will cause the system of equations given by Eq. (4.1-14) to become rank-deficient. This oversampling is generally unavoidable in practice because the exact rank is not known. Even when exactly $2AL$ uniformly spaced (linearly independent) samples are used, the system of equations given by Eq. (4.1-14) will generally be rank-deficient. This is due to the fact that the number of elements in the unknown image $\mathbf{f}$ must be larger than $2AL$ in order to provide increased resolution. Thus the system of equations given by Eq. (4.1-14) will be underdetermined and thus rank-deficient.

The discrete formulation given by Eq. (4.1-14) can be analyzed in much the same way as the continuous case given by Eq. (4.1-1). For this analysis we shall diagonalize the particular degradation matrix $\mathbf{H}$ in order to obtain a set of orthonormal vectors. The singular value decomposition (SVD) is a common method of diagonalizing an arbitrary matrix [4]. Any $m \times n$ matrix $\mathbf{H}$ with $m > n$ can be decomposed in the following way:

$$\mathbf{H} = \mathbf{U}\mathbf{\Lambda}^{1/2}\mathbf{V}^{T}, \tag{4.1-15}$$

where $\mathbf{U}$ and $\mathbf{V}^T$ are orthogonal unitary matrices, i.e.,

$$\mathbf{U}^T\mathbf{U} = \mathbf{V}^T\mathbf{V} = \mathbf{V}\mathbf{V}^T = \mathbf{I}_n, \tag{4.1-16}$$

where the superscript T denotes matrix transpose and $\mathbf{I}_n$ is the n-dimensional identity matrix. Thus, the transpose of a unitary matrix is equivalent to its inverse. The matrix $\mathbf{U}$ consists of the n orthonormal eigenvectors associated with the n largest eigenvalues of $\mathbf{H}\mathbf{H}^T$, and the matrix $\mathbf{V}$ consists of the orthonormal eigenvectors of $\mathbf{H}^T\mathbf{H}$. The diagonal elements of $\mathbf{\Lambda}^{1/2}$ are the nonnegative square roots of the eigenvalues of $\mathbf{H}^T\mathbf{H}$. They are called singular values. Thus,

$$\mathbf{\Lambda}^{1/2} = \text{diag}(\sigma_1, \sigma_2, \ldots, \sigma_n), \tag{4.1-17}$$

where we shall assume for convenience that

$$\sigma_1 \geqslant \sigma_2 \geqslant \cdots \sigma_n \geqslant 0. \tag{4.1-18}$$

Thus, if rank $(\mathbf{H}) = r$, $\sigma_{r+1} = \sigma_{r+2} = \cdots = \sigma_n = 0$.

The decomposition given by Eq. (4.1-15) is the singular value decomposition. There is no loss of generality in assuming that $m \geqslant n$ because in the case where $m < n$, we can compute the SVD of $\mathbf{H}^T$, i.e.,

$$\mathbf{H}^T = \mathbf{U}\mathbf{\Lambda}^{1/2}\mathbf{V}^T, \tag{4.1-19}$$

where the SVD of $\mathbf{H}$ is given by

$$\mathbf{H} = \mathbf{V}\mathbf{\Lambda}^{1/2}\mathbf{U}^T. \tag{4.1-20}$$

The SVD is very important in numerical analysis and linear algebra. Efficient SVD algorithms are readily available [17]. If the matrix $\mathbf{H}$ is rank-deficient, i.e., $r < N$, the pseudoinverse may be used to obtain a solution to Eq. (4.1-14). The pseudoinverse [4, 18] of the matrix $\mathbf{H}$ is defined as

$$\mathbf{H}^\dagger = \mathbf{V}\mathbf{\Lambda}^{-1/2}\mathbf{U}^T, \tag{4.1-21}$$

where the superscript dagger indicates pseudoinverse. The inverse notation of the diagonal matrix denotes replacing the nonzero elements of $\mathbf{\Lambda}^{1/2}$ with the reciprocal values and leaving zero values unchanged.

A matrix $\mathbf{H}$ whose singular values are very small, i.e., numerically zero, can be very closely approximated by a rank-deficient matrix [19]. This matrix is obtained by replacing the small singular values with zero. The pseudoinverse of this rank-deficient matrix can then be taken.

The measurement vector $\mathbf{g}$ can be filtered to eliminate irrelevant noise. The noise reduction filter is accomplished in the following way. We first write Eq. (4.1-14) with the matrix $\mathbf{H}$ replaced with its SVD, i.e.,

$$\mathbf{g} = \mathbf{U}\mathbf{\Lambda}^{1/2}\mathbf{V}^T\mathbf{f}, \tag{4.1-22}$$

and premultiplying by $\mathbf{U}^T$ we obtain

$$\mathbf{U}^T\mathbf{g} = \mathbf{\Lambda}^{1/2}\mathbf{V}^T\mathbf{f}. \tag{4.1-23}$$

We then denote the unitary transformations of the measured and desired images by defining

$$\mathbf{G} \equiv \mathbf{U}^T\mathbf{g}, \tag{4.1-24}$$

and

$$\mathbf{F} \equiv \mathbf{V}^T\mathbf{f}, \tag{4.1-25}$$

where the capital letters $\mathbf{G}$ and $\mathbf{F}$ are used to indicate transformed vectors. Equation (4.1-14) can then be rewritten in the transform domain as

$$\mathbf{G} = \mathbf{\Lambda}^{1/2}\mathbf{F}. \tag{4.1-26}$$

This representation of the image degradation operator indicates the filtering process which takes place. The diagonal matrix $\mathbf{\Lambda}^{1/2}$ is similar to a transfer function. The degradation operators of interest contain zeros or near zeros which effectively eliminate image components.

However, for measurement noise which is additive zero mean white Gaussian, the statistics of the noise in the transform domain are unchanged from those in the spatial domain. Since the transformation is linear and unitary, the mean and the variance are the same in the transform domain. Thus, the transformed components $\mathbf{G}$ of the observed image are strongly dominated by noise when the corresponding diagonal elements of $\mathbf{\Lambda}^{1/2}$ are much less than the standard deviation of the noise. We can then consider $\mathbf{H}$ to be rank-deficient with a pseudorank given by r, where r is the number of singular values which are significantly greater than the standard deviation of the noise. This pseudorank can then be used to determine the number of singular values to be included in the pseudoinverse.

The filtered image is obtained by forming

$$\hat{\mathbf{g}} = \mathbf{U}\hat{\mathbf{G}}, \tag{4.1-27}$$

where $\hat{\mathbf{G}}$ is obtained by setting all those components of $\mathbf{G}$ to zero whose eigenvalues are effectively zero. The resultant vector $\hat{\mathbf{g}}$ will contain no components which are inconsistent with the $\mathbf{H}$ operator's characteristic of eliminating high-order eigenvectors. The unitary matrices $\mathbf{U}^T$ and $\mathbf{V}^T$ form a set of orthogonal vectors, which are similar to the sinusoidal basis functions used in Fourier analysis. Thus, arbitrary input images $\mathbf{f}$ can be decomposed into the row vectors of $\mathbf{V}^T$ and, similarly, the output image vector $\mathbf{g}$ can be decomposed into the row vectors of $\mathbf{U}^T$. The operator $\mathbf{H}$ can then be analyzed as a filter in the transform domain. The ill-conditioned nature of the operator $\mathbf{H}$ will be made evident by the occurrence of filter coefficients which are zero or effectively zero. Each zero of the filter removes a component of the input image **f** which lies in the direction of a row vector of the unitary matrix $\mathbf{V}^T$. We shall refer to the components of **f** lying in the range of row vectors which are not eliminated by the operator as being "in-band" and the remaining components are said to be "out-of-band." Thus the process indicated by Eq. (4.1-27) is performed to remove out-of-band noise components. The noise that cannot be eliminated, that is, the in-band noise, determines the actual signal-to-noise ratio, which is called the in-band signal-to-noise ratio.

Another useful interpretation of the degradation operation given by Eq. (4.1-14) can be obtained from Eq. (4.1-26). The vectors **g** and **f** can be thought of as sets of N Cartesian coordinates for points in an N-dimensional space. For ease of presentation, we shall limit the present discussion to the case where both **g** and **f** contain the same number of elements N. The more general case of unequal dimensions of **f** and **g** follows in much the same way. The unitary transformations given by Eqs. (4.1-22) and (4.1-25) transform each original set of N coordinates into a new set of rotated coordinates. It should be recalled that a unitary transformation preserves the Euclidean distance between the N-dimensional points. The degradation operator as given by Eq. (4.1-26) can now be interpreted in the following way: The original point **f** is transformed into a new set of coordinates given by point **F**. In the transformed coordinate system, several dimensions are eliminated to produce the point **G**. The coordinates of **G** are then transformed back to form the measured coordinates **g** by using Eq. (4.1-27).

Consider a group of points in three-dimensional (3-D) space. The coordinates are transformed and one dimension is eliminated. The resulting two-dimensional (2-D) representation is an isometric projection of the 3-D group of points. Isometric projections are used in graphics to form a 2-D view of a 3-D object. The sparseness constraint is then equivalent to requiring the 3-D group of points to be constrained to lie within a 2-D plane. This is equivalent to the restriction that in order to restore a 3-D image from a 2-D projection, the original image must be constrained to lie in a plane, i.e., must be a 2-D object.

4.2 NUMERICAL EXAMPLE OF THE MATRIX DIAGONALIZATION OF *H*

The concepts of the preceding section can probably best be understood by examining a numerical example. Consider the problem of solving a system of equations as given by Eq. (4.1-14), where the degradation matrix is cyclic and is given by

$$\mathbf{H} = \frac{1}{4}\begin{bmatrix} 3 & 1 & -1 & 1 \\ 1 & 3 & 1 & -1 \\ -1 & 1 & 3 & 1 \\ 1 & -1 & 1 & 3 \end{bmatrix}. \tag{4.2-1}$$

This example corresponds to the deconvolution of a rank-deficient cyclic operator. The singular value decomposition can be performed in this case. However, there are many unitary transformations which can be found to diagonalize a matrix **H**. The SVD method was described in the previous section since it applies to all matrices. Symmetric matrices can be diagonalized with unitary matrices such that $\mathbf{U} = \mathbf{V}$. Cyclic matrices can always be diagonalized by using the discrete Fourier transform (DFT) matrices [16]. Toeplitz matrices are frequently encountered when dealing with shift-invariant systems. Toeplitz systems can be replaced with cyclic systems of equations by appropriately zero padding [20] the **H** matrix and the **g** and **f** vectors. Thus, cyclic matrices are frequently encountered in practice. The DFT is commonly used in treating cyclic systems. Therefore, we shall use the more popular DFT matrix for the diagonalization of this cyclic matrix. In this case, the unitary matrices **U** and $\mathbf{V}^T$ are the inverse discrete Fourier transform (IDFT) and the discrete Fourier transform matrices [16], respectively. The elements on the diagonal are the DFT of the first row of **H** in time-reversed order. The degradation matrix can thus be decomposed into a form similar to that given by Eq. (4.1-15) in the following way:

$$\mathbf{H} = \frac{1}{4}\begin{bmatrix} 1 & 1 & 1 & 1 \\ 1 & j & -1 & -j \\ 1 & -1 & 1 & -1 \\ 1 & -j & -1 & j \end{bmatrix}\begin{bmatrix} 1 & 0 & 0 & 0 \\ 0 & 1 & 0 & 0 \\ 0 & 0 & 0 & 0 \\ 0 & 0 & 0 & 1 \end{bmatrix}\begin{bmatrix} 1 & 1 & 1 & 1 \\ 1 & -j & -1 & j \\ 1 & -1 & 1 & -1 \\ 1 & j & -1 & -j \end{bmatrix}. \tag{4.2-2}$$

The diagonal matrix $\mathbf{\Lambda}^{1/2}$ contains a zero along the diagonal. Thus, the degradation matrix **H** is of rank three. There are four eigenvectors $\mathbf{e}_i$, $i = 1, 2, 3, 4$. The eigenvector $\mathbf{e}_3$ is seen to be eliminated by the matrix **H** while the remaining eigenvectors remain unchanged. Thus,

$$\begin{aligned} \mathbf{e}_1 &= \mathbf{He}_1, \\ \mathbf{e}_2 &= \mathbf{He}_2, \\ \mathbf{0} &= \mathbf{He}_3, \\ \mathbf{e}_4 &= \mathbf{He}_4, \end{aligned} \tag{4.2-3}$$

where **0** is the null vector.

In the present example, the additional constraint that the object be nonnegative will be shown to be sufficient in this case for a unique solution. This can be seen by selecting the constant K so as to eliminate the negative element (-1) in the last row of the image vector. Then $K = -1/4$ as before, and

$$\mathbf{f} = \frac{1}{4}\begin{bmatrix} 5 \\ 7 \\ 1 \\ -1 \end{bmatrix} - \frac{1}{4}\begin{bmatrix} 1 \\ -1 \\ 1 \\ -1 \end{bmatrix} = \begin{bmatrix} 1 \\ 2 \\ 0 \\ 0 \end{bmatrix}. \tag{4.2-13}$$

Thus we find the desired solution as one of the solutions which satisfies the system of equations given by Eq. (4.1-1) and, in addition, is sparse and nonnegative. This solution is not necessarily unique as there may be many sparse nonnegative solutions of the form

$$\mathbf{f} = \begin{bmatrix} 1 \\ 2 \\ 0 \\ 0 \end{bmatrix} + K\begin{bmatrix} 1 \\ -1 \\ 1 \\ -1 \end{bmatrix}. \tag{4.2-14}$$

The number of acceptable nonzero values of the constant K indicates the number of acceptable solutions. In this case the solution given by Eq. (4.2-13) is unique. This can be seen in the following way. If the constant K in Eq. (4.2-14) is a positive number, the resulting solution will have a negative value in the last row. If K is negative, the resulting solution will have a negative value in the second to last row. Thus K must have a value of zero in order to yield a nonnegative solution. Therefore, the only possible solution is the one given by Eq. (4.2-13).

This example serves to demonstrate some important properties of rank-deficient linear systems. First, the constraint that the solution be sparse, i.e., contain at least $N - r$ zeros, is not sufficient for uniqueness. In fact, there are at least $\binom{N}{r}$ such solutions. Second, the constraint of nonnegativity in addition to sparseness is frequently sufficient for uniqueness but this is not guaranteed. The method of linear programming to be presented will assume that both constraints apply. A nonnegative sparse solution will be provided if one exists. If no unique solution or more than one solution exists, then this will be indicated by the method.

The linear programming method also works well when the nonnegativity constraint does not apply, although in this case there are at least $\binom{N}{r}$ solutions. The method tends to find the solution with the largest number of zero values. Thus, a bipolar image must be very sparse, i.e., contain strictly much more than $N - r$ zero values.

4.3 LINEAR PROGRAMMING

Linear programming is the process whereby resources are scheduled or programmed subject to linear constraints. The most common method of solving linear programming problems is the simplex method developed by Dantzig [21]. The general form of a linear program is

$$\text{minimize:} \quad \mathbf{C}^T\mathbf{X} \tag{4.3-1}$$

$$\text{subject to:} \quad \begin{aligned} &\mathbf{A}_1\mathbf{X} < \mathbf{b}_1 \\ &\mathbf{A}_2\mathbf{X} > \mathbf{b}_2 \\ &\mathbf{A}_3\mathbf{X} = \mathbf{b}_3, \end{aligned} \tag{4.3-2}$$

where $\mathbf{X}$ is a vector of unknown variables, and $\mathbf{C}^T\mathbf{X}$ is called the cost function or objective function. The elements of the cost vector $\mathbf{C}$ are called the marginal costs, the matrices $\mathbf{A}_1$, $\mathbf{A}_2$, and $\mathbf{A}_3$ are called the constraint matrices, and the vectors $\mathbf{b}_1, \mathbf{b}_2, \mathbf{b}_3$ are called the right-hand sides.

Linear programming algorithms minimize the objective function subject to equality and inequality constraints. The word linear is used here to indicate that the objective function and the constraints are linear combinations of the variables. It should be noted that linear programming algorithms are generally not linear operations. That is, they do not exhibit the properties of homogeneity and superposition.

If the noise is to be taken into account, we can rewrite Eq. (4.1-14) as

$$\mathbf{g} = \mathbf{Hf} + \mathbf{n}, \tag{4.3-3}$$

where $\mathbf{n}$ denotes a random variable which represents the noise.

In anticipation of using linear programming methods we shall rewrite Eq. (4.3-3) in the form

$$\mathbf{g} = \mathbf{Hf} + \mathbf{n}^+ - \mathbf{n}^-, \tag{4.3-4}$$

where the error vector has been decomposed into two nonnegative vectors. This notational device is needed in order to conform with the nonnegativity constraint of the variables in linear programming. The solution vector $\mathbf{f}$ should be such that the error

$$n_i = n_i^+ - n_i^- = g_i - H_i\mathbf{f} \tag{4.3-5}$$

is small in some sense. Here the subscript indicates the ith element of a vector or the ith row of a matrix. The standard least-squares approach is to minimize the L_2 norm of the error, i.e.,

$$\sum_{i=1}^{M} n_i^2 = \sum_{i=1}^{M} (g_i - H_i\mathbf{f})^2, \tag{4.3-6}$$

where M is the number of equations. We then take the partial derivatives of Eq. (4.3-6) with respect to the elements of **f** and, setting them equal to zero, we obtain the normal equations

$$\mathbf{H}^T(\mathbf{g} - \mathbf{H}\mathbf{f}) = 0, \tag{4.3-7}$$

or more commonly

$$\mathbf{H}^T\mathbf{g} = \mathbf{H}^T\mathbf{H}\mathbf{f}. \tag{4.3-8}$$

Equation (4.3-8) implies the necessity to take the inverse of the matrix $\mathbf{H}^T\mathbf{H}$, which is also rank-deficient when **H** is rank-deficient. Thus, the unconstrained least-squares solution is not unique. The pseudoinverse solution suggests itself here. It has been shown that the pseudoinverse provides the least-sqaures solution with the minimum L_2 norm [18] of the solution vector **f**. However, these unconstrained least-squares approaches do not attempt to restore the eigenvector components of the image **f** which have been eliminated by the operator **H**.

The approach we shall take is to minimize a metric of the error subject to the "constraints" given by Eq. (4.3-4). Thus, as a linear programming problem we can write

$$\text{minimize:} \quad n_1^+ + n_1^- + n_2^+ + n_2^- + \cdots n_M^+ + n_M^- \tag{4.3-9}$$

$$\text{subject to:} \quad \mathbf{g} = \mathbf{H}\mathbf{f} + \mathbf{n}^+ - \mathbf{n}^-. \tag{4.3-10}$$

The objective function Eq. (4.3-9) is the sum of the absolute values of the residual errors. It is called the L_1 norm of the error. Linear programming methods can also be used to minimize the largest residual error value or the sum of the squares of the error [8,9]. These metrics are called the L_∞ and L_2 norms, respectively. It should be noted that the L_2 norm approach in linear programming differs from the conventional least squares in that provision is made for inequality constraints.

A fundamental assumption made in the simplex method of linear programming (4.3-10) is that only r values of the vectors $\mathbf{f}, \mathbf{n}^+$, and $\mathbf{n}^-$ will be nonzero [21]. This sparseness constraint is generally not explicitly present in other restoration methods. We have also included the constraint that the elements of **f** be nonnegative. The nonnegative constraint could be relaxed by rewriting **f** in terms of two nonnegative components, i.e.,

$$\mathbf{f} = \mathbf{f}^+ - \mathbf{f}^-. \tag{4.3-11}$$

The image restoration problem as formulated by Eqs. (4.3-9) and (4.3-10) appears to be one of estimation of the image vector **f**. The problem becomes one of detection if it is known *a priori* that the vector **f** must be one of some finite number J of possible images. In this case, a matched filter will yield the optimum detection scheme [22]. We have seen that the constraint that the image vector **f** must be sparse limits the number of possible solutions to a large but finite number of possibilities. The optimal solution can be found by enumerating all the possible solutions. The norm of the residual of each possible solution can then be computed. The optimal solution is then selected to yield the minimum norm. This is an exhaustive approach which does not use any information gained from the similarity between the J images. Thus, if it is found that the error metric is small for one image, only images similar to this one need be considered further. We shall introduce an iterative method which in general requires much less computational power than an exhaustive search approach.

4.4 NORMS OF THE ERROR

When no measurement error is present, there are many exact solutions to the rank-deficient linear system given by Eq. (4.1-14). The errors introduce random variations in Eq. (4.1-14). This gives rise to inconsistencies in the equations, so that, for example, one equation might be $1 = f_1 + f_2$ but for some linear combinations of the remaining equations we might obtain $2 = f_1 + f_2$. Thus, when measurement error (noise) is considered, an exact solution to Eq. (4.1-14) generally cannot be found. The errors or residual variables which indicate the deviation from the exact solution must be explicitly considered as in Eq. (4.3-3). Some norm or metric of this error is usually taken so as to provide a scalar number indicative of the total deviation from an exact solution to Eq. (4.1-14). An estimate is then found such that this norm of the error is minimized. The norm of the error should not be sensitive to the sign of individual errors. Thus, for example, the sum of the errors is a poor indicator since large negative errors could cancel with large positive errors and the sum would indicate no error. A commonly used norm is the L_2 norm, which yields the least-squares estimate.

The selection of which norm of error is to be minimized is based on the statistics of the error. This can be seen in the following way. The maximum *a posteriori* (MAP) estimate of **f** is found so that the *a posteriori* probability $P_{\hat{\mathbf{f}}/\hat{\mathbf{g}}}(\mathbf{f}/\mathbf{g})$ is a maximum. We write $P_{\hat{\mathbf{f}}/\hat{\mathbf{g}}}(\mathbf{f}/\mathbf{g})$ to denote the probability that the random variable elements of the vectors $\hat{\mathbf{f}}$ and $\hat{\mathbf{g}}$ are equal to the values given in the vectors **f** and **g**. This conditional probability is generally not available, so we can use Bayes' rule and write:

$$P_{\hat{\mathbf{f}}/\hat{\mathbf{g}}}(\mathbf{f}/\mathbf{g}) = P_{\hat{\mathbf{g}}/\hat{\mathbf{f}}}(\mathbf{g}/\mathbf{f}) P_{\hat{\mathbf{f}}}(\mathbf{f}) / P_{\hat{\mathbf{g}}}(\mathbf{g}). \tag{4.4-1}$$

The denominator $P_{\hat{\mathbf{g}}}(\mathbf{g})$ does not involve the vector **f**, which is the variable with respect to which we wish to maximize. Thus the MAP estimate need only consider the expression in the numerator. The probability $P_{\hat{\mathbf{f}}}(\mathbf{f})$ is called the *a priori* probability of $\mathbf{f}$. This probability gives the likelihood of particular objects being imaged. If we assume that all objects are equally likely, then $P_{\hat{\mathbf{f}}}(\mathbf{f})$ is a constant. In this case we find the image $\mathbf{f}$ such that we maximize the conditional probability $P_{\hat{\mathbf{g}}/\hat{\mathbf{f}}}(\mathbf{g}/\mathbf{f})$. The estimate $\mathbf{f}$ is then called the maximum likelihood (ML) estimate.

The individual errors $n_i^+ - n_i^-$ are assumed to be independent, zero mean with a probability density function of the form

$$P_{\hat{g}_i/\hat{f}_i}(g_i/f_i) = P_n(n_i^+ - n_i^-) = C_i \exp\{-K|\mathbf{n}^+ - \mathbf{n}^-|^p\}, \tag{4.4-2}$$

where C, K, and p are appropriately chosen constants. The conditional probability for the vector then has the form

$$P_{\mathbf{n}}(\mathbf{n}^+ - \mathbf{n}^-) = (C)^M \exp\{-K \sum_{i=1}^{M} |n_i^+ - n_i^-|^p\}. \tag{4.4-3}$$

Maximizing $P_{\mathbf{n}}(\mathbf{n}^+ - \mathbf{n}^-)$ is thus equivalent to minimizing the exponent $\sum_{i=1}^{M} |n_i^+ - n_i^-|^p$.

Therefore, we see that the maximum likelihood estimate for Gaussian random errors is obtained by minimizing the L_2 norm or the sum of the squares of the error. The ML estimate for uniformly distributed errors is obtained by minimizing the L_∞ norm or the maximum error. The ML estimate for exponentially distributed or Laplacian random errors is obtained by minimizing the L_1 norm or the sum of the absolute values of the error. The Laplacian random variables are fairly similar to Gaussian ones, except that due to the larger density in the tails of the probability density function, there is greater allowance for a few occasional large errors. Thus, the minimum L_1 norm estimate may deviate from satisfying a few equations by a large margin but the total deviation is still small. This is in contrast to the least-squares or L_2 norm, where the tendency is to emphasize the minimization of large errors at the expense of smaller errors. This is due to the squaring process, which makes large errors larger and small errors smaller. The estimate obtained by the L_1 norm is called robust because of its immunity to occasional large errors. This spiky type of error for the estimate is desirable for the linear programming method of image restoration described here. The sparse estimate of the error is brought about by the nonlinear nature of the linear programming estimate. That is, the estimate $\mathbf{f}$ is constrained to be sparse. Therefore, the in-band noise must also be represented by a sparse vector. The prefiltering of the measurement as described in Section 4.2 will guarantee that the noise is in-band; thus we can write

$$\mathbf{g} = \mathbf{H}(\mathbf{f} + \boldsymbol{\varepsilon}_{\text{in}}), \tag{4.4-4}$$

where we define $\boldsymbol{\varepsilon}_{\text{in}}$ as the in-band noise. The true sparse object is now contaminated so as to contain many elements. The linear programming estimate will replace ε_{in} with a sparse vector $\boldsymbol{\varepsilon}$. This vector will be equivalent to $\boldsymbol{\varepsilon}_{\text{in}}$ in the sense that

$$\mathbf{H}\boldsymbol{\varepsilon} = \mathbf{H}\boldsymbol{\varepsilon}_{\text{in}}. \tag{4.4-5}$$

The inherent nonlinearity of the linear programming estimation process is exhibited by the fact that the resultant error is sparse even when the measurement error is Gaussian. A linear estimation process would produce a Gaussian error for the estimate when the measurement error is Gaussian. It is for this reason that the L_1 norm is the preferred norm for the linear programming method. The estimate obtained by minimizing the L_1 norm will tend to produce a sparse error, whereas the estimate obtained by minimizing the unconstrained L_2 norm will tend to produce an error with many nonzero elements. The solution to Eq. (4.3-3) can contain only $r < N$ nonzero elements for the $N + M$ unknowns in the desired image $\mathbf{f}$ and the noise vectors $\mathbf{n}^+$ and $\mathbf{n}^-$. Thus, a sparse representation of the error is desired to allow the greatest number of nonzero elements in the desired image estimate $\mathbf{f}$.

4.5 NUMERICAL EXAMPLE OF MINIMUM L_1 NORM METHOD

The preceding sections discussed the basic ideas involved in solving an ill-conditioned problem. In this section the linear programming concepts will be made more concrete. The example of Section 4.2 can be written as an approximation due to uncertainties in the left-hand side, i.e.

$$\frac{1}{4}\begin{bmatrix} 5 \\ 7 \\ +1 \\ -1 \end{bmatrix} \approx \frac{1}{4}\begin{bmatrix} 3 & 1 & -1 & 1 \\ 1 & 3 & 1 & -1 \\ -1 & 1 & 3 & 1 \\ +1 & -1 & 1 & 3 \end{bmatrix}\begin{bmatrix} f_1 \\ f_2 \\ f_3 \\ f_4 \end{bmatrix}. \tag{4.5-1}$$

We shall neglect the factor of 1/4 on both sides and rewrite Eq. (4.5-1) as a system of exact linear equations given by:

$$\begin{aligned} 5 &= 3f_1 + f_2 - f_3 + f_4 + n_1^+ - n_1^-, \\ 7 &= f_1 + 3f_2 + f_3 - f_4 + n_2^+ - n_2^-, \\ 1 &= -f_1 + f_2 + 3f_3 + f_4 + n_3^+ - n_3^-, \\ -1 &= f_1 - f_2 + f_3 + 3f_4 + n_4^+ - n_4^-. \end{aligned} \tag{4.5-2}$$

The exact solution should contain zero values for the variables n_i^+ and n_i^-.

The image estimate variables f_i should be nonzero. The variables n_i^+ and n_i^- are the errors between the degraded estimate and the measurement g, as can be seen by writing the previous equations in the form

$$\begin{aligned} n_1^+ - n_1^- &= 5 - 3f_1 - f_2 + f_3 - f_4, \\ n_2^+ - n_2^- &= 7 - f_1 - 3f_2 - f_3 + f_4, \\ n_3^+ - n_3^- &= 1 + f_1 - f_2 - 3f_3 - f_4, \\ n_4^+ - n_4^- &= -1 - f_1 + f_2 - f_3 - 3f_4. \end{aligned} \tag{4.5-3}$$

Note that the right-hand side is the difference between the measurement and the degraded estimate. This quantity is commonly referred to as the residual. The variables n_i^+ and n_i^- are constrained to be nonnegative. In each equation only one of the two variables n_i^+ and n_i^- will be nonzero. Thus the addition of the two residual variables will yield the absolute value of the difference between the degraded estimate and the measurement, that is,

$$\begin{aligned} n_1^+ + n_1^- &= |5 - 3f_1 - f_2 + f_3 - f_4|, \\ n_2^+ + n_2^- &= |7 - f_1 - 3f_2 - f_3 + f_4|, \\ n_3^+ + n_3^- &= |1 + f_1 - f_2 - 3f_3 - f_4|, \\ n_4^+ + n_4^- &= |-1 - f_1 + f_2 - f_3 - 3f_4|. \end{aligned} \tag{4.5-4}$$

An iterative method of obtaining an exact solution would drive the variables on the left-hand side of Eq. (4.5-4) to zero. One approach is to form a scalar quantity or norm of the left-hand side and minimize this scalar quantity. We shall use the L_1 norm, which can be obtained by simply adding the left-hand side variables. Thus the desired solution will be obtained if we minimize the sum

$$n_1^+ + n_1^- + n_2^+ + n_2^- + \cdots + n_M^+ + n_M^-, \tag{4.5-5}$$

subject to the constraints given by (4.3-4).

Linear programming problems are generally solved in two separate phases. The first phase consists of finding a feasible solution of the problem. The variables in this temporary solution are said to be in the basis and are called basic variables. The equations must then be rewritten in such a way that no basic variable appears explicitly in the cost function. The cost function is then examined to determine whether it can be reduced further. If no further reduction is possible, the present feasible solution is optimal. In general, this will not be the case and an iterative process is performed. This iterative process is called phase II. The second phase of linear programming consists of exchanging basic variables for nonbasic variables so as to minimize the cost. At each iteration a new feasible solution is found which reduces the cost function.

This procedure can be demonstrated numerically by rewriting Eq. (4.5-4) in the following way:

$$\begin{aligned} n_1^+ &= 5 - 3f_1 - f_2 + f_3 - f_4 + n_1^-, \\ n_2^+ &= 7 - f_1 - 3f_2 - f_3 + f_4 + n_2^-, \\ n_3^+ &= 1 + f_1 - f_2 - 3f_3 - f_4 + n_3^-, \\ n_4^- &= 1 + f_1 - f_2 + f_3 + 3f_4 + n_4^+, \end{aligned} \tag{4.5-6}$$

where the last equation has been multiplied by -1. At this point we have "solved" the four equations. The nonbasic variables, appearing on the right-hand side, are all assumed zero at this point. Thus an initial feasible solution is given by

$$\begin{aligned} n_1^+ &= 5, \\ n_2^+ &= 7, \\ n_3^+ &= 1, \\ n_4^- &= 1. \end{aligned} \tag{4.5-7}$$

This "solution" equates the residual error to the measurement, and the estimate of the image f is taken to be zero. In order to see if a better solution exists we shall rewrite the cost function given by Eq. (4.5-5) so as to eliminate the basic variables. The basic variables are given in terms of nonbasic variables by Eq. (4.5-6). We substitute these expressions into the left side of Eq. (4.5-5) and obtain the following reformulation of the original problem:

$$\begin{aligned} \text{minimize:} \quad & (5 - 3f_1 - f_2 + f_3 - f_4 + n_1^-) + n_1^- \\ & + (7 - f_1 - 3f_2 - f_3 + f_4 + n_2^-) + n_2^- \\ & + (1 + f_1 - f_2 - 3f_3 - f_4 + n_3^-) + n_3^- \\ & + (1 + f_1 - f_2 + f_3 + 3f_4 + n_4^+) + n_4^+, \end{aligned} \tag{4.5-8}$$

or, collecting terms,

$$\text{minimize:} \quad 14 - 2f_1 - 6f_2 - 2f_3 + 2f_4 + 2n_1^- + 2n_2^- + 2n_3^- + 2n_4^+ \tag{4.5-9}$$

$$\begin{aligned} \text{subject to:} \quad n_1^+ &= 5 - 3f_1 - f_2 + f_3 - f_4 + n_1^-, \\ n_2^+ &= 7 - f_1 - 3f_2 - f_3 + f_4 + n_2^-, \\ n_3^+ &= 1 + f_1 - f_2 - 3f_3 - f_4 + n_3, \\ n_4^- &= 1 + f_1 - f_2 + f_3 + 3f_4 + n_4^+. \end{aligned} \tag{4.5-10}$$

i.e., n_3^+. This exchange is called a pivot operation because of the similarity to the pivot operation used in taking the inverse of a matrix. The simplex method of linear programming continues the pivot operations until no further reduction in the cost function is possible. The solution is then read from the basic variables. The coefficients in the cost function at this point indicate the ambiguity of the solution. That is, if there are coefficients close to or equal to zero for some elements of the vector **f**, then these elements can enter the basis with no change in the norm of the error. Thus, zero coefficients in the final cost vector are an indication that solutions exist with essentially the same value of the L_1 norm of the error.

The preceding example of the minimum L_1 method would continue in the following way. The variable f_2 would become a basic variable and must be eliminated from the right-hand side. This is accomplished by solving the pivot row equations for f_2, i.e.,

$$f_2 = 1 + f_1 - 3f_3 - f_4 + n_3^- - n_3^+. \tag{4.5-16}$$

This expression for f_2 is then substituted into all of the equations so as to leave only nonbasic variables on the right-hand side. Substituting and rewriting Eq. (4.5-14), we obtain

$$\begin{aligned}
\text{minimize:} \quad & 8 - 8f_1 + 16f_3 + 8f_4 - 6n_3^- + 6n_3^+ + 2n_1^- + 2n_2^- + 2n_4^+, \\
\text{subject to:} \quad & n_1^+ = 4 - 4f_1 + 4f_3 + 0f_4 + n_1^- - n_3^- + n_3^+, \\
& n_2^+ = 4 - 4f_1 + 8f_3 + 4f_4 + n_2^- - 3n_3^- + 3n_3^+, \\
& f_2 = 1 + f_1 - 3f_3 - f_4 + n_3^- - n_3^+, \\
& n_4^- = 0 + 0f_1 + 4f_3 + 4f_4 + n_4^+ - n_3^- + n_3^+.
\end{aligned} \tag{4.5-17}$$

Thus, after the first pivot the solution vector is

$$\begin{bmatrix} f_1 \\ f_2 \\ f_3 \\ f_4 \end{bmatrix} = \begin{bmatrix} 0 \\ 1 \\ 0 \\ 0 \end{bmatrix}, \tag{4.5-18}$$

and the error vector is

$$\begin{bmatrix} n_1^+ - n_1^- \\ n_2^+ - n_2^- \\ n_3^+ - n_3^- \\ n_4^+ - n_4^- \end{bmatrix} = \begin{bmatrix} 4 \\ 4 \\ 0 \\ 0 \end{bmatrix}. \tag{4.5-19}$$

Examination of Eq. (4.5-16) indicates that a further reduction in the L_1 norm of the error can be obtained by bringing f_1 into the basis. The variable which would leave the basis is then determined by the ratio test. Note that the ratio is not used on the last equation, since this would result in division by zero. Either the first or second equation can be selected as the pivot row. We shall arbitrarily select the first equation. In order to perform the pivot operation in which we exchange the nonbasic variable f_1 with the basic variable n_1^+, we first solve the pivot equation for f_1:

$$f_1 = 1 + f_3 + \tfrac{1}{4}n_1^- - \tfrac{1}{4}n_3^- + \tfrac{1}{4}n_3^+ - \tfrac{1}{4}n_1^+. \tag{4.5-20}$$

Substituting into (4.5-14) and rewriting, we obtain

$$\begin{aligned} \text{minimize:} \quad & 0 + 8f_3 + 8f_4 + 2n_1^+ + 2n_2^- - 4n_3^+ - 4n_3^- + 2n_4^+ \\ \text{subject to:} \quad & f_1 = 1 + f_3 + \tfrac{1}{4}n_1^+ - \tfrac{1}{4}n_3^- + \tfrac{1}{4}n_3^+ - \tfrac{1}{4}n_3^-, \\ & n_2^+ = 0 - 4f_3 + 4f_4 - n_1^+ - n_1^- + n_2^- + 2n_3^+ - 2n_3^-, \\ & f_2 = 2 - 2f_3 - f_4 + \tfrac{1}{4}n_1^- - \tfrac{1}{4}n_1^+ - \tfrac{3}{4}n_3^+ + \tfrac{3}{4}n_3^-, \\ & n_4^- = 0 + 4f_3 + 4f_4 + n_3^+ - n_3^- + n_4^+. \end{aligned} \tag{4.5-21}$$

The L_1 norm of the error is now zero and the coefficients of the nonbasic variables of f in the cost function are all nonnegative, indicating that no further improvement is possible. The relatively large values of the coefficients in the cost function indicate that no alternative solutions exist for this value of the cost function. The solution and error vectors are therefore

$$f = \begin{bmatrix} 1 \\ 2 \\ 0 \\ 0 \end{bmatrix} \quad \text{and} \quad (n^+ - n^-) = \begin{bmatrix} 0 \\ 0 \\ 0 \\ 0 \end{bmatrix}. \tag{4.5-22}$$

In the presence of noise, the residual vector should consist of random variables. Therefore, its L_1 norm cannot be reduced to zero. If the underlying probability density function is zero mean Gaussian with variance σ, then it is easy to show that [23] the expected value for the absolute value of the error is

$$E(|n_i|) = \sqrt{\frac{2}{\pi}}\,\sigma. \tag{4.5-23}$$

The expected value of the L_1 norm of the error is then

$$E\left(\sum_{i=1}^{N} |n_i|\right) = N\sqrt{\frac{2}{\pi}}\,\sigma, \tag{4.5-24}$$

where N is the number of random noise variables. Thus the iterative process is stopped when the L_1 norm of the error reaches this value.

4.6 COMPUTATION CONSIDERATIONS

There are computational advantages of the simplex algorithm as described in Section 4.5 over a direct linear programming implementation of finding the minimum L_1 norm estimate. A direct realization of finding the minimum L_1 norm estimate by linear programming would be formulated as

$$\text{minimize:} \quad \mathbf{n}^{+}\mathbf{I} + \mathbf{n}^{-}\mathbf{I} \tag{4.6-1}$$

$$\text{subject to:} \quad \mathbf{g} = \mathbf{Hf} + \mathbf{n}^{+}\mathbf{I} - \mathbf{n}^{-}\mathbf{I}. \tag{4.6-2}$$

Thus, for an N by N degradation matrix $\mathbf{H}$, Eq. (4.6-2) would require $3N$ by N elements of computer memory. The corresponding computer memory required to perform the minimum L_1 norm method as described in Section 4.5 is approximately N by N elements. The variables $\mathbf{n}^{+}$ and $\mathbf{n}^{-}$ for $i = 1, 2, 3, \ldots, N$ could be represented by one n-dimensional array. The signs of the elements in the array indicate whether the element is to be associated with $\mathbf{n}^{+}$ or $\mathbf{n}^{-}$. After each pivot operation the pivot column and row are exchanged in memory. Therefore, the memory required remains unchanged throughout the algorithm.

The problem as formulated by Eqs. (4.6-1) and (4.6-2) is not in the standard form for linear programming. Therefore, a feasible solution must be found first. In general, the process of finding a feasible solution is computationally comparable to that of finding an optimum solution starting with a feasible solution. This problem is alleviated altogether in the algorithm described in Section 4.5 since the phase I problem has been eliminated.

A useful technique which has been found to reduce the number of iterations [9] is to select the pivot column so as to maximize the reduction of the value of the L_1 norm of error rather than the rate of reduction. This involves computation of the minimum ratio for all columns with negative costs. The pivot column and row are then selected which yield the largest value for the product of the minimum ratio with the cost. This product is the amount by which the present L_1 norm of the error will be decreased.

Another computational technique that has been found useful arises in the case of a pivot operation between an n_i^{+} variable and an n_i^{-} variable [24]. This would be accomplished very inefficiently by the standard pivot process. A computationally more efficient exchange is accomplished by exchanging the variables in the pivot row equation, negating the equation, and subtracting twice the pivot row equation from the costs.

4.7 SPATIAL RESOLUTION

A point source will produce a diffraction-limited image with an amplitude distribution given by $\sin(Ax)/x$. The minimum distance that two points can be separated and still be distinguished as two separate points is used to measure the resolution of an imaging system. Lord Rayleigh (for example, see reference [25]) specified a lower bound for this distance where the first minimum of $\sin(Ax)/x$ of one point overlaps the maximum $\sin(Ax)/x$ of the second point. This distance for a circular noncoherent imaging system is given by

$$R = 1.22\lambda / D, \tag{4.7-1}$$

where λ is the average wavelength. The utility of this limit for coherent systems is complicated by phase considerations [25]. However, the distance provides a good figure of merit for comparison purposes. An image restoration method is called superresolving if two points spaced closer than this distance can be resolved after processing the image. The computer results presented in Section 4.8 demonstrate that the minimum L_1 norm method is superresolving.

The term visual acuity is used for spatial resolution in physiological optics. Visual acuity is defined as the reciprocal of the angle in minutes subtended by the smallest discernible dimension in the field of view. The usual procedure used to measure a patient's visual acuity involves test targets such as Snellen letters or Landolt C rings. In general, visual acuity is on the order of the diffraction limit of a 2-mm pupil for the human eye. The spatial resolution limit also corresponds to the spacing of the receptor elements in the central retina.

Vernier acuity, on the other hand, is an order of magnitude better than diffraction considerations alone would dictate [26]. Vernier acuity is the ability to detect small displacements. A typical target used to measure vernier acuity is a vertical straight line with a small abrupt displacement in the line. Whereas the average visual acuity for the eye is about 1 minute of an arc, the vernier acuity is just a few arc seconds. The human vision system thus has the ability to resolve two lines spaced an order of magnitude closer together than the diffraction limit of the eye would permit. This type of superresolution performed by the brain appears to be similar to the L_1 norm method of image restoration in that the target must be sparse. This preference for sparse images has also been found to be true when using maximum entropy superresolution methods [5]. Frieden [27] also points out that images with piecewise linear gray level variations can be restored by differentiation. Thus, we see that in some cases the sparseness constraint can be met by a simple linear transformation of the data.

4.8 RESULTS

In this section we will demonstrate the capabilities of the algorithm described in Section 4.5. Figure 4.8-1 contains a test image in the upper

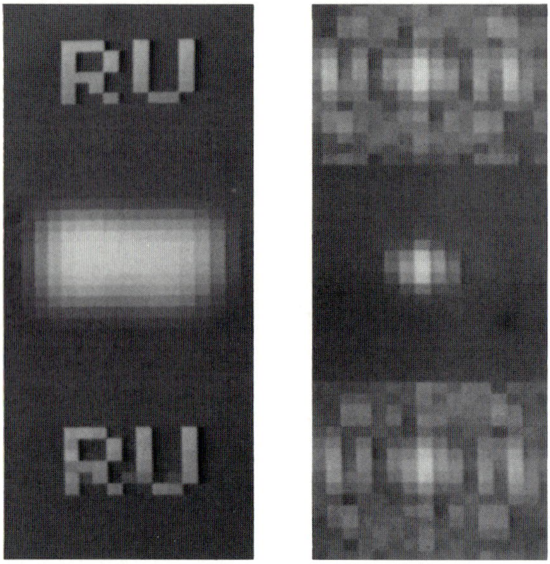

Fig. 4.8-1. Original image (top left), degraded image (middle left), and restored image (bottom left). The images on the right side illustrate the 2-D DFT of the image to the left.

left-hand corner. The test image is composed of the letters R and U. The gray scale decreases linearly from a value of 100 at the top of the image to a value of 50 at the bottom. The image is in a 16×16-pixel by 8-bit format. The degraded image is shown directly beneath the test image. The degradation in this case was obtained by using a seven-point lowpass finite impulse response (FIR) filter designed with a narrow transition band and high stopband attenuation [28]. The cutoff frequency was $\pi/2$ and the stopband attenuation was greater than 42 db. A two-dimensional separable FIR was then used for the degradation process, using full precision arithmetic. This degraded image was quantized to 8 bits and the degradation matrix was also quantized to 8 bits for the restoration. The quantization of the degradation matrix was performed in order to better simulate the typical situation where the degradation matrix is not known exactly. The restored image, using the method described in Section 4.5, is shown below the degraded image. Notice the good recovery of details such as the contrast and the right leg of the letter R. On the right-hand side of Fig. 4.8-1 are the two-dimensional DFTs of the corresponding images on the left. The topmost 2-D DFT indicates that a wide band of spatial frequencies are contained in the test image. The DFT of the degraded image shows that only a small band of low spatial frequencies are passed through the degradation operation. The DFT components in the stopband were found to be below the level of uncertainty due to quantization. The DFT of the restored image on the bottom is seen to extrapolate the DFT of the degraded image very well. Figure 4.8-2 illustrates a very sparse test image at the top. Directly

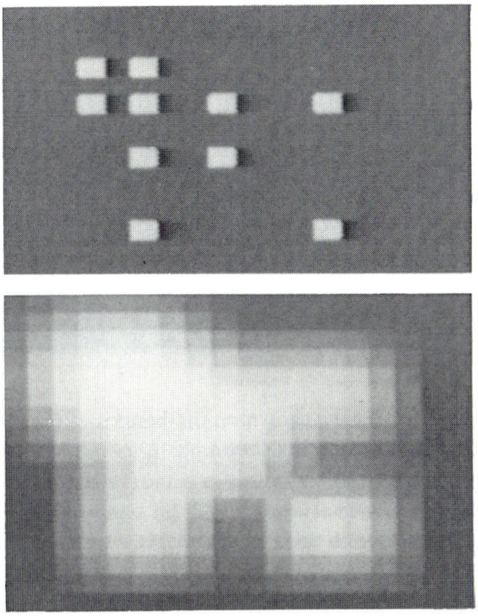

Fig. 4.8-2. Original image (top) degraded image (bottom).

beneath this appears the test image degraded by the same degradation used in Fig. 4.8-1. Figure 4.8-3 illustrates the difference between in-band restoration by the pseudoinverse solution (top view) and restoration by linear programming with prefiltering as given by Eq. (4.1-25) (bottom view). Notice that the high spatial frequencies are recovered by the linear programming method whereas they are not recovered by the pseudoinverse solution. Figure 4.8-4 illustrates the effects of the algorithm on real data. The test target was two wires spaced less than one-third the Rayleigh distance in the horizontal and vertical directions. A B-scan ultrasound image of the target was taken with a commercial ultrasound unit, which was modified to provide the RF signal directly. The restored image is displayed below the measured image. The image of the wires was restored to within the accuracy and resolution of micrometer settings used in positioning the targets. The two artifacts to the right of the image of the two wires are due to reverberations of the ultrasound pulse between the wires.

10. A. Papoulis, A new algorithm in spectral analysis and bandlimited extrapolation. *IEEE Trans. Circuits Syst.* **CAS-22**, 735–742 (1975).
11. M. I. Sezan and H. Stark, Image restoration by the method of convex projections, Part 2—Applications and numerical results. *IEEE Trans. on Med. Imag.* **MI-1**, 95-101 (1982).
12. H. Stark, D. Cahana, and H. Webb, Restoration of arbitrary finite-energy optical objects from spatial and spectral information. *J. Opt. Soc. Amer.* **71**(6), 635-642 (1981).
13. D. Slepian and H. O. Pollak, Prolate spheroidal wave functions, Fourier analysis and uncertainty-I. *Bell Sys. Tech. J.* **40**, 43–63 (1961).
14. C. K. Rushforth and R. W. Harris, Restoration, resolution, and noise. *J. Opt. Soc. Amer.* **58**, 539–545 (1968).
15. D. Slepian, Prolate spheroidal wave functions, Fourier analysis, and uncertainty-V: The discrete case. *Bell Sys. Tech. J.* **57**, 1371-1430 (1978).
16. R. Bellman, "Introduction to Matrix Analysis," New York: McGraw-Hill, 1960.
17. T. F. Chan, An improved algorithm for computing the singular value decomposition Algorithm 581. *ACM Trans. Math. Software,* **8**, 72–83 (1982).
18. M. Z. Nashed, "Generalized Inverses and Applications," New York: Academic Press, 1976.
19. L. Eckert and G. Young, The approximation of one matrix by another of lower rank. *Psychometrika,* **1**, 1936.
20. A. Oppenheim and R. Schafer, "Digital Signal Processing," Englewood Cliffs, New-Jersey: Prentice-Hall, 1975.
21. G. B. Dantzig, "Linear Programming and Extensions," Princeton, NJ: Princeton University Press, 1963.
22. J. M. Wozencraft and I. M. Jacobs, "Principles of Communication Engineering," New York: John Wiley, 1965.
23. A. Papoulis, "Probability, Random Variables, and Stochastic Processes," New York: McGraw-Hill, 1965.
24. I. Barrodale and F. Roberts, Solutions of an overdetermined system of equations in the L_1 norm, Algorithm 478. *Comm. ACM,* **17**, 319 (1973).
25. J. W. Goodman (1968). *Introduction to Fourier Optics,* New York: McGraw-Hill, 1968.
26. G. Westheimer, Spatial Frequency and light-spread descriptions of visual acuity and hyperacuity. *J. Opt. Soc. Amer.* **67**, 207–212 (1977).
27. B. R. Frieden, Image enhancement and restoration, *In* Picture Processing and Digital Filtering (T. S. Huang, ed.), New York: Springer Verlag, 1975.
28. J. H. McClellan, T. W. Parks, and L. R. Rabiner, A computer program for designing linear optimum FIR linear phase digital filters. *IEEE Trans. on Audio and Electro-acoustics* **AU-21**, 506–526 (1973).
29. L. G. Kachian, A polynomial algorithm for linear programming. *Doklady Akad. Nauk, USSR,* **244**(5), 1073-96, translated in *Soviet math, Doklady,* **20**, 191–94 (1979).
30. N. Karmarker, A new polynomial-time algorithm for linear programming. *Combinatorica* **4**, 373-395 (1984).
31. K. Preston and L. Uhr (eds.), "Multi-Computers and Image Processing Algorithms and Programs," New York: Academic Press, 1982.

CHAPTER **5**

The Principle of Maximum Entropy in Image Recovery

*Xinhua Zhuang**

Coordinated Science Laboratory
University of Illinois at
Champaign-Urbana
Urbana, Illinois 61801

Einar Østevold

Norwegian Defense Research
Establishment
N-2007 Kjeller, Norway

Robert M. Haralick

Machine Vision International
Ann Arbor, Michigan 48104

5.1 INTRODUCTION

For many decades it has been recognized, or conjectured, that the notion of entropy defines a kind of measure on the space of probability distributions, such that those of high entropy are in some sense favored over others, all other things being equal. The basis for this was stated first in a variety of intuitive forms: that distributions of higher entropy represent more "disorder," that they are "smoother," "more probable," "less predictable," that they "assume less" and hence "are more natural" according to Shannon's interpretation of entropy as an information measure, etc.

While each of these points of view doubtless reflects an element of truth, none seems explicit enough to lend itself to a quantitative demonstration of the kind we are accustomed to having in other fields of applied mathematics. Accordingly, many approaching this field are disconcerted by what they sense as a kind of vagueness, the underlying theory lacking a solid content.

* Visiting Professor, People's Republic of China.

This has not prevented the fruitful exploitation of this property of entropy. The maximum entropy (ME) principle, briefly speaking, is: when we make inferences based on incomplete information, we should draw them from that probability distribution that has the maximum entropy allowed by the information we do have. Jaynes [1,2] has been a foremost proponent of maximum entropy prior distributions consistent with known information. Essentially all of the known results of statistical mechanics, equilibrium and nonequilibrium, are derivable consequences of this principle. In image restoration and spectrum analysis, the maximum entropy principle takes into account cogent information about multiplicities that orthodox statistics misses because of its failure to admit prior probabilities.

Perhaps Frieden [3] was the first one to correctly pose one type of maximum entropy image restoration problem. In his statistical model, the image space is divided into n resolution cells (or n events). The frequency of occurrence of the ith event is identified with

$$p_i = f_i \Big/ \sum_{i=1}^{n} f_j,$$

where f_i represents the gray tone intensity in the ith cell. Thus, the entropy is defined as

$$H(p_1, \ldots, p_n) = -\sum_{i=1}^{n} p_i \log p_i.$$

Apart from Frieden's statistical model, the maximization of $H(f_1, \ldots, f_n)$,

$$H(f_1, \ldots, f_n) = -\sum_{i=1}^{n} f_i \log f_i, \tag{5.1-1}$$

does have a smoothing influence on the restored image f_i, $i = 1, \ldots, n$, assuming the sum of gray levels $\sum_{i=1}^{n} f_i = t$ is a known constant. To show the smoothing influence we show that as f_i and f_j approach each other while t stays the same, $H(f_1, \ldots, f_n)$ will increase. Assume that $f_i > f_j$; thus for the two pixels to approach each other in value while t continues to be the same, we may introduce the following changes:

$$\nabla f_i = -\varepsilon, \qquad \nabla f_j = \varepsilon, \tag{5.1-2}$$

as $\varepsilon \to 0$. After these changes, we can calculate the new summation in Eq. (5.1-1) denoted by H':

$$\begin{aligned} H' &= H(f_1, \ldots, f_n) + f_i \log f_i + f_j \log f_j \\ &\quad -(f_i - \varepsilon) \log(f_i - \varepsilon) - (f_j + \varepsilon) \log(f_j + \varepsilon) \\ &= (f_1, \ldots, f_n) + \varepsilon \log(f_i/f_j) \\ &> H(f_1, \ldots, f_n), \end{aligned} \tag{5.1-3}$$

where we have used the approximate formula

$$\log(1 + x) = x, \qquad |x| \ll 1.$$

Besides maximizing $H(f_1, \ldots, f_n)$, many other image restoration procedures exert a smoothing influence on the solution. In fact, various linear techniques can be used to improve the resolution of a blurred image of an object. However, the linear techniques often induce erroneous detail such as oscillations or "ringing" around sharp changes in intensity in the image. That is a particular problem with images of complex objects since the superposition of many oscillations can give spurious detail with unpredictable shapes and intensities. Therefore, a "safe" inversion technique is required—one giving restorations free from ringing, such that confidence can be placed in the reality of all extra details revealed by the restoration method. Naturally, interest has centered on nonlinear techniques which incorporate constraints to reduce the artifacts generated in the restoration. Of the nonlinear techniques, maximum entropy is a fundamental method for the solution of the inversion problems in image restoration. A variety of iterative algorithms have been proposed to obtain the maximum entropy solution.

The computing times required by most of the these algorithms are, however, much higher than those required by linear methods, and this has often procluded their use on images containing more than 64×64 pixels. Therefore, designing more efficient algorithms becomes crucial.

Section 5.2 introduces Frieden's statistical model for an "object" in optical image formation and a maximum entropy image restoration algorithm which requires solving $m + 1$ nonlinear equations with $m + 1$ unknowns, where m represents the number of observed data values.

Section 5.3 introduces Burch *et al.* [4] algorithm, which requires solving a optimization problem in $(n + 1)$-dimensional space.

In Section 5.4 we present our own technique [5]. This is a differential equation approach to the ME restoration problem. The problem is reduced to solving a system of differential equations with easily obtained initial conditions. Instead of doing an $(n + 1)$-dimensional search as required by most ME techniques, a one-dimensional search along a well-defined path in the $(n + 1)$-dimensional space is performed. The search path is defined by the Cauchy problem of differential equations. Results are shown for cases with strong deblurring and for cases with high noise levels. Our results are compared with those produced by competing ME techniques. The aspect of computational cost is also discussed.

Section 5.5 is the conclusion.

5.2 FRIEDEN'S APPROACH [3]

Frieden's approach is based on a statistical model for an "object" in optical image formation. The object is defined as an unknown spatial radiance distribution $f_i = f(x_i)$, $i = 1, \ldots, n$, which produces an image irradiance distribution $d_j = d(y_j)$, $j = 1, \ldots, m$, which is collected as data. The connection between f and d is given by

$$d_j = \sum_{i=1}^{n} A_{ji} f_i + e_j, \qquad j = 1, \ldots, m, \tag{5.2-1}$$

where $e_j = e(y_j)$ denotes noise. Here, the object is imagined to be composed of discrete, mathematical "grains" of small intensity Δ which are distributed over the object scene. The scene is subdivided into cells centered on a subdivision of points $\{x_i\}$, $i = 1, \ldots, n$, and the unknown object is assumed to have g_i grains in cell i. Thus $f(x_i) = \Delta \cdot g_i$. Let p_i represent the probability of a grain locating in cell i. Then if a large number of grains are distributed over the object, by the law of large numbers

$$p_i = g_i / g_T,$$

where g_T is the total number of grains in the object; g_T is assumed known by conservation of energy from the image data.

The entropy, from information theory, is defined as

$$\begin{aligned} H(p_1, \ldots, p_n) &\triangleq -\sum_{i=1}^{n} p_i \log p_i \\ &= -\sum_{i=1}^{n} \left(\frac{g_i}{g_T}\right) \log\left(\frac{g_i}{g_T}\right) \\ &= -\sum_{i=1}^{n} \left(\frac{f_i}{\sum_{i=1}^{n} f_i}\right) \log\left(\frac{f_i}{\sum_{i=1}^{n} f_i}\right) \\ &= -\frac{\sum_{i=1}^{n} f_i \log f_i}{\sum_{i=1}^{n} f_i} - \log\left(\sum_{i=1}^{n} f_i\right). \end{aligned} \tag{5.2-2}$$

Since the total spatial radiances $\sum_{i=1}^{n} f_i$ $(= \Delta \cdot g_T)$ is a constant, the principle of maximum entropy then becomes: maximize

$$H(f_1, \ldots, f_n) = -\sum_{i=1}^{n} f_i \log f_i \tag{5.2-3}$$

subject to two constraints: Eq. (5.2-1) and $\sum_{i=1}^{n} f_i = t$, where $t = \Delta \cdot g_T$ is a constant. Since Eq. (5.2-1) contains the noise term, a weighted sum of the image entropy and the observation noise entropy is maximized in Frieden's approach.

The noise component may take either positive or negative values. Hence, the definition of noise entropy requires some care. The difficulty is overcome by applying additional *a priori* knowledge to the problem by assuming that a constant B is known such that

$$B = \max_j(-e_j). \tag{5.2-4}$$

In practice, B may be roughly estimated by setting it equal to 2σ, where σ is the standard deviation of the noise. According to Frieden, the quality of the restoration does not depend critically on B precisely satisfying Eq. (5.2-4); however, the results are best when Eq. (5.2-4) is indeed satisfied. We may now define a new nonnegative sequence

$$e_j' = e_j + B \geqslant 0 \tag{5.2-5}$$

and the noise entropy as

$$H_e = -\sum_{j=1}^{n} e_j' \log e_j'. \tag{5.2-6}$$

The restoration problem is solved by maximizing the weight sum

$$H(f_1, \ldots, f_n) + \rho H_e, \tag{5.2-7}$$

where the constant ρ is used to emphasize one of $H(f_1, \ldots, f_n)$ and H_ρ with respect to the other. Frieden recommends a value of $\rho = 20$ as about optimum for a wide range of object and noise situations.

Maximization of Eq. (5.2-7) is carried out subject to satisfying the following two equalities:

$$d_j = \sum_{i=1}^{n} A_{ji} f_i + e_j' - B, \tag{5.2-8}$$

$$\sum_{i=1}^{n} f_i = t. \tag{5.2-9}$$

The maximization can be carried out with Lagrange multipliers. We find the $m + n$ unknowns f_i $(i = 1, \ldots, n)$, e_j' $(j = 1, \ldots, m)$ that maximize the functional

$$\begin{aligned} K \triangleq\ & H(f_1, \ldots, f_n) + \rho H_e \\ & + \sum_{j=1}^{m} \lambda_j \left(\sum_{i=1}^{n} A_{ji} f_i + e_j' - B - d_j \right) + \mu \left(\sum_{i=1}^{n} f_i - t \right), \end{aligned} \tag{5.2-10}$$

where λ_j, $j = 1, \ldots, m$, and μ and $m + 1$ Lagrange multipliers.

Let $\hat{f}_i$ and $\hat{e}_j$ denote the optimum values of f_i and e'_j. Clearly, letting $\mathbf{f} \triangleq [f_1, \ldots, f_n]^T$, $\hat{\mathbf{f}} \triangleq [\hat{f}_1, \ldots, \hat{f}_n]^T$, $\mathbf{e} \triangleq [e_1, \ldots, e_m]^T$, and $\hat{\mathbf{e}} \triangleq [\hat{e}_1, \ldots, \hat{e}_m]^T$ we should have

$$\partial K / \partial f_i |_{f=\hat{f}} = 0, \tag{5.2-11}$$

$$\partial K / \partial e'_j |_{e'=\hat{e}} = 0. \tag{5.2-12}$$

Performing these differentiations, we obtain

$$\hat{f}_i = \exp\left(-1 + \mu + \sum_{j=1}^{m} \lambda_j A_{ji} \right) \tag{5.2-13}$$

and

$$\hat{e}_j = \exp(-1 + \lambda_j / \rho). \tag{5.2-14}$$

These solutions must satisfy the constraints given by Eqs. (5.2-8) and (5.2-9):

$$\sum_{i=1}^{n} A_{ji} \hat{f}_i + \hat{e}_j - B = d_j, \qquad j = 1, \ldots, m, \tag{5.2-15}$$

and

$$\sum_{i=1}^{n} \hat{f}_i = t. \tag{5.2-16}$$

The restored f_i is obtained by first substituting Eqs. (5.2-13) and (5.2-14) in Eqs. (5.2-15) and (5.2-16). The resulting $m+1$ equations are solved for the $m+1$ unknowns λ_j, $j = 1, \ldots, m$, and μ. These equations are highly nonlinear but, according to Frieden, are always solvable by an $(m+1)$-dimensional Newton–Raphson method, provided B is large enough.

5.3 BURCH, GULL, AND SKILLING'S APPROACH [4]

Again let the required restored image have pixel values represented by nonnegative numbers $f_1, \ldots, f_n$ and the observed image data be given by

$$d_j = \sum_{i=1}^{n} A_{ji} f_i + e_j, \qquad j = 1, \ldots, m, \tag{5.3-1}$$

where the e_j's represent independent zero mean, σ_j^2 variance noise terms.

Burch *et al.* attack the problem in a different way. They define

$$Q(f_1, \ldots, f_n) = \frac{1}{2} \sum_{j=1}^{m} \frac{1}{\sigma_j^2} \left(\sum_{i=1}^{n} A_{ji} f_i - d_j \right)^2. \tag{5.3-2}$$

Instead of the m constraint equations such as Eq. (5.3-1) Burch *et al.* use a single chi-squared, χ^2, statistic

$$Q(f_1, \ldots, f_n) = m/2. \tag{5.3-3}$$

The motivation for this constraint comes about from the central limit theorem [6], which states

$$\lim_{m \to 0} \frac{1}{m} \sum_{j=1}^{m} \frac{e_j^2}{\sigma_j^2} = 1. \tag{5.3-4}$$

Thus, provided that m is large, we would expect the true values of $f_1, \ldots, f_n$ to satisfy Eq. (5.3-3). This approach avoids a mathematically unwieldy proliferation of Lagrange multipliers and allows the development of an iterative technique applicable to large images. Instead of the functional K in Eq. (5.2-10), the following functional is maximized subject to $Q(f_1, \ldots, f_n) = m/2$ in Burch *et al.*'s approach:

$$J = -\sum_{i=1}^{n} f_i \log\left(\frac{f_i}{eA}\right) - \lambda Q(f_1, \ldots, f_n), \tag{5.3-5}$$

where A is a predetermined parameter and e the natural exponent. Expanding (5.3-5), we obtain

$$J = -\sum_{i=1}^{n} f_i \log f_i - \lambda Q(f_1, \ldots, f_n) + (1 + \log A) \sum_{i=1}^{n} f_i$$

$$= H(f_1, \ldots, f_n) - \lambda Q(f_1, \ldots, f_n) + (1 + \log A) \sum_{i=1}^{n} f_i \tag{5.3-6}$$

$$= H(f_1, \ldots, f_n) - \lambda Q(f_1, \ldots, f_n) + \mu \sum_{i=1}^{n} f_i, \tag{5.3-7}$$

where $\mu = (1 + \log A)$. If $f_1^*, \ldots, f_n^*$ maximize Eq. (5.3-7) subject to $Q(f_1, \ldots, f_n) = m/2$, then $f_1^*, \ldots, f_n^*$ would maximize $H(f_1, \ldots, f_n)$ subject to two constraints:

$$Q(f_1, \ldots, f_n) = m/2$$

and

$$\sum_{i=1}^{n} f_i = \sum_{i=1}^{n} f_i^*.$$

Noticing that

$$H(p_1, \ldots, p_n) = \frac{H(f_1, \ldots, f_n)}{\sum_{i=1}^{n} f_i} \log\left(\sum_{i=1}^{n} f_i\right)$$

we observe that the same $f_1^*, \ldots, f_n^*$ would maximize the entropy $H(p_1, \ldots, p_n)$ subject to the same two constraints. Here, $p_i = f_i / \sum_{i=1}^n f_i$ as before. The stationary point equation of Eq. (5.3-6) or Eq. (5.3-7) is

$$\nabla J = 0 \tag{5.3-8}$$

or

$$\log A - \log f_i - \lambda \frac{\partial}{\partial f_i} Q(f_1, \ldots, f_n) = 0 \tag{5.3-9}$$

or

$$(\mu - 1) - \log f_i - \lambda \frac{\partial}{\partial f_i} Q(f_1, \ldots, f_n) = 0. \tag{5.3-10}$$

Normally, we first attempt to solve an unconstrained maximization of J or its stationary point equation $\nabla J = 0$. The simplest such algorithm is steepest ascent, in which one performs a line search along the direction of ∇J. This is well known to be inefficient. The standard way to improve a steepest ascent algorithm is to use some variant of the conjugate gradient technique. Instead of using ∇J itself as the search direction, one uses a combination of this with previous gradient(s) chosen to give quick convergence if J is exactly quadratic in $f_1, \ldots, f_n$. Unfortunately, J in maximum entropy is highly nonquadratic, and conjugate gradients will also not be powerful.

Instead of searching along just one direction at a time, Burch *et al.* use a subspace of three directions, namely

$$(\mathbf{e}_1)_i \triangleq f_i \frac{\partial S}{\partial f_i}, \qquad \text{where for convenience let } S = -\sum_{i=1}^n f_i \log\left(\frac{f_i}{eA}\right), \tag{5.3-11}$$

$$(\mathbf{e}_2)_i \triangleq f_i \frac{\partial Q}{\partial f_i}, \tag{5.3-12}$$

$$(\mathbf{e}_3)_i \triangleq \sum_{j=1}^n f_i \frac{\partial^2 Q}{\partial f_i \, \partial f_j} f_j \frac{\partial J}{\partial f_j}, \tag{5.3-13}$$

where, according to the required stationary point equation $\nabla J = 0$ ($J = S - \lambda Q$), λ is necessarily given by

$$\lambda = \left(\frac{\sum_{i=1}^n f_i (\partial S / \partial f_i)^2}{\sum f_i (\partial Q / \partial f_i)^2} \right)^{1/2}. \tag{5.3-14}$$

Here the first direction $\mathbf{e}_1$ helps S to be maximized through $\partial S / \partial f_i$. The extra factor f_i has the operational characteristic of letting the algorithm concentrate on intense areas of the image instead of areas already known to be faint and requiring little further attention. Likewise, the second direction $\mathbf{e}_2$ helps the desired value of Q to be reached, with a similar factor of f_i. These factors can be understood by using the entropy curvature $-\nabla\nabla S = -(\partial^2 S/\partial f_i\, \partial f_j) = \text{diag}[1/f_1, \ldots, 1/f_n]$ as the metric tensor defining lengths and angles in f space. The third direction $\mathbf{e}_3$ corresponds to the initial correction which a conjugate gradient maximization of J would make to the steepest ascent direction ∇J, provided $-\nabla\nabla S$ was being used as the metric.

Within this three-dimensional subspace, the iterative change to be made to f can be expressed as

$$\delta\mathbf{f} = x^1\mathbf{e}_1 + x^2\mathbf{e}_2 + x^3\mathbf{e}_3, \tag{5.3-15}$$

where suitable coefficients x^i $(i = 1, 2, 3)$ are to be determined. Burch *et al.* begin by constructing quadratic models for S and Q with respect to variables x^1, x^2, x^3 from the local gradients and curvatures, thus modeling

$$\begin{aligned} \tilde{S}(x) &= S + \nabla S \cdot \delta\mathbf{f} + \frac{1}{2}(\delta\mathbf{f})^T \nabla\nabla S(\delta\mathbf{f}) \\ &= S + \sum_{i=1}^{3} (\nabla S, \mathbf{e}_i) x^i - \frac{1}{2}(\delta\mathbf{f})^T \,\text{diag}[1/f_1, \ldots, 1/f_n](\delta\mathbf{f}) \\ &= S + \sum_{i=1}^{3} (\nabla S, \mathbf{e}_i) x^i - \frac{1}{2}\sum_{i,j=1}^{3} (\mathbf{e}_i^T \,\text{diag}[1/f_1, \ldots, 1/f_n]\mathbf{e}_j) x^i x^j, \end{aligned} \tag{5.3-16}$$

$$\begin{aligned} \tilde{Q}(x) &= Q + (\nabla Q, \delta f) + \frac{1}{2}(\delta\mathbf{f})^T \nabla\nabla Q(\partial\mathbf{f}) \\ &= Q + \sum_{i=1}^{3} (\nabla Q, \mathbf{e}_i) x^i + \frac{1}{2}(\delta\mathbf{f})^T A' D A(\delta\mathbf{f}) \\ &= Q + \sum_{i=1}^{3} (\nabla Q, \mathbf{e}_i) x^i + \frac{1}{2}\sum_{i,j=1}^{3} (\mathbf{e}_i^T \mathbf{A}^T \mathbf{D} \mathbf{A} \mathbf{e}_j) x^i x^j, \end{aligned} \tag{5.3-17}$$

where $\mathbf{A} = (A_{ji})_{m \times n}$, $\mathbf{D} = \text{diag}[d_1, \ldots, d_m]$.

In keeping with using $-\nabla\nabla S$ as the metric tensor in $\mathbf{f}$ space, Burch *et al.* define the length squared of the increment $\delta\mathbf{f}$ as

$$\begin{aligned} l^2 &= (\delta\mathbf{f})^T(-\nabla\nabla S)(\delta f) \\ &= \sum_{i,j=1}^{3} (\mathbf{e}_i^T \,\text{diag}[1/f_1, \ldots, 1/f_n]\mathbf{e}_j) x^i x^j. \end{aligned} \tag{5.3-18}$$

Starting with a flat iterate $f_j = A$, the control of the algorithm now passes into the subspace, where highly sophisticated processing can be performed on the 26 defining scalars S, Q, $(\nabla S, \mathbf{e}_i)$, $(\mathbf{e}_i^T \operatorname{diag}[1/f_1, \ldots, 1/f_n]\mathbf{e}_j)$, $(\nabla Q, \mathbf{e}_i)$, $(\mathbf{e}_i^T \mathbf{A}^T \mathbf{D} \mathbf{A} \mathbf{e}_j)$, $i, j = 1, 2, 3$.

The first "control" step is to rotate the coordinate system within the subspace to diagonalize the symmetric matrix

$$(\mathbf{e}_i^T \operatorname{diag}[1/f_1, \ldots, 1/f_n]\mathbf{e}_j)_{3\times 3}.$$

This enables the algorithm to be protected against accidental linear dependence of the search directions by discarding eigenvectors corresponding to unusually small eigenvalues of $(\mathbf{e}_i^T \operatorname{diag}[1/f_1, \ldots, 1/f_n]\mathbf{e}_j)_{3\times 3}$.

Assume $\mathbf{R} = [\mathbf{r}_1, \mathbf{r}_2, \mathbf{r}_3]$ is a 3×3 rotational matrix, i.e., orthonormal with $\det(\mathbf{R}) = 1$ such that

$$\mathbf{R}^T(\mathbf{e}_i^T \operatorname{diag}[1/f_1, \ldots, 1/f_n]\mathbf{e}_j)_{3\times 3}\mathbf{R} = \operatorname{diag}[\lambda_1, \lambda_2, \lambda_3] \qquad (5.3\text{-}19)$$

or after rearrangement

$$([\mathbf{e}_1, \mathbf{e}_2, \mathbf{e}_3]\mathbf{R})^T \operatorname{diag}[1/f_1, \ldots, 1/f_n)([\mathbf{e}_1, \mathbf{e}_2, \mathbf{e}_3]\mathbf{R})$$
$$= \operatorname{diag}[\lambda_1, \lambda_2, \lambda_3](\operatorname{diag}[1/\sqrt{f_1}, \ldots, 1/\sqrt{f_n}][\mathbf{e}_1, \mathbf{e}_2, \mathbf{e}_3]\mathbf{R})^T. \qquad (5.3\text{-}20)$$

$$(\operatorname{diag}[1/\sqrt{f_1}, \ldots, 1/\sqrt{f_n}][\mathbf{e}_1, \mathbf{e}_2, \mathbf{e}_3]\mathbf{R}) = \operatorname{diag}[\lambda_1, \lambda_2, \lambda_3]. \quad (5.3\text{-}21)$$

It is clear that

$$\begin{aligned}
\lambda_1 &= \|\operatorname{diag}[1/\sqrt{f_1}, \ldots, 1/\sqrt{f_n}][\mathbf{e}_1, \mathbf{e}_2, \mathbf{e}_3]\mathbf{R}[1, 0, 0]^T\|^2 \geqslant 0,\\
\lambda_2 &= \|\operatorname{diag}[1/\sqrt{f_1}, \ldots, 1/\sqrt{f_n}][\mathbf{e}_1, \mathbf{e}_2, \mathbf{e}_3]\mathbf{R}[0, 1, 0]^T\|^2 \geqslant 0,\\
\lambda_3 &= \|\operatorname{diag}[1/\sqrt{f_1}, \ldots, 1/\sqrt{f_n}][\mathbf{e}_1, \mathbf{e}_2, \mathbf{e}_3]\mathbf{R}[0, 0, 1]^T\|^2 \geqslant 0.
\end{aligned} \qquad (5.3\text{-}22)$$

Thus, from Eq. (5.3-22) it is clear that there is a zero eigenvalue $\lambda_i = 0$ if and only if

$$\operatorname{diag}[1/\sqrt{f_1}, \ldots, 1/\sqrt{f_n}][\mathbf{e}_1, \mathbf{e}_2, \mathbf{e}_3]\mathbf{r}_i = 0 \qquad (5.3\text{-}23)$$

or

$$[\mathbf{e}_1, \mathbf{e}_2, \mathbf{e}_3]\mathbf{r}_i = 0 \qquad (5.3\text{-}24)$$

or the search directions $\mathbf{e}_1$, $\mathbf{e}_2$, $\mathbf{e}_3$, are linearly dependent. Moreover, it is easy to verify that $\mathbf{r}_i$ is the eigenvector of

$$([\mathbf{e}_1, \mathbf{e}_2, \mathbf{e}_3]^T \operatorname{diag}[1/f_1, \ldots, 1/f_n][\mathbf{e}_1, \mathbf{e}_2, \mathbf{e}_3])_{3\times 3}$$

corresponding to zero eigenvalue $\lambda_i = 0$.

Let $\mathbf{y} = \mathbf{R}\mathbf{x}$. Then $\tilde{S}(\mathbf{x})$, $\tilde{Q}(\mathbf{x})$ becomes $\tilde{S}(\mathbf{y})$, $\tilde{Q}(\mathbf{y})$, respectively, where

$$\tilde{S}(\mathbf{y}) = S + [(\nabla S, \mathbf{e}_i)]_{3\times 1}^T \mathbf{R}^T \mathbf{y} - \tfrac{1}{2}\mathbf{y}^T \operatorname{diag}[\lambda_1, \lambda_2, \lambda_3]\mathbf{y}, \qquad (5.3\text{-}25)$$

$$\tilde{Q}(\mathbf{y}) = Q + [(\nabla Q, \mathbf{e}_i)]_{3\times 1}^T \mathbf{R}^T \mathbf{y} + \tfrac{1}{2}\mathbf{y}^T \mathbf{R}^T (\mathbf{e}_i^T \mathbf{A}^T \mathbf{D} \mathbf{A} \mathbf{e}_j)_{3\times 3} \mathbf{R}\mathbf{y}. \qquad (5.3\text{-}26)$$

If none of three eigenvalues λ_1, λ_2, λ_3 is small, letting $\mathbf{z} = \mathrm{diag}[\sqrt{\lambda_1}, \sqrt{\lambda_2}, \sqrt{\lambda_3}]\mathbf{y}$, $\tilde{S}(\mathbf{u})$, $\tilde{Q}(\mathbf{y})$ becomes $\tilde{S}(\mathbf{z})$, $\tilde{Q}(\mathbf{z})$, respectively, where

$$\tilde{S}(\mathbf{z}) = S + [(\nabla S, \mathbf{e}_i)]^T_{3\times 1}\mathbf{R}^T \mathrm{diag}[1/\sqrt{\lambda_1}, 1/\sqrt{\lambda_2}, 1/\sqrt{\lambda_3}]\mathbf{z} - \tfrac{1}{2}\mathbf{z}^T\mathbf{z}, \quad (5.3\text{-}27)$$

$$\begin{aligned}\tilde{Q}(\mathbf{z}) = Q &+ [(\nabla Q, \mathbf{e}_i)]^T_{3\times 1}\mathbf{R}^T \mathrm{diag}[1/\sqrt{\lambda_1}, 1/\sqrt{\lambda_2}, 1/\sqrt{\lambda_3}]\mathbf{z} \\ &+ \tfrac{1}{2}\mathbf{z}^T \mathrm{diag}[1/\sqrt{\lambda_1}, 1/\sqrt{\lambda_2}, 1/\sqrt{\lambda_3}]\mathbf{R}^T(\mathbf{e}_i^T\mathbf{A}^T\mathbf{D}\mathbf{A}\mathbf{e}_j)_{3\times 3}\mathbf{R} \\ &\cdot \mathrm{diag}[1/\sqrt{\lambda_1}, 1/\sqrt{\lambda_2}, 1/\sqrt{\lambda_3}]\mathbf{z}. \end{aligned} \quad (5.3\text{-}28)$$

Introduce the second rotational matrix $\mathbf{V}$ such that

$$\begin{aligned}&\mathbf{V}^T \mathrm{diag}[1/\sqrt{\lambda_1}, 1/\sqrt{\lambda_2}, 1/\sqrt{\lambda_3}]\mathbf{R}^T(\mathbf{e}_i^T\mathbf{A}^T\mathbf{D}\mathbf{A}\mathbf{e}_j)_{3\times 3}\mathbf{R} \\ &\quad\cdot \mathrm{diag}[1/\sqrt{\lambda_1}, 1/\sqrt{\lambda_2}, 1/\sqrt{\lambda_3}]\mathbf{V} = \mathrm{diag}[\mu_1, \mu_2, \mu_3]. \end{aligned} \quad (5.3\text{-}29)$$

Letting $\mathbf{u} = \mathbf{V}\mathbf{z}$, $\tilde{S}(\mathbf{z})$ and $\tilde{Q}(\mathbf{z})$ become $\tilde{S}(\mathbf{u})$ and $\tilde{Q}(\mathbf{u})$, respectively, where

$$\tilde{S}(\mathbf{u}) = S + [(\nabla S, \mathbf{e}_i)]^T_{3\times 1}\mathbf{R}^T \mathrm{diag}[1/\sqrt{\lambda_1}, 1/\sqrt{\lambda_2}, 1/\sqrt{\lambda_3}]\mathbf{V}^T\mathbf{u} - \frac{1}{2}\mathbf{u}^T\mathbf{u}, \quad (5.3\text{-}30)$$

$$\tilde{Q}(\mathbf{u}) = Q + [(\nabla Q, \mathbf{e}_i)]^T_{3\times 1}\mathbf{R}^T \mathrm{diag}[1/\sqrt{\lambda_1}, 1/\sqrt{\lambda_2}, 1/\sqrt{\lambda_3}]\mathbf{V}^T\mathbf{u} + \frac{1}{2}\sum_{i=1}^{3}\mu_i u_i^2. \quad (5.3\text{-}31)$$

If λ_3 is very small but λ_1, λ_2 are not so small, we let $y_3 = 0$ and

$$[z_1, z_2]^T = \mathrm{diag}[\sqrt{\lambda_1}, \sqrt{\lambda_2}][y_1, y_2]^T;$$

then $\tilde{S}(y)$, $\tilde{Q}(y)$ become $\tilde{S}(z)$, $\tilde{Q}(z)$, where

$$\begin{aligned}\tilde{S}(\mathbf{z}) &= S + [(\nabla S, \mathbf{e}_i)]^T_{3\times 1}\mathbf{R}^T\begin{bmatrix}\mathrm{diag}[1/\sqrt{\lambda_1}, 1/\sqrt{\lambda_2}][z_1, z_2]^T \\ 0\end{bmatrix} - \frac{1}{2}\sum_{i=1}^{2} z_i^2 \\ &= S + \sum_{i=1}^{2} S_i z_i - \frac{1}{2}\sum_{i=1}^{2} z_i^2, \end{aligned} \quad (5.3\text{-}32)$$

$$\tilde{Q}(\mathbf{z}) = Q + \sum_{i=1}^{2} Q_i z_i + \frac{1}{2}\sum_{i,j=1}^{2} M_{ij} z_i z_j, \quad (5.3\text{-}33)$$

where S_i, Q_i, M_{ij} all could be explicitly determined.

Introduce the second rotational matrix $\mathbf{V}_{2\times 2}$ such that

$$V^T M V = \mathrm{diag}[\mu_1, \mu_2]. \quad (5.3\text{-}34)$$

Letting $\mathbf{u} = V\mathbf{z}$, $\tilde{S}(\mathbf{z})$, $\tilde{Q}(\mathbf{z})$ become $\tilde{S}(\mathbf{u})$, $\tilde{Q}(\mathbf{u})$:

$$\tilde{S}(\mathbf{u}) = S + [S_1, S_2]^T\mathbf{V}^T\mathbf{u} - \tfrac{1}{2}\mathbf{u}^T\mathbf{u}, \tag{5.3-35}$$

$$\tilde{Q}(\mathbf{u}) = Q + [Q_1, Q_2]^T\mathbf{V}^T\mathbf{u} + \frac{1}{2}\sum_{i=1}^{2} \mu_i\mu_i^2. \tag{5.3-36}$$

In any other cases, the treatment is similar.

Now we attempt to maximize S subject to $Q = m/2$ within the **u**-subspace. Here without loss of generality we could write

$$\tilde{S}(\mathbf{u}) = S + \sum S_i u_i - \frac{1}{2}\sum u_i^2, \tag{5.3-37}$$

$$\tilde{Q}(\mathbf{u}) = Q + \sum Q_i u_i + \frac{1}{2}\sum \mu_i u_i^2. \tag{5.3-38}$$

Since the symmetric matrix $(\mathbf{e}_i^T\mathbf{A}^T\mathbf{D}\mathbf{A}\mathbf{e}_j)_{3\times 3}$ is nonnegative, eigenvalues μ_i are nonnegative, too. Thus, most likely $\tilde{Q}(\mathbf{u})$ will have a minimum value $Q - \frac{1}{2}\sum Q_i^2\mu_i$. If the minimum value exceeds the desired value $m/2$, it is impossible to satisfy the constraint $\tilde{Q} = m/2$. Accordingly, Burch *et al.* set the more modest constraint $\tilde{Q} = \tilde{Q}_0$,

$$\tilde{Q}_0 = \max\left(Q - \frac{1}{3}\sum Q_i^2/\mu_i, m/2\right) \tag{5.3-39}$$

which is always accessible within the **u** subspace.

Maximizing $\tilde{S}$ subject to $\tilde{Q} = \tilde{Q}_0$ is equivalent to maximizing $\tilde{S} - \tilde{\lambda}\tilde{Q}$ subject to $\tilde{Q} = \tilde{Q}_0$. It leads to

$$u_i = (S_i - \tilde{\lambda}Q_i)/(1 + \tilde{\lambda}\mu_i) \tag{5.3-40}$$

A binary search could be used to give the value of $\tilde{\lambda}$ which attains the aim of $\tilde{Q}(\mathbf{u}) = \tilde{Q}_0$.

However, the quadratic model for S will only be accurate locally, within some distance limit: $u^Tu \leqslant l_0^2 \simeq 0.2\sum f_i$, the factor 0.2 being a dimensionless coefficient suggested by Burch *et al.* from their practical experience. As easily seen, the following holds.

$$\begin{aligned} l^2 &\triangleq \mathbf{x}^T(e_i^T \operatorname{diag}[1/f_1, \ldots, 1/f_n]e_j)_{3\times 3}\mathbf{x} \\ &= \mathbf{y}^T \operatorname{diag}[\lambda_1, \lambda_2, \lambda_3]\mathbf{y} \\ &= \mathbf{z}^T\mathbf{z} \\ &= \mathbf{u}^T\mathbf{u}. \end{aligned} \tag{5.3-41}$$

So $\tilde{S}$ must be maximized subject to $\tilde{Q} = \tilde{Q}_0$ and an overrriding constraint $l^2 \leqslant l_0^2$. To satisfy the distance constraint, the algorithm normally uses the same values of $\tilde{\lambda}$ and **u** as before, but the binary search is redirected toward the current value Q [see Eq. (5.3-38)] whenever $\tilde{\lambda}$ gives an increment with l^2 too large. This gives extra safety without changing the model parameters S, S_i, Q, q_i, μ_i.

When there is no sufficiently close increment, the distance constraint must be introduced explicitly via a second multiplier P, giving

$$\tilde{W} = \tilde{S} - \tilde{\lambda}\tilde{Q} - \tilde{\lambda}Pl^2 \qquad (P \geqslant 0), \tag{5.3-42}$$

since otherwise the first constraint $\tilde{Q} = \tilde{Q}_0$ could be drastically overriden by the second constraint $l^2 \leqslant l_0^2$. $\tilde{W}$ is maximized at

$$u_i = \frac{S_i - \tilde{\lambda}Q_i}{1 + \tilde{\lambda}(\mu_i + P)}. \tag{5.3-43}$$

P can be interpreted as a modification of $\tilde{Q}$ which increases its curvature components from μ_i to $\mu_i + P$, giving a revised model

$$\tilde{Q}_P = \tilde{Q} + \frac{1}{2}P\sum u_i^2. \tag{5.3-44}$$

With P invoked, $\tilde{\lambda}$ is searched toward $\tilde{Q} = \tilde{Q}_0$, but is redirected toward $\tilde{Q}_P = Q$ whenever l^2 is too large. For sufficiently large P, Burch *et al.* claimed that the binary search could always reach a result obeying

$$\tilde{Q}_0 \leqslant \tilde{Q} < \tilde{Q}_P \leqslant Q \tag{5.3-45}$$

and the smallest such P is used to give the final result **x**.

All that remains is to increment **f** by the multiples x_i of the search directions while protecting against the very occasional stray nonnegative value.

Burch *et al.* also claimed that, in any case, the results are not sensitive to the precise value of the parameter A.

5.4 A DIFFERENTIAL EQUATION APPROACH TO MAXIMUM ENTROPY IMAGE RESTORATION [5]

In this section we develop a new algorithm for solving the maximum entropy image restoration problem. The problem is reduced to solving a system of ordinary differential equations with appropriate initial conditions. The choice of initial conditions closely relates to the satisfaction of constraints, and we show how the initial conditions are determined. The algorithm does not involve any optimization method. Instead of searching in the $(n + 1)$-dimensional space as required for most ME algorithms, our approach relies on solving a one-dimensional search along a self-defined and easily mastered path. Moreover, an efficient algorithm is developed to handle the search. The computer restoration verifies the theory.

In subsection A of this section, the basic concepts for the maximum entropy restoration technique are given. This part includes the derivation of a system of differential equations defining a branch of solutions over which the one-dimensional search is performed. The section concludes with an analysis of the existence interval for λ, the search parameter. Subsection B deals mainly with the problem of selecting initial conditions so that the constraints imposed on the original formulation of the problem can be satisfied. A more comprehensive analysis of this subject is carried out in subsection C. The algorithm for solving the system of differential equations for the maximum entropy reconstruction problem is outlined in subsection D. Subsection E deals with the problem of adjusting a parameter in order to satisfy the constraint on the total image intensity embedded in the original problem formulation. In subsection F experiments in image reconstruction using the algorithm outlined in subsection E are described and results are presented.

A. *Maximum Entropy, Stationary Point Equations, and Cauchy Problem of Differential Equations*

Let the required reconstructed image have pixel values represented by positive numbers $f_1, \ldots, f_n$ which are to be determined, and on which the entropy

$$H(p_1, \ldots, p_n) = \sum_{i=1}^{n} p_i \log p_i, \qquad p_i = \frac{f_i}{\sum_{i=1}^{n} f_i}, \qquad i = 1, \ldots, n, \tag{5.4-1}$$

is defined. The entropy depends only on the distribution of gray levels in the image and not on the total intensity $\sum_{i=1}^{n} f_i$.

Let the observed image data be given by

$$d_j = \sum_{i=1}^{n} A_{ji} f_i + e_j, \qquad j = 1, \ldots, m, \tag{5.4-2}$$

where the e_j's represent independent zero mean, σ_j^2 variance noise terms. We define

$$\begin{aligned} Q(f_1, \ldots, f_n) &= \frac{1}{2} \sum_{j=1}^{m} \frac{1}{\sigma_j^2} \left(\sum_{i=1}^{n} A_{ji} f_i - d_j \right)^2 \\ &= \frac{1}{2} (\mathbf{Af} - \mathbf{d})^T \operatorname{diag}[1/\sigma_1^2, \ldots, 1/\sigma_m^2](\mathbf{Af} - \mathbf{d}) \frac{1}{2} \|\mathbf{Af} - \mathbf{d}\|_{\mathbf{D}}^2, \end{aligned} \tag{5.4-3}$$

where $\mathbf{A} \triangleq [A_{ji}]_{m\times n}$, $\mathbf{f} \triangleq [f_1, \ldots, f_n]^T$, $\mathbf{d} \triangleq [d_1, \ldots, d_m]^T$, and $\mathbf{D} \triangleq \mathrm{diag}[1/\sigma_1^2, \ldots, 1/\sigma_m^2]$. Typical least-squares approaches would try to determine those values $f_1, \ldots, f_n$ which minimize $Q(f_1, \ldots, f_n)$. Rather than do this, we seek those $f_1, \ldots, f_n$ which maximize the entropy subject to the constraint

$$Q(f_1, \ldots, f_n) = m/2. \tag{5.4-4}$$

The motivation for this constraint comes about from the central limit theorem [6], which states

$$\lim_{m\to\infty} \frac{1}{m} \sum_{j=1} \frac{e_j^2}{\sigma_j^2} = 1.$$

Thus, provided that m is large, we would expect the true value of $f_1, \ldots, f_n$ to satisfy Eq. (5.4-4). The condition $Q = m/2$ now determines the set of feasible images $\mathbf{f}$ which pass the given statistical test for consistency with the actual image data $\{d_1, \ldots, d_m\}$.

Although any of the feasible images is acceptable as a reconstruction, the maximum entropy criterion selects that particular $\mathbf{f}$ which has the least configurational information, i.e., the one where the pixel values are least separated. Hence it can be looked on as a smoothing criterion. Of all reconstructions which fit the actual data $\{d_1, \ldots, d_m\}$, the maximum entropy solution thus gives uniquely safe and unprejudicial results.

Formally we are to maximize the entropy $H(p_1, \ldots, p_n)$ given the constraints $Q(f_1, \ldots, f_n) = m/2$ and $\sum_{i=1}^n f_i = t$ $(t > 0)$. We introduce the second constraint for two reasons. One is that the total intensity has a status different from individual pixel values. It does not contribute to the shape of the gray tone intensity function of the image $\mathbf{f}$. For the second reason we must look to the equality

$$H(p_1, \ldots, p_n) = \frac{H(f_1, \ldots, f_n)}{\sum_{i=1}^n f_i} - \log \sum_{i=1}^n f_i,$$

where $H(f_1, \ldots, f_n) = -\sum_i f_i \log f_i$. Introducing the second constraint $\sum_i f_i = t$, we obtain a linear relation between $H(f_1, \ldots, f_n)$ and $H(p_1, \ldots, p_n)$ enabling us to treat $H(f_1, \ldots, f_n)$ instead of $H(p_1, \ldots, p_n)$, the first one being much more tractable. Anyway, this constraint is not very strict, and it may be varied to obtain any required total intensity.

From the relation between $H(p_1, \ldots, p_n)$ and $H(f_1, \ldots, f_n)$ it is easily seen that the following three problems are equivalent to each other:

Problem 1. $\max H(p_1, \ldots, p_n)$

subject to $\quad Q = m/2, \quad \sum_i f_i = t.$

Problem 2. $\max H(f_1, \ldots, f_n)$

subject to $\quad Q = m/2, \quad \sum_i f_i = t.$

Problem 3. $\max[H(f_1, \ldots, f_n) + \mu \sum_i f_i - \lambda Q(f_1, \ldots, f_n)]$

subject to $\quad Q = m/2, \quad \sum_i f_i = t.$

Instead of solving problem 3, we solve the following problem:

Problem 4. $\max H[(f_1, \ldots, f_n) + \mu \sum_i f_i - \lambda Q(f_1, \ldots, f_n)]$

subject to $\quad Q = m/2.$

In problem 4 the total intensity is treated like a "free boundary condition." No specific value is assigned to it in advance. In fact, problem 4 is a very reasonable formulation of the maximum entropy image reconstruction problem. If $f_1^*, \ldots, f_n^*$ is a solution of problem 4, then it is also a solution of problems 1, 2, 3, with $t = \sum_i f_i^*$. Usually we are satisfied with the solution of problem 4. If not, we could adjust the parameter μ so that the solution of problem 4 has the required total intensity (see subsection F).

To solve problem 4 we need to consider the related unconstrained maximization problem.

Problem 5. $\max[H(f_1, \ldots, f_n) + \mu \sum_i f_i - \lambda Q(f_1, \ldots, f_n)].$ (5.4-5)

Let

$$J(f_1, \ldots, f_n; \mu, \lambda) = H(f_1, \ldots, f_n) + \mu \sum_i f_i - \lambda Q(f_i, \ldots, f_n)$$

Then problem 5 is to find maximal points of $J(f_1, \ldots, f_n; \mu, \lambda)$. The function $J(f; \mu, \lambda)$ is defined and continuous in the closed domain $\bar{E}$, where $E \triangleq \{f_i | f_i > 0, i = 1, \ldots, n\}$. When $\lambda \geqslant 0$ the continuous function J has a unique maximal point, which is finite since J tends to minus infinity as $\|\mathbf{f}\| \to \infty$ and is strictly concave due to the negative definiteness of its Jacobian $\nabla^2 J$:

$$\nabla^2 J = -\mathrm{diag}[1/f_1, \ldots, 1/f_n] - \lambda \mathbf{A}^T \mathbf{D} \mathbf{A} < 0. \tag{5.4-6}$$

As known, if the maximal point $\mathbf{f}^0$ is an inner point, i.e., $\mathbf{f}^0 \in E$, then the following stationary point equations hold:

$$\nabla J = -[\log f_1^0, \ldots, \log f_n^0]^T + (\mu - 1)\mathbf{h} - \lambda \mathbf{a}^T \mathbf{D}(\mathbf{A}\mathbf{f}^0 - \mathbf{d}) = 0, \tag{5.4-7}$$

where for abbreviation $\mathbf{h} \triangleq [1, \ldots, 1]_{1\times n}^T$.

Conversely, if ∇J equals zero at a point $\mathbf{f}^0 \in E$ as $\lambda \geqslant 0$, then $\mathbf{f}^0$ must be a maximal point, as easily seen from Eq. (5.4-6).

In the following we will prove that Eq. (5.4-7) is always solvable as $\lambda \geqslant 0$. Thus, problem 5 is actually equivalent to Eq. (5.4-7) as $\lambda \geqslant 0$. The solution of Eq. (5.4-7) gives that unique maximal point required by problem 5 as $\lambda \geqslant 0$.

To prove that Eq. (5.4-7) is always (uniquely) solvable as $\lambda \geqslant 0$, all we need is to find such a path $\mathbf{f}(\lambda; \mu)$ in E, defined for $0 \leqslant \lambda < \infty$, that the gradient ∇J is identically equal to zero along it:

$$\nabla J(\mathbf{f}(\lambda; \mu); \mu, \lambda) \equiv 0, \qquad 0 \leqslant \lambda < \infty. \tag{5.4-8}$$

It is clear that Eq. (5.4-8) is equivalent to the following Eqs. (5.4-9) and (5.4-10).

$$\frac{d}{d\lambda} \nabla J(f(\lambda; \mu); \mu, \lambda) \equiv 0, \qquad 0 \leqslant \lambda < \infty, \tag{5.4-9}$$

$$\nabla J(\mathbf{f}(0; \mu); \mu, 0) = 0. \tag{5.4-10}$$

It is also clear that Eq. (5.4-9) is equivalent to the following differential equations for $\mathbf{f}(\lambda; \mu)$ (by working out the derivative $d\nabla J/d\lambda$):

$$\nabla^2 J(\mathbf{f}; \mu, \lambda)\, d\mathbf{f}/d\lambda = \partial Q(\mathbf{f}), \qquad 0 \leqslant \lambda < \infty, \tag{5.4-11}$$

and Eq. (5.4-10) directly gives $\mathbf{f}(0; \mu)$ as

$$\mathbf{f}(0; \mu) = \exp(\mu - 1)\mathbf{h}, \tag{5.4-12}$$

where $\mathbf{h}$ equals $[1, \ldots, 1]^T_{1\times n}$.

Thus, the stationary point equation is always (uniquely) solvable as $\lambda \geqslant 0$ if the following Cauchy problem of differential equations defines a solution curve in E for $0 \leqslant \lambda < \infty$:

$$\begin{aligned} \nabla^2 J\, d\mathbf{f}/d\lambda &= \nabla Q, \\ \mathbf{f}|_{\lambda=0} &= \exp(\mu - 1)\mathbf{h}. \end{aligned} \tag{5.4-13}$$

This is proved in Theorem 1 of the Appendix. Having shown that the unconstrained maximization problem 5 has the form of Eq. (5.4-13), we are ready to consider the variety of computing schemes available to solve it.

To solve the maximum entropy image reconstruction problem, the only thing left is to select suitable initial conditions for the Cauchy problem, Eq. (5.4-13), in order for the constraint $Q = m/2$ to be reached by the solution curve. For this we need to know more about the solution curve of Eq. (5.4-13). We discuss this in the next section.

B. Properties of Eq. (5.4-13)

In this subsection we define the sets giving the μ's for which a maximum entropy solution satisfying the constraint $Q = m/2$ can be found. Naturally, we are interested in the behavior of $Q(\mathbf{f}(\lambda; \mu))$ along the solution curve.

It is easy to see that

$$\begin{aligned} dQ(\mathbf{f}(\lambda;\mu))/d\lambda &= (\nabla Q)^T d\mathbf{f}/d\lambda \\ &\equiv (\nabla Q)^T(\nabla^2 J)^{-1}(\nabla Q). \end{aligned} \tag{5.4-14}$$

In the following we prove that $Q(\mathbf{f}(\lambda;\mu))$ decreases strictly monotonically as λ increases unless the solution curve $\mathbf{f}(\lambda;\mu)$ shrinks into a single point $\exp(\mu-1)\mathbf{h}$. As a matter of fact, along the solution curve there are only two possibilities: either

$$\mathbf{f}(\lambda;\mu) \equiv \exp(\mu-1)\mathbf{h} \qquad \text{and} \qquad \nabla Q(\exp(\mu-1)\mathbf{h}) = 0, \tag{5.4-15}$$

or

$$\nabla Q(\mathbf{f}(\lambda;\mu)) \neq 0, \qquad 0 \leqslant \lambda < \infty. \tag{5.4-16}$$

Thus, when $\mathbf{f}(\lambda;\mu)$ does not shrink into a single point, the derivative $dQ(\mathbf{f}(\lambda;\mu))/d\lambda$ will always be negative because of (5.4-14), (5.4-16), and $(\nabla^2 J)^{-1} < 0$. As for (5.4-15) and (5.4-16) we can proceed as follows.

Suppose there exists a point $\mathbf{f}^* = \mathbf{f}(\lambda^*;\mu)$ along the solution curve such that

$$\nabla Q(\mathbf{f}^*) = 0. \tag{5.4-17}$$

Then $\mathbf{f}(\lambda;\mu)$ will coincide with the solution curve defined by the following Cauchy problem due to the uniqueness of the solution to

$$\begin{aligned} \nabla^2 J\, d\mathbf{f}/d\lambda &= \nabla Q, \\ \mathbf{f}|_{\lambda=\lambda^*} &= \mathbf{f}^*. \end{aligned} \tag{5.4-18}$$

However, the solution curve of Eq. (5.4-18) obviously is a single point $\mathbf{f}^*$ due to Eq. (5.4-17) and the initial condition in Eq. (5.4-18). thus it holds

$$\mathbf{f}(\lambda;\mu) \equiv \mathbf{f}^*, \qquad 0 \leqslant \lambda < \infty. \tag{5.4-19}$$

Now it is not hard to see that Eq. (5.4-15) is valid since $\mathbf{f}(0;\mu) = \exp(\mu-1)\mathbf{h}$. Thus, we have proved

Theorem 5.4-1. The derivative $dQ(\mathbf{f}(\lambda;\mu))/d\lambda$ is always negative unless $\nabla Q(\exp(\mu-1)\mathbf{h}) = 0$.

In order for the solution curve $\mathbf{f}(\lambda;\mu)$ to satisfy the constraint $Q = m/2$, only those μ's need be considered which satisfy either

$$Q(\exp((\mu-1)\mathbf{h}) = m/2 \tag{5.4-20}$$

or

$$Q(\exp(\mu-1)\mathbf{h}) > m/2 \qquad \text{and} \qquad \nabla Q(\exp(\mu-1)\mathbf{h}) \neq 0. \tag{5.4-21}$$

If Eq. (5.4-20) holds, then $\exp(\mu - 1)\mathbf{h}$ solves problem 4. Otherwise, we need to select suitable μ's by means of Eq. (5.4-21) and then compute $\mathbf{f}(\lambda; \mu)$ by means of Eq. (5.4-13).

Let the sets U, V, W be defined as follows:

$$U = \{\mu : Q(\exp(\mu - 1)\mathbf{h}) = m/2\}, \tag{5.4-22}$$

$$V = \{\mu : Q(\exp(\mu - 1)\mathbf{h}) > m/2\}, \tag{5.4-23}$$

$$W = \{\mu : \nabla Q(\exp(\mu - 1)\mathbf{h}) \neq 0\}. \tag{5.4-24}$$

From the previous discussion it is clear that only those μ's which belong to U or $V \cap W$ can produce meaningful results.

C. Computing the Sets U, V, and W

In this subsection, we will investigate further the sets U, V, and W in order to determine initial values for which a maximum entropy solution satisfying the constraint can be found.

For abbreviation we let

$$\alpha = \exp(\mu - 1), \tag{5.4-25}$$

$$\mathbf{r} = \left[\sum_{i=1}^{n} A_{1i}/\sigma_1, \ldots, \sum_{i=1}^{n} A_{mi}/\sigma_m\right]^T, \tag{5.4-26}$$

$$\mathbf{S} = [d_1/\sigma_1, \ldots, d_m/\sigma_m]^T. \tag{5.4-27}$$

Then a direct computation yields

$$Q(\alpha\mathbf{h}) = a\alpha^2 + b\alpha + c, \tag{5.4-28}$$

where three coefficients a, b, c are as follows:

$$a = \tfrac{1}{2}\|\mathbf{r}\|^2 \tag{5.4-29}$$

$$b = -(\mathbf{r}, \mathbf{s}), \tag{5.4-30}$$

$$c = \tfrac{1}{2}\|\mathbf{s}\|^2. \tag{5.4-31}$$

By using the Schwartz inequality it is easily verified that $b^2 \leqslant 4ac$. In the following we prove that the expectation $E\{c\} \geqslant m/2$ and $(1/m)(c - \frac{1}{2}\|g\|^2) \to \frac{1}{2}$ with probability one as $m \to \infty$, where $\mathbf{g} \triangleq [\sum_{i=1}^{n} A_{1i}f_i/\sigma_1, \ldots, \sum_{i=1}^{m} A_{mi}f_i/\sigma_m]^T$ and the appearing f_i's are true image intensity values.

Let

$$\mathbf{u} = [e_1/\sigma_1, \ldots, e_m/\sigma_m]^T, \tag{5.4-32}$$

where the e_j's represent independent zero mean, σ_j^2 variance noise terms, as known in subsection A. From Eq. (5.4-2) we have

$$\mathbf{s} = \mathbf{g} + \mathbf{u} \tag{5.4-33}$$

and hence

$$\begin{aligned} E\{c\} &= \tfrac{1}{2}E\{\|\mathbf{g} + \mathbf{u}\|^2\} \\ &= \tfrac{1}{2}\|\mathbf{g}\|^2 + (\mathbf{g}, E\{\mathbf{u}\}) + \tfrac{1}{2}E\{\|\mathbf{u}\|^2\} \\ &= \tfrac{1}{2}\|\mathbf{g}\|^2 + m/2 \geqslant m/2. \end{aligned} \tag{5.4-34}$$

From Eq. (5.4-34) it is clear that $E\{c\} = m/2$ if and only if $\mathbf{g} = 0$.

As proved in Theorem 2 in the Appendix, if $\|g\|/m \to 0$ and $m \to \infty$, then with probability one

$$(1/m)(g, u) \to 0 \qquad \text{as } m \to \infty \tag{5.4-35}$$

and

$$(1/m)(c - \tfrac{1}{2}\|g\|^2) \to 0 \qquad \text{as } m \to \infty. \tag{5.4-36}$$

On the basis of this fact, if m is large and $\|g\|/m$ is small, then c can be rather safely replaced by $m/2 + \|g|^2/2$, which is never less than $m/2$. Hence in computing the sets U, V, and W, we assume $c \geqslant m/2$. In particular, we are interested in determining a nonempty set u or in determining the set intersection $V \cap W$ should $U = \varnothing$.

1. *Computing the Set U*

According to Eqs. (5.4-22), (5.4-25), and (5.4-28) the set U is determined as

$$U = \{1 + \log\alpha : a\alpha^2 + b\alpha + (c - m/2) = 0 \text{ with } \alpha > 0\}. \tag{5.4-37}$$

Case 1. $a > 0$, $b^2 > 4a(c - m/2)$, $b < 0$:

$$U = \begin{cases} \{1 + \log[(-b \pm \sqrt{b^2 - 4a(c - m/2)})/(2a)\}, & c > m/2, \\ \{1 + \log(-b/a)\}, & c = m/2. \end{cases} \tag{5.4-38}$$

Case 2. $a = 0$ ($\Rightarrow b = 0$), $c = m/2$:

$$U = \{(\infty, -\infty)\}. \tag{5.4-39}$$

In all other cases $U = \varnothing$, an empty set. Here the second case occurs very rarely. Thus in general U is empty or has at most two elements. This corresponds to no equientropy solution or at most two equientropy solutions for problem 4, respectively.

2. Computing the Set V When U Is Empty

The set V is determined as

$$V = \{1 + \log \alpha : a\alpha^2 + b\alpha + (c - m/2) > 0 \text{ with } \alpha > 0\}. \quad (5.4\text{-}40)$$

Since the set U is empty, the following cases exhaust all possibilities:

Case 1. $a > 0,\ b^2 - 4a(c - m/2) < 0.$
Case 2. $a > 0,\ b^2 - 4a(c - m/2) \geq 0,\ b \geq 0.$
Case 3. $a = 0\ (\Rightarrow b = 0),\ c > m/2.$

It is easy to see that all three cases lead to the same V:

$$V = \{(-\infty, \infty)\}. \quad (5.4\text{-}41)$$

3. Computing the Set W

The set W is determined as:

$$W = \{1 + \log \alpha : \alpha(\mathbf{A}^T\mathbf{DAh}) \neq \mathbf{A}^T\mathbf{Dd} \text{ with } \alpha > 0\}. \quad (5.4\text{-}42)$$

From Eq. (5.4-42) it is clear that W equals $\{(-\infty, \infty)\}$ unless the $(n \times 1)$-vector $\mathbf{A}^T\mathbf{Dd}$ has the same direction as the $(n \times 1)$-vector $\mathbf{A}^T\mathbf{DAh}$. The exceptional case occurs very rarely. Thus in general there holds

$$W = \{(-\infty, \infty)\}. \quad (5.4\text{-}43)$$

In summary, there are only two possibilities in general. Either U is not empty and thus problem 4 has at most two equientropy solutions. Or $V \cap W = \{(-\infty, \infty)\}$ and thus for any μ there hold

$$Q(\exp(\mu - 1)\mathbf{h} > m/2, \quad (5.4\text{-}44)$$

$$\Delta Q(\exp(\mu - 1)\mathbf{h}) \neq 0. \quad (5.4\text{-}45)$$

As known from Theorem 5.4-1, Eq. (5.4-45) assures

$$dQ(\mathbf{f}(\lambda; \mu))/d\lambda < 0 \qquad \text{as } \lambda \geq 0, \quad (5.4\text{-}46)$$

where $\mathbf{f}(\lambda; \mu)$ is the unique solution of the Cauchy problem, Eq. (5.4-13). Intuitively, from Eq. (5.4-44) we would like to choose such a μ that $Q(\exp(\mu - 1)\mathbf{h})$ is as close to $m/2$ as possible. When $-b/2a > 0$, we can simply choose that

$$\mu = 1 + \log(-b/2a), \quad (5.4\text{-}47)$$

because of $[dQ(\alpha\mathbf{h})/d\alpha]_{\alpha=-b/2a} = 0$.

From Eqs. (5.4-26), (5.4-27) and (5.4-29), (5.4-30) it is apparent that the condition, $-b/2a > 0$, is equivalent to $(r, s) > 0$. Notice that d_j and f_i are related by

$$d_j = \sum_{i=1}^{n} A_{ji} f_i + e_j, \qquad j = 1, \ldots, m,$$

where all f_i are nonnegative, and the matrix $(A_{ji})_{n \times m}$ represents the blurring operation which occurred in the imaging process. The latter implies $A_{ji} \geqslant 0$ in most practical cases. Thus, $(\mathbf{r}, \mathbf{s}) > 0$ will hold whenever the noise is not too large.

D. Algorithm for Solving Differential Equations

In this subsection the algorithm for solving the system of differential equations (5.4-13) is presented. As mentioned in the introduction, we do a search along the solution curve by varying the parameter λ, starting with $\lambda_0 = 0$. A satisfactory solution is found when $Q(\mathbf{f}(\lambda; \mu))/(m/2)$ is sufficiently near 1. For each λ a large system of linear equations must be solved. This is done by making extensive use of the properties of the coefficient matrix and applying the Gauss–Seidel algorithm [7], which is an iterative technique providing fast convergence.

Suppose that the U is empty, i.e., problem 4 has no equientropy solution, and moreover an appropriate μ has been chosen from the intersection $V \cap W$. From differential equations (5.4-13) we have the following discretized iterative equations:

$$\begin{cases} \mathbf{f}^0 = \exp(\mu - 1)\cdot\mathbf{h}, \\ k \geqslant 0: \quad (\mathbf{F}^k + \lambda_k \mathbf{A}^T\mathbf{D}\mathbf{A})(\mathbf{f}^{k+1} - \mathbf{f}^k) = -(\lambda_{k+1} - \lambda_k)\mathbf{A}^T\mathbf{D}(\mathbf{A}\mathbf{f}^k - \mathbf{d}), \end{cases} \tag{5.4-48}$$

where $\mathbf{F}^k = \operatorname{diag}[1/f_1^k, \ldots, 1/f_n^k]$, $\lambda_0 = 0$, and $\lambda_{k+1} > \lambda_k$. When all $|\lambda_{k+1} - \lambda_k|$ are small enough, $\mathbf{f}^k$ will approximate $\mathbf{f}(\lambda_k; \mu)$ very well. Equation (5.4-48) can be written as follows:

$$\begin{cases} \mathbf{f}^0 = \exp(\mu - 1)\mathbf{h}, \\ k \geqslant 0: \quad (\mathbf{F}^k + \lambda_k \mathbf{A}^T\mathbf{D}\mathbf{A})\mathbf{f}^{k+1} = (2\lambda_k - \lambda_{k+1})\mathbf{A}^T\mathbf{D}\mathbf{A}\mathbf{f}^k \\ \qquad\qquad + \mathbf{h} + (\lambda_{k+1} - \lambda_k)\mathbf{A}^T\mathbf{D}\mathbf{d}. \end{cases} \tag{5.4-49}$$

This is a large linear system of equations. Fortunately, the coefficient matrix $(\mathbf{F}^k + \lambda_k \mathbf{A}^T\mathbf{D}\mathbf{A})$ is positive definite and symmetric. The Gauss–Seidel iterative scheme (see [7]) can then be used to solve Eq. (5.4-49) efficiently.

Let $\mathbf{P}$ represent the matrix $(\mathbf{F}^k + \lambda_k \mathbf{A}^T\mathbf{D}\mathbf{A})$. Let $\mathbf{b}$ represent the vector $(2\lambda_k - \lambda_{k+1})\mathbf{A}^T\mathbf{D}\mathbf{A}\mathbf{f}^k + \mathbf{h} + (\lambda_{k+1} - \lambda_k)\mathbf{A}^T\mathbf{D}\mathbf{d}$. Further let $\mathbf{x}$ be the $(k+1)$st estimate $\mathbf{f}^{k+1}$ of $\mathbf{f}(\lambda_{k+1}; \mu)$. For $m \geqslant 1$ we have the following Gauss–Seidel iterative equations for solving $\mathbf{Px} = \mathbf{b}$:

$$\begin{cases} \mathbf{x}^{(0)} = \mathbf{f}^k, \\ x_i^{(m)} = \dfrac{1}{P_{ii}}\left[v_i - \sum_{j<i} P_{ij}x_j^{(m)} - \sum_{j>i} P_{ij}x_j^{(m-1)}\right], \qquad i = 1, \ldots, n. \end{cases} \tag{5.4-50}$$

It is proved (see [7]) that the convergence and the limit point are independent of the choice of $\mathbf{x}^{(0)}$. Thus $\lim \mathbf{x}^{(m)} = \mathbf{f}^{k+1}$. We choose $\mathbf{x}^{(0)} = \mathbf{f}^k$ since $\mathbf{f}^{k+1}$ should be near $\mathbf{f}^k$ when $|\lambda_{k+1} - \lambda_k|$ is small. This choice of $\mathbf{x}^{(0)}$ greatly speeds up the convergence of (5.4-50). Experiments indicate that in most cases only one iteration for each k is enough to obtain $\mathbf{f}^{k+1}$ from $\mathbf{f}^k$.

In the sequel we let

$$\begin{cases} \mathbf{L} = \mathbf{A}^T\mathbf{D}\mathbf{A}, \\ \mathbf{p} = \mathbf{A}^T\mathbf{D}\mathbf{d}, \\ \mathbf{g}^k = \mathbf{L}\mathbf{f}^k, \qquad k = 0, 1, 2, \ldots. \end{cases} \tag{5.4-51}$$

Then from Eq. (5.4-49) it follows

$$k \geqslant 0: \qquad (\mathbf{F}^k + \lambda_k\mathbf{f}^{k+1} = (2\lambda_k - \lambda_{k+1})\mathbf{g}^k + \mathbf{h} + (\lambda_{k+1} - \lambda_k)\mathbf{p}, \tag{5.4-52}$$

$$k \geqslant 1: \qquad \mathbf{g}^{k+1} = \left(2 - \frac{\lambda_{k+1}}{\lambda_k}\right)\mathbf{g}^k + \frac{1}{\lambda_k}\left[\frac{f_1^k - f_1^{k+1}}{f_1^k}, \ldots, \frac{f_n^k - f_n^{k+1}}{f_n^k}\right]^T + \left(\frac{\lambda_{k+1}}{\lambda_k} - 1\right)\mathbf{p}. \tag{5.4-53}$$

The initial values $\mathbf{f}^0$, $\mathbf{g}^0$, $\mathbf{g}^1$ for Eqs. (5.4-52)-(5.4-53) are as follows:

$$\begin{aligned} \mathbf{f}^0 &= \exp(\nu - 1)\mathbf{h}, \\ \mathbf{g}^0 &= \mathbf{L}\mathbf{f}^0 = \exp(\mu - 1)\left[\sum_{i=1}^n L_{1i}, \ldots, \sum_{i=1}^n L_{ni}\right]^T, \\ \mathbf{g}^1 &= \mathbf{L}\mathbf{f}^1. \end{aligned} \tag{5.4-54}$$

Thus Eq. (5.4-54), which involves matrix calculations, is used to find $\mathbf{g}^0$, $\mathbf{g}^1$, and the recursive formula (5.4-53) involving only vector calculations is used for $\mathbf{g}^k$, $k \geqslant 2$.

Finally we summarize this subsection by giving the following easily implemented algorithm for solving the maximum entropy image reconstruction problem.

Step 1. If $U \neq \varnothing$ and $\mu \in U$, then there is an equientropy solution $\mathbf{f}^* = \exp(\mu - 1)\mathbf{h}$. Stop.

Step 2. Choose an appropriate $\mu \in V \cap W$ such that $Q(\exp(\mu - 1)\mathbf{h})$ is as small as possible.

Step 3. Set $k = 0$, $\lambda_k = 0$, $\mathbf{f}^k = \exp(\mu - 1)\mathbf{h}$, $\mathbf{g}^k = \mathbf{L}\mathbf{f}^k$, and $Q^k = Q(\mathbf{f}^k)$.

Step 4. If $|Q^k/(m/2) - 1| \leqslant \varepsilon$, then Stop. Here ε is a prescribed small positive number.

Step 5. Solve Eq. (5.4-49) for $\mathbf{f}^{k+1}$ by using Gauss–Seidel iterative scheme, Eq. (5.4-50).

Step 6. Compute $Q^{k+1} = Q(\mathbf{f}^{k+1})$.

Step 7. If k is large enough, Stop.

Step 8. If $|Q^{k+1}/(m/2) - 1| \leqslant \varepsilon$, then Stop.

Step 9. If $k \geqslant 1$, then compute $\mathbf{g}^{k+1}$ by Eq. (5.4-53), else $\mathbf{g}^{k+1} = \mathbf{L}\mathbf{f}^{k+1}$.

Step 10. If $Q^{k+1} > m/2$, then set $\delta\lambda > 0$, else set $\delta\lambda < 0$.

Step 11. Set $k = k + 1$, $\lambda_k = \lambda_{k-1} + \delta\lambda$.

Step 12. Goto Step 5.

In the experiments we consider only cases where the blurring given by the matrix **A** can be represented as a convolution between the undergraded image and a (regular) mask. By making full use of the fact that very few of the elements in the matrix **L** are nonzero for moderate mask sizes, it has been possible to construct a fast algorithm without resorting to Fourier transforms. If the blurring happens in the imaging processes, then the mask usually is small. For very large masks, however, the situation is different, and the possibility of saving time by use of Fourier transforms should be investigated. Equation (5.4-49), however, does not have a convolution form, so implementation of Fourier transforms is not straightforward.

E. Adjusting μ to Get the Required Total Intensity

Assume that the set U is empty and we have parameters μ_0 and λ_0 such that $\lambda_0 \geqslant 0$ and $|Q(\mathbf{f}(\lambda_0; \mu_0))/(m/2) - 1| \leqslant \varepsilon$. We are looking for a differentiable curve $\lambda = \lambda(u)$ passsing through (μ_0, λ_0), along which the Q values, $Q(\mathbf{f}(\lambda(\mu); \mu))$, are constant. In this case, adjusting the parameter μ will not affect the satisfaction of the constraint $|Q/(m/2) - 1| \leqslant \varepsilon$. Now along the required curve we should have two relations:

$$dQ(\mathbf{f}(\lambda(\mu); \mu))/d\mu = 0, \tag{5.4-55}$$

$$\nabla J(\mathbf{f}(\lambda(\mu); \mu); \mu, \lambda(\mu)) = 0, \tag{5.4-56}$$

the latter because $f(\lambda(\mu); \mu)$ is the stationary point of the function $J(\mathbf{f}; \mu, \lambda(\mu))$ as known before.

From (5.4-56) it is easy to derive that

$$\nabla^2 J(\mathbf{f}(\lambda(\mu); \mu); \mu, \lambda(\mu)) \frac{d\mathbf{f}(\lambda(\mu); \mu)}{d\mu} + h - \frac{d\lambda(\mu)}{d\mu} \nabla Q(\mathbf{f}(\lambda(\mu); \mu)) = 0$$

or

$$\frac{d\mathbf{f}(\lambda(\mu);\mu)}{d\mu} \equiv [\nabla^2 J(\mathbf{f}(\lambda(\mu);\mu))]^{-1}\left[\frac{d\lambda(\mu)}{d\mu}\nabla Q(\mathbf{f}(\lambda(\mu);\mu)) - h\right]. \tag{5.4-57}$$

Under the assumption $U = \varnothing$, we have that

$$V \cap W = \{(-\infty, \infty)\}, \tag{5.4-58}$$

which implies that

$$\nabla Q(\exp(\mu - 1))\mathbf{h} \neq 0. \tag{5.4-59}$$

As stated before [see Eqs. (5.4-15) and (5.4-16)], Eq. (5.4-56) implies that

$$\nabla Q(\mathbf{f}(\lambda(\mu);\mu)) \neq 0 \tag{5.4-60}$$

everywhere.

From (5.4-55) we have

$$[\nabla Q(\mathbf{f}(\lambda(\mu);\mu))]^T d\mathbf{f}(\lambda(\mu);\mu)/d\mu \equiv 0. \tag{5.4-61}$$

Combining (5.4-59), (5.4-60), and (5.4-61), we obtain

$$d\lambda(\mu)/d\mu \equiv [\nabla Q]^T[\nabla^2 J]^{-1}\mathbf{h}/[\nabla Q]^T[\nabla^2 J]^{-1}[\nabla Q], \tag{5.4-62}$$

where the differential curve passes through (μ_0, λ_0), i.e., $\lambda(\mu_0) = \lambda_0$.

Then, multiplying both sides of (5.4-57) by $\mathbf{h}^T$, it follows

$$\begin{aligned}\frac{d\sum_{i=1}^n f_i(\lambda(\mu);\mu)}{d\mu} &\equiv \mathbf{h}^T[\nabla^2 J]^{-1}\nabla Q\cdot\frac{d\lambda}{d\mu} - \mathbf{h}^T[\nabla^2 J]^{-1}\mathbf{h}\\ &\equiv \frac{([\nabla Q]^T[\nabla^2 J]^{-1}\mathbf{h})^2}{[\nabla Q]^T[\nabla^2 J]^{-1}[\nabla Q]} - \mathbf{h}^T[\nabla^2 J]^{-1}\mathbf{h}.\end{aligned} \tag{5.4-63}$$

Finally, we obtain the following Cauchy problem of $n+2$ differential equations for adjusting μ to ge the required total intensity:

$$\begin{aligned}[\nabla^2 J]\frac{d\mathbf{f}(\lambda(\mu);\mu)}{d\mu} &= \frac{d\lambda(\mu)}{d\mu}[\nabla Q] - \mathbf{h},\\ \frac{d\lambda(\mu)}{d\mu} &= \frac{[\nabla Q]^T[\nabla^2 J]^{-1}\mathbf{h}}{[\nabla Q]^T[\nabla^2 J]^{-1}[\nabla Q]},\\ \frac{d\sum_{i=1}^n f_i(\lambda(\mu);\mu)}{d\mu} &= \frac{([\nabla Q]^T[\nabla^2 J]^{-1}\mathbf{h})^2}{[\nabla Q]^T[\nabla^2 J]^{-1}[\nabla Q]} - \mathbf{h}^T[\nabla^2 J]^{-1}\mathbf{h},\\ \mathbf{f}|_{\mu=\mu_0} &= \mathbf{f}(\lambda_0, \mu_0),\\ \lambda|_{\mu=\mu_0} &= \lambda_0,\\ \sum_{i=1}^n f_i|_{\mu=\mu_0} &= \mathbf{h}^T\mathbf{f}(\lambda_0;\mu_0).\end{aligned} \tag{5.4-64}$$

The iterative scheme for solving (5.4-64) is as follows.

Given: μ_0, λ_0, $\mathbf{f}^0 = \mathbf{f}(\lambda_0; \mu_0)$, $t_0 = \mathbf{h}^T\mathbf{f}_0$, and $k \geqslant 0$:

$$\frac{\lambda_{k+1} - \lambda_k}{\mu_{k+1} - \mu_k} = \frac{[\nabla Q(\mathbf{f}^k)]^T[\nabla^2 J(\mathbf{f}^k; \mu_k, \lambda_k)]^{-1}\mathbf{h}}{[\nabla Q(\mathbf{f}^k)]^T[\nabla^2 J(\mathbf{f}^k; \mu_k, \lambda_k)]^{-1}[\nabla Q(\mathbf{f}^k)]},$$

$$[\nabla^2 J](\mathbf{f}^k; \mu_k, \lambda_k)](\mathbf{f}^{k+1} - \mathbf{f}^k) = (\lambda_{k+1} - \lambda_k)[\nabla Q(\mathbf{f}^k)] - (\mu_{k+1} - \mu_k)\mathbf{h},$$

$$\frac{t_{k+1} - t_k}{\mu_{k+1} - \mu_k} = \frac{([\nabla Q(\mathbf{f}^k)]^T[\nabla^2 J(\mathbf{f}^k; \mu_k, \lambda_k)]^{-1}\mathbf{h})^2}{[\nabla Q(\mathbf{f}^k)]^T[\nabla^2 J(\mathbf{f}^k; \mu_k, \lambda_k)]^{-1}[\nabla Q(\mathbf{f}^k)]} - \mathbf{h}^T[\nabla^2 J(\mathbf{f}^k; \mu_k, \lambda_k)]^{-1}\mathbf{h}. \quad (5.4\text{-}65)$$

F. Experiments

Three source images were used in the experiments. The first one (Fig. 5.4-1) is a picture of a girl. The image size is 128×128 pixels, and the

Fig. 5.4-1. The 128×128 source image of a girl's face.

resolution in gray level is 8 bits. Five different degraded images were produced from this image. These images, which are input images for the algorithm, will be referred to as image 1 through image 5. Image 1 and image 2 were created by first convolving the source image with a 5×5 box filter (all kernel elements equal to unit). Random noise with zero mean and standard deviations of 4 and 40, respectively, was added to the blurred image. This corresponds to signal-to-noise ratio (S/N) of 1000 and 10, respectively, for the resulting images. To get a stronger blurring, this source image was convolved with a 7×7 mask with the center 5×5 elements all equal to zero and the 24 border elements equal to unity. Noise with zero mean and standard deviations of 4 and 40, respectively, was added to the blurred image to produce image 3 and image 4. To test the ability of the algorithm to reconstruct an image when we have incomplete measurements, image 5 was created. It is identical to image 1, but the gray level is known only for every tenth pixel. For convenience that is done in a regular manner, producing the line pattern which can be seen in Fig. 5.4-2, where the five input images are shown. Data for these images are listed in Table 5.4-1.

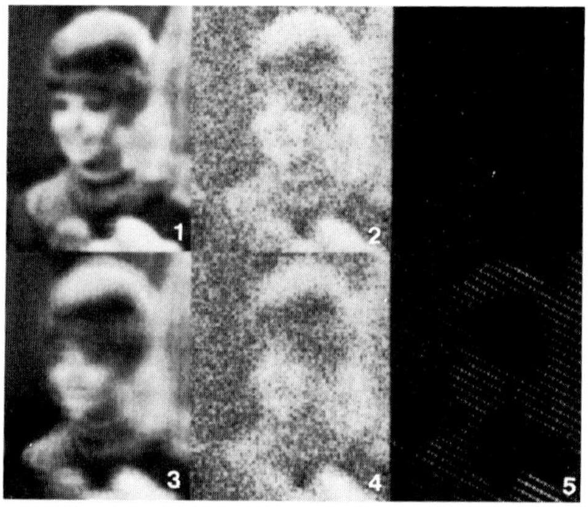

Fig. 5.4.-2. Five degraded images. Images 1 to 4 show the source image of Fig. 5.4-1; (1) blurred with a 5×5 box filter and having additive Gaussian noise of 4; (2) blurred with a 5×5 box filter and having additive Gaussian noise of 40; (3) blurred with a 7×7 box filter and having additive Gaussian noise of 4; (4) blurred with a 7×7 box filter and having additive Gaussian noise of 40. Image 5 is image 1 sampled every tenth pixel by row and every tenth pixel by column.

Table 5.4-1
Input Images Created from Girl Source Image

Image	Size of mask	Standard deviations of noise	(S/N)	Sampling interval
Image 1	5×5	4.0	1000	1
Image 2	5×5	40.0	10	1
Image 3	7×7	4.0	1000	1
Image 4	7×7	40.0	10	1
Image 5	5×5	4.0	1000	10

The second source image (Fig. 5.4-3a) contains text and has size 480 columns by 100 lines. It was blurred by a 7×7 Gaussian filter with standard deviation of 5. Noise with a standard deviation of 4 was added to produce the input image shown in Fig. 5.4-3b. The third source image (Fig. 5.4-4a) is a 128×128 checkerboard image. The corresponding input image was produced by the same degradation process as was applied to the text image.

As stated by Eq. (5.4-25), the initial value for the solution vector $\mathbf{f}$ is $\mathbf{f}^0 = \alpha\mathbf{h}$, where $\alpha = \exp(\mu - 1)$. For fastest possible convergence α is chosen so that $Q(\alpha\mathbf{h})$ is the smallest possible (provided $U = \emptyset$); that is, we choose $\alpha > 0$ so that $dQ(\alpha, \ldots, \alpha)/d\alpha = 0$ [see Eq. (5.4-20)].

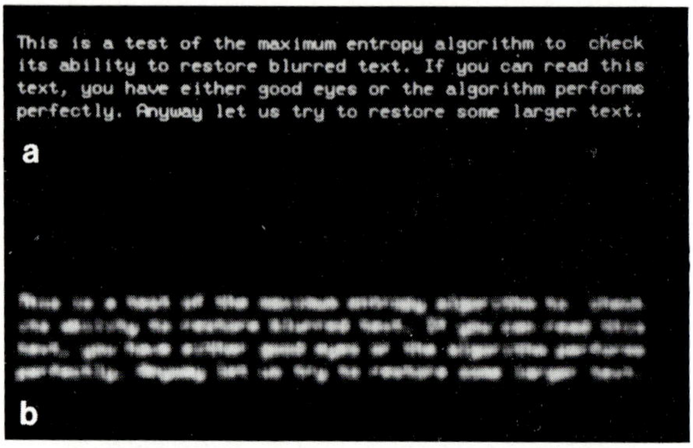

Fig. 5.4-3. (a) A 100×480 text source image. (b) The text source image blurred by a 7×7 Gaussian filter having a standard deviation of 5 and additive Gaussian noise with standard deviation 4.

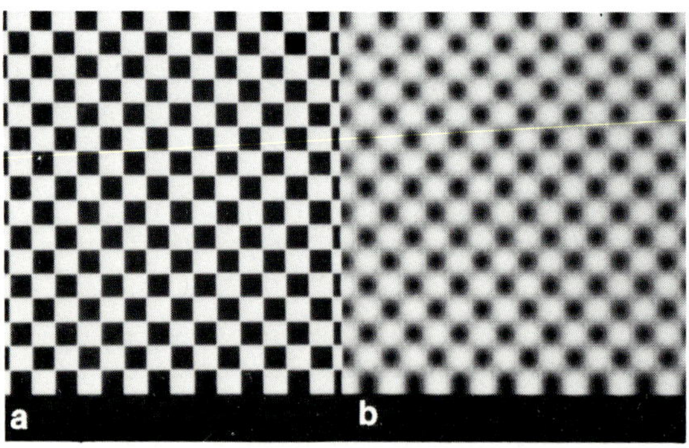

Fig. 5.4.-4. (a) A 128×128 checkerboard image. (b) The checkerboard image blurred by a 7×7 Gaussian filter having a standard deviation of 5 and additive Gaussian noise with standard deviation 4.

For each reconstruction the algorithm was run until the resulting Q was sufficiently near $Q = m/2$ (m is the number of data points, i.e., pixels). The number of iterations for each λ and step size in λ, $\delta\lambda = |\lambda_{k+1} - \lambda_k|$, were determined empirically, and the chosen values represent a reasonable, but in no way the best, choice. The optimal choice of $\delta\lambda$ also varies strongly with noise level. If the relative reduction in Q in one step is below a user-specified limit, the step size is doubled for the next step. This can save a great deal of time compared to keeping $\delta\lambda$ constant for a whole search. Indeed, in some cases it is quite necessary in order to be able to reach a solution in a reasonable time. One Gauss–Seidel iteration was enough to obtain f^{k+1} from f^k with satisfactory accuracy.

Table 5.4-2 gives an overview of some of the experiments run for the five input images generated from the Girl source image. K is the number of

Table 5.4-2
Data for Reconstruction of Girl Image

Image	K	$\delta\lambda_1$	$\delta\lambda_2$	λ-final	Q-init	Q-final	$m/2$	CPU
Image 1	1	0.2	0.2	0.2	769041	52588		1.36
Image 1	6	0.2	0.4	1.4	769041	10340		10.15
Image 1	12	0.2	0.4	3.8	769041	8381	8192	20.10
Image 2	1	4.0	4.0	4.0	15970	13539		1.36
Image 2	5	4.0	4.0	20.0	15970	8875		8.40
Image 2	9	4.0	4.0	36.0	15970	8379	8192	15.05
Image 3	1	0.2	0.3	0.2	643424	73208		2.50
Image 3	6	0.2	0.4	1.6	643424	11845		18.10
Image 3	10	0.25	0.5	4.0	643424	8423	8192	30.15
Image 4	1	4.0	4.0	4.0	14715	12814		2.50
Image 4	4	4.0	4.0	16.0	14715	9473		12.40
Image 4	8	4.0	4.0	32.0	14715	8412	8192	24.00
Image 5	1	0.5	0.5	0.5	77151	47400		1.45
Image 5	8	0.5	1.0	5.5	77151	3777		14.30
Image 5	15	0.5	1.0	12.5	77151	825	820	26.20

steps. $\delta\lambda_1$ is the initial step size in λ and $\delta\lambda_2$ is the final step size. Q-init is the initial Q value computed from the flat (constant) image which is the starting image. CPU is the VAX 11/780 CPU time in minutes and seconds. The corresponding images are shown in Figs. 5.4.5 to 5.4-9 for image 1 through image 5. The upper left image is the input image, upper right is the result after one step in λ, lower left is the intermediate result, and lower right is the final result.

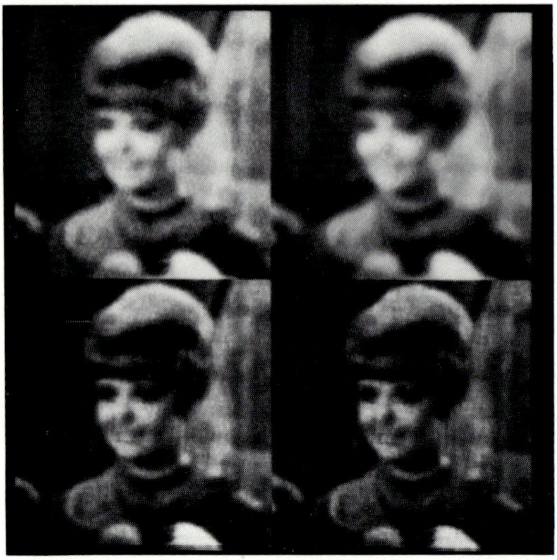

Fig. 5.4-5. Restoration results for image 1. Image 1 is shown at the upper left. The restoration is shown after 1 step in λ at the upper right, after 6 steps in λ at the lower left, and after 12 steps in λ at the lower right.

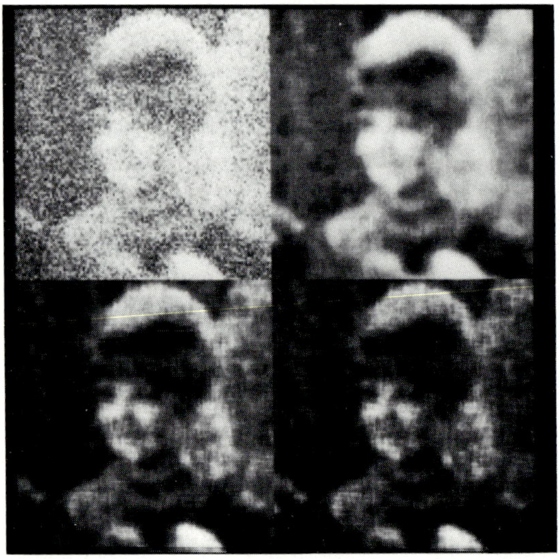

Fig. 5.4-6. Restoration results for image 2. Image 2 is shown at the upper left. The restoration is shown after 1 step in λ at the upper right, after 6 steps in λ at the bottom left, and after 12 steps in λ at the lower right.

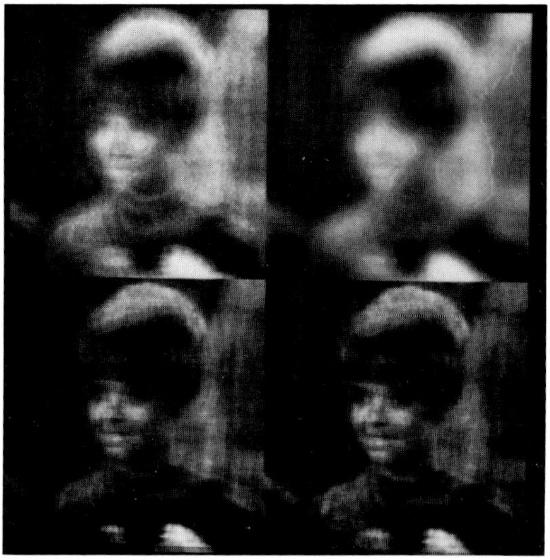

Fig. 5.4-7. Restoration results for image 3. Image 3 is shown at the upper left. The restoration is shown after 1 step in λ at the upper right, after 6 steps in λ at the lower left, and after 12 steps in λ at the lower right.

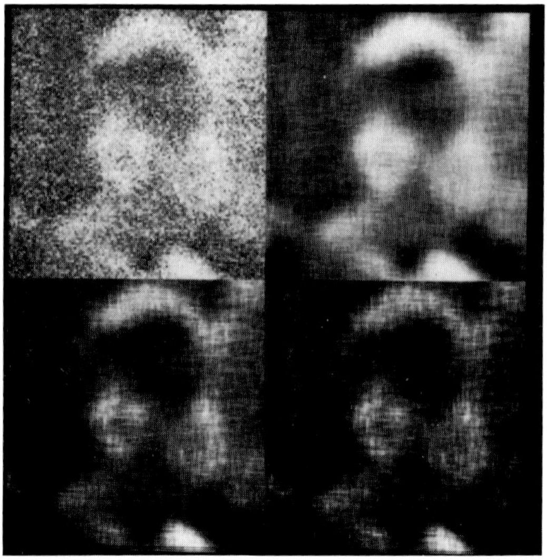

Fig. 5.4-8. Restoration results for image 4. Image 4 is shown at the upper left. The restoration is shown after 1 step in λ at the upper right, after 6 steps in λ at the lower left, and after 12 steps in λ at the lower right.

Fig. 5.4-9. Restoration results for image 5. Image 5 is shown at the upper left. The restoration is shown after 1 step in λ at the upper right, after 6 steps in λ at the lower left, and after 12 steps in λ at the lower right.

Table 5.4.-3 shows some of the experimental data for the text image and checkerboard image. The corresponding reconstructed images are shown in Figs. 5.4-10 and 5.4-11, respectively.

Table 5.4-3
Reconstruction Data for Text and Checkerboard Images

Image	K	$\delta\lambda_1$	$\delta\lambda_2$	λ-final	Q-init	Q-final	$m/2$	CPU
Text	22	0.50	256.0	768.5	827968	24574	24000	3.13.00
Check	6	0.25	0.5	2.0	1935043	50248		18.00
Check	25	0.25	32768.0	98308.0	1935043	9297	8192	1.15.00

The overall results indicate very good performance in reconstruction of blurred images even in cases where we have very few measurements. The performance in the reconstruction of extremely noisy images seems to be a little poorer. The experiments indicate that the algorithm runs better in deblurring than in noise removal. In fact, if no noise was present, it would be possible (in principle at least) to obtain a perfect reconstruction (which in this case is a pure deconvolution or deblurring). In practice, however,

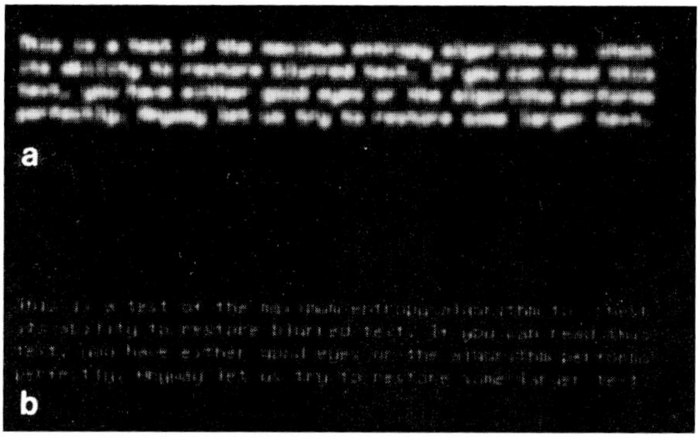

Fig. 5.4-10. Reconstruction of the text image. (a) Noisy blurred source image; (b) reconstructed image.

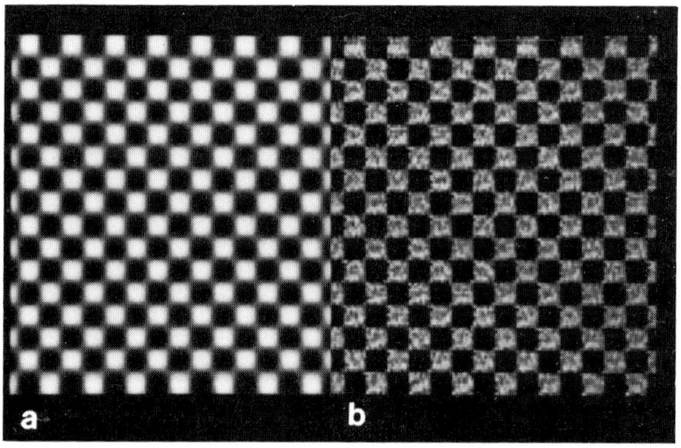

Fig. 5.4-11. Reconstruction of the degraded checkerboard image of Fig. 5.4-4. (a) Reconstruction after 6 steps in λ; (b) reconstruction after 25 steps in λ.

noise is always present, and even very small amounts of noise can cause the reconstructed data to fluctuate somewhat. The present algorithm, however, applies noise statistics [see Eq. (5.4-4)], which is the best way (statistically speaking) to deal with the noise.

Figures 5.4-5 to 5.4-9 indicate that in many cases it is not necessary to search all the way until Q reaches Q_0. The final results would not be very much different from the intermediate results obtained after 12 steps in λ as displayed in the figures. Thus, by accepting the intermediate results as the final reconstruction, a great deal of CPU time can be saved.

Table 5.4-3 reveals the importance of varying the step size $\delta\lambda$. The initial step size cannot be increased significantly from its present values, so a constant step size would in many cases make practical applications impossible.

The checkerboard image was the only case where the result did not indisputably improve with the number of iterations. The average contrast between black and white squares did increase, but pixel values for the two regions overlap after 25 iterations (see Fig. 5.4-11), whereas this was not the case for just 6 iterations with a final λ of only 2.0 (shown in the same figure).

The experiments were run on a VAX-11/780 computer. Data for the experiments as well as CPU time consumed are listed in Tables 5.4-2 and 5.4-3. Though the general problem complexity is the square of the number of pixels, the computational complexity can in many cases (for instance, when the degradation matrix has a simple convolution form) be reduced to linear in the number of pixels. This is confirmed through experiments on a 256×256 image. For many applications CPU time is crucial. Taking into account the size of the problem, the CPU times listed in Tables 5.4-2 and 5.4-3 are very satisfactory. It is, however, evident that this time can be reduced even more if a better technique for finding an optimal or near-optimal $\delta\lambda$ for each step in the search along the path is developed.

5.5 CONCLUSION

We have briefly reviewed Frieden's approach and the Burch *et al.* approach to maximum entropy restoration. We have given a detailed description of a new differential equation approach to maximum entropy image restoration. We have performed and discussed an initial set of experiments with the differential equation approach.

The experimentation phase has not yet been completed. More combinations of the most essential parameters need to be investigated. To fully verify the potential of the algorithm, this more extensive experimentation should include:

(1) Testing more combinations of noise and blurring,

(2) Optimizing the search in λ,

(3) Possibly evaluating a better criterion for termination of the search process,

(4) Trying more source images, and

(5) Investigating possible time savings by use of Fourier transforms (for large masks).

A final assessment of the algorithm must wait for this work to be completed. However, the experiments run so far indicate that the algorithm presented in this chapter is a very prospective tool in image reconstruction. In the most general case this reconstruction will require either an enormous storage capacity (the matrix $\mathbf{L} = \mathbf{A}^T\mathbf{D}\mathbf{A}$ has N^2 elements, where N is number of pixels in an image) or a substantial amount of CPU time (if all elements of $\mathbf{L}$ have to be computed every time they are used). However, in many practical applications simplifications are possible and the reconstruction can be done at a reasonable cost in CPU time without requiring excessive storage capacity.

APPENDIX

Theorem 1. A unique solution curve $\mathbf{f}(\lambda;\mu)$ is defined in E for $0 \leqslant \lambda < \infty$ by the following Cauchy problem of differential equations:

$$\begin{aligned} \nabla^2 J\, d\mathbf{f}/d\lambda &= \nabla Q, \\ \mathbf{f}\big|_{\lambda=0} &= \exp(\mu - 1)\mathbf{h}. \end{aligned} \tag{5.A-1}$$

Proof. Since the coefficient matrix $\nabla^2 J$ is negative definite and both $\nabla^2 J$ and ∇Q are analytic in E, Eq. (5.A-1) does define a unique solution curve $\mathbf{f}(\lambda;\mu)$ in E for $0 \leqslant \lambda < \alpha$ where α is some positive number or plus infinity. Thus all we need now is to prove that the solution curve could be extended whenever $\alpha < \infty$.

Suppose $\alpha < \infty$. We are to prove first that the solution curve $\mathbf{f}(\lambda;\mu)$ tends to a point $\mathbf{f}^* \in E$ as λ approaches α from below and the limit point $\mathbf{f}^*$ satisfies the stationary point equations, Eq. (5.4-7). Then the solution curve $\mathbf{f}(\lambda;\mu)$ is extended beyond α by the solution curve of the following Cauchy problem:

$$\begin{aligned} \nabla^2 J\, d\mathbf{f}/d\lambda &= \nabla Q, \\ \mathbf{f}\big|_{\lambda=\alpha} &= \mathbf{f}^*. \end{aligned} \tag{5.A-2}$$

Along the solution curve $\mathbf{f}(\lambda, \mu)$ we have

$$\nabla J(\mathbf{f}(\lambda;\mu);\mu,\lambda) \equiv 0, \qquad 0 \leqslant \lambda < \alpha. \tag{5.A-3}$$

Premultiplying Eq. (5.A-3) with $\mathbf{f}^T(\lambda;\mu)$ yields

$$\begin{aligned} \mathbf{f}^T\,\nabla J &\equiv -\sum_i f_i \log f_i + (\mu - 1)\sum_i f_i - \lambda \mathbf{f}^T(\mathbf{A}^T\mathbf{D}\mathbf{A})\mathbf{f} + \lambda \mathbf{f}^T\mathbf{A}^T\mathbf{D}\mathbf{d} \\ &\equiv 0, \qquad 0 \leqslant \lambda < \alpha, \end{aligned} \tag{5.A-4}$$

where for abbreviation the arguments in $\mathbf{f}(\lambda; \mu)$ are dropped. Equation (5.A-4) indicates that the norm $\|\mathbf{f}(\lambda; \mu)\|$ must be bounded as $\lambda \to \alpha - 0$. Otherwise $\mathbf{f}^T(\lambda; \mu)\, \nabla J(\mathbf{f}(\lambda; \mu); \mu, \lambda)$ would go to $-\infty$ because of $\mathbf{f}^T(\lambda; \mu)(\mathbf{A}^T\mathbf{D}\mathbf{A})\mathbf{f}(\lambda; \mu) \geqslant 0$ and $-\sum f_i \log f_i + (\mu - 1)\sum f_i + \lambda \mathbf{f}^T\mathbf{A}^T\mathbf{D}\mathbf{A} \to -\infty$. The boundedness of $\|\mathbf{f}(\lambda; \mu)\|$ assures that each component $f_i(\lambda; \mu)$ has a positive lower bound as $\lambda \to \alpha - 0$. Otherwise the corresponding component of $\nabla J(\mathbf{f}(\lambda; \mu); \mu, \lambda)$ would go to ∞, as easily seen from Eq. (5.4-7). Therefore, not only the entire solution curve but also its limit point(s) as $\lambda \to \alpha - 0$ belong to E. Now it is clear that each limit point will satisfy Eq. (5.4-7) and hence coincide with that unique maximal point of J at α. If we denote the unique limit point by $\mathbf{f}^*$, then Eq. (5.A-2) does define a unique solution curve in E for some small interval $(\alpha - \delta, \alpha + \delta)$ with $\alpha - \delta \geqslant 0$ which coincides with the maximal point path of J as $\lambda \in (\alpha - \delta, \alpha + \delta)$. Hence the solution curve given by Eq. (5.A-2) coincides with $\mathbf{f}(\lambda; \mu)$ as $\lambda \in (\alpha - \delta, \alpha)$ and extends $\mathbf{f}(\lambda; \mu)$ beyond α.

Theorem 2. Assume that $\|\mathbf{g}\|/m \to 0$ as $m \to \infty$. Then with probability one,

$$\frac{1}{m}(\mathbf{g}, \mathbf{u}) \to 0 \qquad \text{as } m \to \infty, \tag{5.A-5}$$

$$\frac{1}{m}\left(c - \frac{1}{2}\|\mathbf{g}\|^2\right) \to \frac{1}{2} \qquad \text{as } m \to \infty. \tag{5.A-6}$$

Proof. Let $x_m = (1/m)(\mathbf{g}, \mathbf{u})$. It is easy to verify by means of $E\{u_i\} = 0$, $E\{u_i u_j\} = \delta_{ij}$, and the assumption $\|\mathbf{g}\|/m \to 0$ as $m \to \infty$ that

$$E\{x_m\} = 0, \tag{5.A-7}$$

$$\operatorname{Var}\{x_m\} = E\{x_m^2\} = \left(\frac{\|\mathbf{g}\|}{m}\right)^2 \to 0 \qquad \text{as } m \to \infty. \tag{5.A-8}$$

The latter will lead to $x_m \to 0$ as $m \to \infty$ with probability one by Chebychev's inequality (see [6], p. 151). This validates the first part of the theorem. To prove the second part we need to write $(1/m)(c - \frac{1}{2}\|\mathbf{g}\|^2)$ as

$$(1/m)(c - \tfrac{1}{2}\|\mathbf{g}\|^2) = x_m + (1/2m)\|\mathbf{u}\|^2. \tag{5.A-9}$$

The components u_j of $\mathbf{u}$ are independent, identically distributed random variables with zero mean and variance 1; thus $(1/m)\|\mathbf{u}\|^2 \to 1$ as $m \to \infty$ with probability one by the central limit theorem [6]. Combining this fact with $x_m \to 0$ as $m \to \infty$ with probability one, we obtain $(1/m)(c - \frac{1}{2}\|\mathbf{g}\|^2) \to \frac{1}{2}$ as $m \to \infty$ with probability one. This completes the second part of the theorem.

REFERENCES

1. E. T. Jaynes, Prior probabilities. *IEEE Trans. on Systems, Science and Cybernetics* **SSC-4**(3), 227–241 (1968).
2. E. T. Jaynes, The well-posed problem. *Foundation of Physics* **3**(4), 477–492 (1973).
3. B. R. Frieden, Restoring with maximum likelihood and maximum entropy. *J. Opt. Soc. Amer.* **G2**, 511–518 (1972).
4. S. F. Burch, S. F. Gull, and J. K. Skilling, Image restoration by a powerful maximum entropy method. *Computer Vision, Graphics, and Image Processing* **23**, 113–128 (1983).
5. X. Zhuang, E. Østevold, and R. M. Haralick, A differential equation approach to maximum entropy image reconstruction (to appear in *IEEE Trans. on Acoustics, Speech, and Signal Processing*).
6. W. Feller, "An Introduction to Probability Theory and Its Application, Vol. II," Second edition, John Wiley and Sons, Inc. 1971.
7. J. N. Franklin, "Matrix Theory," Prentice-Hall, Inc., 1968.

CHAPTER 6

The Unique Reconstruction of Multidimensional Sequences from Fourier Transform Magnitude or Phase

Monson H. Hayes

School of Electrical Engineering
Georgia Institute of Technology
Atlanta, Georgia 30332

6.1 INTRODUCTION

This chapter is concerned with the problem of recovering a signal given partial information about its Fourier transform. Although there is a one-to-one correspondence between a signal and its Fourier transform, there are many applications in which some of the Fourier domain information is either degraded or missing. In order to restore the signal, therefore, it is desirable to either restore or recover the missing spectral information. The primary problem of concern in this chapter is recovery of the phase of the Fourier transform given the magnitude and recovery of the Fourier transform magnitude given the phase. Clearly, in the absence of any underlying signal model or constraints, the loss of either phase or magnitude information of a complex function is irreversible. What is surprising, however, is that under some fairly general conditions it is possible to recover a signal from the phase of its Fourier transform or from its magnitude. In fact, for a useful class of signals it is almost always possible to recover one from the other; i.e., the phase and the magnitude constitute redundant pieces of information.

The emphasis in this chapter will be on the uniqueness question. In other words, we will be concerned with the conditions (signal constraints) which are sufficient to uniquely recover a signal given knowledge of only the magnitude or the phase of its Fourier transform. The signals of primary interest are two-dimensional, such as images. In order to motivate some of the uniqueness results, however, one-dimensional signals will also be considered. Before beginning the development of the theory surrounding the unique recovery of signals from Fourier transform phase or Fourier transform magnitude, we will survey some of the applications in which these problems are of importance.

The recovery of spectral phase information from only spectral magnitude (or intensity) information, generally referred to as the *phase retrieval* problem, arises in a variety of different contexts and applications [1–4]. In x-ray crystallography, for example, the molecular structure of a crystal is to be inferred from the observed diffraction pattern of x-rays [1]. Although the diffraction pattern is related to the scattering density of the crystal by a Fourier transformation, only the intensity (squared magnitude) of the diffraction pattern may be measured. Since knowledge of the phase of the diffracted wave is indispensable for determination of the crystal structure, the phase retrieval problem is particularly important. A similar problem arises in optical and electron microscopy when, for example, the index of refraction of a thin object or the height distribution of a surface is to be determined from the intensity of the wave distribution in the image plane or some other plane in the microscope [3]. Again, in order to determine the object structure, phase information of the wave is required. In some applications, however, it is only the wave intensity in the image plane of an optical system which is of interest. Nevertheless, the retrieval of phase information may still be an important problem. Imaging through a turbulent atmosphere, for example, may reduce the resolution of objects well below the diffraction limits of the telescope. With interferometric techniques, however, it is possible to obtain diffraction-limited information about the Fourier transform of the object [5]. Therefore, if it were possible to recover the phase of the Fourier transform of an object from only its magnitude, it might then be possible to obtain a diffraction-limited image through a turbulent atmosphere.

The reconstruction of a signal from the phase of its Fourier transform, referred to here as the *magnitude retrieval problem*, was initially motivated on theoretical grounds in order to address the apparent importance of phase information in the representation of signals [6]. However, in a number of applications the ability to reconstruct a signal from the phase of its Fourier transform would be useful. For example, in the class of problems referred to as *blind deconvolution*, a desired signal is to be recovered from an observation which is the convolution of the desired signal with some unknown signal [7]. Since little is usually known about either the desired signal or the distorting signal, deconvolution of the two signals is generally a very difficult problem. However, in the special case in which the distorting signal is known to have a zero-phase Fourier transform, the spectral phase of the desired signal is undistorted. Such a situation occurs, at least approximately, in long-term exposure to atmospheric turbulence or when images are blurred by severely defocused lenses with circular aperture stops [8]. In this case, except for phase reversals, the phase of the observed signal is approximately the same as the phase of the desired signal and, therefore, it is of interest to consider reconstructing the signal from phase information alone. As another example, a situation frequently encountered in such applications as seismology, ocean acoustics, radar, and sonar is that of an unknown signal (wavelet) propagating nondispersively in a reverberatory or multipath environment. Two receivers placed in such an environment will record the arrival of this wavelet many times with different attenuation factors. However, it may be shown that under ideal conditions the arrival times and attenuation factors may be determined from the phase of the cross-spectrum of the two signals [9]. Finally, we note that there are examples for which phase information is more readily accessible than magnitude information while knowledge of both is desired. In paleomagnetism, for example, the history of the direction (phase) of the magnetic field of the earth may be determined from core samples from the earth. Since the magnetic field strength (magnitude) cannot be obtained from these samples, it would be of interest to consider methods for recovering the magnetic field strength from information on the magnetic field direction. As another example, consider the generation of wind flow maps. Since measurements of wind direction are relatively easy to obtain from wind vanes or cloud movements, one can consider the recovery of wind velocity or strength from the wind direction vectors.

In each of the applications above as well as in the discussions which follow, there are two distinct sets of problems. These are the problems of synthesis and reconstruction. The *synthesis* problem is concerned with the determination of a signal, such as an image, which is consistent with the known constraints. A phase-only Fourier synthesis, for example, is concerned with the determination of a signal which has a Fourier transform with a given phase which, at the same time, satisfies some other constraints. As discussed in the following section, it is often possible to synthesize a signal in a straightforward manner from a partial Fourier domain description in such a way that much of the information in the original signal is accurately represented in the synthesized signal even though the synthesized signal may not be an accurate representation on a point-by-point basis.

The *reconstruction* problem, on the other hand, is concerned with the *unique recovery* of a signal given a partial Fourier domain description along with a signal model or a set of signal constraints. Here there are the separate issues related to the conditions under which exact reconstruction is theoretically possible and the development of practical algorithms for performing the reconstruction.

Before we begin our treatment of the phase and magnitude retrieval problems it is appropriate at this point to specify the types of signals of interest in this chapter and to define what is meant by the Fourier and z-transforms of these signals.

A one-dimensional (1-D) sequence is a function of a single integer variable n and will be denoted by $x(n)$. Unless otherwise specified, $x(n)$ will be assumed to be a real-valued function of n. The z-transform of $x(n)$ is a function of the complex variable z and is defined by [10]

$$X(z) = \mathscr{Z}[x(n)] = \sum_{n=-\infty}^{\infty} x(n)z^{-n}. \tag{6.1-1}$$

The Fourier transform of $x(n)$ is a function of the real frequency variable ω and is equal to the z-transform evaluated on the unit circle $z = e^{j\omega}$

$$X(e^{j\omega}) = \mathscr{F}[x(n)] = \sum_{n=-\infty}^{\infty} x(n)\, e^{-jn\omega}. \tag{6.1-2}$$

Being a complex function of the real variable ω, $X(e^{j\omega})$ may be written in polar form as

$$X(e^{j\omega}) = |X(e^{j\omega})| \exp[\phi_x(\omega)], \tag{6.1-3}$$

where $|X(e^{j\omega})|$ is the magnitude and $\phi_x(\omega)$ is the phase of the Fourier transform of $x(n)$.

Whereas a 1-D signal is a function of a single integer variable, a two-dimensional (2-D) sequence is a function of two integer variables, m and n, and will be denoted by $x(m, n)$. As is the case for 1-D sequences, unless otherwise specified, 2-D sequences will be assumed to be real-valued for all m and n. The two-dimensional z-transform of $x(m, n)$, denoted by $X(z_1, z_2)$, is defined by [11]

$$X(z_1, z_2) = \mathscr{Z}[x(m, n)] = \sum_{m=-\infty}^{\infty} \sum_{n=-\infty}^{\infty} x(m, n) z_1^{-m} z_2^{-n}, \tag{6.1-4}$$

where z_1 and z_2 are complex variables. The 2-D Fourier transform of $x(m, n)$ is a function of two frequency variables, μ and ν, and is equal to the z-transform of $x(m, n)$ evaluated on the unit bicircle, $z_1 = e^{j\mu}$ and $z_2 = e^{j\nu}$,

$$X(e^{j\mu}, e^{j\nu}) = \mathscr{F}[x(m, n)] = \sum_{m=-\infty}^{\infty} \sum_{n=-\infty}^{\infty} x(m, n)\, e^{-j[m\mu + n\nu]}. \tag{6.1-5}$$

Expressed in polar form, $X(e^{j\mu}, e^{j\nu})$ is given by

$$X(e^{j\mu}, e^{j\nu}) = |X(\mu, \nu)| \exp[j\phi_x(\mu, \nu)], \tag{6.1-6}$$

where $|X(\mu, \nu)|$ is the magnitude and $\phi_x(\mu, \nu)$ is the phase of the 2-D Fourier transform of $x(m, n)$. From Eq. (6.1-6) we note that the phase retrieval problem is concerned with the recovery of $x(m, n)$ given only the spectral magnitude function $|X(\mu, \nu)|$, whereas the magnitude retrieval problem is concerned with the recovery of $x(m, n)$ from only spectral phase, $\phi_x(\mu, \nu)$.

6.2 FOURIER SYNTHESIS FROM PARTIAL INFORMATION

In this section we look at the Fourier synthesis of an image given an incomplete characterization in the Fourier domain. In particular, we will examine the effect of combining phase information with an estimate of the magnitude function and the effect of combining magnitude information with an estimate of the phase information. We will also look briefly at the effects of coarse quantization of the phase.

It has been observed in a number of different contexts and applications that many of the important features of a signal are represented in the phase of its Fourier transform [12]. Specifically, it is well known that a *phase-only synthesis* of a signal, formed by combining the phase of the Fourier transform of the signal with either a constant or an ensemble average magnitude, contains a number of similarities to the true signal. One of the first contexts in which the similarity between a signal and its phase-only synthesis was noted is in the field of x-ray crystallography [13]. Since the x-ray diffraction pattern of a crystal is related to the electron density distribution in the crystal by a Fourier transform, if both the magnitude and phase of the diffracted wave could be measured, then an inverse Fourier transform of the diffraction pattern would yield the desired structure. However, since only the intensity of the diffraction pattern may be measured, the only available information which is obtained by measurement is that which is contained in the autocorrelation function (Patterson diagram) of the electron density distribution. Therefore, although the autocorrelation function contains peaks at positions corresponding to the interatomic vectors between the atoms in the crystal, phase information is necessary to uniquely recover the structure. In order to determine the importance of phase information in crystal structure determination, Srinivasan [13] performed a number of different experiments in the Fourier synthesis of crystal structures by combining the correct phase with various other magnitude functions. For example, shown in Fig. 6.2-1A is a contour diagram of the electron density

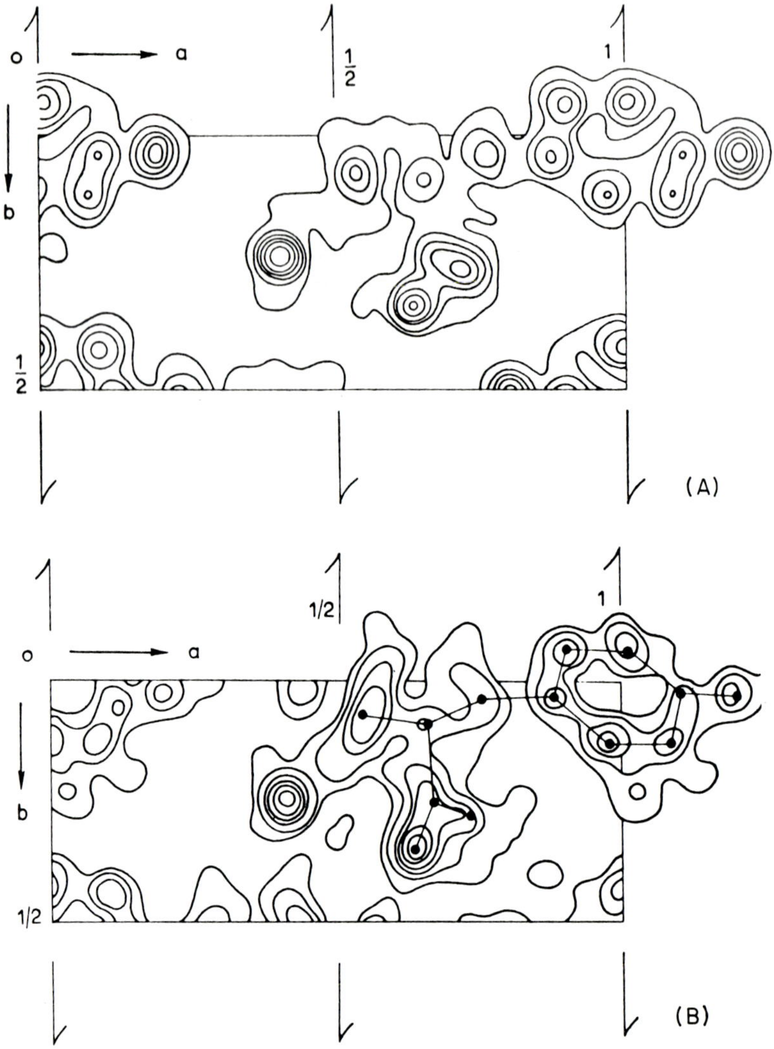

Fig. 6.2-1. Contour diagrams of electron density distribution of L-tyrosine HCl. See text for details. [After Ramachandran and Srinivasan [1]. Copyright 1970 by John Wiley & Sons, Inc.]

distribution of L-tyrosine HCl that was obtained by combining the correct phase and magnitude information. The contours in this plot correspond to regions of constant electron density with the peaks representing the locations of the atomic positions. Shown in Fig. 6.2-1B is the contour plot obtained by combining the correct phase with a magnitude function which is inversely

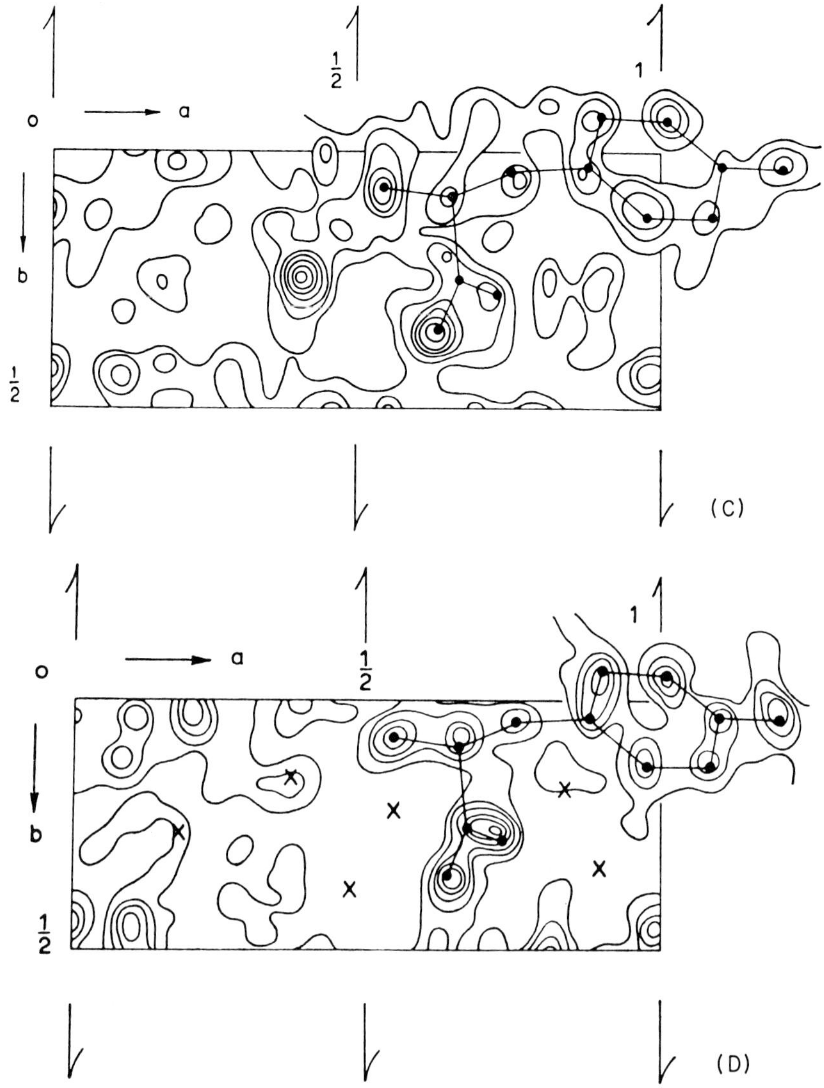

(*continued*)

proportional to frequency (a property known to be characteristic of the diffracted wave). In this case, although the magnitude information has been essentially discarded, the resulting synthesis still contains many of the important features and properties of the correct structure (the correct atomic structure is overlaid in solid lines). Shown in Fig. 6.2-1C is the synthesis obtained by randomly permuting the correct magnitudes, and again many of the important properties are preserved. Finally, in Fig. 6.2-1D is the result obtained by combining the correct phase with the magnitude of a totally different structure which has atomic locations at the positions indicated by the crosses. It is apparent that the resulting structure most closely resembles the structure corresponding to the phase used in the synthesis. The conclusion which may be drawn from this series of experiments is that, at least for the signals which typically arise in x-ray crystallography, most of the essential information of interest is contained in the phase.

The similarity between a signal and its phase-only synthesis has also been noted in the context of image processing by investigators performing a similar set of experiments [6, 12, 14]. In particular, it has been observed that when only phase information is used to synthesize an image, the result contains many of the important features of the original image. Shown in Fig. 6.2-2a, for example, is an original image and in Fig. 6.2-2b is the image formed by combining the phase of the Fourier transform of the image with a constant magnitude

$$y(m, n) = \mathscr{F}^{-1}[\exp\{j\phi_x(\mu, \nu)\}]. \tag{6.2-1}$$

Clearly, the essence of the image is preserved with this phase-only synthesis. An even better phase-only synthesis may be obtained, however, by combining the phase of the Fourier transform of the image with a magnitude function which is "more representative" of images. For example, shown in Fig. 6.2-3a is the phase-only synthesis obtained by using an ensemble average magnitude function which was derived by averaging the Fourier transform magnitudes of several different images. In Fig. 6.2-3b, on the other hand, is the phase-only synthesis obtained by simply using the magnitude of the Fourier transform of another image.

Although much of the useful information in a signal appears to be represented in the phase of its Fourier transform, the phase-only syntheses examined above assumed perfect phase information, i.e., no quantization errors and no additive noise. It is remarkable, however, that even with a coarsely quantized representation of the phase, a reasonable phase-only synthesis may be obtained. For example, shown in Fig. 6.2-4a is a phase-only synthesis formed from one bit of phase information and a constant magnitude. The one-bit phase function is given by

$$Q[\phi_x(\mu, \nu)] = \begin{cases} 0 & \text{if } |\phi_x(\mu, \nu| \leq \pi/2, \\ \pi & \text{otherwise.} \end{cases} \tag{6.2-2}$$

As a result, the phase-only synthesis in Eq. (6.2-1) is given simply by

$$y(m, n) = \mathscr{F}^{-1}[Y(\mu, \nu)] \tag{6.2-3}$$

Fig. 6.2-2 (a)

Fig. 6.2-2 (b)

Fig. 6.2-2. (a) Original image and (b) its phase-only synthesis using a constant magnitude.

Fig. 6.2-3 (a)

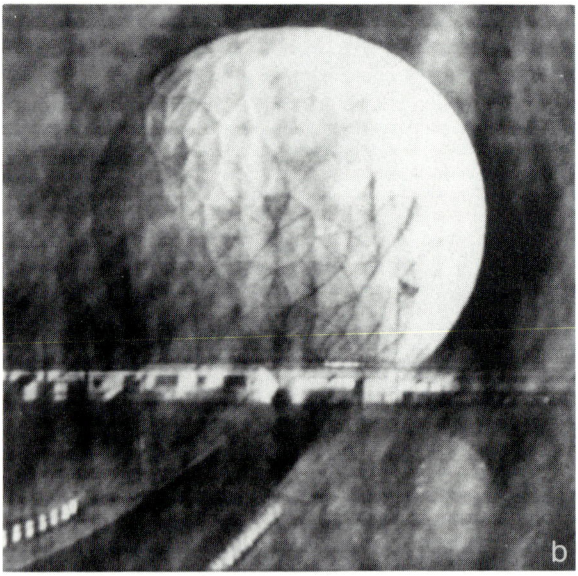

Fig. 6.2-3 (b)

Fig. 6.2-3. Phase-only syntheses using (a) an ensemble average magnitude function and (b) the magnitude function from another image.

Fig. 6.2-4 (a)

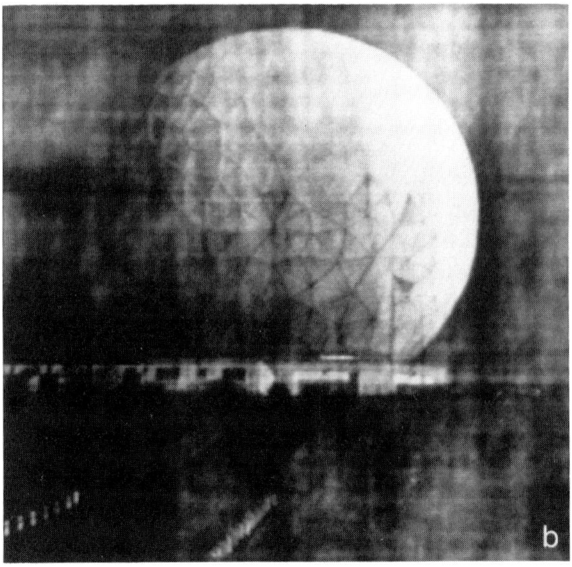

Fig. 6.2-4 (b)

Fig. 6.2-4. One bit phase-only syntheses using (a) a constant magnitude and (b) an ensemble average magnitude function.

where

$$Y(\mu, \nu)] = \begin{cases} 1 & \text{if } |\phi_x(\mu, \nu)| \leq \pi/2, \\ -1 & \text{otherwise.} \end{cases} \tag{6.2-4}$$

An improved synthesis may be obtained by using an ensemble average magnitude function (as opposed to a constant) as the result shown in Fig. 6.2-4b indicates. What these figures suggest is the possibility that an image may be recoverable from only one bit of phase information (see Section 6.6).

Having observed the importance of phase information in the representation of images, it is of interest to investigate the relative importance of magnitude information in signal representation. Therefore, a set of experiments similar to those described above was performed to examine the importance of magnitude information [14]. Shown in Fig. 6.2-5a, for example, is an original image, $x(m, n)$, and shown in Fig. 6.2-5b is a magnitude-only synthesis formed by combining the magnitude of the Fourier transform of $x(m, n)$ with zero phase

$$y(m, n) = \mathscr{F}^{-1}[|X(\mu, \nu)|]. \tag{6.2-5}$$

Clearly, there is no useful information portrayed with this synthesis. Combining the magnitude with an ensemble average phase produces the magnitude-only synthesis shown in Fig. 6.2-6a, and, again, the synthesis is not useful in representing $x(m, n)$. Finally, shown in Fig. 6.2-6b is the synthesis produced by combining the magnitude of the Fourier transform of $x(m, n)$ with the phase of another image. As we would expect, the image produced by this synthesis bears a resemblance to the image whose Fourier phase was used.

The importance of phase information in the representation of signals has also been observed in a number of other applications such as speech processing [6], digital filtering [15], and acoustical and optical holography [16–19]. In speech processing, for example, a set of experiments similar to those described above was performed with similar results [6]. Specifically, if the Fourier transform of a sentence of speech is computed and the phase is set equal to zero, the intelligibility is lost in the signal obtained by an inverse Fourier transform. However, if the phase is preserved and the magnitude is set equal to a constant, an inverse Fourier transform leads to an intelligible synthesis of the original sentence of speech.

The conclusion to be drawn from this discussion is that there is, apparently, a significant amount of information conveyed in the phase of the Fourier transform of signals which are of interest in such fields as image processing, x-ray crystallography, microscopy, and astronomy. Furthermore, this information does not rely on the precise representation of the phase.

Fig. 6.2-5 (a)

Fig. 6.2-5 (b)

Fig. 6.2-5. (a) Original image and (b) its magnitude-only synthesis using zero phase.

Fig. 6.2-6 (a)

Fig. 6.2-6 (b)

Fig. 6.2-6. Magnitude-only synthesis using (a) an ensemble average phase and (b) the phase of another image.

6.3 THE ALGEBRA OF POLYNOMIALS IN TWO OR MORE VARIABLES

In this section we establish the algebraic notation and background necessary for our analysis of the reconstruction problem. In particular, z-transforms of both one- and two-dimensional signals are interpreted in terms of pole/zero representations and some important results from algebra are presented which are concerned with the factorization of polynomials in one or more variables.

Unless otherwise noted, the signals considered in this chapter are assumed to have finite duration or support. Therefore, for convenience it will be assumed without any loss in generality that a finite-duration 1-D sequence is *causal* so $x(n) = 0$ for all $n < 0$. In the general case, any finite-duration sequence which is not causal may always be shifted in order to satisfy this constraint. In addition, if it is known that $x(n)$ is zero outside the interval $[0, N\text{-}1]$, then $x(n)$ will be said to have support $R(N)$. Similarly, for 2-D sequences it will be assumed that $x(m, n)$ is nonzero only for a finite number of values of m and n. In addition, it will be assumed, without any loss in generality, that a sequence with finite support has first quadrant support so $x(m, n) = 0$ if $m < 0$ or if $n < 0$. Again, in the general case, any sequence with finite support may always be shifted (delayed) in order to satisfy this constraint. If it is known that $x(m, n)$ is zero outside the rectangular region consisting of all points (m, n) for which $0 \leqslant m < M$ and $0 \leqslant n < N$, then $x(m, n)$ will be said to have support $R(M, N)$.

It is important to note that a one-dimensional sequence with finite support $R(N)$ has a z-transform which is a polynomial in z^{-1} of degree $N - 1$. Since the *fundamental theorem of algebra* states that any polynomial in a single variable may always be uniquely factored into a product of first-order polynomials [20], it follows that any 1-D sequence with support $R(N)$ has a z-transform which may be written in the form

$$X(z) = \sum_{n=0}^{N-1} x(n)z^{-n} = \alpha z^{-n_0} \prod_{k=1}^{p} A_k(z) = \alpha z^{-n_0} \prod_{k=1}^{p} (1 - a_k z^{-1}). \quad (6.3\text{-}1)$$

The polynomials $A_k(z) = (1 - a_k z^{-1})$ are the irreducible factors of $X(z)$ and the constants a_k are the *zeros* of $X(z)$. As a result, any $1 - \mathrm{D}$ signal with finite support is uniquely characterized (to within a scale factor) by the distribution of the zeros of its z-transform. In other words, a 1-D sequence is uniquely defined to within a scale factor by the coefficients a_k in Eq. (6.3-1) for $k = 1, \ldots, p$ and the value of n_0.

The z-transform of a 2-D sequence $x(m, n)$ with support $R(M, N)$ is a polynomial in the two complex variables z_1 and z_2. Therefore, $X(z_1, z_2)$

may always be written uniquely (to within factors of zero degree) as a product of polynomials which are irreducible over the field of complex numbers [20]

$$X(z_1, z_2) = \sum_{m=0}^{M-1} \sum_{n=0}^{N-1} x(m, n) z_1^{-m} z_2^{-n} = \alpha z_1^{-n_1} z_2^{-n_2} \prod_{k=1}^{p} A_k(z_1, z_2), \quad (6.3\text{-}2)$$

where α is a real number and n_1 and n_2 are nonnegative integers. The polynomials $A_k(z_1, z_2)$ are the irreducible factors of $X(z_1, z_2)$. Therefore, for both 1-D and 2-D signals, the z-transform may always be written as a product of irreducible factors. There are some important differences, however, between the 1-D and 2-D cases. For example, whereas the *zero set* for a one-dimensional sequence consists of the finite collection of isolated points $\{a_k\}$ in Eq. (6.3-1), the zeros of the z-transform of a 2-D sequence consists of *zero contours.* In particular, the solutions to the equation $X(z) = 0$ are points while the solutions to the equation $X(z_1, z_2) = 0$ are continuous functions of z_1 and z_2. Another important difference follows from the fact that the fundamental theorem of algebra is not valid for polynomials in two or more variables. In particular, whereas this theorem guarantees that the irreducible factors of 1-D polynomials will always be linear (i.e., first-order) factors, the irreducible factors of 2-D polynomials may be of arbitrarily large degree. For example, with $a(n)$ an arbitrary 1-D sequence with support $R(N)$ and with $\delta(n-k)$ denoting a point source at index $n = k$, the z-transform of the sequence $x(m, n)$ given by

$$x(m, n) = a(m) + \delta(n - k) \quad (6.3\text{-}3)$$

is irreducible (nonfactorable) for any nonzero value of k [21]. Therefore, $p = 1$ in the expansion given by Eq. (6.3-2) and

$$X(z_1, z_2) = A(z_1) + z_2^{-k} \quad (6.3\text{-}4)$$

As another example, let $x(m, n)$ be an arbitrary 2-D sequence with support $R(M, N)$. Assuming $x(0, 1) \neq 0$, then with $a \neq 0$ the sequence

$$w(m, n) = x(m, n) + a\delta(m, n - N) \quad (6.3\text{-}5)$$

has a z-transform which satisfies *Eisenstein's criterion* and, therefore, is guaranteed to be irreducible [22]. Therefore $p = 1$ in the expansion given in Eq. (6.3-2). The term $a\delta(m, n - N)$ has been referred to as a reference function and the signal $w(m, n)$ has similarities to off-axis holography.

Since polynomials in two variables of arbitrarily large degree may be irreducible, it is of interest to determine the likelihood that an arbitrary 2-D signal with support $R(M, N)$ will have an irreducible z-transform. More specifically, given an arbitrary signal $x(m, n)$ with support $R(M, N)$, is it more likely that $X(z_1, z_2)$ is reducible (factorable) or irreducible? In order to address this question, note that it is possible to make a correspondence between the polynomials which are z-transforms of sequences having support $R(M, N)$ and an ordered sequence (vector) of $\rho = M \times N$ real numbers. This correspondence consists of arranging the coefficients $x(m, n)$ of the polynomial $X(z_1, z_2)$ into a vector in some fixed order, e.g., lexicographically. Since the set of reducible polynomials may be identified in a one-to-one manner with a subset S of the ρ-dimensional Euclidean space R^ρ, i.e., the set of all vectors of length ρ, then it is possible to determine the Lebesque measure of the set S. It has been shown that this set has measure zero [23]. Therefore, it follows that "almost all" polynomials in two or more variables are irreducible.

Although a set of measure zero is, in some sense, a *sparse* set, it is important to note that sets with measure zero may have distinctly different geometric properties. In R^2, for example, the set S_1 of all ordered pairs $\mathbf{x} = (x_1, x_2)$ which have integer coefficients is a set of measure zero. This set consists of an infinite collection of isolated points. Another set of measure zero is the set S_2 of all ordered pairs $\mathbf{x}(x_1, x_2)$, where $x_1 = f(x_2)$ is a continuous differentiable function. This set consists of a curve in R^2. Finally, the set S_3 of all ordered pairs $\mathbf{x} = (x_1, x_2)$ which have rational coefficients is a set of measure zero. This set differs from the previous two sets since it is *dense* in R^2, i.e., the closure of the set is R^2. As a result, for any point $\mathbf{x} = (x_1, x_2)$ not in the set S_3 it is possible to find a point $\mathbf{y} = (y_1, y_2)$ in S_3 which is arbitrarily close to $\mathbf{x}$.

Since the set of reducible polynomials is a set of measure zero, it is of interest to determine whether geometrically it more closely resembles the set S_1, S_2, or S_3. In particular, if it is dense in the set of all polynomials then it would be possible to find, for any given irreducible polynomial, a reducible polynomial which is arbitrarily close to it. Practically speaking, this would imply that the irreducibility property of a given polynomial is extremely sensitive to noise since arbitrarily small perturbations of the coefficients may lead to a reducible polynomial. Fortunately, this is not the case. In particular, it has been shown that the set of reducible polynomials is contained in a nontrivial algebraic set and, as a result, geometrically resembles the set S_2 described above [24]. As a very simple example to illustrate this, consider the set of all 2-D sequences which have support $R(M, N)$ where $M = N = 1$. In this case, a general sequence is given by

$$x(m, n) = a\delta(m, n) + b\delta(m - 1, n) + c\delta(m, n - 1) + d\delta(m - 1, n - 1). \tag{6.3-6}$$

It readily follows that $X(z_1, z_2)$ will be reducible if and only if $a \times d = b \times c$. In the space R^4 defined by the four coefficients a, b, c, and d, it follows that the subset S of reducible polynomials is a hyperquadratic surface.

6.4 THE MAGNITUDE RETRIEVAL PROBLEM

In this section we consider the uniqueness of a discrete two-dimensional signal in terms of the phase of its Fourier transform. More specifically, assuming that the phase of the Fourier transform of a 2-D signal, $x(m, n)$, is known exactly for all frequencies, we consider sets of signal constraints which are sufficient to allow for a unique reconstruction of $x(m, n)$ from its phase. That we need to consider constraints at all follows from the observation that there are an infinite number of signals which have a Fourier transform with the same phase as $x(m, n)$. In particular, suppose we let $\phi_x(\mu, \nu)$ be the phase of the Fourier transform of the signal $x(m, n)$ and let $A(\mu, \nu)$ be any real symmetric function of μ and ν. In this case the signal

$$\begin{aligned} y(m, n) &= \mathscr{F}^{-1}[A(\mu, \nu) \exp\{j\phi_x(\mu, \nu)\}] \\ &= \frac{1}{4\pi^2} \int_{-\pi}^{\pi} \int_{-\pi}^{\pi} A(\mu, \nu) \exp\{j[\phi_x(\mu, \nu) + m\mu + n\nu]\} \, d\mu \, d\nu \end{aligned} \tag{6.4-1}$$

cannot be distinguished from $x(m, n)$ on the basis of phase information alone.

Before examining the 2-D uniqueness question we first look at the uniqueness question for one-dimensional signals. This provides us with a geometric understanding of the issues involved in signal reconstruction from phase. The one-dimensional results are then extended in a straightforward manner to two-dimensional signals. Assuming that a unique reconstruction is possible, several approaches for recovering a signal from only phase information are then briefly outlined.

A. *One-Dimensional Signals*

Let $x(n)$ be a one-dimensional sequence with support $R(N) = [0, N-1]$. The z-transform of $x(n)$ is a polynomial of degree $N-1$ in the complex variable z^{-1} and is, therefore, uniquely defined (to within a scale factor) by the distribution of its zeros in the z-plane, i.e., by the set of numbers $\{a_1, \ldots, a_p\}$ and the value of n_0 in Eq. (6.3-1). However, the phase of the Fourier transform of $x(n)$ does not uniquely define the sequence since it is always possible to find another sequence which has a Fourier transform with the same phase by simply convolving $x(n)$ with a zero-phase sequence $g(n)$:

$$y(n) = x(n) * g(n) \equiv \sum_{k=-\infty}^{\infty} x(k)g(n-k). \tag{6.4-2}$$

Conversely, if two sequences $x(n)$ and $y(n)$ have Fourier transforms with the same phase, then $G(z) = Y(z)/X(z)$ is the z-transform of a zero-phase sequence and $x(n)$ and $y(n)$ are related by convolution with $g(n)$ as shown in Eq. (6.4-2). At this point, therefore, it is of interest to characterize the class of zero-phase sequences by examining their properties.

By definition, a zero-phase sequence $g(n)$ must have a Fourier transform which is *real.* As a result, it follows that $g(n)$ must be conjugate symmetric [10]:

$$g(n) = g^*(-n). \tag{6.4-3}$$

Therefore, its z-transform must have the following property:

$$G(z) = G^*(1/z^*). \tag{6.4-4}$$

Thus, the singularities of $G(z)$ occur in conjugate reciprocal pairs so that if $G(z)$ has a zero (pole) at $z = a$ then $G(z)$ must also have a zero (pole) at $z = 1/a^*$. For example, the following is the z-transform of a zero-phase sequence which has four zeros and two poles:

$$G(z) = z\frac{1 + 2z^{-1} + 5z^{-2} + 2z^{-3} + z^{-4}}{1 + 5z^{-1} + z^{-2}}. \tag{6.4-5}$$

It follows that if $g(n)$ is a zero-phase sequence then its z-transform can be factored in the form (assuming a rational function)

$$G(z) = \frac{B(z)B^*(1/z^*)}{C(z)C^*(1/z^*)}. \tag{6.4-6}$$

where $B(z)$ and $C(z)$ are polynomials in z^{-1}.

Now consider the effect of convolving a sequence $x(n)$ with a zero-phase sequence $g(n)$. Since $Y(z) = X(z)G(z)$, the effect of convolving with $g(n)$ is to

(a) Add or delete conjugate reciprocal poles or zeros in $X(z)$, or
(b) Replace a zero (pole) at $z = a$ with a pole (zero) at $z = 1/a^*$.

As a result, observe that if $x(n)$ has a z-transform with all of its singularities (poles and zeros) inside the unit circle, then convolution with a zero-phase sequence will produce a sequence which has at least one singularity outside the unit circle. Similarly, if all of the singularities are outside the unit circle, then convolution with a zero-phase sequence will produce a sequence with at least one singularity inside the unit circle. A sequence which has a z-transform with all of its singularities *inside* the unit circle is called a *minimum phase sequence* and one with all of its singularities outside the unit circle is called a *maximum phase sequence.* We therefore have the following theorem [10]:

Theorem 1. Any minimum phase sequence is uniquely defined (to within a scale factor) by the phase of its Fourier transform.

It should be emphasized that this result asserts that if we restrict our set of admissible solutions to the set of minimum phase sequences or to the set of maximum phase sequences, then there exists only one sequence with a given phase. In other words, whereas there exists an infinite number of sequences with a given phase function, there exists only one minimum phase sequence and only one maximum phase sequence having this phase. Given a phase function, the minimum phase and maximum phase sequence may be found with a Hilbert transform [10].

Let us now restrict our attention to the class of sequences which are finite in length. As previously noted, finite-length sequences have z-transforms which contain only zeros. If a finite-length sequence $x(n)$ is convolved with a zero-phase sequence $g(n)$, then

(a) Poles are added to $X(z)$, or
(b) Conjugate reciprocal zeros are added to or deleted from $X(z)$.

Poles are added to $X(z)$ only when $G(z)$ contains conjugate reciprocal poles, e.g., one at $z = a$ and one at $z = 1/a^*$. If $X(z)$ does not have a zero either at $z = a$ or at $z = 1/a^*$ then $Y(z) = X(z)G(z)$ will contain a pair of conjugate reciprocal poles. If $X(z)$ has a zero at $z = a$ but not at $z = 1/a^*$ then, effectively, the zero at $z = a$ is *flipped* about the unit circle and replaced by a pole at $z = 1/a^*$. In other words, a factor of $(1 - az^{-1})$ in $X(z)$ is replaced by the factor $1/(z^{-1} - a^*)$. However, if $X(z)$ contains a pair of conjugate reciprocal zeros which correspond to the conjugate reciprocal poles in $G(z)$ then these zeros will not be found in $Y(z)$. Finally, any conjugate reciprocal zeros in $G(z)$ will be added to $X(z)$.

The important point of these observations is that if a pole is added to $X(z)$ then the resulting sequence no longer has finite support. Therefore, if we restrict our attention to the class of sequences which have finite support with z-transforms having no conjugate reciprocal zeros, then there may exist, at most, one sequence (within this class) having a given phase function. Thus, we have the following result [25].

Theorem 2. A finite-length sequence having a z-transform with no conjugate reciprocal zeros is uniquely defined (to within a scale factor) by the phase of its Fourier transform.

It should be noted that the exclusion of conjugate reciprocal zeros is equivalent to the constraint that $X(z)$ cannot be factored into a product of two polynomials

$$X(z) = P(z)Q(z), \tag{6.4-7}$$

where $Q(z)$ is the z-transform of a zero-phase sequence:

$$Q(z) = Q^*(1/z^*). \tag{6.4-8}$$

That such a restriction is placed in $X(z)$ is not surprising when we observe that the phase of the Fourier transform of the sequence $x(n)$ whose z-transform is given by Eq. (6.4-7) is equal to

$$\phi_x(\omega) = \phi_p(\omega) + \phi_q(\omega), \tag{6.4-9}$$

where $\phi_p(\omega)$ is the phase of the Fourier transform of $p(n)$ and $\phi_q(\omega)$ is the phase of the Fourier transform of $q(n)$. Since $\phi_q(\omega) = 0$, the factor $Q(z)$ is transparent based on phase information alone. In other words, measuring the phase of $X(e^{j\omega})$ provides no information about $Q(e^{j\omega})$.

In the next section we extend the results presented in Theorem 2 to the case of two-dimensional signals.

B. Two-Dimensional Signals

In the previous section it was shown that without any additional information or constraints a 1-D sequence is not uniquely defined by the phase of its Fourier transform. However, by placing constraints on the type and/or location of the singularities in the z-transform of a signal a unique specification may be possible. As we show in this section, the same results apply for two-dimensional signals.

By following the development in the previous section for 1-D signals, it follows that two 2-D sequences, $x(m, n)$ and $y(m, n)$, have Fourier transforms with the same phase if and only if one is related to the other by convolution with a zero-phase sequence:

$$y(m, n) = x(m, n) ** g(m, n) \equiv \sum_{k=-\infty}^{\infty} \sum_{l=-\infty}^{\infty} x(k, l)g(m-k, n-l). \tag{6.4-10}$$

A 2-D zero-phase sequence has a Fourier transform which is real, which implies that its z-transform has the property

$$G(z_1, z_2) = G^*(1/z_1^*, 1/z_2^*). \tag{6.4-11}$$

Unlike the case of one-dimensional signals, however, it is not necessarily true that a 2-D zero-phase sequence may be factored in a form similar to Eq. (6.4-6). In fact, since almost all 2-D polynomials are irreducible, all that can be stated in general is that if $g(m, n)$ is a zero-phase array then $G(z_1, z_2)$ may be written as

$$G(z_1, z_2) = \prod_{k=1}^{p} \frac{B_k(z_1,z_2)}{C_k(z_1,z_2)}, \tag{6.4-12}$$

where

$$B_k(z_1, z_2) = \pm B_k^*(1/z_1^*, 1/z_2^*), \tag{6.4-13}$$

$$C_k(z_1, z_2) = \pm C_k^*(1/z_1^*, 1/z_2^*). \tag{6.4-14}$$

Factors of the form given in Eqs. (6.4-13) and (6.4-14) will be said to be *symmetric.*

Proceeding as we did in the previous section, let us now restrict our attention to the class of sequences $x(m, n)$ which have finite support. Convolving $x(m, n)$ with a zero-phase sequence which has a rational z-transform of the form given in Eq. (6.4-12), we can make the following observations. If the denominator factors $C_k(z_1, z_2)$ are not canceled by identical factors in $X(z_1, z_2)$, then $Y(z_1, z_2) = X(z_1, z_2)G(z_1, z_2)$ will be a rational function of z_1 and z_2 and, therefore, will be the z-transform of a sequence which does not have finite support. If, on the other hand, the denominator factors are canceled by factors of $X(z_1, z_2)$, then, by definition, $X(z_1, z_2)$ contains symmetric factors. Finally, since $X(z_1, z_2)$ is a polynomial, any numerator factors $A_k(z_1, z_2)$ will not be canceled by $X(z_1, z_2)$ and will show up as symmetric factors in $Y(z_1, z_2)$. Therefore, as we did for 1-D sequences, if we restrict our attention to that class of 2-D sequences which has finite support and z-transforms with no symmetric factors, then there may exist, at most, one sequence within this class having a given phase function. We thus have the following 2-D version of Theorem 2 [26]:

Theorem 3. A 2-D sequence with finite support which has a z-transform with no symmetric factors is uniquely defined (to within a scale factor) by the phase of its Fourier transform.

As a special case of this theorem, note that if $x(m, n)$ has a z-transform of the form

$$X(z_1, z_2) = \alpha z_1^{k_1} z_2^{k_2} P(z_1, z_2), \tag{6.4-15}$$

where k_1 and k_2 are integers and $P(z_1, z_2)$ is a polynomial which is irreducible and nonsymmetric, then $x(m, n)$ satisfies the constraints of the theorem. Since "almost all" polynomials in two variables are irreducible and thus of the form given in Eq. (6.4-14), it follows that "almost all" 2-D sequences are uniquely defined to within a scale factor by the phase of their Fourier transform.

One of the limitations of Theorems 2 and 3 is that the phase of the Fourier transform of $x(n)$ or of $x(m, n)$ is assumed to be known for all frequencies. Fortunately, however, it is possible to generalize these theorems to show that uniqueness still holds for the case in which the phase of the discrete Fourier transform (DFT) of $x(n)$ or $x(m, n)$ is known [26]. The importance of this generalization lies in the fact that it allows for the development of practical numerical algorithms for reconstructing signals from their phase [26–28].

C. Reconstruction Techniques

A number of different algorithms have been proposed for reconstructing a signal from the phase of its Fourier transform [26–28]. Since some of these techniques are treated elsewhere in this volume, in this section we present only a brief overview of two phase-only reconstruction algorithms. The first is a closed form solution which requires finding the solution to a set of linear equations. The second is an iterative procedure which provides the flexibility for incorporating additional signal constraints. In the presentation which follows, in order to ensure that there is a unique solution to the reconstruction problem, it will be assumed that $x(m, n)$ has support $R(M, N)$ and that its z-transform $X(z_1, z_2)$ is an irreducible polynomial. For convenience, it will also be assumed that $x(0, 0) = \alpha_0$ is nonzero and known. In this case, α_0 along with the phase of the $2M \times 2N$-point DFT of $x(m, n)$, $\phi_x(k, l)$, uniquely specify $x(m, n)$ [26].

One algorithm for reconstructing $x(m, n)$ from the phase of its Fourier transform follows from the definition of $\phi_x(\mu, \nu)$. Specifically, it may be easily shown that $x(m, n)$ must satisfy the following equation for any values of μ and ν [26]:

$$\sum_{(m,n)\neq(0,0)}^{M-1} \sum^{N-1} x(m, n) \sin[\phi_x(\mu, \nu) + m\mu + n\nu] = -\alpha_0 \sin \phi_x(\mu, \nu) \qquad (6.4\text{-}16)$$

provided $\phi_x(\mu, \nu) \neq \pm\pi/2$. For the case in which $\phi_x(\mu, \nu) = \pm\pi/2$, $x(m, n)$ must satisfy the equation

$$\sum_{(m,n)\neq(0,0)}^{M-1} \sum^{N-1} x(m, n) \cos(m\mu + n\nu) = -\alpha_0. \qquad (6.4\text{-}17)$$

Substituting the values of the phase of the $2M \times 2N$-point DFT of $x(m, n)$ into Eq. (6.4-16) leads to a set of Q linear equations in the P unknowns of $x(m, n)$, where $P = MN$ and $Q = 4MN$ (due to the symmetry of the phase, half of these equations are redundant and may be eliminated). If we arrange the elements of $x(m, n)$ into a vector $\mathbf{v}_x$ of length P, these equations, when augmented with the equation $x(0, 0) = \alpha_0$, may be written as

$$\mathbf{A}\mathbf{v}_x = -\alpha_0 \mathbf{b}, \tag{6.4-18}$$

where $\mathbf{A}$ is a matrix with P columns and Q rows and $\mathbf{b}$ is a vector of length Q. As a result of the assumptions made on $x(m, n)$, it follows from the results of Theorem 3 that the columns of $\mathbf{A}$ must be linearly independent (otherwise multiple solutions would exist). Therefore, the matrix $\mathbf{S} = \mathbf{A}^T\mathbf{A}$ is nonsingular and the desired sequence may be reconstructed from

$$\mathbf{v}_x = -\alpha_0(\mathbf{A}^T\mathbf{A})^{-1}\mathbf{A}^T\mathbf{b}. \tag{6.4-19}$$

Although this algorithm has been used successfully in reconstructing 2-D sequences of moderate size ($P \leqslant 256$), its application in practice is limited by the computational difficulties encountered in solving linear equations when the number of unknowns becomes large. Therefore, it is of interest to consider alternative solutions to the reconstruction problem. One such algorithm is an iterative approach which is similar in style to the iterative algorithm first proposed by Gerchberg and Saxton [29]. This algorithm may be formally derived by writing $x(m, n)$ as the solution of a nonlinear fixed point equation of the form $x(m, n) = \mathbf{F}_\phi[x(m, n)]$, where $\mathbf{F}_\phi[\cdot]$ is a nonlinear operator which imposes the known constraints on $x(m, n)$, that is, the constraint that $x(m, n)$ has support $R(M, N)$ with $x(0, 0) = \alpha_0$ and the constraint that the phase of the $2M \times 2N$-point DFT of $x(m, n)$ is equal to $\phi_x(k, l)$. Then, using the method of successive approximations, the solution to this fixed point equation is iteratively determined [26, 27]. More specifically, consider the *windowing* operator $\mathbf{D}[\cdot]$ which imposes the known time (space) domain constraints:

$$\mathbf{D}[y(m, n)] = \begin{cases} y(m, n) & \text{if } (m, n) \in R(M, N) \text{ and } (m, n) \neq (0, 0), \\ \alpha_0 & \text{if } (m, n) = (0, 0), \\ 0 & \text{otherwise.} \end{cases} \tag{6.4-20}$$

In addition, consider the signal-dependent operator $\mathbf{C}_\phi[y(m, n)]$, where $\mathbf{C}_\phi = \mathbf{W}^{-1}\mathbf{\Phi}\mathbf{W}$ is a cascade of three operators. The first, $\mathbf{W}[\cdot]$, is the DFT operator defined by

$$\mathbf{W}[y(m, n)] = \sum_{m=0}^{M-1} \sum_{n=0}^{N-1} y(m, n) \exp\left[-j\frac{2\pi}{M}mk - j\frac{2\pi}{N}nl\right] \equiv Y(k, l). \tag{6.4-21}$$

The second operator, $\Phi[\cdot]$, replaces the phase of $Y(k, l)$ with the known phase, $\phi_x(k, l)$ of $X(k, l)$:

$$\Phi[Y(k, l)] = |Y(k, l)| \exp[j\phi_x(k, l)]. \qquad (6.4\text{-}22)$$

Finally, $\mathbf{W}^{-1}[\cdot]$ is the inverse DFT operator. Therefore, $x(m, n)$ satisfies the nonlinear fixed point equation

$$x(m, n) = \mathbf{F}_\phi[x(m, n)] = \mathbf{C}_\phi \mathbf{D}[x(m, n)]. \qquad (6.4\text{-}23)$$

Even more important, however, as a consequence of Theorem 3, it follows that $x(m, n)$ is the unique solution to Eq. (6.4-23). Having $x(m, n)$ as the unique solution to a fixed-point equation, an iterative approach based on the method of successive approximations may be used to solve the fixed-point equation for $x(m, n)$ [14, 26, 30]. This iterative algorithm is given by

$$x_{p+1}(m, n) = \mathbf{F}_\phi[x_p(m, n)], \qquad (6.4\text{-}24)$$

where $x_p(m, n)$ is the pth approximation or estimate of the fixed point $x(m, n)$. Note that this fixed-point iteration involves the successive application of the known signal constraints. Specifically, in the time (space) domain the finite support constraint is incorporated, whereas in the frequency domain the known spectral phase information is imposed. A block diagram which more clearly illustrates the iteration is shown in Fig. 6.4-1.

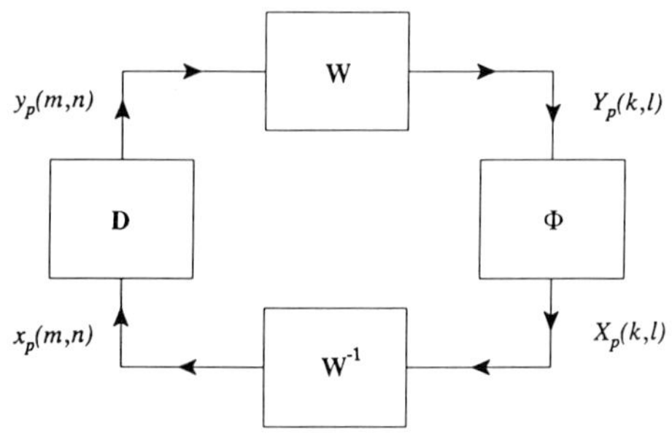

Fig. 6.4-1. Block diagram of iterative approach to solution of Eq. (6.4-23).

A fundamental question of considerable practical and theoretical importance concerns the conditions under which the estimate $x_p(m, n)$ converges to the correct sequence $x(m, n)$ as $p \to \infty$. By defining the distance between two sequences using the standard Euclidean metric

$$d(\mathbf{x}, \mathbf{y}) = \sum_{m=0}^{M-1} \sum_{n=0}^{N-1} [x(m, n) - y(m, n)]^2 \qquad (6.4\text{-}25)$$

it is straightforward to verify that $\mathbf{F}_\phi[\cdot]$ is a *strictly nonexpansive operator*

[26,30], that is,

$$d[\mathbf{F}_\phi \mathbf{x}, \mathbf{F}_\phi \mathbf{y}] < d(\mathbf{x}, \mathbf{y}) \qquad \text{provided } \mathbf{x} \neq \mathbf{y} \tag{6.4-26}$$

for any two sequences $x(m, n)$ and $y(m, n)$. As a result, it may be shown that for any initial estimate $x_0(m, n)$, the iteration converges to the desired solution

$$\lim_{p\to\infty} x_p(m, n) = x(m, n) \tag{6.4-27}$$

provided only that the constraint operators $\mathbf{D}[\cdot]$ and $\Phi[\cdot]$ define a unique fixed-point solution to Eq. (6.4-23) [26, 30]. This theoretical result has been confirmed experimentally by running many different types of examples. In all cases it has been observed that the iteration given in Eq. (6.4-24) converges to the correct signal.

6.5 THE PHASE RETRIEVAL PROBLEM

In this section we consider the reconstruction of a discrete 2-D signal from the magnitude of its Fourier transform. Just as we demonstrated for the magnitude retrieval problem, without any additional information or constraints, a signal is not uniquely specified by the magnitude of its Fourier transform [14, 26]. For example, a signal may always be shifted (delayed) or multiplied by a factor of -1 without changing the magnitude of its Fourier transform. Thus, any signal $y(m, n)$ which is related to $x(m, n)$ by

$$y(m, n) = \pm x(m + k, n + l) \tag{6.5-1}$$

for any integers k and l has a Fourier transform with the same magnitude as $x(m, n)$. In addition, rotating a signal by 180° [or reflection about the point $(0, 0)$] will not change the magnitude of its Fourier transform, so $x(m, n)$ and

$$y(m, n) = x(-m, -n) \tag{6.5-2}$$

have the same Fourier transform magnitude. In most applications, such ambiguities are of no concern since it is the relative amplitudes of the various samples within the signal that are of primary importance. Therefore, two signals $x(m, n)$ and $y(m, n)$ will be said to be *equivalent* if they are related to one another by any combination of the relations in Eq. (6.5-1) and (6.5-2), and this equivalence relation will be denoted by

$$y(m, n) \sim x(m, n) \tag{6.5-3}$$

Excluding these trivial ambiguities, it is still not possible to uniquely define a signal in terms of its spectral magnitude because it is always possible to convolve a signal with an arbitrary *all-pass* signal (one which has a Fourier transform with unit modulus) to obtain another signal with the same spectral magnitude. In the following section, the uniqueness issue is addressed for the case of one-dimensional signals. Having developed the geometric properties of the problem, the two-dimensional uniqueness question is then described.

A One-Dimensional Signals

Let $x(n)$ be a one-dimensional sequence with support $R(N)$. With $X(e^{j\omega})$ the Fourier transform of $x(n)$, it follows that the inverse Fourier transform of $|X(e^{j\omega})|^2$ is the autocorrelation of $x(n)$:

$$r_x(n) = x(n) * x(-n) = \mathscr{F}^{-1}[|X(e^{j\omega})|^2]. \qquad (6.5\text{-}4)$$

Therefore, if $x(n)$ is uniquely defined by the magnitude of its Fourier transform, it must also be uniquely defined by its autocorrelation. It is well known that without any additional information this is not the case. More specifically, note that for any sequence $x(n)$ it is always possible to find another sequence which has a Fourier transform with the same magnitude by simply convolving $x(n)$ with an *all-pass* sequence $g(n)$ where $|\mathscr{F}[g(n)]| = 1$:

$$y(n) = x(n) * g(n)\,; \quad |G(e^{j\omega})| = 1. \qquad (6.5\text{-}5)$$

Conversely, if two sequences, $x(n)$ and $y(n)$, have Fourier transforms with the same magnitude, then $G(z) = Y(z)/X(z)$ is the z-transform of an all-pass sequence and $x(n)$ and $y(n)$ are related by convolution with $g(n)$ as given in Eq. (6.5-5).

As we did for zero-phase sequences, it will be important in the phase retrieval problem to look at the z-transform properties of an all-pass sequence. Note that if $g(n)$ is an all-pass sequence, then by definition $|G(e^{j\omega})|^2 = 1$, which [since $g(n)$ is real] implies that

$$G(e^{j\omega})G^*(e^{-j\omega}) = 1. \qquad (6.5\text{-}6)$$

Therefore, $g(n) * g^*(-n) = \delta(n)$ and

$$G(z)G^*(1/z^*) = 1. \qquad (6.5\text{-}7)$$

It follows that the singularities of $G(z)$ must occur in conjugate reciprocal pole/zero pairs, so that if $G(z)$ has a zero (pole) at $z = a$ then $G(z)$ must also have a pole (zero) at $z = 1/a^*$. Therefore, if $g(n)$ is an all-pass sequence then its z-transform must be of the form (assuming a rational function):

$$G(z) = z^{-n_0} \frac{A(z)}{A^*(1/z^*)} = z^{-(n_0+p)} \prod_{k=1}^{p} \frac{1 - a_k z^{-1}}{z^{-1} - a_k^*}, \tag{6.5-8}$$

where $A(z)$ is a polynomial in z. Conversely, any rational all-pass signal has a z-transform which may be written in the form given in Eq. (6.5-8). Therefore, it follows that the effect of convolving a sequence $x(n)$ with an all-pass sequence $g(n)$ is to

(a) Add or delete conjugate reciprocal pole/zero pairs (all-pass factors),

or

(b) Replace a zero (pole) at $z = a$ with a zero (pole) at $z = 1/a^*$.

The transformation in (b) has been referred to as *zero flipping*. Specifically, with $x(n)$ a 1-D finite-length sequence with support $R(N)$, its z-transform is a product of $(N - 1)$ linear factors of the form $(1 - a_k z^{-1})$ as given in Eq. (6.3-1). If any of these factors are replaced with a factor of the form $(z^{-1} - a_k^*)$, then the corresponding sequence $y(n)$ will still have support $R(N)$ and the magnitude of its Fourier transform will be unchanged. This follows from the fact that $Y(z) = X(z)G_k(z)$, where

$$G_k(z) = (1 - a_k z^{-1})/(z^{-1} - a_k^*) \tag{6.5-9}$$

is the transform of an all-pass signal. Since the effect of $H_k(z)$ on $X(z)$ is to *flip* the zero of $X(z)$ at $z = a_k$ about the unit circle, it follows that if an arbitrary number of zeros of $X(z)$ are flipped about the unit circle the result will be a signal of the same duration with a Fourier transform with the same magnitude as $x(n)$. Clearly, the same results apply to poles. Specifically, if $x(n)$ has a z-transform containing poles and if an arbitrary number of these poles are flipped about the unit circle, then the result will be a signal having a Fourier transform with the same magnitude as $x(n)$. Therefore, in order to guarantee a unique solution, it is necessary to place constraints on $x(n)$ in order to resolve the ambiguity about the side of the unit circle on which each pole and each zero lies.

Note that if $x(n)$ has a z-transform with all of its singularities (poles and zeros) inside the unit circle, i.e., $x(n)$ is minimum phase, then convolution with an all-pass sequence will always produce a sequence which has at least one singularity outside the unit circle. Similarly, if all of the singularities are outside the unit circle, i.e., $x(n)$ is maximum phase, then convolution with an all-pass sequence will produce a sequence with at least one singularity inside the unit circle. In other words, convolution with an all-pass sequence does not preserve the minimum and maximum phase property. We therefore have the following theorem [10]:

Theorem 4. Any minimum phase sequence or maximum phase sequence is uniquely defined (to within a sign and a delay) by the magnitude of its Fourier transform.

Again it should be emphasized that this result asserts that if we restrict our set of admissible solutions to the set of minimum phase sequences or to the set of maximum phase sequences, then there exists only one sequence with a given magnitude. Given a magnitude function, the minimum phase and maximum phase sequence may be found with a Hilbert transform [10].

In the previous section we saw that a finite-length constraint along with the exclusion of zero-phase factors is sufficient to guarantee that a one-dimensional signal is uniquely defined by the phase of its Fourier transform. Due to the possibility of zero flipping, however, a similar set of constraints (i.e., finite length and the exclusion of all-pass factors) will not guarantee that a one-dimensional signal is uniquely defined in terms of the magnitude of its Fourier transform. Nevertheless, consider the signal $x(n) = \delta(n) + a\delta(n-1)$ which has a z-transform with only one zero. Given the magnitude of the Fourier transform of $x(n)$, i.e.,

$$|X(e^{j\omega})|^2 = 1 + a^2 + 2a \cos \omega, \tag{6.5-10}$$

the only other finite-length signals which have the same magnitude are delayed versions of the signal $x(n)$ and delayed versions of the "time-reversed" signal $\tilde{x}(n) = a\delta(n) + \delta(n-1)$. In other words, for a one-dimensional signal with a z-transform, $X(z)$, containing only *one factor*, the signal is uniquely defined to within a linear phase shift and a time reversal by the magnitude of its Fourier transform. This result may be stated more formally as follows [26]:

Theorem 5. A finite-length signal which has an irreducible z-transform is uniquely defined (to within a sign, a time shift, and a time reversal) by the magnitude of its Fourier transform.

Since the only one-dimensional signals which have irreducible z-transforms are those which are of length $N = 1$ or $N = 2$, this result is not particularly useful. The significance of this result, however, lies in the fact that we will find the same constraint in the case of two-dimensional signals. In two dimensions, however, the class of signals which have z-transforms with only *one factor* are much richer.

B. Two-Dimensional Signals

In the previous section it was shown that without any additional information or constraints a 1-D sequence is not uniquely defined by the magnitude of its Fourier transform. The ambiguity arises as a result of the possibility of zero flipping and pole flipping. Therefore, additional information must be provided which specifies the side of the unit circle on which each singularity resides. In contrast with the case for known Fourier transform phase, a finite-length constraint does not provide the information necessary for a unique characterization. Nevertheless, due to the irreducibility of almost all polynomials in two or more variables, an important and useful uniqueness theorem may be developed. In particular, it may be shown that almost all 2-D sequences with finite support will be uniquely defined by the magnitude of their Fourier transform.

By following the development in the previous section for 1-D signals, it follows that two 2-D sequences $x(m, n)$ and $y(m, n)$ have Fourier transforms with the same magnitude if and only if one is related to the other by convolution with an all-pass sequence:

$$y(m, n) = x(m, n) ** g(m, n), \tag{6.5-11}$$

where $|\mathscr{F}[g(m, n)]| = 1$. As in the case of 1-D sequences, any 2-D all-pass sequence $g(m, n)$ has a z-transform with the following property:

$$G(z_1, z_2)G^*(1/z_1^*, 1/z_2^*) = 1. \tag{6.5-12}$$

Therefore, it follows that any all-pass sequence which has a rational z-transform must have a z-transform of the following form:

$$G(z_1, z_2) = \frac{H(z_1, z_2)}{H^*(1/z_1^*, 1/z_2^*)} = \prod_{k=1}^{p} \frac{H_k(z_1, z_2)}{H_k^*(1/z_1^*, 1/z_2^*)}, \tag{6.5-13}$$

where $H_k(z_1, z_2)$ are the irreducible factors of $H(z_1, z_2)$.

Just as in the one-dimensional case, multiple solutions to the 2-D phase retrieval problem generally exist due to the possibility of *zero contour flipping*. Specifically, with $x(m, n)$ a 2-D sequence with support $R(M, N)$, its z-transform may be written as a product of p irreducible polynomials $A_k(z_1, z_2)$ as in Eq. (6.3-2). If any of these irreducible factors are replaced with the factor $A_k^*(1/z_1^*, 1/z_2^*)$, then the resulting 2-D sequence $y(m, n)$ will also have finite support and the magnitude of its Fourier transform will be proportional to that of $x(m, n)$. As in the 1-D case, this follows from the fact that $Y(z_1, z_2) = X(z_1, z_2)H_k(z_1, z_2)$, where

$$H_k(z_1, z_2) = A_k^*(1/z_1^*, 1/z_2^*)/A_k(z_1, z_2) \tag{6.5-14}$$

is the z-transform of an all-pass 2-D signal which *flips* the zero contour $A_k(z_1, z_2)$ about the unit bicircle. Note, however, that if $X(z_1, z_2)$ is irreducible, then the only other sequences which have the same Fourier transform magnitude as $x(m, n)$ are those which are related to $x(m, n)$ by a linear shift, multiplication by -1, or by reflecting the sequence about the origin [replacing $X(z_1, z_2)$ with $X^*(1/z_1^*, 1/z_2^*)$]. More specifically, we have the following two-dimensional version of Theorem 5 [26]:

Theorem 6. A 2-D sequence with finite support which has an irreducible z-transform is uniquely defined (to within a sign, a time shift, and a time reversal) by the magnitude of its Fourier transform.

Since most polynomials in two or more variables are irreducible, Theorem 6 implies that almost all 2-D sequences with finite support are uniquely defined by the magnitude of their Fourier transform. It should be pointed out that this theorem assumes that the Fourier transform magnitude is known for all frequencies. As is the case for phase, however, it is possible to generalize this theorem to show that uniqueness still holds if the magnitude of the DFT of $x(m, n)$ is known [26].

Finally, it should be pointed out that a slightly more general result may be derived which allows the z-transform of a 2-D sequence to contain an arbitrary number of symmetric factors [26]. Since symmetric factors are invariant under rotations, zero contour flipping of these factors will not change the sequence.

C. Reconstruction Techniques

Because of its importance in the phase retrieval problem for wave amplitudes and coherence functions, an algorithm for reconstructing a 2-D sequence from the magnitude of its Fourier transform has been the objective of many research efforts and the subject of a considerable number of published papers. Although in almost all cases a 2-D sequence with finite support is uniquely defined (to within some trivial ambituities) by the magnitude of its Fourier transform, reconstrucing the sequence remains a very difficult problem in the absence of additional information or constraints. It appears that there are at least two problems which have made the development of a reconstruction algorithm difficult. The first is the nonlinear relationship between the coefficients of a 2-D sequence and the magnitude of its Fourier transform. Recall, for example, that the magnitude of the Fourier transform of a sequence $x(m, n)$ may be used to obtain the autocorrelation $r_x(m, n)$ of $x(m, n)$. Therefore, with magnitude information alone it is always possible to define a set of second-order equations which relate the known values of the autocorrelation of $x(m, n)$ at various lags with the unknown coefficients of the desired 2-D sequence, e.g.,

$$r_x(m, n) = \sum_{k=-\infty}^{\infty} \sum_{l=-\infty}^{\infty} x(m + k, n + l)x(k, l). \qquad (6.5\text{-}15)$$

One possible solution to the phase retrieval problem may thus consist of attempting to solve these nonlinear equations for $x(m, n)$. Although there exist techniques for finding solutions to a set of simultaneous nonlinear equations in more than one unknown, when the number of equations and the number of unknowns become large, these algorithms are not generally practical.

The second problem which has contributed to the difficulty in developing a practical phase retrieval algorithm is that, without any phase information, it is not generally possible to obtain a very accurate estimate of the unknown sequence from only the magnitude of its Fourier transform. Recall, for example, that if the Fourier transform magnitude of an an image is combined with either zero phase or an ensemble average phase function, then the result obtained with an inverse Fourier transform is an image which generally does not contain any recognizable features (see Figs. 6.2-5 and 6.2-6). In contrast, a phase-only image which has a Fourier transform with the correct phase and a constant magnitude contains many of the important characteristics of the original image (see Fig. 6.2-2).

Nevertheless, a number of different phase retrieval algorithms have been proposed [3, 4, 14, 26, 31–33]. One of these is an iterative algorithm which is similar in style to the phase-only iteration described in the previous section [26]. Specifically, again assuming that $x(m, n)$ has support $R(M, N)$ and that $X(z_1, z_2)$ is irreducible, then $x(m, n)$ satisfies the fixed-point equation

$$x(m, n) = \mathbf{F}_A[x(m, n)] = \mathbf{C}_A\mathbf{D}'[x(m, n)], \tag{6.5-16}$$

where $\mathbf{D}'$ is the operator which imposes the known time (space) domain constraints and $\mathbf{C}_A$ is the operator which imposes the known frequency domain constraints. In particular, $\mathbf{D}'[\cdot]$ is the window operator given by

$$\mathbf{D}'[y(m, n)] = \begin{cases} y(m, n) & \text{if } (m, n) \in R(M, N), \\ 0 & \text{otherwise.} \end{cases} \tag{6.5-17}$$

The operator $\mathbf{C}_A$, on the other hand, is a cascade of three operators, $\mathbf{C}_A = \mathbf{W}^{-1}\mathbf{A}\mathbf{W}$, where $\mathbf{W}$ and $\mathbf{W}^{-1}$ denote the forward and inverse DFT operators, respectively, and $\mathbf{A}[\cdot]$ is a nonlinear signal-dependent operator which replaces the magnitude of the DFT of a sequence with the known magnitude $|X(k, l)|$ of the DFT of $x(m, n)$:

$$\mathbf{A}[Y(k, l)] = |X(k, l)| \exp[j\phi_y(k, l)]. \tag{6.5-18}$$

As with the phase-only iteration, use of the method of successive approximations leads to the iterative algorithm

$$x_{p+1}(m, n) = \mathbf{F}_A[x_p(m, n)], \tag{6.5-19}$$

where $x_p(m, n)$ is the pth approximation or estimate of the desired solution $x(m, n)$. The basic structure of this iteration is the same as that of the phase-only iteration shown in Fig. 6.4-1. Specifically, the algorithm involves repeated Fourier transformation between the time (space) and frequency domains, where in each domain the known information about the desired signal is imposed on the current estimate. In the time domain, for example, the finite support constraint is imposed, whereas in the frequency domain the sequence is constrained to have a given Fourier transform magnitude. Unlike the phase-only iteration, however, it has been observed that the magnitude-only iteration will not generally converge to the correct sequence even if the desired solution satisfies the constraints of Theorem 6 [26].

6.6 SUMMARY AND OTHER PROBLEMS

In this chapter we consider the problem of reconstructing a signal from either the phase or the magnitude of its Fourier transform. In both of these problems the Fourier transform of the signal is incompletely specified, so the goal is to find signal constraints that will allow for a unique reconstruction. It was shown that if a one-dimensional signal is constrained to be finite in length and to have no zero-phase factors (conjugate reciprocal zeros), then the signal is uniquely defined by the phase of its Fourier transform. Similarly, if a two-dimensional signal is constrained to have finite support and to have no zero-phase factors (symmetric factors in its z-transform), then it is uniquely defined by the phase of its Fourier transform. In both the one-dimensional and the two-dimensional case, assuming that a signal is uniquely defined by the phase of its Fourier transform and assuming that this phase information is known exactly, then there exist a number of different algorithms which will reconstruct the signal from this phase information.

For the phase retrieval problem, on the other hand, it was shown that a finite-length constraint is not sufficient to guarantee that a one-dimensional signal is uniquely defined by the magnitude of its Fourier transform due to the possibility of zero flipping. However, since almost all polynomials in two or more variables are irreducible, it was shown that a finite-support constraint is sufficient (in most cases) to ensure that a two-dimensional signal is uniquely defined by the magnitude of its Fourier transform. In spite of this uniqueness property, it is generally very difficult to reconstruct a two-dimensional signal from magnitude information alone.

Because of the importance of the phase retrieval problem in many applications and the difficulties surrounding the reconstruction of two-dimensional signals from magnitude information alone, a number of people have considered including additional signal constraints. For example, many two-dimensional signals of interest are nonnegative. Therefore, positivity constraints are often used in signal reconstruction algorithms. As another example, since many of the important features of an image are preserved when it is synthesized from only one bit of phase information, the possibility of including one bit of phase information has been considered [34, 35]. For one-dimensional signals it has been shown that the ambiguity introduced by zero flipping is eliminated with the addition of one bit of phase information along with a causality constraint. In particular, it has been shown that a causal one-dimensional signal is uniquely defined by the magnitude of its Fourier transform and one bit of phase [34]. Although one bit of phase information is not required for a unique solution to the two-dimensional phase retrieval problem, it has been observed that with this phase information an iterative algorithm similar to that described in Section 6.5 may be used to successfully reconstruct the signal in most cases [34]. Carrying this idea one step further, the uniqueness of a 2-D signal in terms of one bit of phase information and no magnitude information has also been considered [36]. In particular, conditions have been developed which are sufficient for a 2-D signal to be uniquely defined by one bit of phase information, and an iterative algorithm for reconstructing the signal from this phase information has been studied.

REFERENCES

1. G. N. Ramachandran and R. Srinivasan, "Fourier Methods in Crystallography," Wiley-Interscience, New York, 1970.
2. W. O. Saxton, "Computer Techniques for Image Processing in Electron Microscopy," Academic Press, New York, 1978.
3. H. A. Ferwerda, The phase reconstruction problem for wave amplitudes and coherence functions. *In* "Inverse Source Problems in Optics," (edited by H. P. Baltes) Springer-Verlag, Berlin, 1978, Chapter 2.
4. L. S. Taylor, The phase retrieval problem. *IEEE Trans. Ant. and Prop.* **AP-29** (2), 386–391 (1981).
5. J. R. Fienup, Space object imaging through the turbulent atmosphere. *Opt. Eng.* 529–534 (1979).
6. A. V. Oppenheim, J. S. Lim, G. Kopek, and S. C. Pohlig, Phase in speech and pictures. *Proc. IEEE Int. Conf. on Acoust., Speech, Sig. Proc.* 632–637 (1979).
7. T. G. Stockham, T. M. Cannon, and R. B. Ingebretson, Blind deconvolution through digital signal processing. *Proc. IEEE* **63**, 678–692 (1975).
8. H. C. Andrews and B. R. Hunt, "Digital Image Restoration," Prentice-Hall, Englewood-Cliffs, New Jersey, 1977.
9. Y. T. Li and A. L. Kurkjian, Arrival time determination using iterative signal reconstruction from the phase of the cross spectrum. *IEEE Trans. Acoust., Speech, Sig. Proc.* **ASSP-31**, 502–504 (1983).

10. A. V. Oppenheim and R. W. Schafer, "Digital Signal Processing," Prentice-Hall, Englewood Cliffs, New Jersey, 1975.
11. D. E. Dudgeon and R. M. Mersereau, "Multidimensional Digital Signal Processing," Prentice-Hall, Englewood Cliffs, New Jersey, 1984.
12. A. V. Oppenheim and J. S. Lim, The importance of phase in signals. *Proc. IEEE* **69**, 529–541 (1981).
13. R. Srinivasan, The significance of the phase synthesis. *Proc. Indian Acad. Sci.* **53A**, 252–261 (1961).
14. M. H. Hayes, Signal reconstruction from phase or magnitude, Sc.D Thesis, Massachusetts Institute of Technology, Cambridge, Massachusetts, 1981.
15. T. S. Huang, J. W. Burnett, and A. G. Deczky, The importance of phase in image processing filters. *IEEE Trans. Acoust., Speech, Sig. Proc.* **ASSP-23** (6), 529–542 (1975).
16. L. Larimore, H. M. A. El-Sum, and A. E. Metherell, Acoustical holograms using phase information only. *Applied Optics* **8**, 1533–1536 (1969).
17. L. B. Lesem, P. M. Hirsch, and J. A. Jordan Jr., The kinoform: A new wavefront reconstruction device. *IBM J. Res. Dev.* 150–155 (1969).
18. A. F. Metherell, The relative importance of phase and amplitude in acoustical holography, *In* "Acoustical Holography, Vol. 2," (A. F. Metherell and L. Larimore, eds.), Plenum Press, New York, 1970, Chapt. 14.
19. J. Powers, J. Landry, and G. Wade, Computed reconstructions from phase-only and amplitude-only holograms. *In* "Acoustical Holography, Vol. 2," (A. F. Metherell and L. Larimore, eds.), Plenum Press, New York, 1970, Chapt. 13.
20. A. Mostowski and M. Stark, "Introduction to Higher Algebra," Pergamon Press, New York, 1964.
21. Yu. M. Bruck and L. G. Sodin, On the ambiguity of the image reconstruction problem. *Optics Comm.* **30** (3), 304–308, 1979.
22. M. A. Fiddy, B. J. Brames, and J. C. Dainty, Enforcing irreducibility for phase retrieval in two dimensions. *Optics Letters* **8** (7), 96–98 (1983).
23. M. H. Hayes and J. H. McClellan, Reducible polynomials in two or more variables. *Proc. IEEE* **70** (2), 197–198 (1982).
24. J. L. C. Sanz and T. S. Huang, Stability of unique Fourier-transform phase reconstruction. *J. Opt. Soc. Am.* **73** (11), 1442–1445 (1983).
25. M. H. Hayes, J. S. Lim, and A. V. Oppenheim, Signal reconstruction from the phase or magnitude of its Fourier transform. *IEEE Trans. Acoust., Speech, Sig. Proc.* **ASSP-28**, 670–680 (1980).
26. M. H. Hayes, The reconstruction of a multidimensional sequence from the phase or magnitude of its Fourier transform. *IEEE Trans. Acoust., Speech, Sig. Proc.* **ASSP-30** (2), 140–154 (1982).
27. A. V. Oppenheim, M. H. Hayes, and J. S. Lim, Iterative procedures for signal reconstruction from Fourier transform phase. *Optical Engineering* **21** (1), 122–127 (1982).
28. A. Levi and H. Stark, Signal reconstruction from phase by projections onto convex sets. *J. Opt. Soc. Am.* **73** (6), 810–822 (1983).
29. R. Gerchberg and W. O. Saxton, A practical algorithm for the determination of phase from image and diffraction plane pictures. *Optik 35*, 237–246 (1972).
30. V. T. Tom, T. F. Quatieri, M. H. Hayes, and J. H. McClellan, Convergence of iterative nonexpansive signal reconstruction algorithms. *IEEE Trans. Acoust., Speech, Sig. Proc.* **ASSP-29** (5), 1052–1058 (1981).
31. A. Levi and H. Stark, Image restoration by the method of generalized projections with application to restoration from magnitude. *J. Opt. Soc. Am.*, **1** (9), 932–943 (1983).
32. J. R. Fienup, Phase retrieval algorithms: A comparison. *Applied Optics*, **21** (15), 2758–2769 (1982).

33. M. H. Hayes and T. F. Quatieri, Recursive phase retrieval using boundary conditions. *J. Opt. Soc. Am.* **73**, 1427–1433 (1983).
34. P. L. Van Hove, M. H. Hayes, J. S. Lim, and A. V. Oppenheim, Signal reconstruction from signed Fourier transform magnitude. *IEEE Trans. Acoust., Speech, Sig. Proc.* **ASSP-31** (5), 1286–1293 (1983).
35. A. V. Oppenheim, J. S. Lim, and S. R. Curtis, Signal synthesis and reconstruction from partial Fourier domain information. *JOSA* **73** (11), 1413–1420 (1983).
36. S. R. Curtis, J. S. Lim, and A. V. Oppenheim, Signal reconstruction from one bit of phase information. *Proc. 1984 Int. Conf. Acoust., Speech, Sig. Proc.* 12A5.1–12A5.4 (1984).

CHAPTER 7

Phase Retrieval and Image Reconstruction for Astronomy

J. C. Dainty

Blackett Laboratory
Imperial College
London SW7 2BZ, England

J. R. Fienup

Environmental Research
Institute of Michigan
Ann Arbor, Michigan 48107

7.1 INTRODUCTION

In optical astronomy, the traditional way to obtain an object map or object intensity $f(x, y)$† is to use a single large telescope to record a conventional long-exposure image. This image can be enhanced, if necessary, by using the standard techniques of digital image processing. Although this may, at first sight, seem to be the most obvious method of estimating the object intensity, it suffers from two drawbacks. First, for earth-based observation, atmospheric turbulence limits the resolution obtainable to approximately 1 arc-sec (at the best observing sites), compared to the limit set by diffraction, $\alpha_{\min} \sim \lambda / D$, where λ is the wavelength and D the telescope diameter. This limit is approximately 0.02 arc-sec for a 5-m telescope at a wavelength of 400 nm. Second, ignoring atmospheric turbulence, the diameter D of a single large telescope limits the resolution (as above) and it is difficult to imagine a telescope with a diameter greater than about 50 m in the foreseeable future.

The solution to the problem of limited resolution lies with a family of techniques which can be described as *interferometric imaging*. The basis of

† In astronomy, it is customary to use angular variables (α, β) instead of the spatial variables (x, y) used in this chapter.

these is the van Cittert–Zernike theorem [1, Sections 5.6 and 7.4], which, in the form of use in astronomy, states that the spatial coherence function $F(u, v)$ in the far field of a thermal light source is proportional to the Fourier transform of the object intensity $f(x, y)$. The spatial coherence of starlight can be measured by a variety of techniques, including speckle and Michelson interferometry.

Some of these interferometric methods are particularly suited to long-baseline and synthetic aperture techniques, in which the "diameter" D is potentially 1 km or more, providing the possibility of sub-milliarcsecond resolution in the future. Synthetic aperture techniques are, of course, widely used for imaging in radio astronomy.

Interferometric imaging techniques in optical astronomy all suffer from one major disadvantage: it is difficult, and sometimes impossible, to get an accurate estimate of the *phase* of $F(u, v)$. We are then faced with the problem of reconstructing the object intensity $f(x, y)$ either from $|F(u, v)|$ alone or with limited phase information. In both cases, the modulus (and phase information, if any) is known only within a finite region of the spatial frequency (u, v) plane, typically in a band extending from 0 to $\sqrt{u_{\max}^2 + v_{\max}^2} = D/(z\lambda)$, where z is the telescope focal length. In addition, $f(x, y)$ is known to be a nonnegative function and of fine extent.

In Section 7.2 we summarize the uniqueness aspects of phase retrieval from modulus-only data (this is discussed in Chapter 6 in more detail). Section 7.3 describes several algorithms for modulus-only data, for both special and general objects. Iterative algorithms have proved to be the most successful in practice, and Section 7.4 describes these and gives detailed instructions on how to obtain satisfactory reconstructions with a particular iterative approach. In the final section, we describe a number of solutions to the phase problem in optical astronomy that are specific to speckle and pupil plane (Michelson) interferometry. All of these utilize the (limited) phase information provided by these particular observing techniques.

Other reviews of stellar speckle interferometry can be found in [2–4].

7.2 UNIQUENESS OF PHASE RETRIEVAL FROM MODULUS DATA

In interferometric imaging, an attempt is made to measure the object's Fourier transform

$$F(u, v) = |F(u, v)| \exp[i\psi(u, v)]$$
$$= \iint_{-\infty}^{\infty} f(x, y) \exp[-i2\pi(ux + vy)]\, dx\, dy, \qquad (7.2\text{-}1)$$

where $f(x, y)$ is the object's brightness or intensity distribution, $|F(u, v)|$ the Fourier modulus, and $\psi(u, v)$ the Fourier phase. In the extreme case of severe atmospheric turbulence, which is ordinarily the case at optical wavelengths, only $|F(u, v)|$ can be measured faithfully and all information about the phase, $\psi(u, v)$, is lost. Then two questions arise: first, how can one reconstruct $f(x, y)$, or equivalently reconstruct $\psi(u, v)$, from $|F(u, v)|$ given the *a priori* information that a physical brightness distribution is a real, nonnegative function? This first question is answered in Sections 7.3 and 7.4, which discuss various reconstruction algorithms. The second question is, given an image produced by a reconstruction algorithm, how does one know whether it is unique? That is, might not there be other real, nonnegative solutions having the same Fourier modulus?

First, it is necessary to recognize that $f(x, y)$, $f(x - x_0, y - y_0)$, and the image rotated by 180° (reflected through the origin), $f(-x - x_0, -y - y_0)$, all have the same Fourier modulus. This fundamental ambiguity does not cause much concern since it is the appearance of the object that is usually sought rather than its position. The position and the 180° rotational ambiguity can easily be resolved in many cases from a very low resolution image of the object. If the above fundamental ambiguities are the only ambiguities, we refer to the object as being unique; only ambiguous solutions having forms different from the object are considered ambiguous.

Until less than one decade ago, there was much pessimism that the phase retrieval problem could be solved or that the solution would be useful, because the available theory at the time, which was for one-dimensional functions, indicated that there were hopelessly many ambiguous solutions [5]. It can be shown that Fourier transforms of one-dimensional functions of finite support are analytic functions of exponential type. They are characterized by their complex zeros, i.e., the locations of the zeros of the (Fourier) function analytically extended into the complex plane, by

$$F(z) = \int_{-\infty}^{\infty} f(x)\, e^{-i2\pi zx}\, dx, \tag{7.2-2}$$

where $z = u + iu'$, u being the physical real axis and u' the imaginary axis. If the set of complex zeros, for which $F(z) = 0$, are $\{z_n\}$, then the Fourier transform can be expressed as the Hadamard product

$$F(z) \propto \prod_{n=1}^{\infty} (1 - z/z_n). \tag{7.2-3}$$

Given $|F(u)|^2 = F(u)F^*(u)$, the complex zeros of its analytic extension into the complex plane, $|F(z)|^2 = F(z)F^*(z^*)$, can be computed, but they include both the zeros $\{z_n\}$ of $F(z)$ and the zeros of $F^*(z^*)$, which are $\{z_n^*\}$.

If $F(z)$ has N complex zeros, then not knowing which of the $2N$ complex zeros of $|F(z)|^2$ go with $F(z)$ and which go with $F^*(z^*)$ leads to a 2^N-fold ambiguity. (Half of these solutions are the mirror images of the other half, reducing the ambiguity to 2^{N-1}-fold.) Ordinarily, N is comparable to the space-bandwidth product of $f(x)$, and so 2^N is an enormous ambiguity, making the one-dimensional problem seem hopeless. Only for the case of objects known to consist of two separated parts is the one-dimensional phase retrieval problem likely to be unique and therefore interesting [6–8].

The first indications that the situation is much different in two dimensions came from empirical reconstruction experiments [9,10]: images that were reconstructed looked like the simulated objects used to compute the Fourier modulus data. Those results gave hope that the two-dimensional problem may be soluble and unique, giving impetus to attempts to extend the one-dimensional theory of analytic functions to two dimensions in such a way that predictions concerning uniqueness could be made. Unfortunately, the theoretical level of understanding of the two-dimensional problem, while progressing [11–16], has not matched that available for the one-dimensional case.

If the object is approximated by an array of delta functions or by an array of numbers in a computer, then its Fourier transform reduces to a discrete Fourier transform which, with a change in variables, can be written as a polynomial of a complex variable. When this discrete approximation is made, one can understand the vast difference between the one-dimensional and two-dimensional phase retrieval problems [16]. It can be shown in the one-dimensional case that the zeros of the polynomial are analogous to the complex zeros for the continuous (analytic functions) case and that there are 2^{N-1} ambiguous solutions, where N is the number of factors into which the polynomial can be factored. Since the fundamental theorem of algebra states that a polynomial of degree N has N prime (irreducible) factors, one has a high degree of ambiguity for the one-dimensional case. As pointed out by Bruck and Sodin [16], it is known that polynomials of two complex variables are rarely factorable. Hence the two-dimensional phase retrieval problem is usually unique (ignoring noise issues). Chapter 6 describes the uniqueness issues for discrete functions in some detail.

Although most two-dimensional discrete objects are unique, one would like to know what conditions ensure uniqueness. As described in Section 6.3, objects whose Fourier transforms satisfy Eisenstein's criterion are unconditionally unique [17]. The type of support that the object must have (i.e., the set of points or the shape within which the object is nonzero) to satisfy this criterion is the only known support which guarantees unconditional uniqueness. The uniqueness of such objects can also be demonstrated by the existence of a recursive algorithm (a closed-form solution) that can reconstruct such objects from their Fourier modulus via their autocorrelation functions [18]. This recursive algorithm can also be used to reconstruct in closed form objects having other supports as well, such as triangular supports having nonzero corners. But for these other objects the support must be known *a priori*, since there may be other solutions having different supports [18]. Objects that are unique among the class of objects having that given support include all objects for which the convex hull of their support is a polygon having no parallel sides [19,20]. Another such class of objects are those satisfying the holography condition: the object consists of an extended part plus a delta function (unresolved component) separated from the extended part by a distance greater than the diameter of the extended part. Then the extended part of the object is found as an isolated term in the autocorrelation function. Note, however, that objects satisfying the holography condition are not necessarily unique among all objects—they are only guaranteed to be unique among the class of objects satisfying the holography condition.

Although it remains to be proved definitively, the general feeling is that the physical case of interest, the two-dimensional continuous case, mirrors the discrete case and will, except for special objects, usually (but not always) be unique. It does not appear that the addition of small amounts of noise will change that assessment [21].

7.3 ALGORITHMS FOR PHASE RETRIEVAL FROM MODULUS

Since most phase retrieval/image reconstruction algorithms are performed on sampled data in the computer, we will employ the discrete Fourier transform

$$F(u, v) = |F(u, v)| \exp[i\psi(u, v)] = \mathscr{F}[f(x, y)]$$
$$= \sum_{x=0}^{N-1} \sum_{y=0}^{N-1} f(x, y) \exp[-i2\pi(ux + vy)/N] \qquad (7.3\text{-}1)$$

and its inverse

$$f(x, y) = N^{-2} \sum_{u=0}^{N-1} \sum_{v=0}^{N-1} F(u, v) \exp[i2\pi(ux + vy)/N], \qquad (7.3\text{-}2)$$

where u, v, x, and $y = 0, 1, 2, \ldots, N-1$ (for simplicity, square arrays are assumed). Any computation of the discrete Fourier transform is, of course, accomplished by using the fast Fourier transform (FFT) method. Note that in order to compute $|F(u, v)|^2$ without aliasing it is required that $f(x, y)$ be zero outside $0 \leq x \leq M-1$ and $0 \leq y \leq M-1$, where $M \leq N/2$. The autocorrelation of $f(x, y)$ is given by

$$r_f(x, y) = \sum_{x'=0}^{N-1} \sum_{y'=0}^{N-1} f(x', y')f(x' - x, y' - y) \tag{7.3-3}$$

$$= \mathscr{F}^{-1}[|F(u, v)|^2]. \tag{7.3-4}$$

The problem of reconstructing $\psi(u, v)$ or $f(x, y)$ from $|F(u, v)|$ or $r_f(x, y)$ has received much attention, especially within the past decade. Section 7.3.A describes several reconstruction methods that are tailored to special types of objects. Most of these methods work only for objects consisting of a collection of delta-function-like points or at least containing a pointlike component. Although they work only for such special cases, they are of interest because in astronomy the existence of pointlike stars is common and these methods tend to be simpler than the methods for more general objects. Section 7.3.B describes more general methods designed to work on most types of astronomical objects (objects of finite extent). Section 7.4 describes in more detail the iterative Fourier transform algorithm, which has been the most general and most successful reconstruction algorithm to date.

A. *Algorithms for Special Types of Objects*

1. Holography

The classic reconstruction method, discussed briefly in Section 7.2, is holography [22]: when the object includes a delta-function component (an unresolved point called the reference point) sufficiently separated from the rest of the object, then the object can be found as one term in the autocorrelation. Although holography was originally conceived for applications employing coherent light, the same principles apply for incoherent light if one has measured the modulus of the Fourier transform of the object [23,24].

2. Recursive Reconstruction Using Latent Reference Points

The closed-form recursive method [18] mentioned in Section 7.2 is akin to holography. It requires the existence of two or more reference points, making it more restrictive than the holography condition, but these reference points can be very close to the extended part of the object, making it less restrictive than the holography condition. These reference points satisfy the holography condition only for the far edge of the object, which is immediately reconstructed from the autocorrelation. After those edge points are solved (including the reference points), then at least one of the autocorrelation values can be expressed as a sum of products of the previously solved points plus the product of an unknown point with a reference point. This single linear equation in a single unknown may be solved trivially. The rest of the points in the object are solved recursively in a similar manner. The reference points are referred to as latent since they cannot be used to solve for the interior points in the object until the exterior points are first reconstructed. This algorithm works only for particular types of sampled objects, including those for which the convex hull of their support is a polygon having no parallel sides† [20]. As mentioned in Section 7.2, one must know the support *a priori* to employ this algorithm. Except in special cases, given just the support of the autocorrelation one cannot find the support of the object [25]: one can only narrow down the possibilities.

3. *Difference Fourier Synthesis*

Some of the phase retrieval methods from x-ray crystallography [26] can be used for the astronomical application. One of these is difference Fourier synthesis [27]. Suppose the object consists of a collection of pointlike stars and one has an estimate of the object that contains some, but not all, of its stars. One creates a new image by (1) Fourier transforming the estimate of the object, (2) replacing the computed Fourier modulus by the true Fourier modulus while keeping the computed Fourier phase, and (3) inverse Fourier transforming the result. Under the assumptions mentioned above, it can be shown that the new image will contain the stars missing in the object's estimate, but at half their true value. So by taking the difference between the new image and the estimate of the object one can identify the locations and relative brightnesses of the missing stars and use that information to form a new estimate of the object. This algorithm must be applied iteratively because extraneous terms are produced in the new image along with the desired missing stars. This method was successful for pointlike objects but not for extended objects [27]. It is tantalizingly close to the iterative transform algorithm described in Section 7.4, which is applicable to a much more general class of objects.

4. *Products of Autocorrelations*

If the object consists of a collection of pointlike stars, then the autocorrelation $r_f(x, y)$ also consists of a number of pointlike terms. First one computes

† Note that Fig. 6(a) in [18] has two solutions rather than the single solution claimed. It would have a single solution if either point C′ or point C″ was zero.

the product $r_f(x - x_0, y - y_0)r_f(x, y)$, where the point (x_0, y_0) is one for which $r_f(x_0, y_0)$ is nonzero. Next one computes the triple product $r_f(x - x_1, y - y_1)r_f(x - x_0, y - y_0)r_f(x, y)$, where the point (x_1, y_1) is one for which $r_f(x_1 - x_0, y_1 - y_0)r_f(x_1, y_1)$ is nonzero. If the object points satisfy certain nonredundancy conditions, then the support of the triple product is identical to the support of the object (or the object rotated by 180°). Furthermore, the values of the object can be reconstructed in closed form by some very simple equations involving taking the cube root of (all but three of) the points in the triple product [25]. The nonredundancy conditions that the object must satisfy are that (1) the vector separations between three pairs of six distinct points must not sum to zero, (2) the vector separations between two different pairs of points must not be equal, and (3) the vector separation between one pair of points must not equal twice the vector separation between another pair of points. It is likely that a randomly distributed collection of stars would satisfy the nonredundancy conditions unless the number of stars in the image is large.

B. *Methods Proposed for General Objects*

The phase retrieval problem for general objects of finite support can be approached in a number of ways. Brute-force searching through all possible objects or all possible phases is, of course, impossible. For example, if there are K different values for each of the M^2 pixels of $f(x, y)$ that must be searched to see which objects agree with the given Fourier modulus, then the number of possibilities to be investigated is K^{M^2}, an overwhelming number, even for $K = 2$. If each of the particles in the universe were a computer that could search 10^9 objects per second, then it would take the age of the universe to search all possible binary objects for $M = 19$. Clearly, a very intelligent algorithm is needed to search through the possible objects to find one consistent with the Fourier modulus data.

Consider taking the modulus squared of both sides of Eq. (7.3-1) or simply use Eq. (7.3-3). In either case one has a system of N^2 nonlinear equations—one for each value of $|F(u, v)|^2$ or of $r_f(x, y)$—in M^2 unknowns, the values of $f(x, y)$ within its support. One could equally well treat the Fourier phases as the unknowns. Obviously, there is a high degree of redundancy in the set of equations since there are more equations than unknowns. Nevertheless, for large values of M and N the solution of a set of such nonlinear equations is a formidable task.

The classic approach to such a problem is to define an error metric for any given estimate for the object (or for the phase) and use a Newton-Raphson or a gradient search method to seek a solution, that is, an estimate for which the error metric is zero. For example, the error metric could be

$$E_r = \sum_{x=0}^{N-1} \sum_{y=0}^{N-1} [r_g(x, y) - r_f(x, y)]^2, \tag{7.3-5}$$

where $r_g(x, y)$ is the autocorrelation of $g(x, y)$, an estimate of $f(x, y)$. If $E_r = 0$ for a $g(x, y)$ that satisfies the object-domain constraints, then $g(x, y)$ is a solution. The solution is necessarily equal to $f(x, y)$ only if the phase retrieval problem is unique for $f(x, y)$, but at the reconstruction stage of the game this is not a problem since we are only concerned with finding a solution (or solutions).

The error metric E_r can be thought of as a function in an M^2-dimensional space where the values $g(x, y)$ are the coordinates of the space. The collection of the M^2 values of $g(x, y)$ can be thought of as an M^2-dimensional vector coordinate in that space. The problem is to find, within the subset of that space for which $g(x, y)$ satisfies the object-domain nonnegativity and support constraints, the point at which E_r is at a global minimum ($E_r = 0$ if no noise is present). One can seek the global minimum by starting with an estimate $g_0(x, y)$, computing the gradient of E_r at that point, and using that information to step down the side of the hill toward a lower value of E_r at a new estimate $g_1(x, y)$. Further steps are taken iteratively until the global solution is found or until stagnation occurs at a local minimum.

1. *Newton–Raphson*

The Newton–Raphson method [28] involves the local linearization of the problem by expressing $r_g(x, y)$ in Eq. (7.3-5) in a Taylor series expansion about the current estimate $g_k(x, y)$:

$$g(x, y) = g_k(x, y) + \Delta g(x, y), \tag{7.3-6}$$

and keeping only up to first-order terms:

$$r_g(x, y) \simeq r_{g_k}(x, y) + \sum_{x'} \sum_{y'} \frac{\partial r_g(x, y)}{\partial g(x', y')} \Delta g(x', y'). \tag{7.3-7}$$

Inserting this into the expression for E_r, E_r is minimized by setting equal to zero the partial derivatives of E_r with respect to $\Delta g(x_0, y_0)$ for each value of (x_0, y_0). This gives a system of M^2 linear equations in M^2 unknowns (the Δg's), which can be solved by inversion of an M^2 by M^2 matrix. A new estimate is given by Eq. (7.3-6) and the procedure is repeated. The matrix inversion takes on the order of M^6 operations, making this method computationally very intensive despite the fact that only a small number of iterations are required.

2. Gradient Search

Although requiring several times as many iterations as the Newton–Raphson method, gradient search algorithms are attractive because the amount of computation per iteration is far less. Ordinarily the computation of the gradient itself, if done with finite differences, would be very time-consuming, requiring one or two times M^2 FFTs. However, by a proper choice of the error metric it is possible to compute the entire gradient with only two FFTs [29]. If the error metric is defined in the Fourier domain by

$$E_F^2 = \sum_{u=0}^{N-1} \sum_{v=0}^{N-1} [|G(u, v)| - |F(u, v)|]^2, \tag{7.3-8}$$

where $G(u, v)$ is the Fourier transform of the current estimate $g(x, y)$, then it can be shown that the bulk of the computation for the gradient can be accomplished by the first three steps of the iterative transform algorithm, which is described in Section 7.4.

In general, the kth iteration of a gradient search method would proceed by computing a desired direction, $d_k(x, y)$, in which to go from a given object estimate, $g_k(x, y)$, and take a step of size h_k in that direction:

$$g_k''(x, y) = g_k(x, y) + h_k d_k(x, y) \tag{7.3-9}$$

For the steepest-descent gradient search method, $d_k(x, y)$ is just the negative of the partial derivative of E_F^2 with respect to $g(x, y)$. For the conjugate gradient method, $d_k(x, y)$ is a linear sum of the negative of the gradient and $d_{k-1}(x, y)$. After stepping to $g_k''(x, y)$ a test is made to see if it violates the object-domain constraints. It usually does, and a new estimate $g_{k+1}(x, y)$ is taken to be the function in object space satisfying the object-domain constraints that is closest to $g_k''(x, y)$. [In the language of Chapter 8, $g_k''(x, y)$ is projected onto the space of functions satisfying the constraints.] As for most applications, the steepest-descent method is very slow to converge, and the conjugate gradient method is much better [29]. Although the gradient search methods have not proved to converge as quickly as the algorithm described in Section 7.4, they have not yet been fully optimized and deserve further study.

3. Tracking of 1-D Complex Zeros

The method of tracking 1-D complex zeros [10] starts with one-dimensional radial cuts through the Fourier modulus. By the projection-slice theorem a radial cut through the 2-D Fourier transform is the 1-D Fourier transform of the projection of the object along a direction perpendicular to the direction of the radial cut. As mentioned in Section 7.2, there are 2^N

different solutions for a given 1-D radial cut, all of which can be computed by using different combinations of the conjugated or nonconjugated zeros $\{z_n, z_n^*\}$. Of these solutions only a small fraction will correspond to nonnegative projected images, and only these need be retained since the object brightness distribution is known to be nonnegative. As the angle of the cut is rotated slightly, the complex zeros should change in a continuous fashion, and so the complex zeros for one cut can be associated with the complex zeros for other cuts. By finding the complex zeros that yield nonnegative projected images over a range of cut angles, one can identify the set of zeros $\{z_n\}$ associated with each cut of the Fourier transform and solve the 1-D phase retrieval problem for each of the cuts, thereby solving the 2-D phase retrieval problem. This method has the disadvantage of having to compute 2^N different images for a number of cuts, which is computationally very burdensome. Application of this method gave one of the first indications that the 2-D problem is more likely to be unique than the 1-D problem [10].

4. *Making 1-D Complex Zeros Consistent between Orthogonal Cuts*

Employing a philosophy similar to that of the method described above, this method [30] computes the complex zeros of 1-D cuts through the 2-D Fourier modulus squared, and then from the complex zeros computes all 2^N different phase solutions for each cut. In this case the cuts are each line and each column of the 2-D Fourier transform. At each point in the Fourier domain a true solution must have a phase that is consistent with the phase of both orthogonal cuts. By checking consistency in phase at each point among all 2^N possible phases for each cut, it is possible to find what 2-D phase functions are consistent across the entire Fourier domain. A disadvantage of this method is the enormous computational burden of computing 2^N solutions for each of $2N$ cuts and comparing all the potential combinations.

5. *Sampling Theorem Methods*

The use of the sampling theorem is based on knowing that the object is of finite extent. If the object is zero outside a square of length L centered at the origin, then the Shannon sampling theorem gives

$$F(u, v) = \sum_p \sum_q F(p\,\Delta u, q\,\Delta v)\,\mathrm{sinc}\left(\frac{u - p\,\Delta u}{\Delta u}\right)\mathrm{sinc}\left(\frac{v - q\,\Delta v}{\Delta v}\right), \tag{7.3-10}$$

where $\Delta u = \Delta v = L^{-1}$. In the sampling theorem method [31,32], the samples $F(p\,\Delta u, q\,\Delta v)$ are estimated by $|F(p\,\Delta u, q\,\Delta v)|\exp[i\phi(p\,\Delta u, q\,\Delta v)]$, where

ϕ is an estimate of the unknown phase. This is inserted into the right-hand side of Eq. (7.3-10), which is evaluated at the half-integer coordinates, and the squared modulus is computed. Since adequate sampling of $|F(u, v)|^2$ is twice the sampling required for $F(u, v)$, and it is assumed that $|F(u, v)|^2$ was adequately sampled, the computed squared modulus values can be compared with the measured value of $|F(u, v)|^2$. This leads to a system of N^2 nonlinear equations in N^2 unknowns, which can be solved, for example, by the Newton–Raphson algorithm. This differs from the Newton–Raphson method described in Section 7.3.B.1—the set of nonlinear equations to be solved are different—but the computational complexity is the same.

A computationally more efficient version of the sampling theorem method has been devised [33]. Letting $v = 0$, Eq. (7.3-10) reduces to

$$F(u, 0) = \sum_p F(p\,\Delta u, 0)\,\mathrm{sinc}\left(\frac{u - p\,\Delta u}{\Delta u}\right). \tag{7.3-11}$$

Since the half-integer values of u along the $v = 0$ axis depend only on the sampled values along the $v = 0$ axis, the phase of $F(u, 0)$ can be determined by solving the one-dimensional problem, which involves only N nonlinear equations in N unknowns. In a similar fashion, given $F(p\,\Delta u, 0)$ for each integer value of p the phase of $F(p\,\Delta u, v)$ can be determined by solving a system of N nonlinear equations in N unknowns. The total required computation to solve $(N + 1)$ sets of N nonlinear equations in N unknowns, when solved by the Newton–Raphson method, involves $(N + 1)N^3$ operations times the number of iterations per 1-D solution, which is a great savings compared with the fully 2-D form described earlier. A disadvantage of this method is that each of the 1-D problems that is solved is just one of the 2^N ambiguous solutions that are always present for the 1-D problem. Consequently, this method will fail unless one starts with an initial estimate very close to the true solution or until one finds a way of incorporating more constraints in the algorithm.

A third variation on the sampling theorem method is the crude phase estimation method [34]. If one approximates the sinc(u) kernels in Eq. (7.3-10) by a function that is unity at $u = 0$, a constant β at $u = \pm 1/2$, and zero for all other integer and half-integer values of u, then Eq. (7.3-10) simplifies to such an extent that the phase can be expressed in closed form as a function of the modulus of Eq. (7.3-10). The closed-form expression involves an arc cosine function which leads to twofold sign ambiguity, which can be resolved by a consistency check over a closed 2-D path in Fourier space [34]. While this method is computationally simple, it unfortunately yields only a very crude estimate of the Fourier phase owing to the crudeness with which the sinc function is approximated.

7.4 ITERATIVE TRANSFORM ALGORITHM

The iterative transform algorithm is a descendant of the Gerchberg–Saxton algorithm [35–37]. It bounces back and forth between the object domain, where *a priori* knowledge about the object—its nonnegativity and bounds on its support—is applied, and the Fourier domain, where the measured Fourier modulus data are applied. The iterative transform algorithm differs from the Gerchberg–Saxton algorithm both in the type of data and constraints available and in the more powerful object-domain operations applied for the hybrid input–output version of the algorithm. The iterative transform algorithm has been the most successful method for reconstructing nonnegative objects from the modulus of their Fourier transforms because it works for the most general types of objects, uses all the available data and constraints, is not highly sensitive to noise, and is not as computationally burdensome as most other methods. In Section 7.4.A the basic iterative algorithm is described. Additional details that will be helpful in the practical implementation of the algorithm are included in Section 7.4.B. In some difficult cases the basic iterative transform algorithm fails to converge to a solution. Distinct modes of stagnation have been identified and methods for overcoming them have been devised, as described in Section 7.4.C. Section 7.4.D shows an example of a simulation of noisy stellar speckle interferometry data, Fourier modulus estimation, and iterative transform reconstruction.

A. *The Basic Iterative Algorithm*

1. *The Error-Reduction Algorithm*

The simplest version of the iterative transform algorithm follows the philosophy of the Gerchberg–Saxton algorithm [35–37], and it is known as the error-reduction algorithm [9,29,38]. It can be viewed in a number of different ways: in terms of the method of successive approximations [39], as a form of steepest-descent gradient search [29], or as a projection onto sets in a Hilbert space (the Fourier modulus constraint being onto a nonconvex set, however, so convergence is not assured). The latter point of view is described in more detail in Chapter 8.

For the most general problem, the error-reduction algorithm consists of the following four steps (for the kth iteration): (1) Fourier transform $g_k(x, y)$, an estimate of $f(x, y)$, yielding $G_k(u, v)$; (2) make the minimum changes in $G_k(u, v)$ which allow it to satisfy the Fourier-domain constraints to form $G'_k(u, v)$, an estimate of $F(u, v)$; (3) inverse Fourier transform

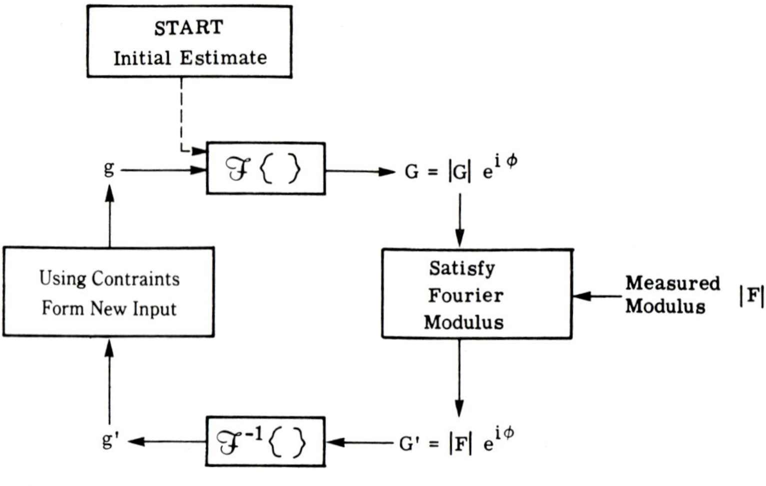

Fig. 7.4-1. Block diagram of the iterative transform algorithm.

$G'_k(u, v)$, yielding $g'_k(x, y)$, the corresponding image; and (4) make the minimum changes in $g'_k(x, y)$ that allow it to satisfy the object-domain constraints to form $g_{k+1}(x, y)$, a new estimate of the object. For phase retrieval for a nonnegative object from the Fourier modulus $|F(u, v)|$, these four steps are

Step 1: $$G_k(u, v) = |G_k(u, v)| \exp[i\phi_k(u, v)] = \mathscr{F}[g_k(x, y)], \tag{7.4-1}$$
Step 2: $$G'_k(u, v) = |F(u, v)| \exp[i\phi_k(u, v)], \tag{7.4-2}$$
Step 3: $$g'_k(x, y) = \mathscr{F}^{-1}[G'_k(u, v)], \tag{7.4-3}$$
Step 4: $$g_{k+1}(x, y) = \begin{cases} g'_k(x, y), & (x, y) \notin \gamma, \\ 0, & (x, y) \in \gamma, \end{cases} \tag{7.4-4}$$

where γ is the set of points at which $g'_k(x, y)$ violates the object-domain constraints and where g_k, G'_k, and ϕ_k are estimates of f, F, and the phase ψ of F, respectively. The algorithm is typically started by using an array of random numbers for $g_0(x, y)$ or $\phi_0(u, v)$. Figure 7.4-1 shows a block diagram of the iterative transform algorithm.

For the astronomy problem, the object-domain constraints are the object's nonnegativity and a (usually loose) support constraint. Then the points in the set γ are those outside the assumed support and those within the support for which $g'_k(x, y) < 0$. The diameter of the object can be computed since it is just half the diameter of the autocorrelation; however, the exact support of the object in general cannot be determined uniquely from the support of the autocorrelation [25], and so the support constraint cannot be applied tightly. For other problems, one may not have a nonnegativity constraint, but may have *a priori* knowledge of a tighter support constraint [40].

For the problem of phase retrieval from two intensity measurements [i.e., $|F(u, v)|^2$ and $|f(x, y)|^2$], $g'_k(x, y) = |g'_k(x, y)| \exp[i\theta'_k(x, y)]$ is complex-valued, and step 4 becomes

$$g_{k+1}(x, y) = |f(x, y)| \exp[i\theta_{k+1}(x, y)] = |f(x, y)| \exp[i\theta'_k(x, y)], \tag{7.4-5}$$

where $|f(x, y)|$ is the known modulus of the complex-valued object and θ_k is an estimate of the phase of the object. With this modulus constraint in the object domain, the error-reduction algorithm is precisely the Gerchberg-Saxton algorithm. In this chapter we consider only the problem of phase retrieval from a single intensity measurement, the squared Fourier modulus.

A measure of the convergence of the algorithm to a solution (a Fourier transform pair satisfying all the constraints in both domains) is the normalized root-mean-squared (NRMS) error metric in the Fourier domain,

$$E_F = \left[\frac{\sum_u \sum_v [|G(u, v)| - |F(u, v)|]^2}{\sum_u \sum_v |F(u, v)|^2}\right]^{1/2}, \tag{7.4-6}$$

or in the object domain,

$$E_o = \left[\frac{\sum\sum_{(x,y)\in\gamma} |g'_k(x, y)|^2}{\sum_x \sum_y |g'_k(x, y)|^2}\right]^{1/2}, \tag{7.4-7}$$

where γ is defined as in Eq. (7.4-4). Unless otherwise noted, the summations are performed over all points in image or Fourier space.

2. *Proof of Convergence for the Error-Reduction Algorithm*

It can be shown that the error-reduction algorithm "converges" in the weak sense that the squared error cannot increase with an increasing number of iterations [29]. This can be shown by considering an unnormalized squared error at the kth iteration,

$$e_{Fk}^2 = N^{-2} \sum_u \sum_v [|G_k(u, v)| - |F(u, v)|]^2 \tag{7.4-8}$$

$$= N^{-2} \sum_u \sum_v |G_k(u, v) - G'_k(u, v)|^2, \tag{7.4-9}$$

which by Parseval's theorem is

$$e_{Fk}^2 = \sum_x \sum_y |g_k(x, y) - g'_k(x, y)|^2. \tag{7.4-10}$$

Now compare this with the corresponding expression for the object-domain error

$$e_{ok}^2 = \sum\sum_{(x,y)\in\gamma} [g'_k(x, y)]^2 \tag{7.4-11}$$

$$= \sum_x \sum_y |g_{k+1}(x, y) - g'_k(x, y)|^2. \tag{7.4-12}$$

Both $g_k(x, y)$ and $g_{k+1}(x, y)$ by definition satisfy the object-domain constraints. Also, at any point (x, y), by definition $g_{k+1}(x, y)$ is the nearest value to $g'_k(x, y)$ that satisfies the object-domain constraints. Therefore for all points,

$$|g_{k+1}(x, y) - g'_k(x, y)| \leqslant |g_k(x, y) - g'_k(x, y)|, \tag{7.4-13}$$

and therefore from Eqs. (7.4-10) and (7.4-12)

$$e_{ok}^2 \leqslant e_{Fk}^2. \tag{7.4-14}$$

Similarly, by Parseval's theorem

$$e_{ok}^2 = N^{-2} \sum_u \sum_v |G_{k+1}(u, v) - G'_k(u, v)|^2. \tag{7.4-15}$$

Since both $G'_k(u, v)$ and $G'_{k+1}(u, v)$ satisfy the Fourier-domain constraint, having modulus $|F(u, v)|$, and by definition $G'_{k+1}(u, v)$ is the nearest value to $G_{k+1}(u, v)$ that satisfies the Fourier-domain constraint, then

$$|G_{k+1}(u, v) - G'_{k+1}(u, v)| \leqslant |G_{k+1}(u, v) - G'_k(u, v)|. \tag{7.4-16}$$

Therefore from Eqs. (7.4-9) and (7.4-15),

$$e_{F,k+1}^2 \leqslant e_{ok}^2, \tag{7.4-17}$$

and combining this with Eq. (7.4-14) gives the desired result

$$e_{F,k+1}^2 \leqslant e_{ok}^2 \leqslant e_{Fk}^2. \tag{7.4-18}$$

That is, the error can only decrease, or at worst stay the same, at each successive iteration of the error-reduction algorithm.

3. *The Hybrid Input–Output Algorithm*

Although it works well for the problem of phase retrieval from two intensity measurements, the error-reduction algorithm usually converges very slowly for the problem of phase retrieval from a single intensity measurement being considered here [29]. Several modifications of the iterative transform algorithm were made and tested, and most of them converged faster than the error-reduction algorithm [29]. To date, the most successful version is the hybrid input–output algorithm, which replaces step

4 of the algorithm by [8,29]

$$g_{k+1}(x, y) = \begin{cases} g'_k(x, y) & (x, y) \notin \gamma \\ g_k(x, y) - \beta g'_k(x, y), & (x, y) \in \gamma, \end{cases} \tag{7.4-19}$$

where β is a constant feedback parameter. Values of β between 0.5 and 1.0 work well. When using the hybrid input-output algorithm, $g_k(x, y)$ is no longer an estimate of $f(x, y)$; it is instead the input function used to drive the output $g'_k(x, y)$ [which *is* an estimate of $f(x, y)$] to satisfy the constraints. Hence only the object-domain error E_o is meaningful. When using using the hybrid input-output algorithm, even E_o does not always correlate with image quality as well as one would like. For this reason one may prefer to perform a number of cycles of iterations, where one cycle consists of, say, 20–50 iterations of the hybrid input-output algorithm followed by 5–10 iterations of the error-reduction algorithm, and note E_o only at the end of a cycle.

For a more complete description of the various versions of the iterative algorithm, see [29]. Additional details concerning the implementation of the algorithm are given in Section 7.4B. A description of the algorithm as it applies to a number of different problems is given in [39].

B. Additional Aspects of the Iterative Transform Algorithm

In the past, some researchers have had varying success in applying the iterative transform algorithm to phase retrieval from a single intensity measurement. In this section, a number of additional aspects of making the iterative algorithm work are given as an aid to the practical implementation of the algorithm.

The information one must have available is an estimate $|\hat{F}(u, v)|$ of the modulus $|F(u, v)|$ of the Fourier transform of the object. Although the iterative transform reconstruction algorithm is not hypersensitive to noise, care must be taken to obtain the best possible estimate of the Fourier modulus, which may involve considerable compensation of the raw data [41], depending on how they are collected. In many circumstances one can estimate the expected value of the NRMS error of the data:

$$E_{|\hat{F}|} = \left[\frac{\sum_u \sum_v [|\hat{F}(u, v)| - |F(u, v)|]^2}{\sum_u \sum_v |F(u, v)|^2}\right]^{1/2}. \tag{7.4-20}$$

As described later, this is useful for deciding when one is close enough to a solution.

As the iterations progress, the NRMS error in the object domain, E_0, given by Eq. (7.4-7), should be computed. E_0 is a measure of how close the current Fourier transform pair is to a solution. Note that the denominator of Eq. (7.4-7) is a constant that must be computed only once. When E_o goes significantly below $E_{|\hat{F}|}$ one has a solution consistent with the measured data and constraints to within the limits of the error in the given data. It is unlikely that E_o will ever go to zero because (1) diffraction by the telescope aperture causes side lobes in the image that extend beyond the support of the object and (2) noise in $|\hat{F}(u, v)|$ almost always results in a Fourier modulus that is inconsistent with either the nonnegativity constraint or any reasonable support constraint. Furthermore, in the presence of noise there will ordinarily exist an output image $g'(x, y)$ that is in better agreement with the noisy data than the true image is. Consequently, for the case of noisy data, a "solution" is not found until E_o is decreased to a level somewhat less than $E_{|\hat{F}|}$.

For the astronomy problem one has a nonnegativity constraint in the object domain. Furthermore, one can compute upper bounds on the support of the object in any of several ways [25]. The simplest way is to use a rectangle that is half the size, in each dimension, of the smallest rectangle that encloses the autocorrelation. If the actual support of the object is known *a priori*, it should, of course, be used. Any other types of *a priori* information should be used during the iterations if available. The support constraint can in general be defined by a binary mask which is unity within the support and zero outside; then the computations of Eqs. (7.4-4), (7.4-7), and (7.4-19) can be performed arithmetically without the use of logic, which is advantageous when using array processors.

There are many ways to pick an initial input to the algorithm. Although claims have been made that the crude phase estimation method described in Section 7.3.B.5 offers a superior starting point [34], others have found that random numbers do as well or better [42]. Having an initial input close to the true solution reduces the number of iterations required and might help to avoid some stagnation problems. If one of the reconstruction methods described in Section 7.5 has yielded an image, then that image would be an appropriate starting input. One can either view the other reconstruction method as a means for supplying starting inputs for the iterative transform algorithm or view the iterative transform algorithm as a means for "cleaning up" images reconstructed by the other method. If no other initial estimate for the object is available, one should use random numbers in the object domain or for the Fourier phase, giving an unbiased start to the algorithm. In the object domain, a convenient starting guess $g_0(x)$ can be formed by filling the support mask with random numbers. Another method [29] is to threshold the autocorrelation (at, say, 0.005 its maximum value), demagnify that by a factor of 2 in each dimension by discarding every other row and every other column, and finally fill the resulting shape with random numbers. (Note that this shape does not necessarily contain the support of the object [25].)

The algorithm can be made to converge faster and avoid a stagnation problem (see Section 7.4.C.3) if the support mask is chosen to be somewhat smaller than the correct support for the first cycle or two of iterations. Since it is the incorrect support, the smaller support mask is inconsistent with Fourier modulus and stagnation will eventually occur when it is used. Nevertheless, the smaller support mask helps to force most of the energy of the output, $g'(x, y)$, into a confined region in fewer iterations. After this has happened the support mask should be enlarged to the correct support constraint for the object. This enlargement of the mask could be done in more than one step if desired. When the algorithm has nearly finished reconstructing the object, it is often beneficial to make the support mask even larger than the correct support for the object. This helps to ensure that no parts of the object are being inadvertently truncated by the support constraint. The progression from a smaller support mask to a larger one also helps to avoid having edges of the output image biased toward falling right at the edges of the support mask.

Recall from Section 7.4.A that the heart of the algorithm consists of several cycles of iterations, where one cycle consists of K_1 iterations of the hybrid input-output algorithm [Eqs. (7.4-1)-(7.4-3) and (7.4-19)] followed by K_2 iterations of the error-reduction algorithm [Eqs. (7.4-1)-(7.4-4)]. Our experience has shown that values of K_1 from 20 to 100, of K_2 from 5 to 10, and of the feedback parameter β from 0.5 to 1.0 (use, say, 0.7) work well.

The discrete Fourier transforms are computed with the FFT algorithm. The sampling in the Fourier domain should be fine enough to ensure that the object domain array size is at least twice the width and height of the object itself, which is equivalent to achieving the Nyquist sampling rate for $|F(u, v)|^2$.

A straightforward method for evaluating Eq. (7.4-2) is to compute the phase from the real and imaginary parts of $G_k(u, v)$, then combine it with $|F(u, v)|$ to form $G'_k(u, v)$, and finally compute the real and imaginary parts of $G'_k(u, v)$ (which are required by the FFT) from its modulus and phase. Alternatively, one can employ

$$G'_k(u, v) = G_k(u, v)|F(u, v)|/[|G_k(u, v)| + \delta], \tag{7.4-21}$$

where δ is a very small number used to prevent overflow problems in the rare event that $|G_k(u, v)| = 0$. (For some computers one can use $\delta = 0$ with no ill effects.)

If all goes well, the iterative transform algorithm will converge to a solution after a small number of cycles of iterations. If there are multiple solutions, the iterative transform algorithm is capable of finding any one of them, depending on the starting input [43,44]. Confidence that the solution is the one and only true solution can be increased by performing two or more trials of the algorithm, each time using different random numbers for the initial input.

In some cases the iterative transform algorithm will stagnate before reaching a solution. The algorithm can be considered to have stagnated if the error E_o is greater than $E_{|\hat{F}|}$ and has failed to decrease after three additional cycles. While some objects can be reconstructed very easily, requiring only one or two cycles, other more difficult objects can require many cycles comprising well over 1000 iterations. Consequently, one should not too readily jump to the conclusion that the algorithm has stagnated. It often occurs that very slow progress is made for many iterations, but then the algorithm suddenly finds its way and rapid progress is made in just a few iterations.

If the iterative transform algorithm does stagnate, then one can start over with a different set of random numbers for the initial input; alternatively, one can try other simple alterations of the algorithm. One can use a larger value for β, say 1.2, causing larger changes to be made; but this should not be carried out for too many iterations since it causes the algorithm to become unstable. Use of a different support constraint mask may also be helpful. For certain special modes of stagnation special methods have been devised for overcoming the stagnation, as described below.

C. *Methods for Overcoming Stagnation*

Three particular modes of stagnation have been identified: those characterized by (1) simultaneous twin images, (2) stripes, and (3) truncated support. Methods have been devised to jump each of these hurdles. They are briefly described in this section; further details can be found in [45]. (See Chapter 8 for another view of stagnation problems.)

1. *Simultaneous Twin Images*

Since both $f(x, y)$ and its twin image $f(-x, -y)$ have the same Fourier modulus, the iterative transform algorithm is equally likely to reconstruct either one. When the support of $f(x, y)$ is symmetric, the algorithm often outputs a partially reconstructed image having features of both $f(x, y)$ and $f(-x, -y)$ as illustrated in Fig. 7.4-2. In many cases the algorithm stagnates with such an output image. (It is not a solution because $E_o > E_{|\hat{F}|}$.) If the algorithm has stagnated on an output image having a strong centrosymmetric component, then one would suspect this mode of stagnation. A method for

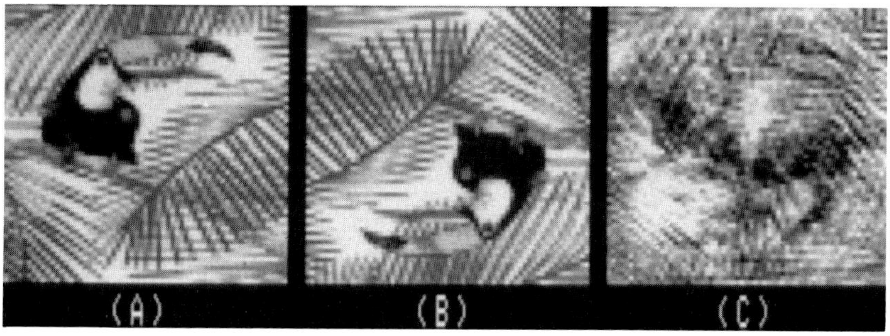

Fig. 7.4-2. Example of stagnation with simultaneous twin images. (A) Object $f(x, y)$; (B) twin image $f(-x, -y)$; (C) stagnated output image partially reconstructed from Fourier modulus, having features of both twins.

circumventing this stagnation problem is to use a reduced-area temporary support constraint for a few iterations. The temporary support should be highly asymmetric and should include one or two edges of the object's support but not the opposite edge. After a few iterations with the reduced-area support constraint, the symmetry in the output image may be sufficiently broken that on further iterations with the correct support constraint the iterative transform algorithm can converge to either $f(x, y)$ or its twin.

For example, to overcome the simultaneous twin images stagnation problem exhibited in Fig. 7.4-2C, a triangular support covering the lower right half of the true square support was used for 10 error reduction iterations. When further iterations of the algorithm were performed with the correct square support, it converged to the solution.

2. *Stripes*

The algorithm frequently stagnates in a mode in which the output image is a faithful representation of the object but with a pattern of stripes superimposed. This can be recognized by the fact that the stripes extend (although with less contrast) beyond the support of the object. The near-sinusoidal nature of the stripes is an indication that the phase errors occur in a symmetric pair of fairly well-defined areas in the Fourier domain. It has been observed that, if the iterative transform algorithm is applied more than once to the same Fourier modulus data but with different random initial inputs, then the stripes (if there are any) of each of the reconstructed images tend to have different frequencies and orientations [46], which means that the errors occur in different areas of the Fourier domain. Figures 7.4-3A to 7.4-3D show an example of this.

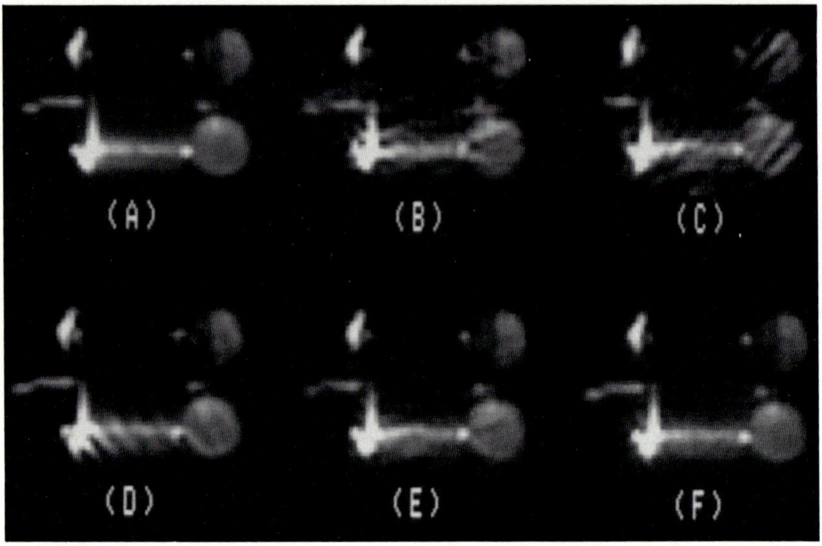

Fig. 7.4-3. Example of stagnation with stripes and overcoming it by the voting method. (A) Object; (B–D) images partially reconstructed by the iterative transform algorithm, each from a different random starting input and each having a different set of stripes superimposed on an otherwise good-quality image; (E) output of the voting method; (F) output image after further iterations.

One method for solving this stagnation problem is the voting method [45]. The iterative transform algorithm is applied three times, each time with a different random initial input to produce three different striped images, as shown in Figs. 7.4-3B to 7.4-3D. (If one of them has no stripes, then, of course, the problem is solved.) By cross-correlations the relative translations and orientations of the three images are determined and corrrected so that they are made to be in perfect registration. The Fourier transforms of the three images are compared. At each point the two complex Fourier values that are closest are averaged and the third (assumed to have a phase error) is discarded. The resulting composite Fourier function is inverse transformed to produce the corresponding output image. Then further iterations of the iterative transform algorithm are performed. For example, Fig. 7.4-3E shows the output of the voting method applied to the images shown in Figs. 7.4-3B to 7.4-3D, and Fig. 7.4-3F shows the result of further iterations.

A second method for overcoming stagnation with stripes is the patching method [45]. It requires that only two striped images be produced and made to be in registration. For each of these images the area of the support of the object is zeroed out, leaving just the stripes in the area outside the object's support. These stripes are Fourier transformed, and the resulting Fourier modulus arrays are used to determine what areas in the Fourier domains have phase errors associated with the stripes. It is assumed that those areas are different for each of two striped images. A new composite Fourier transform is set equal to the Fourier transform of the first striped image where it is not affected by the phase errors and to the Fourier transform of the second striped image where the first is affected by the phase errors. Both methods of overcoming the problem of stagnation with striped images have been shown to be effective [45].

3. *Truncated Support*

Since $f(x - x_0, y - y_0)$ has the same Fourier modulus as $f(x, y)$, the location of the object's support is arbitrary. Frequently the image partially reconstructed by the algorithm is not in perfect registration with the support constraint. Then enforcing the support constraint causes an inadvertent truncation of part of the desired image, causing the algorithm to stagnate. In addition to the enlarging support method described in Section 7.4.B, a method of combating this stagnation problem is to dynamically translate either the support constraint or the image.

The amount of translation to be used can be determined as follows. Compute the total energy of the output image $g'_k(x, y)$ (i.e., square and sum) over the area of the support constraint for the current position of the support constraint and for the support constraint translated by one or two pixels in every direction. The support constraint should be translated to the position for which the energy is maximized. Alternatively, compute the cross-correlation of $g'_k(x, y)$ with the support mask, and translate according to the peak of the cross-correlation. This can be done occasionally or at every iteration. This method would be particularly effective if, just prior to performing it, a support constraint larger than the usual support were used for a few iterations; that would give the truncated part of the image a chance to establish itself.

D. *Iterative Reconstruction Example*

In this section, the results of a computer experiment in which stellar speckle interferometry [47] was realistically simulated—including the effects of atmospheric turbulence and photon noise—and reconstruction from noisy Fourier modulus data by the iterative transform algorithm are shown [46]. (See Section 7.5.A for a description of stellar speckle interferometry.)

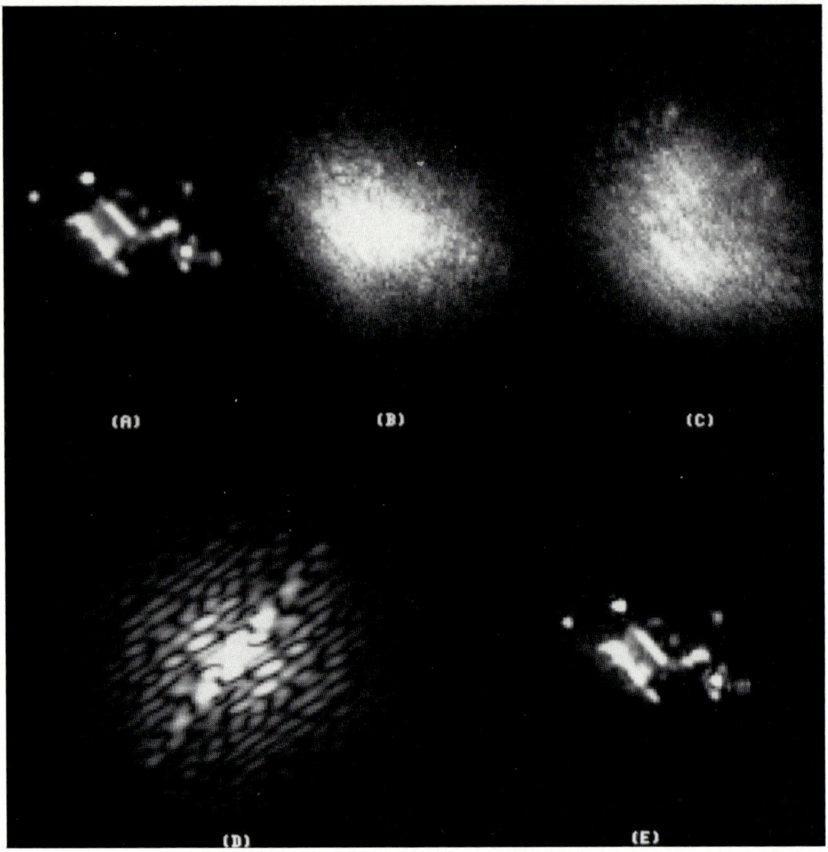

Fig. 7.4-4. Image reconstruction example. (A) Undegraded object; (B, C) examples of degraded images simulated to include the effects of atmospheric turbulence and photon noise; (D) Fourier modulus estimate computed from the degraded images; (E) image reconstructed by using the iterative transform algorithm.

An undegraded object, a low-pass filtered, digitized photograph of a satellite, shown in Fig. 7.4-4A, was convolved with 156 different point spread functions to produce 156 different blurred images. Each of the point spread functions represented a different realization of the effects of the turbulent atmosphere. The blurred images were then subjected to a Poisson noise process to simulate the effects of photon noise. For this example, there were approximately 300,000 photons per degraded image (or on the order of 100 photons per pixel over the extent of the object), which is realistic for objects of this type when imaged through a telescope of diameter 1.6 m. Two of the resulting 156 degraded images are shown in Figs. 7.4-4B and 7.4-C. The degraded images were then processed by Labeyrie's [47] method as modified by Goodman and Belsher [48]. The estimate of the modulus of the Fourier transform of the object is given by [46]

$$F(u, v) = W(u, v)\left[\frac{\sum_{m=1}^{M}|I_m(u, v)|^2 - N_p}{\sum_{m=M+1}^{2M}|T_m(u, v)|^2}\right]^{1/2}, \qquad (7.4\text{-}22)$$

where $I_m(u, v)$ is the Fourier transform of the mth degraded image, N_p is the total number of photons detected (it is subtracted in order to compensate for a noise bias term that arises in the power spectrum due to photon noise [48]), $T_m(u, v)$ is the Fourier transform of the mth point-spread function [to provide compensation for the modulation transfer function (MTF) of the speckle interferometry process], and the weighting factor $W(u, v)$ is the MTF due to the telescope aperture. In practice, the denominator of this expression would be obtained by making measurements on a reference star through an atmosphere having the same statistics as that which blurred the images or by using a model of the effects of atmospheric turbulence. The term $W(u, v)$ was included in order to restore the natural MTF due to the telescope aperture, which was removed by the denominator of Eq. (7.4-22). Figure 7.4-4D shows the resulting Fourier modulus estimate.

The object was reconstructed (or, equivalently, the Fourier phase was retrieved) using the hybrid input-output algorithm alternately with the error-reduction algorithm. The result, shown in Fig. 7.4-E, agrees very well with the original object shown in Fig. 7.4-4A, despite the noise present in the Fourier modulus data. Good reconstructed images were also obtained when only one-tenth as many photons were assumed to be available [46].

For this reconstruction example, Fig. 7.4-5 shows E_o versus the number of iterations and Fig. 7.4-6 shows the output image at various stages. The

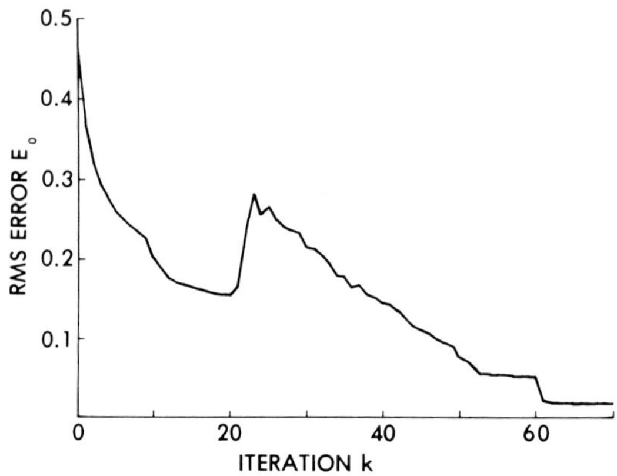

Fig. 7.4-5 Normalized root-mean-squared (NRMS) error metric, E_o, versus iteration number for the example.

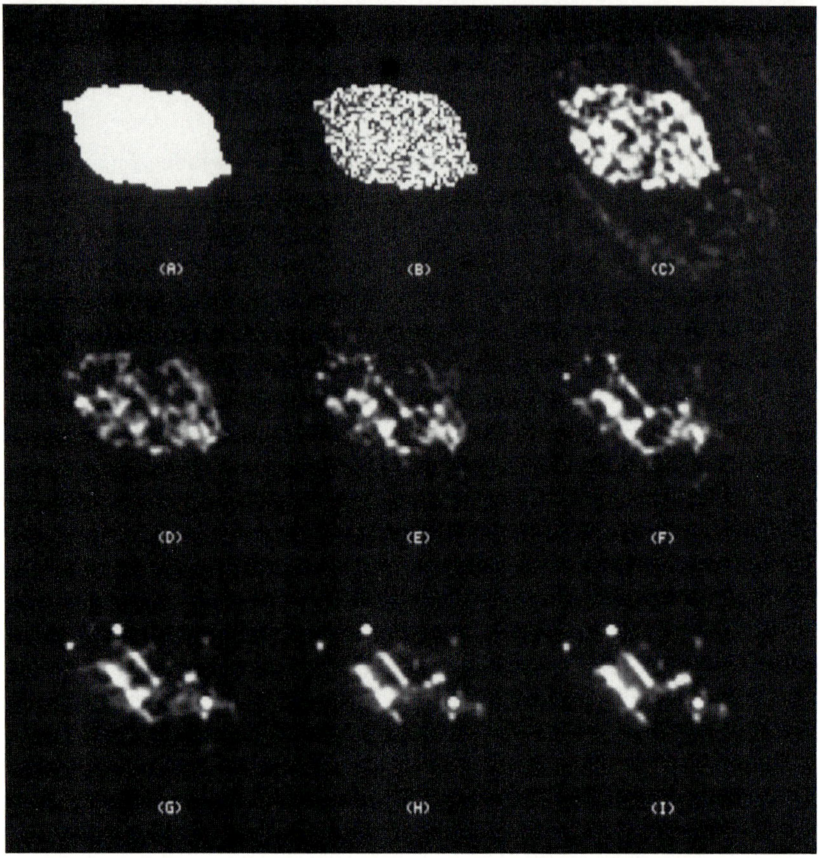

Fig. 7.4-6. Details of the image reconstruction example. (A) Mask array defining the support constraint for the first 10 iterations; (B) initial input to the iterative transform algorithm; (C–I) output images—number of iterations: (C) 1, (D) 5, (E) 20, (F) 25, (G) 35, (H) 45, (I) 55.

starting input used, shown in Fig. 7.4-6B, was the randomized, demagnified autocorrelation described in Section 7.4.B, using a threshold value of 0.0004 times the peak of the autocorrelation. For the first 10 iterations, the error-reduction algorithm was used and the mask defining the support constraint was chosen to be the region over which the autocorrelation function, spatially demagnified by a factor of 2, exceeded 0.0004 of its maximum value, providing a fairly tight diameter constraint (it is just the support of Fig. 7.4-6B, which is shown in Fig. 7.4-6A). For iterations 11 to 20 the error-reduction algorithm was used, and the mask for these and the remaining iterations was chosen to be a square of length 64 pixels, which is larger than the actual object extent of about 60×40 pixels (embedded in an array of 128×128 pixels). The error E_0 decreased rapidly at first, and the visual appearance of the output image improved substantially from the first iteration, shown in Fig. 7.4-6C, to the fifth iteration, shown in Fig. 7.4-6D. The error decreased suddenly at the tenth iteration since some positive-valued points that were inside the second mask but outside the first mask were no longer counted as contributing to E_0. By the twentieth iteration the error-reduction algorithm was converging very slowly, and the output image, shown in Fig. 7.4-6E, is only slightly improved over that in Fig. 7.4-6D. For interations 21 to 60 the hybrid input–output algorithm, with $\beta = 1$, was used. At first E_0 increased sharply, although, as shown in Fig. 7.4-6F, the output image at iteration 25 appeared no worse than at iteration 20; then E_0 decreased fairly rapidly until stagnation at $E_0 \simeq 0.05$ at about iteration 55. Output images at iterations 35, 45, and 55 are shown in Figs. 7.4-6G, H, and I, respectively. For iterations 61 to 70 the error-reduction algorithm was used, for which E_0 dropped suddenly from 0.05 to 0.02, although the visual appearance of the reconstructed image remained the same as for iteration 55.

This final value of E_0 is somewhat less than 0.03, the normalized rms error of the Fourier modulus estimate itself.

Reconstruction experiments do not always proceed as smoothly as the one described above. When stagnation occurs, the methods described in Sections 7.4.B and 7.4.C should be employed.

7.5 SOLUTIONS SPECIFIC TO MEASUREMENT TECHNIQUES

We have assumed so far in this chapter that only the modulus of the Fourier transform $|F(u, v)|$ of the object intensity $f(x, y)$ is known and have combined these data with the *a priori* knowledge that the object is nonnegative and of finite extent to achieve a reconstruction of the object. Astronomical observing techniques based on either speckle (image plane) or Michelson (pupil plane) interferometry can yield data in addition to the modulus of the Fourier transform: for example, speckle interferometry gives, in addition to a diffraction-limited estimate of the modulus, a low-resolution image of the object, i.e., an estimate of both the modulus and phase of the transform at low spatial frequencies. These additional data may in some cases be sufficient to determine the object intensity without any ambiguity. Since the different methods are specific to the observing technique, they are described separately under the headings of speckle interferometry and Michelson interferometry.

A. Object Recovery from Speckle Interferometry Data

1. Basic Speckle Technique

Speckle interferometry was invented by A. Labeyrie [47] in 1970 as a means of obtaining *diffraction-limited* resolution (but not images) from large telescopes, despite the presence of atmospheric turbulence or "seeing." The method is illustrated in Fig. 7.5-1 for an unresolved star, binary stars of two separations, and a resolved star. A large number of short-exposure (~10 ms) photographs are taken, each through a different realization of the atmosphere, typical examples being shown in row B (see also Fig. 7.5-2). For a binary star, each component produces an identical speckle pattern and a "double-speckle" effect may be visible in each short-exposure image in favorable circumstances. The optical diffraction pattern, or squared modulus of the Fourier transform, of each photograph is shown in row C—the signal-to-noise ratio is low for a single record and may be improved by adding many such diffraction patterns (row D). The unresolved object has a diffraction halo of relatively large spatial extent, the binaries give fringes of period inversely proportional to their separation, and the resolved object gives a diffraction halo whose diameter is inversely proportional to the diameter of the object. By taking a further Fourier transform of each ensemble-averaged diffraction pattern, we obtain the average spatial autocorrelation of the diffraction-limited images of each object (row E).

In mathematical terms, one estimates either the average spatial autocorrelation $\langle g(x, y) \star g(x, y)\rangle$ or the power† spectrum $\langle |G(u, v)|^2\rangle$ of the image intensity $g(x, y)$, where $\langle\ \rangle$ denotes the ensemble average, from a finite number of frames. It can be shown that the image power spectrum is related to that of the object, $|F(u, v)|^2$, through the speckle transfer function $\langle |T(u, v)|^2\rangle$:

$$\langle |G(u, v)|^2\rangle = |F(u, v)|^2 \langle |T(u, v)|^2\rangle, \tag{7.5-1}$$

where

$$\langle |T(u, v)|^2\rangle \simeq |\langle T(u, v)\rangle|^2 + T_D(u, v)/N_s, \tag{7.5-2}$$

and all spatial frequencies are referred to those of the image (formed by a telescope of focal length z using light of wavelength λ). In Eq. (7.5-2), $\langle T(u, v)\rangle$ is the average (seeing-limited) transfer function, $T_D(u, v)$ is the *diffraction-limited* transfer function of the telescope, and N_s is the average number of speckles per frame $[\simeq 2.3(D/r_0)^2]$, where r_0 is the atmospheric seeing parameter.

† Strictly speaking, this should be called the average *energy* spectrum [3, p. 260].

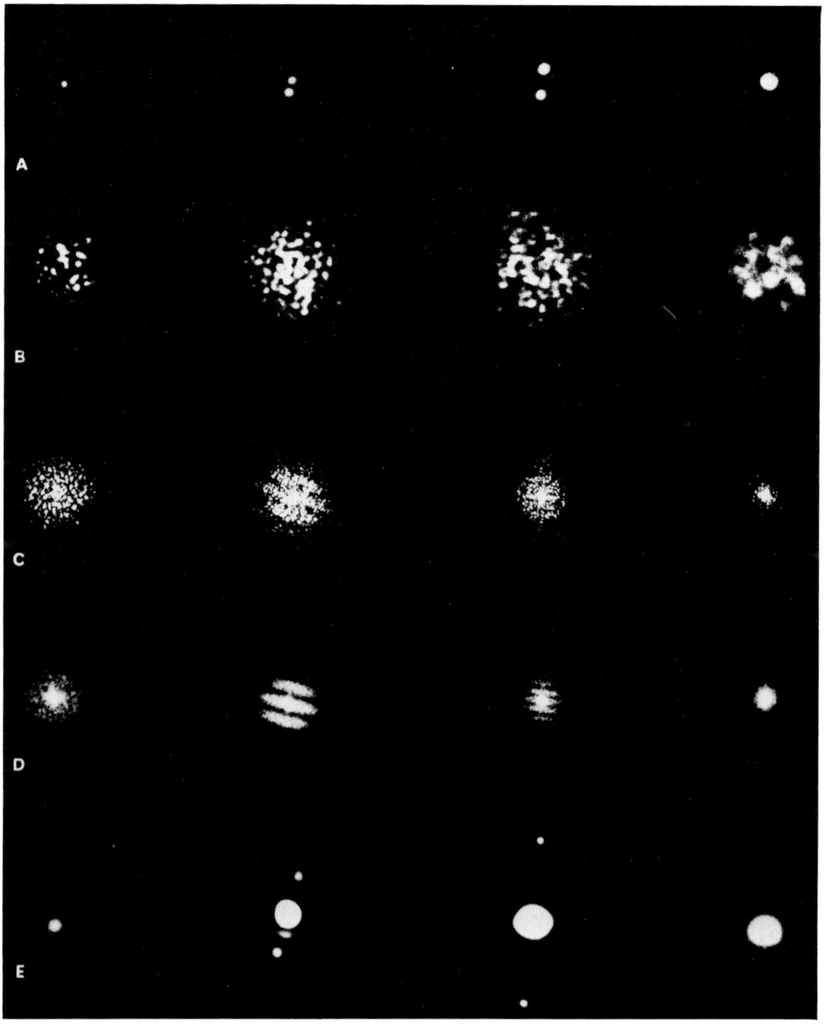

Fig. 7.5-1. Laboratory simulation showing the principles of stellar speckle interferometry. (A) Objects; (B) typical short-exposure photographs; (C) diffraction patterns of row B; (D) sum of 20 diffraction patterns; (E) diffraction patterns of row D (not at the same scale as row A). (Courtesy of A. Labeyrie, CERGA.)

The key feature of speckle interferometry is that the speckle transfer function $\langle |T(u, v)|^2 \rangle$ is nonzero up to the diffraction limit of the telescope, as given by Eq. (7.5-2), allowing the object power spectrum $|F(u, v)|^2$, or equivalently the object autocorrelation $f(x, y) \star f(x, y)$, to be estimated up to the diffraction limit of the telescope. Note that the basic technique does

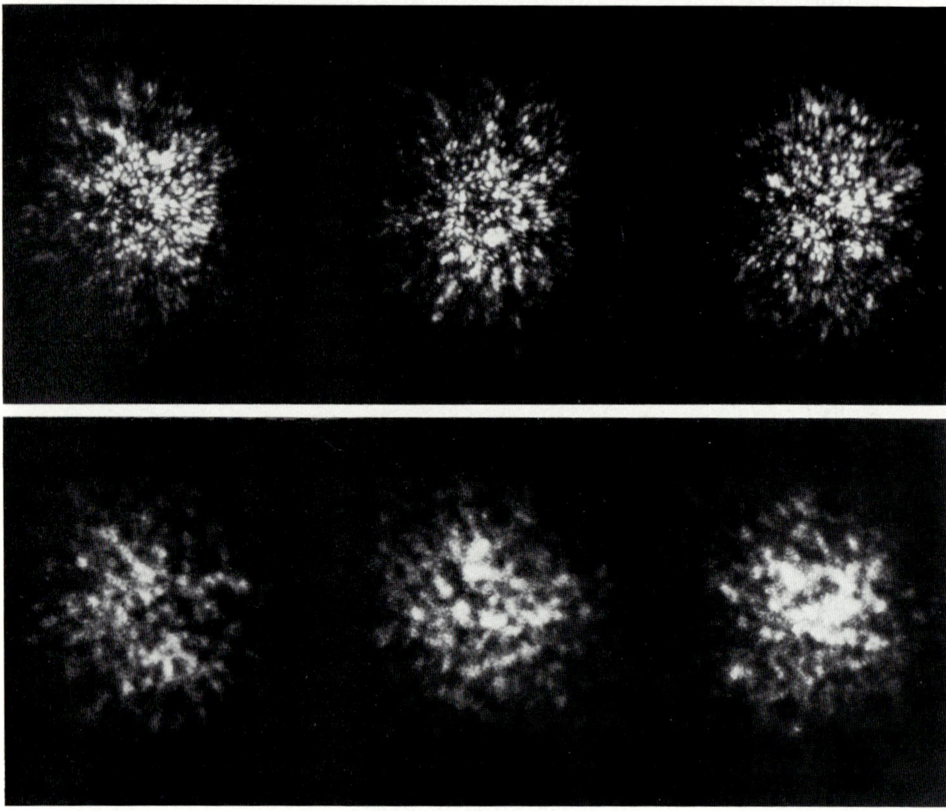

Fig. 7.5-2. Short-exposure photographs of an unresolved point source (upper row) and a resolved star, α-Orionis (lower row), taken on a 4-m telescope. The exposure time and filter bandwidth are 10^{-2} s and 10 nm, respectively. (Courtesy of B. L. Morgan and R. J. Scaddan, Imperial College, London.)

not give the object intensity itself: to address this problem, one must use one of the phase retrieval algorithms described earlier or one of the speckle *imaging* techniques described below.

2. Shift-and-Add Method

The first speckle imaging technique to be suggested used the notion that the individual speckles in a short-exposure image resemble diffraction-limited images of the object. Figure 7.5-2 compares short-exposure images of a point source and a resolvable star, and there is a clear qualitative difference between the speckle structure in the two cases. By averaging the centroided versions of selected bright speckles, Lynds *et al.* [49] were able to reconstruct a map of the red giant α-Orionis. The procedure was made a little more systematic and generalized by Bates and Cady [50], who suggested that the brightest speckles in the image be *shifted* to the origin and the entire images *added* (hence "shift-and-add").

There are two problems with this method that make it unlikely to be of practical significance for most objects. First, to be able to implement the method it is necessary to identify individual speckles, and this can be done only if there are more than approximately 10^5 detected photons per frame, implying an object brighter than about 5th magnitude in the visible. Second, each speckle is obviously *not* an image of the object and there is no linear transfer function for this technique, making it difficult to reconstruct reliable quantitative object maps. The Bates and Cady extension can give good results if the image contains a very bright unresolved component [51].

3. *Cross-Power Spectrum Method*

This method, first suggested by Knox and Thompson [52], makes use of the fact the *phase difference* between two values of $F(u, v)$, separated by $(\Delta u, \Delta v) < r_0/\lambda z$ in spatial frequency, is present in the original speckle data $g(x, y)$ and can be extracted by computing the cross-power spectrum $\langle G(u, v)G^*(u + \Delta u, v + \Delta v)\rangle$. A straightforward analysis shows that

$$\begin{aligned}\text{phase}\{\langle G(u, v)G^*(u + \Delta u, v + \Delta v)\rangle\} \\ = \text{phase}\{F(u, v)\} - \text{phase}\{F(u + \Delta u, v + \Delta v)\} \\ + \text{phase}\{\langle T(u, v)T^*(u+\Delta u, v+\Delta v)\rangle\}. \qquad (7.5\text{-}3)\end{aligned}$$

The phase of a complex quantity has meaning only if its modulus is nonzero, so that from Eq. (7.5-3) there are two conditions to be met if phase differences of $F(u, v)$ are to be computed from the phase of the cross-spectrum:

(i) $|\langle T(u, v)T^*(u + \Delta u, v + \Delta v)\rangle| \neq 0$,
(ii) $\text{phase}\{\langle T(u, v)T^*(u + \Delta u, v + \Delta v)\rangle\}$ either known or zero.

If we assume that the complex amplitude in the telescope pupil has complex Gaussian statistics and that its autocorrelation function has a Gaussian shape, i.e., the "seeing" transfer function $T_s(u, v)$ is Gaussian, then it can be shown [3, p. 303] that

$$\langle T(u, v)T^*(u+\Delta u, v+\Delta v)\rangle \simeq \langle |T(u, v)|^2\rangle |T_s(\Delta u/2, \Delta v/2)|^2. \qquad (7.5\text{-}4)$$

That is, the Knox–Thompson transfer function is simply the product of the normal speckle transfer function [Eq. (7.5-2)] at (u, v) and the squared modulus of the seeing transfer function at $(\Delta u/2, \Delta v/2)$. It follows that the modulus of the Knox–Thompson transfer function is nonzero up to the diffraction limit of the telescope provided that $\Delta u/2$ and $\Delta v/2$ are less than $(r_0/\lambda z)$, the width of the seeing transfer function. It also follows that

$$\text{phase}\{\langle T(u, v)T^*(u + \Delta u, v + \Delta v)\rangle\} \simeq 0.$$

The problem of finding the phase of $F(u, v)$ given the phase differences for frequencies separated by $(\Delta u, \Delta v)$ is identical to that of finding a wave front from sheared interferograms [53]. Simple "bootstrapping" of the phase differences, starting at the origin of the spatial frequency plane, leads to cumulative errors at higher frequencies and in two dimensions yields phase values at a given frequency that depend on the path taken in the frequency plane: averaging these values is not the best way to deal with this problem. A least-squares solution is straightforward to implement, using an iterative technique to solve the usual matrix equation (the matrix that relates the phase differences to the phase values is sparse) [54].

The assumptions leading to Eq. (7.5-4) are not particularly realistic and it remains to be seen how sensitive this technique is to the atmospheric statistics.

4. *Triple Correlation Method*

The average triple correlation $\langle g^{(3)}(\mathbf{x}_1, \mathbf{x}_2)\rangle$ of the image intensity $g(\mathbf{x})$, or equivalently its Fourier transform, the bispectrum $\langle G^{(3)}(\mathbf{u}_1, \mathbf{u}_2)\rangle$, contains information on the phase of $G(\mathbf{u})$ and hence on the phase of the object transform [55,56]. Here we use the notation $\mathbf{x} = (x, y)$ and $\mathbf{u} = (u, v)$. The triple correlation and bispectrum are, for a real function $g(\mathbf{x})$, defined by

$$g^{(3)}(\mathbf{x}_1, \mathbf{x}_2) = \iint_{-\infty}^{\infty} g(\mathbf{x})g(\mathbf{x}+\mathbf{x}_1)g(\mathbf{x}+\mathbf{x}_2)\, d\mathbf{x} \tag{7.5-5}$$

and

$$G^{(3)}(\mathbf{u}_1, \mathbf{u}_2) = G(\mathbf{u}_1)G(\mathbf{u}_2)G(-\mathbf{u}_1 - \mathbf{u}_2). \tag{7.5-6}$$

The average bispectrum of the image is related to that of the object through a bispectrum transfer function $\langle T^{(3)}(\mathbf{u}_1, \mathbf{u}_2)\rangle$,

$$\langle G^{(3)}(\mathbf{u}_1, \mathbf{u}_2)\rangle = F^{(3)}(\mathbf{u}_1, \mathbf{u}_2)\langle T^{(3)}(\mathbf{u}_1, \mathbf{u}_2)\rangle. \tag{7.5-7}$$

If it is assumed that complex amplitude in the telescope pupil has Gaussian statistics, then it can be shown [55] that the bispectrum transfer function is real and nonzero over the diffraction-limited portion of the

$(\mathbf{u}_1, \mathbf{u}_2)$ volume: it follows that the phase of the image bispectrum equals that of the object bispectrum.

To see how the phase of the object spectrum (which is Hermitian) can be calculated recursively from that of the object bispectrum, consider a one-dimensional sampled bispectrum $F^{(3)}_{l,m}$:

$$F^{(3)}_{l,m} = F_l F_m F_{-l-m}. \tag{7.5-8}$$

Denoting the phase of the object spectrum by ψ_n and that of the bispectrum by $\beta_{l,m}$, the phase factors satisfy the equation

$$\exp[i\psi_n] = \exp[i(\psi_m + \psi_{n-m} - \beta_{n-m,m})], \tag{7.5-9}$$

where the substitution $n = l + m$ has been made. Note that both ψ_0 and ψ_1 may be set equal to zero as the object is real and its absolute position is not of interest.

Now consider Eq. (7.5-9) with $m = 1$ and $n = 2, 3, 4, \ldots$:

$$\exp[i\psi_n] = \exp[i(\psi_1 + \psi_{n-1} - \beta_{n-1,1})], \qquad (\psi_1 = 0). \tag{7.5-10}$$

Clearly, ψ_2, $\psi_3, \ldots$ can be found by simple recursion with $n = 2, 3, \ldots$. In fact, the process can be repeated for different values of m to yield a more accurate estimate of the phase at each spatial frequency. The method is in some ways similar to, but more general than, the cross-power spectrum technique described in Section 7.5.A.3.

The triple correlation method is illustrated in Fig. 7.5-3 for a one-dimensional projection of the double star ω-Leonis. For the more interesting case of two-dimensional objects, the reconstruction of the object spectrum from the bispectrum is computationally quite reasonable—the problem lies in computing the average bispectrum, which is a four-dimensional function. Nevertheless, the triple correlation method appears to be the most promising technique of speckle imaging at the present time.

5. *Exponential Filter Method*

For discrete objects, the object spectrum can be written as a polynomial or z-transform, as discussed in Section 6.1. For modulus-only reconstruction in one dimension, the possibility of ambiguities arises because the zeros of the power spectrum of the object include both the zeros of the original object and their inverses: in the one-dimensional case it is not in general possible to distinguish the correct zeros from their inverses, given only the Fourier modulus. The cross-spectrum method eliminates this ambiguity by rotating the "inverse" zeros by a known amount [57], thus enabling them to be distinguished from the correct ones even in the one-dimensional case. The triple correlation method also allows the zeros of the object spectrum

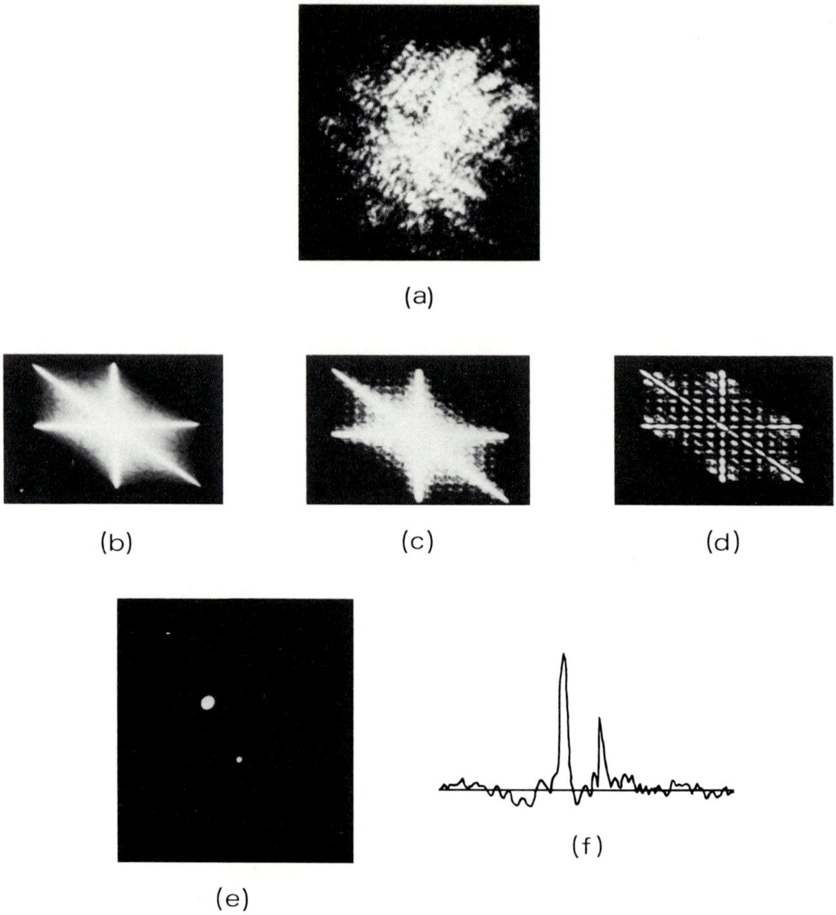

Fig. 7.5-3. Illustration of the triple correlation method in one dimension on the double star ω-Leonis. (a) Typical short-exposure speckle photograph; (b) bispectrum transfer function; (c) average bispectrum of speckle image calculated from 300 exposures; (d) object bispectrum found from (b) and (c); (e) reconstructed image; (f) scan through reconstructed image. (Courtesy of A. W. Lohmann and G. Weigelt, University of Erlangen.)

to be identified correctly from the zeros of the triple correlation [56]. The exponential filter method provides a third way of distinguishing the correct zeros from their inverses: in this case the correct zeros are moved out radially in the complex plane by a known amount.

In this method, one records the instantaneous image intensity $g(x, y)$ in the usual way and computes the average power spectrum $\langle|G(u, v)|^2\rangle$. A second power spectrum, $\langle|G'(u, v)|^2\rangle$ of $g(x, y)p(x, y)$, where $p(x, y)$ is a special modifying function, is also computed. Walker [58] showed that if the modifying function has the property

$$p(x_1 + x_2, y_1 + y_2) = p(x_1, y_1)p(x_2, y_2),$$

then

$$\langle|G'(u, v)|^2\rangle = |F'(u, v)|^2\langle|T'(u, v)|^2\rangle, \tag{7.5-11}$$

where

$$F'(u, v) = F(u, v) * P(u, v)$$

and $\langle|T'(u, v)|^2\rangle$ is a transfer function that can be measured by observing a point source. The exponential, $p(x, y) = \exp(-2\pi ax)$, has the required property. From the data we have thus extracted the power spectrum of the object, $|F(u, v)|^2$, and that of the exponentially filtered object, $|F'(u, v)|^2$; these together, in principle, contain sufficient information to reconstruct the object intensity.

In practice, in the two-dimensional case, an iterative algorithm similar to that described in Section 7.4 is used to obtain the object intensity $f(x, y)$ from these two power spectra, which are both used alternately as Fourier constraints [59]. Figure 7.5-4 shows an example of reconstructions obtained in a computer simulation with this algorithm.

6. Phase Averaging Method

The instantaneous image intensity $g(x, y)$ is related to the object intensity $f(x, y)$ by the usual convolution formula, which in Fourier space becomes a product:

$$G(u, v) = F(u, v)T(u, v), \tag{7.5-12}$$

where $T(u, v)$ is the instantaneous transfer function of the atmosphere/telescope combination. Taking the ensemble average of the logarithm of Eq. (7.5-12) and equating imaginary parts, we obtain

$$\langle\text{phase}\{G(u, v)\}\rangle = \text{phase}\{F(u, v)\} + \langle\text{phase}\{T(u, v)\}\rangle + 2\pi q, \tag{7.5-13}$$

where the phase is the unwrapped value in the interval $-\infty$ to $+\infty$ and q is an integer. McGlamery [60] suggested that if $\langle\text{phase}\{T(u, v)\}\rangle$ is known (or zero), then the phase of the object transform could be found from the average phase of the image transform. Using the central limit theorem, it is possible to show that, for spatial frequencies $r_0/\lambda z < (u, v) < (D - r_0)/\lambda z$, the quantity $T(u, v)$ is a circular complex Gaussian process, with the result that $\langle\text{phase}\{T(u, v)\}\rangle = 0$.

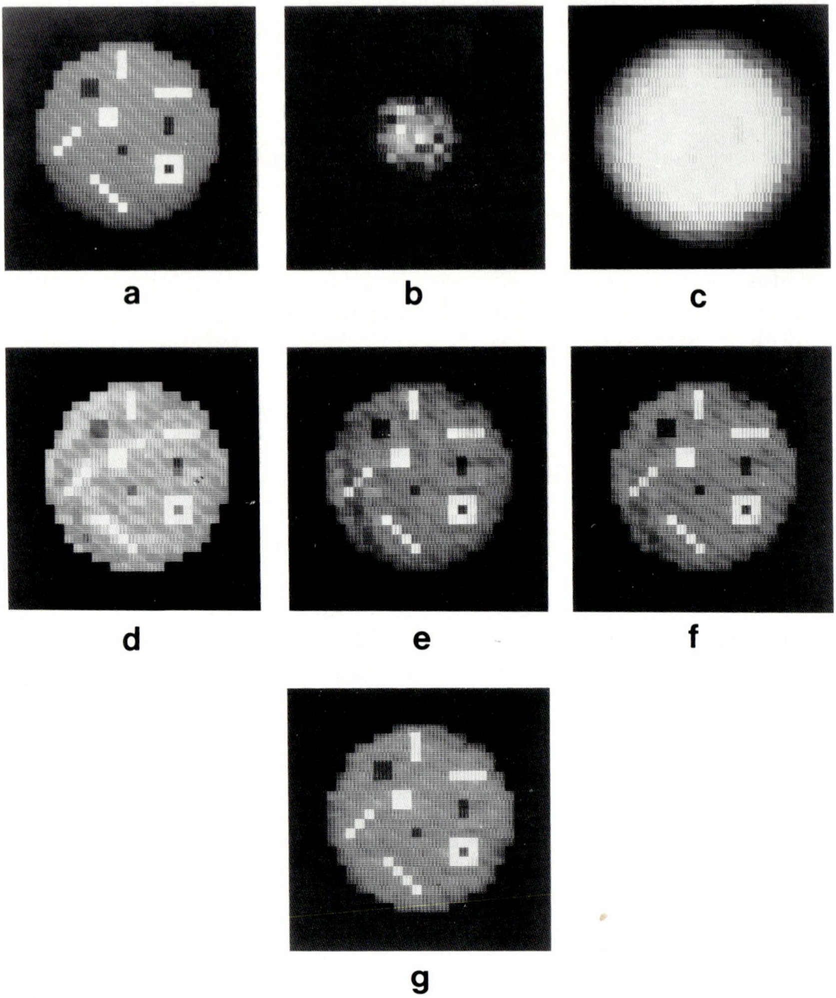

Fig. 7.5-4. Illustration of the exponential filter method. (a) Test object; (b) a short-exposure point spread function; (c) a short-exposure image; (d–g) reconstructions using, respectively, 10, 25, 50, and 100 simulated point and object exposures. (Courtesy of J. W. Walker, Kings College, London.)

The real problem with this phase averaging method is determining, using continuity of $G(u, v)$, the unwrapped phase (the phase in the interval $-\infty$ to $+\infty$) from the phase of $G(u, v)$, which is necessarily in the interval $-\pi$ to $+\pi$. This unwrapping has to be done before the averaging process. When the modulus $|G(u, v)|$ is small, the rms absolute error σ in the unwrapped phase is given approximately by

$$\sigma \simeq \frac{1}{\sqrt{2\bar{N}}\,|G(u, v)|}, \tag{7.5-14}$$

where $\bar{N}$ is the average number of detected photons per frame and $G(u, v)$ is normalized to unity at zero spatial frequency. In addition to having a low *average* value of its modulus [because of the form of the speckle transfer function, Eq. (7.5-2)], the continuous function $G(u, v)$, since it is the Fourier transform of a speckle pattern, also contains many points whose modulus is zero, which can be expected to cause problems in phase unwrapping algorithms [61].

Mertz [62] has suggested that it should be possible to use continuity in *time*, as well as spatial frequency, to help unwrap the phase, and such a proposal may be easy to implement using the time-ordered photon event detectors that are becoming available.

7. *Non-redundant Aperture Methods*

As is well known, the Fourier transform of any intensity point spread function is equal to the autocorrelation of the complex amplitude distribution over the pupil. A given spatial frequency component is therefore the sum (or integral) of products of values of the complex amplitude at all combinations of two points in the pupil separated by a fixed amount. If a special mask of subapertures is used in the pupil, such that no vector separation occurs more than once (i.e., a *nonredundant* array of subapertures), then each spatial frequency component of the point spread function equals the product of two particular values of the complex amplitude in the pupil and hence yields the phase difference between the wave front at these two points. Thus, the Fourier transform of the image intensity, when the imaging aperture is a nonredundant array of subapertures, has a simple, direct interpretation in terms of the wave front in the telescope aperture, and this has been the basis of several possible methods of diffraction-limited imaging from (speckle-free) image plane data. However, in all cases each frame of data has to have an excellent signal-to-noise ratio so that an accurate phase difference can be found.

Three methods of utilizing short-exposure images taken through nonredundant apertures have been put forward. Rhodes and Goodman [63] suggested using two or more masks to form two or more short-exposure images with the same atmospheric distortion—this procedure could be used to eliminate telescope aberrations as well. Brown [64] suggested that even a single image from a single mask could be processed by using a suitable "sharpness" criterion for the reconstructed object. Finally, there is the possibility of phase averaging analogous to that described in Section 7.5.A.6.

A variation on the use of a nonredundant aperture has been proposed by Greenaway [65]. The basic idea is to reconfigure a redundant aperture into a nearly nonredundant one by using a suitable interferometer and then applying phase closure techniques. This allows the full aperture to be utilized, but it is not clear how one constructs the optical system required to reconfigure the aperture.

8. *Phase Diversity Method*

As seen in Section 7.5.A.5, the availability of more than one type of measurement opens up additional possibilities for reconstructing an image. A class of such techniques are those that employ phase diversity in the pupil of the telescope [66]. Two images are collected simultaneously: the usual aberrated image with optical transfer function $T(u, v)$ and Fourier transform

$$G(u, v) = F(u, v)\,T(u, v), \tag{7.5-15}$$

and a "phase-diversity" image with optical transfer function $T_{\mathrm{d}}(u, v)$ and Fourier transform

$$G_{\mathrm{d}}(u, v) = F(u, v)\,T_{\mathrm{d}}(u, v). \tag{7.5-16}$$

The phase-diversity transfer function $T_{\mathrm{d}}(u, v)$ includes a known, purposely induced phase aberration in addition to the same unknown atmospherically induced aberrations included in $T(u, v)$. For this problem both the object and the atmospheric aberrations are unknown.

To solve this problem, Gonsalves [66] introduced the error metric

$$E = \iint [|G(u, v) - \hat{F}(u, v)\hat{T}(u, v)|^2 + |G_{\mathrm{d}}(u, v) - \hat{F}(u, v)\hat{T}_{\mathrm{d}}(u, v)|^2]\, du\, dv \tag{7.5-17}$$

(where $\hat{X}$ is an estimate of X). By the method of calculus of variations, the function $\hat{F}(u, v)$ that minimizes E can be determined and inserted in Eq. (7.5-17) to yield an expression for the error metric that is independent of $\hat{F}(u, v)$:

$$E = \iint \left[\frac{|G(u, v)\hat{T}_{\mathrm{d}}(u, v) - G_{\mathrm{d}}(u, v)\hat{T}(u, v)|^2}{|\hat{T}(u, v)|^2 + |\hat{T}_{\mathrm{d}}(u, v)|^2} \right] du\, dv. \tag{7.5-18}$$

This error metric depends only on the estimates of the transfer functions, which in turn depend only on the estimate of the phase aberrations. One can iteratively search for the estimate of the phase aberrations (usually

parameterized by, say, the coefficients of a Zernike polynomial expansion) that minimize E, by the Fletcher–Powell gradient search method, for example. Having found the aberrations, $T(u, v)$ can be computed and $F(u, v)$ can be determined from Eq. (7.5-15) or, better, by Wiener filtering.

A problem with this method is that, employing only a single pair of frames of data, it works well only for bright objects.

B. Object Recovery from Michelson Interferometry Data

1. Pupil Plane Interferometry

In pupil plane interferometry, the spatial coherence function of starlight is measured by simple two-beam interference either directly in the pupil of a large telescope or with two (or more) separated telescopes in a long-baseline arrangement.

The spatial coherence function $C(\Delta\xi, \Delta\eta)$ is defined by

$$C(\Delta\xi, \Delta\eta) \equiv \langle V(\xi, \eta, t) V^*(\xi + \Delta\xi, \eta + \Delta\eta, t) \rangle, \qquad (7.5\text{-}19)$$

where $V(\xi, \eta, t)$ is the analytic signal representation of the field in the (ξ, η) plane. The real part of $C(\Delta\xi, \Delta\eta)$ can be measured by interfering portions of the wave front located at (ξ, η) and $(\xi + \Delta\xi, \eta + \Delta\eta)$ and measuring the resultant time-average† intensity:

$$\begin{aligned} &\langle | V(\xi, \eta, t) + V(\xi + \Delta\xi, \eta + \Delta\eta, t)|^2 \rangle \\ &\qquad = I(\xi, \eta) + I(\xi + \Delta\xi, \eta + \Delta\eta) + 2\,\mathrm{Re}[C(\Delta\xi, \Delta\eta)]. \qquad (7.5\text{-}20) \end{aligned}$$

The imaginary part is found by retarding one wave front by $\pi/2$ prior to interference. In practice, the two wave fronts are combined either through a beamsplitter arrangement to yield both the real and imaginary parts of $C(\Delta\xi, \Delta\eta)$ simultaneously, or at an angle so that the fringe position gives the phase (if it is possible to follow individual fringes).

According to the van Cittert–Zernike theorem, the coherence function $C(\Delta\xi, \Delta\eta)$ is proportional to the Fourier transform $F(u, v)$ of the object intensity $f(x, y)$; if we take the object coordinates referred to image space, the spatial frequencies (u, v) of the object transform are related to separation in the pupil plane $(\Delta\xi, \Delta\eta)$ given by

$$u = \Delta\xi/\lambda z, \qquad v = \Delta\eta/\lambda z,$$

where z is the focal length of the telescope.‡

† Or ensemble average; the distinction is unimportant for ergodic light sources such as stars.
‡ It is customary in astronomy to use angular frequencies $\Delta\xi/\lambda$ and $\Delta\eta/\lambda$.

Although historically this method was first proposed and demonstrated by Fizeau and Stephan, it is normally known as Michelson interferometry, following Michelson's measurements of the satellites of Jupiter and the red giant α-Orionis, the latter with a specifically constructed two-beam interferometer mounted on the 100-in. Mt. Wilson telescope. The main attraction of the implementation of Michelson interferometry with two small telescopes is that the baseline, which in terms of resolution is equivalent to the diameter of a single-aperture telescope, can potentially be very large, perhaps 1 km or more. It is also relatively easily adapted to use in space. The major disadvantages of the two-telescope approach are the poor coverage of the (u, v) plane and the lack of reliable phase information.

2. *Wave Front Rotational Shearing Method*

This method is equivalent to an array of Michelson interferometers working in parallel and is an alternative to speckle (image plane) methods for single large telescopes. Mertz [67] was the first to point out that a rotational shearing interferometer can be used to measure the two-dimensional Fourier transform of any incoherent source distribution with a variable magnification that depends on the amount of rotational shear. This interferometer has been adapted by Roddier and Roddier [68] for astronomical observation.

In a rotational shearing interferometer, the time-varying complex amplitude in the telescope pupil, $V_1(\xi, \eta, t)$, is divided by a beamsplitter into two beams, one of which is rotated by an angle θ with respect to the other before recombination at the same or another beamsplitter. Variable magnification is achieved by varying θ, but to see the principle of operation we shall let $\theta = 180°$. The output intensity $I_2(\xi, \eta)$ of the interferometer in this case is

$$\begin{aligned} I_2(\xi, \eta) &= \langle |V_1(\xi, \eta, t) + V_1(-\xi, -\eta, t)|^2 \rangle \\ &= 2I_1(\xi, \eta) + 2\,\mathrm{Re}[C(2\xi, 2\eta)] \end{aligned} \tag{7.5-21}$$

for a uniform pupil intensity. The imaginary part of the spatial coherence could be found separately by phase-shifting one beam prior to recombination or, more simply, by adding tilt fringes to form a sideband hologram [68].

In applying this method to imaging through turbulence, it is important to note that the fringe *visibility*, from which the spatial coherence is found, remains constant from frame to frame and the turbulence affects only the detailed form of the fringes. The major disadvantage of the technique is that each frame must have sufficient signal-to-noise ratio to enable the fringe visibility to be determined, and this requires that the individual fringes are clearly defined. This is a significant drawback despite the fact that at high light levels pupil plane interferometers have a higher signal-to-noise ratio than the image plane technique of speckle interferometry.

3. Phase Closure

Phase retrieval in the two (or more) telescope arrangement originally proposed by Michelson is very much more complicated than for the single-aperture case, as there are almost certainly discontinuities of the measurements in the pupil plane because the full aperture is sparsely filled. One approach is to combine simultaneously the outputs from a nonredundant coherent array and use the image plane (speckle) data as described in Section 7.5.A.7. Another technique, originally suggested and used in radio astronomy [69], is to form simultaneously all possible pair correlations from N apertures by using the Michelson technique to determine the Fourier phase, and then use the principle of *phase closure* to eliminate some or (given enough redundant measurements) all the (unknown) atmospheric phase disturbances.

Consider any three apertures and let the (constant) phases across them due to atmospheric turbulence or other phase errors be α_1, α_2, and α_3. The measurement of instantaneous fringe positions for each aperture pair yields phase differences β_{12}, β_{23}, and β_{13}, which are related to the true object phase differences ψ_{12}, ψ_{23}, and ψ_{13} by

$$\beta_{12} = \psi_{12} + \alpha_1 - \alpha_2,$$

$$\beta_{23} = \psi_{23} + \alpha_2 - \alpha_3,$$

$$\beta_{13} = \psi_{13} + \alpha_1 - \alpha_3.$$

By forming the sum

$$\beta_{12} + \beta_{23} - \beta_{13} = \psi_{12} + \psi_{23} - \psi_{13},$$

the atmospheric phase terms are eliminated. If one assumes that two object phases are known, then the third can be calculated. The arbitrary choice of the first two phases simply affects the position of the reconstructed object on the sky and does not affect its morphology. By repeating this with a carefully chosen set of aperture spacings (which include redundant spacings) it is possible to determine the Fourier phase of the object at all spatial frequencies covered by the array. Even if sufficient redundant spacings are not present to ensure a complete reconstruction, the number of unknown phases is greatly reduced by the phase closure method.

7.6 CONCLUSIONS

The methods for reconstructing stellar objects $f(x, y)$ from interferometric data in optical astronomy fall into two classes: those that require only the Fourier modulus $|F(u, v)|$ and those that use the additional, but limited, phase data that are available in certain observational techniques.

It appears that object reconstruction from the Fourier modulus alone is usually unique, although it is not difficult to construct specific counter-examples that do not yield a unique solution. The main problem is how to find this solution. The iterative transform algorithm described in Section 7.4 is the most computationally efficient and successful algorithm to date. Using array processors of modest cost, a complete iteration on a 128×128 data set need take only about 1 s and convergence to a solution can be obtained in a few minutes if stagnation problems are avoided (see Section 7.4.C).

Particular observing techniques, such as speckle interferometry, can provide not only the Fourier modulus but also Fourier phase information. Clearly, if this additional information is available, it should be used in the reconstruction process. The most promising techniques for doing this at the moment are the cross-spectrum, triple correlation, and exponential filter methods described in Sections 7.5.A.3–5 respectively.

The planned construction of several giant 10–20-m diameter ground-based optical telescopes in the near future should provide the stimulation for further advances in phase retrieval and object reconstruction in astronomy.

ACKNOWLEDGMENTS

One author (J.R.F.) thanks the U.S. Air Force Office of Scientific Research for its financial support under contract F49620-82-K-0018.

REFERENCES

1. J. W. Goodman (1984). "Statistical Optics," Wiley, New York.
2. R. H. T. Bates (1982). *Astronomical speckle imaging. Physics Reports* (*Reviews Sect. of Phys. Lett.*) **90**, 203–297.
3. J. C. Dainty (1984). "Laser Speckle and Related Phenomena," Springer Verlag, Heidelberg. 2nd edition, pp. 255–320.
4. J. C. Dainty (1984). Progress in image reconstruction in astronomy. *Proc. SPIE*, **492**.
5. A. Walther (1963). The question of phase retrieval in optics. *Optica Acta* **10**, 41–49.
6. A. H. Greenaway (1977). Proposal for phase recovery from a single intensity distribution. *Opt. Lett.* **1**, 10–12.
7. T. R. Crimmins and J. R. Fienup (1981). Ambiguity of phase retrieval for functions with disconnected support. *J. Opt. Soc. Am.* **71**, 1026–1028.

8. T. R. Crimmins and J. R. Fienup (1983). Uniqueness of phase retrieval for functions with sufficiently disconnected support. *J. Opt. Soc. Am.* **73**, 218–221.
9. J. R. Fienup (1978). Reconstruction of an object from the modulus of its Fourier transform. *Opt. Lett.* **3**, 27–29.
10. P. J. Napier and R. H. T. Bates (1974). Inferring phase information from modulus information in two-dimensional aperture synthesis. *Astron. Astrophys. Suppl.* **15**, 427–430.
11. W. Lawton (1980). A numerical algorithm for 2-D wavefront reconstruction from intensity measurements in a single plane. *Proc. SPIE* **231**, 94–98.
12. M. Nieto-Vesperinas (1980). Dispersion relations in two dimensions: Application to the phase problem. *Optik (Stuttgart)* **56**, 377–384.
13. I. Manolitsakis (1982). Two-dimensional scattered fields: A description in terms of the zeros of entire functions. *J. Math. Physics* **23**, 2291–2298.
14. R. Barakat and G. Newsam (1984). Necessary conditions for a unique solution to two-dimensional phase recovery. *J. Math. Phys.* **25**, 3190–3193.
15. J. L. C. Sanz and T. S. Huang (1983). Unique reconstruction of a band-limited multi-dimensional signal from its phase or magnitude. *J. Opt. Soc. Am.* **73**, 1446–1450.
16. Yu. M. Bruck and L. G. Sodin (1979). On the ambiguity of the image reconstruction problem. *Opt. Commun.* **30**, 304–308.
17. M. A. Fiddy, B. J. Brames, and J. C. Dainty (1983). Enforcing irreducibility for phase retrieval in two dimensions. *Opt. Lett.* **8**, 96–98.
18. J. R. Fienup (1983). Reconstruction of objects having latent reference points. *J. Opt. Soc. Am.* **73**, 1421–1426.
19. B. J. Brames (1985). Uniqueness and other aspects of the optical phase problem. Ph.D. Thesis, University of Rochester.
20. T. R. Crimmins (1986). Phase retrieval for discrete functions with support constraints: summary, *in* OSA Topical Meeting on Signal Recovery and Synthesis II, Honolulu, Hawaii.
21. J. L. C. Sanz, T. S. Huang, and F. Cukierman (1983). Stability of unique Fourier-transform phase reconstruction. *J. Opt. Soc. Am.* **73**, 1142–1145.
22. E. N. Leith and J. Upatnieks (1962). Reconstructed wavefronts and communication theory. *J. Opt. Soc. Am.* **52**, 1123–1130.
23. J. W. Goodman (1970). Analogy between holography and interferometric image formation. *J. Opt. Soc. Am.* **60**, 506–509.
24. C. Y. C. Liu and A. W. Lohmann (1973). High resolution image formation through the turbulent atmosphere. *Opt. Commun.* **8**, 372–377.
25. J. R. Fienup, T. R. Crimmins, and W. Holsztynski (1982). Reconstruction of the support of an object from the support of its autocorrelation. *J. Opt. Soc. Am.* **72**, 610–624.
26. G. H. Stout and L. H. Jensen (1968). "X-ray Structure Determination," Macmillan, London.
27. J. E. Baldwin and P. J. Warner (1978). Phaseless aperture synthesis. *Mon. Not. R. Astr. Soc.* **182**, 411–422.
28. B. R. Frieden and D. G. Currie (1976). On unfolding the autocorrelation function. *J. Opt. Soc. Am.* **66**, 1111A.
29. J. R. Fienup (1982). Phase retrieval algorithms: A comparison. *Appl. Opt.* **21**, 2758–2769.
30. H. V. Deighton, M. S. Scivier, and M. A. Fiddy (1985). Solution of the two-dimensional phase retrieval problem. *Opt. Lett.* **10**, 250–251.
31. H. H. Arsenault and S. Lowenthal (1969). La restitution de la phase a de measures d'eclairement. *C.R. Acad. Sc. Paris* **269**, 518–521.
32. H. H. Arsenault and K. Chalasinska-Macukow (1983). The solution to the phase retrieval problem using the sampling theorem. *Opt. Commun.* **47**, 380–386.
33. K. Chalasinska-Macukow and H. H. Arsenault (1985). Fast iterative solution to exact equations for the two-dimensional phase-retrieval problem. *J. Opt. Soc. Am. A* **2**, 46–50.

34. R. H. T. Bates, *et al.* (1982). Fourier phase problems are uniquely solvable in more than one dimension. I: underlying theory. *Optik* **61**, 247–262; (1982) II: one-dimensional considerations. *Optik* **62**, 131–142; (1982) III: computational examples for two dimensions. *Optik* **62**, 219–230.
35. R. W. Gerchberg and W. O. Saxton (1972). A practical algorithm for the determination of phase from image and diffraction plane pictures. *Optik* **35**, 237–246.
36. R. W. Gerchberg (1974). Super-resolution through error energy reduction. *Optica Acta* **21**, 709–720.
37. W. O. Saxton (1978). "Computer Techniques for Image Processing in Electron Microscopy," Academic Press, New York.
38. J. R. Fienup (1979). Space object imaging through the turbulent atmosphere. *Opt. Eng.* **18**, 529–534.
39. J. R. Fienup (1981). Reconstruction and synthesis applications of an iterative algorithm. *Proc. SPIE* **373**, 147–160.
40. J. R. Fienup (1984). Phase retrieval from a single intensity distribution, in "Optics in Modern Science and Technology," Conference Digest for ICO-13, August 20-24, Sapporo, Japan, pp. 606–609; J. R. Fienup (1987). Reconstruction of a complex-valued object from the modulus of its Fourier transform using a support constraint. *J. Opt. Soc. Am. A* **3** (forthcoming).
41. J. R. Fienup and G. B. Feldkamp (1980). Astronomical imaging by processing stellar speckle interferometry data. *Proc. SPIE* **243**, 95–102.
42. R. J. Sault (1983). The Fourier phase problem, *FLEURS Internal Report No. 83/1*, University of Sydney.
43. P van Toorn, A. H. Greenaway, and A. M. J. Huişer (1984). Phaseless object reconstruction. *Optica Acta* **7**, 767–774.
44. J. R. Fienup (1984). Experimental evidence of the uniqueness of phase retrieval from intensity data. *In* "Indirect Imaging," Proceedings of URSI/IAU Symposium, Sydney, Australia, (J. A. Roberts, ed.), pp. 99–109, Cambridge University Press, Cambridge.
45. J. R. Fienup and C. C. Wackerman (1986). Phase retrieval stagnation problems and solutions. *J. Opt. Soc. Am. A* **2** (forthcoming).
46. G. B. Feldkamp and J. R. Fienup (1980). Noise properties of images reconstructed from Fourier modulus. *Proc. SPIE* **231**, 84–93.
47. A. Labeyrie (1970). Attainment of diffraction limited resolution in large telescopes by Fourier analysing speckle patterns in star images. *Astron. and Astrophys.* **6**, 85–87.
48. J. W. Goodman and J. F. Belsher (1976). Fundamental limitations in linearly invariant restoration of atmospherically degraded images. *Proc SPIE* **75**, 141–154.
49. C. R. Lynds, S. P. Worden, and J. W. Harvey (1976). Digital image reconstruction applied to Alpha Orionis. *Astrophys. J.* **207**, 174–180.
50. R. H. T. Bates and F. M. Cady (1980). Towards true imaging by wideband speckle interferometry. *Opt. Commun.* **32**, 365–369.
51. B. R. Hunt, W. R. Fright, and R. H. T. Bates (1983). Analysis of the shift-and-add method for imaging through turbulent media. *J. Opt. Soc. Am.* **73**, 456.
52. K. T. Knox and B. J. Thompson (1974). Recovery of images from atmospherically degraded short-exposure photographs. *Astrophys. J. Letters* **193**, L45–L48.
53. M. Rimmer (1974). Method for evaluating lateral shearing interferograms. *Appl. Opt.* **13**, 623–629.
54. B. R. Hunt (1979). Matrix formulation of the reconstruction of phase values from phase differences. *J. Opt. Soc. Am.* **69**, 393–399.
55. A. W. Lohmann, G. Weigelt, and B. Wirnitzer (1983). Speckle masking in astronomy: triple correlation theory and applications. *Appl. Opt.* **22**, 4028–4037.

56. H. Bartelt, A. W. Lohmann, and B. Wirnitzer (1984). Phase and amplitude recovery from bispectra. *Appl. Opt.* **23**, 3121–3129.
57. B. J. Brames and J. C. Dainty (1981). Method for determining object intensity distributions in stellar speckle interferometry. *J. Opt. Soc. Am.* **71**, 1542–1545.
58. J. G. Walker (1981). The phase retrieval problem: A solution based on zero location by exponential apodization. *Optica Acta* **28**, 735–738.
59. J. G. Walker (1982). Computer simulation of a method for object reconstruction from stellar speckle interferometry data. *Appl. Opt.* **21**, 3132–3137.
60. B. L. McGlamery (1970). Image restoration techniques applied to astronomical photography. *In* "Astronomical Use of Television-Type Image Sensors" (V. R. Boscarino, ed.), pp. 167–192, NASA SP-256(NTIS N71-28522).
61. M. S. Scivier, T. J. Hall, and M. A. Fiddy (1984). Phase unwrapping using the complex zeros of a bandlimited function and the presence of ambiguities in two dimensions. *Opt. Acta* **31**, 619–623.
62. L. Mertz (1984). Phase estimation with a few photons. *Appl. Opt.* **23**, 1638–1641.
63. W. T. Rhodes and J. W. Goodman (1973). Interferometric technique for recording and restoring images degraded by unknown aberrations. *J. Opt. Soc. Am.* **63**, 647–657.
64. T. M. Brown (1978). Reconstruction of turbulence degraded images using nonredundant aperture arrays. *J. Opt. Soc. Am.* **68**, 883–889.
65. A. H. Greenaway (1982). Diffraction-limited pictures from single turbulence degraded images in astronomy. *Opt. Commun.* **42**, 157–161.
66. R. A. Gonsalves (1982). Phase retrieval and diversity in adaptive optics. *Opt. Eng.* **21**, 829–832.
67. L. Mertz (1965). "Transformations in Optics," Wiley, New York.
68. C. Roddier and F. Roddier (1979). *In* "Image Formation from Coherence Functions in Astronomy," pp. 175–185. D. Reidel, Amsterdam.
69. R. C. Jennison (1958). A phase sensitive interferometer technique for the measurement of the Fourier transforms of spatial brightness distributions of small angular extent. *Mon. Not. R. Astr. Soc.* **118**, 276–284.

CHAPTER **8**

Restoration from Phase and Magnitude by Generalized Projections

*Aharon Levi**

Henry Stark

Electrical, Computer, and Systems Engineering Department
Rensselaer Polytechnic Institute
Troy, New York 12180

8.1 INTRODUCTION

The problems of concern in this chapter are those in which the phase or magnitude of a complex function is known and it is required to recover the entire function. We already have seen in the previous two chapters several approaches to problems of this type. In this chapter we stress solutions by *iterative methods* based on *mathematical projections.* Projection methods can be fruitfully applied to other image recovery problems as well, as in tomographic image recovery from limited view data (Chapter 11).

Typically, in problems in which the phase or magnitude is to be recovered, we are faced with the following situation: we assume an "image" $f(\mathbf{x})$ with Fourier transform $F(\boldsymbol{\omega}) = M(\boldsymbol{\omega}) \exp[j\phi(\boldsymbol{\omega})]$ and—from measurements or other means—we know $M(\boldsymbol{\omega})$ [or $\phi(\boldsymbol{\omega})$] and we wish to recover $\phi(\boldsymbol{\omega})$ [or $M(\boldsymbol{\omega})$]. Usually we have available some *a priori* information, for example, that $f(\mathbf{x})$ has compact finite support or that $f(\mathbf{x})$ is nonnegative or that $f(\mathbf{x})$ is bounded in energy. Problems of this type occur in radio astronomy, in

* Present address: Electronics Division, Rafael, P.O. Box 2250, Haifa 31021, Israel.

which it is required to recover the phase of the measured coherence function (see Chapter 7); in the study of materials, in which it is desired to recover the phase of the x-ray or electron diffraction pattern to estimate the structure of matter; and in situations where the phase of the signal is so scrambled due to noise, e.g., atmospheric turbulence, that only the magnitude is available for processing.

Fundamental to image recovery by projections is the notion that each constraint applied to $f(\mathbf{x})$ forces $f(\mathbf{x})$ to belong to a corresponding set. Thus given m constraints on $f(\mathbf{x})$, we seek a point, i.e., a function that lies at the intersection of m sets. The problem of recovery from partial information can then be viewed as that of finding a procedure that will, in systematic fashion, find a point in the intersection of m sets. From this point of view almost all problems of restoration from partial information can be divided into two types: (1) those in which all constraints are convex, i.e., each constraint requires $f(\mathbf{x})$ to belong to an associated *convex* set; and (2) those in which at least one constraint is not convex. The restoration from phase problem is an example of the former. Restoration from magnitude, a much more difficult problem, belongs to the latter. Problems involving only convex constraints can be treated by the method of projections onto convex sets [1,2] (see also Chapters 2 and 11 of this book).

The restoration from magnitude problem can be treated by the method of generalized projections [3] (MGP). These algorithms and their applications to the aforementioned problems are the main subjects of this chapter.

8.2 THE GERCHBERG–SAXTON AND RELATED ALGORITHMS

A. General

The Gerchberg–Saxton (GS) algorithm is frequently used for restoration problems. It was first derived by Gerchberg and Saxton [4] for the restoration of a signal from its given magnitude and the given magnitude of its Fourier transform. The GS algorithm was adopted by Gerchberg [5] and Papoulis [6] for problems of extrapolation of a space-limited signal from a given portion of its Fourier transform (or a band-limited signal from a given portion of itself). Hayes *et al.* [7] used it for the problem of recovering a signal from Fourier phase, and Fienup [8] described the application of the method, together with some generalizations, to the recovery-from-magnitude-only problem. The algorithm converges for problems of extrapolation [6,9] and restoration from phase [10]. In the case of restoration from magnitude only, a property of "error" reduction can be established [8]. In some cases the generalized version of the algorithm leads to projection operators and therefore is a special case of the method of projections (e.g., the problem of restoration from magnitude), but in general the algorithm may include transformations which are not projection operators.

B. The Gerchberg–Saxton Algorithm

The algorithm is characterized by two steps in each iteration. In step 1 (see Fig. 8.2-1) the known information about the signal in the Fourier

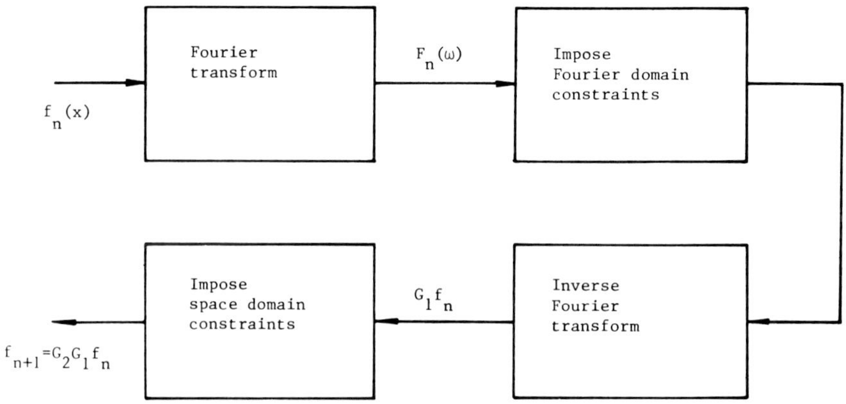

Fig. 8.2-1. Block diagram description of Gerchberg-Saxton algorithm.

(space) domain is imposed on the last estimate of the signal. In step 2, the known information about the signal in the space (Fourier) domain is imposed on the partially corrected estimate from step 1. Let f_n be the nth iterate; then the result of step 1 is $G_1 f_n$, the result of step 2 is $G_2(G_1 f_n)$, and therefore we can write

$$f_{n+1} = G_2 G_1 f_n = G f_n, \tag{8.2-1}$$

where G_1 and G_2 are two transformations representing step 1 and step 2, respectively, and $G \triangleq G_2 G_1$. This algorithm will be used throughout this chapter for comparison with algorithms based on the method of projections.

8.3 THE METHOD OF PROJECTIONS ONTO CONVEX SETS

A. General

The method of projection onto convex sets (POCS) is a recursive algorithm for finding a point in the intersection of m given closed convex sets. The method was developed by a Russian group (Bregman [11] and Gubin *et al.* [12]) and was extended and adapted to signal processing by Youla [1,2]. The method is described in detail in Chapter 2 of this book. In this section a summary of the method is given and the problem of optimization of the relaxation parameters is discussed.

An underlying notion in POCS is that of a projection onto a convex set: given an element $f \in \mathscr{H}$, where $\mathscr{H}$ is a Hilbert space, the projection $g \triangleq P_i f$ of f onto the closed convex set $\mathscr{C}_i$ is that unique element g (proof that g is unique can be found, for example, in [1]) in $\mathscr{C}_i$ that satisfies

$$\operatorname*{Inf}_{x \in \mathscr{C}_i} \|f - x\| = \|f - g\|, \tag{8.3-1}$$

where $\|x\|$ is the norm or "length" and is generally taken as the L_2-space distance. Thus g is that element in $\mathscr{C}_i$ that is closest to f. The synthesis of the operator P_i is obviously a central objective; fortunately, in practice it is often a straightforward task.

In applying the method of POCS in image processing, the following viewpoint is adopted: nearly every known property of an unknown image f is viewed as a constraint that restricts the image to lie in a well-defined closed convex set $\mathscr{C}_i$. Thus for m properties there are m sets $\mathscr{C}_i$, $i = 1, \ldots, m$, and the image must lie in the intersection

$$\mathscr{C}_0 \triangleq \bigcap_{i=1}^{m} \mathscr{C}_i. \tag{8.3-2}$$

The image restoration problem is then to find a point, i.e., an image that lies in $\mathscr{C}_0$. Such an image will satisfy all the constraints and will therefore represent a solution to the problem. If $\mathscr{C}_0$ contains only a single point, the solution is unique; if $\mathscr{C}_0$ is empty, then there is no solution. In practice, the latter might occur with noisy data or inaccurate *a priori* information.

Let P_i denote the projector for the set $\mathscr{C}_i$ and let T_i be the "relaxed" projector defined by

$$T_i \triangleq 1 + \lambda_i(P_i - 1), \qquad 0 < \lambda_i < 2, i = 1, 2, \ldots, m, \tag{8.3-3}$$

where λ_i are constants called relaxation parameters (RPs). Observe that the fixed points of P_i are also those of T_i.† Let us define the composition T by

$$T \triangleq T_m T_{m-1} \cdots T_1; \tag{8.3-4}$$

then it was shown [1] that if $\mathscr{C}_0$ is nonempty the POCS algorithm given by

$$f_{n+1} = Tf_n, \qquad f_0 \text{ arbitrary} \tag{8.3-5}$$

converges weakly to a point of $\mathscr{C}_0$.

† A fixed point operator P is that function for which $Pf = f$.

Gubin *et al.* [12] and Youla [1,2] give several conditions under which the convergence is strong, for example, in the case of finite discrete sequences. In the algorithm given by Eq. (8.3-5) the relaxation parameter λ_i associated with each projection operator P_i, $i = 1, 2, \ldots, m$, is fixed and does not change from iteration to iteration. Before addressing the problem of the optimization of the RPs we must answer the basic question: Is the algorithm still convergent when the RPs are changed independently in each iteration? The answer to this question is positive, and the more general algorithm can be stated as follows. Let $T_{i,n}$, $i = 1, 2, \ldots, m$, $n = 1, 2, 3, \ldots$, be given by

$$T_{i,n} = 1 + \lambda_{i,n}(P_i - 1), \qquad 0 < \lambda_{i,n} \leq \lambda_0 < 2, \tag{8.3-6}$$

where $\lambda_{i,n}$ is the RP associated with the projection operator P_i at the nth iteration. Let T_n be the combined operator

$$T_n = T_{m,n} T_{m-1,n} \cdots T_{1,n}. \tag{8.3-7}$$

The following theorem is by Levi [13].

Theorem 8.3-1. Let $\mathscr{C}_0$ be nonempty. Then for every choice of $\lambda_{i,n}$, provided that $0 < \lambda_{i,n} \leq \lambda_0 < 2$, the algorithm

$$f_{n+1} = T_n f_n, \qquad f_0 \text{ arbitrary}, \tag{8.3-8}$$

converges weakly to a point of $\mathscr{C}_0$.

We shall use Theorem 8.3-1 in subsequent sections.

B. Optimization of the General Two-Step POCS Algorithm

When Eq. (8.3-5) is specialized to the two-step-per-cycle algorithm, the general two-step-per-cycle realization of the POCS algorithm is given by

$$f_{1n} \triangleq T_1 f_n, \qquad f_{n+1} \triangleq T_2 f_{1n}, \qquad f_0 \text{ arbitrary}, \tag{8.3-9}$$

or as a composition†

$$f_{n+1} = T_2 T_1 f_n, \qquad f_0 \text{ arbitrary}. \tag{8.3-10}$$

The estimate f_{1n} is an intermediate result obtained in each iteration after the first step, i.e., after applying T_1 on f_n. In what follows we consider the algorithm in Eq. (8.3-10) with‡

$$T_1 \triangleq 1 + \lambda_{1n}(P_1 - 1), \tag{8.3-11a}$$

$$T_2 \triangleq 1 + \lambda_{2n}(P_2 - 1), \tag{8.3-11b}$$

† Equation (8.3-10) has the form of Eq. (8.3-5) with $T_2 T_1 \triangleq T$.

‡ We replace $\lambda_{i,n}$ by λ_{in} whenever the meaning is clear.

and we attempt to develop some theory and procedures for determining the relaxation parameters $\lambda_{1n}, \lambda_{2n}$ for improving the convergence rate. The first subscript identifies the λ with the operator which it "relaxes" and the second refers to the iteration number. Thus λ_{in} is the relaxation parameter associated with P_i at the nth iteration. Note that in general T_1 and T_2 are functions of n, but we omit the subscript n for simplicity.

In general, it is a very difficult problem to choose $\lambda_{1n}, \lambda_{2n}$ for every iteration n, $n = 1, \ldots, N$, such that at some specified iteration, say the Nth, the error will be a minimum. Instead we consider two different criteria for choosing the RPs. In this chapter we present mostly the results, but more discussion and proofs can be found in [3], [13], and [14].

To understand the two criteria for choosing RPs, define the error after the nth iteration by

$$e_n \triangleq f_n - f, \tag{8.3-12}$$

where $f \in \mathscr{C}_0$ is a valid solution and define the intermediate-step error e_{1n} by

$$e_{1n} \triangleq f_{1n} - f \qquad (\text{recall } f_{1n} \triangleq T_1 f_n). \tag{8.3-13}$$

It is easy to show that the sequence of normed per-cycle errors satisfy

$$\|e_n\| \leqslant \|e_{1,n-1}\| \leqslant \|e_{n-1}\| \leqslant \|e_{1,n-2}\| \leqslant \cdots. \tag{8.3-14}$$

The proof is straightforward and is based on the following results: (1) since f is a fixed point of T_i, $i = 1, 2$, we can write $f = T_1 f = T_2 f$; and (2) projection operators are nonexpansive and hence $\|T_i f_n - T_i f\| \leqslant \|f_n - f\|$, $i = 1, 2$. We define our two criteria as follows: (1) find $\lambda_{1n}, \lambda_{2n}$ that sequentially minimize the normed errors after each step, i.e., so-called *per-step optimization*, and (2) find λ_{1n} and λ_{2n} that minimize the normed errors after each iteration (cycle), i.e., so-called *per-cycle optimization*. In (1) we seek to sequentially minimize $\|e_{1n}\|$ for a given f_n, following which we minimize $\|e_{n+1}\|$ for a given f_{1n}. In (2) we seek to minimize $\|e_{n+1}\|$ for a given f_n. Intuitively we expect criterion (2) to give better results because it considers the minimization for a whole iteration.

C. Per-Step Optimization of the Relaxation Parameters

To simplify the notation we omit the subscript n on the λ's. From Eqs. (8.3-11) and (8.3-12), we obtain†

$$\|e_{1n}\|^2 = \|f_n - f + \lambda_1(P_1 - 1)f_n\|^2 = \|f_n - f\|^2 + 2\lambda_1 \operatorname{Re}[(f_n - f, (P_1 - 1)f_n)] + \lambda_1^2\|(P_1 - 1)f_n\|^2 \tag{8.3-15}$$

† Re(f) denotes the real part of f and (f, g) denotes the usual L_2 inner product of f with g. See Eqs. (8.4-5) and (8.4-6).

from which it is easily verified by differentiation that $\|e_{1n}\|^2$ is a minimum for any fixed f_n when

$$\lambda_1 = \lambda_{1,\text{opt}} = \frac{\text{Re}[(f - f_n, (P_1 - 1)f_n)]}{\|(P_1 - 1)f_{1n}\|^2}. \tag{8.3-16}$$

Similarly, if f_{1n} is fixed, $\|e_{n+1}\|^2$ is a minimum when

$$\lambda_2 = \lambda_{2,\text{opt}} = \frac{\text{Re}[(f - f_{1n}, (P_2 - 1)f_{1n})]}{\|(P_2 - 1)f_{1n}\|^2}. \tag{8.3-17}$$

More generally, with P denoting either P_1 or P_2, λ_{opt} denoting either $\lambda_{1,\text{opt}}$ or $\lambda_{2,\text{opt}}$, and f_n denoting the previous estimate f_n or f_{1n}, respectively, we can write

$$\lambda_{\text{opt}} = \frac{\text{Re}[(f - f_n, (P - 1)f_n)]}{\|(P - 1)f_n\|^2}, \tag{8.3-18}$$

which combines the results of Eqs. (8.3-16) and (8.3-17). In [13] a proof is given for the important result that

$$\lambda_{\text{opt}} \geqslant 1 \tag{8.3-19}$$

and in particular, when $\mathscr{C}$ is a linear subspace, that

$$\lambda_{\text{opt}} = 1. \tag{8.3-20}$$

In general, we cannot evaluate Eq. (8.3-18) because f is not known.‡ However, certain reasonable approximations are possible, as we shall see later.

D. Per-Cycle Optimization of the Relaxation Parameters

Here we consider minimizing the error after the $(n + 1)$th iteration with respect to λ_1 and λ_2 simultaneously. Thus we seek to minimize $\|e_{n+1}\|$ by appropriate choice of λ_1, λ_2, where

$$e_{n+1} = f_{n+1} - f = T_2 T_1 f_n - f. \tag{8.3-21}$$

The optimum RPs in this section will be denoted by $\lambda_{1m}, \lambda_{2m}$ to differentiate them from $\lambda_{1,\text{opt}}, \lambda_{2,\text{opt}}$ computed under the constraints of Section 8.3.C, i.e., the per-step optimization.

First we observe that regardless of what value λ_1 has, we must still compute λ_{2m} from Eq. (8.3-17), namely

$$\lambda_{2m} = \frac{\text{Re}[(f - T_1 f_n, P_2 T_1 f_n - T_1 f_n)]}{\|P_2 T_1 f_n - T_1 f_n\|^2}, \tag{8.3-22}$$

‡ In fact, if $\mathscr{C}_0$ is not a singleton set, f is not unique and e_n in Eq. (8.3-12) is not well defined.

since λ_{2m} will minimize $\|e_{n+1}\|$ for any given f_n and λ_1. However, Eq. (8.3-22) cannot be used directly to evaluate λ_{2m} because (1) we have as yet not determined λ_{1m} on which λ_{2m} depends and (2) we do not know f.

We note that $\|e_{n+1}\|^2$ can be written as

$$\|e_{n+1}\|^2 = \|[1 + \lambda_{2m}(P_2 - 1)]T_1 f_n - f\|^2. \tag{8.3-23}$$

By inserting Eq. (8.3-22) into Eq. (8.3-23) and rearranging terms we obtain

$$\|e_{n+1}\|^2 = \|T_1 f_n - f\|^2 - \lambda_{2m} \operatorname{Re}[(f - T_1 f_n, P_2 T_1 f_n - T_1 f_n)]. \tag{8.3-24}$$

We cannot count on determining λ_{1m} by differentiating Eq. (8.3-24) and isolating λ_{1m} because P_2 is, in general, a nonlinear operator. However, in some cases, notably the restoration from phase (RFP) problem, we can compute a λ_1 which approximates λ_{1m}. This can be done either by finding an upper bound for $\|e_{n+1}\|^2$ and finding a λ_1 that minimizes that upper bound, or by computing an approximate value for $\|e_{n+1}\|^2$ and computing the λ_1 which minimizes that approximate value. In [3] it is shown that the value of λ_1 that minimizes an upper bound for $\|e_{n+1}\|^2$ is found by minimizing the expression

$$I \triangleq -\lambda_1(2 - \lambda_1)\|P_1 f_n - f_n\|^2 - \|P_2 T_1 f_n - T_1 f_n\|^2. \tag{8.3-25}$$

In the general case, where P_2 is nonlinear, I can be minimized by straightforward scan through the λ_1. Then, having obtained an approximate λ_{1m}, we compute λ_{2m} from Eq. (8.3-22). In the special case when P_2 is linear, it is possible to find closed-form solutions for λ_{1m} and λ_{2m}. This is demonstrated in Section 8.4 when dealing with optimization of the RPs in the restoration-from-phase problem.

8.4 APPLICATION OF POCS TO THE PROBLEM OF RESTORATION FROM PHASE

A. General

In the phase restoration problem it is assumed that the object $f(x)$ with Fourier transform (FT) $F(\omega)$ has compact support $[-a, a]$ and the phase $\phi(\omega) \triangleq \arg[F(\omega)]$ is known either at all frequencies or at a discrete set. Our goal is to find an algorithm for reconstructing $f(x)$ from the information given above. However, in general, $f(x)$ is not uniquely determined by the phase of its FT. Nevertheless, it is well known that if $f(x)$ is *minimum phase*, i.e., no zeros in the right half-plane or—for sequences—no zeros inside the unit circle, then the phase uniquely determines the magnitude. However, the restriction of minimum phase is not necessary for uniqueness. A necessary condition for a signal $f(x)$ to be completely specified by the phase $\phi(\omega)$ of its Fourier transform (to a scale factor) is the following restriction on its z-transform $\tilde{F}(z)$: $\tilde{F}(z)$ cannot contain symmetric factors, i.e., factors of the form

$$A(z) = \pm z^{-2n_0} A(z^{-1}) \tag{8.4-1}$$

or

$$z^{n_0} A(z) = \pm z^{-n_0} A(z^{-1}). \tag{8.4-2}$$

In the space domain these relations translate into

$$a(n_0 + x) = \pm a(n_0 - x), \tag{8.4-3}$$

where $a(x)$ and $A(z)$ form a z-transform pair. Thus a symmetric factor in the z-domain is equivalent to a sequence in the space domain that is even-symmetric or odd-symmetric about a shift n_0. Therefore the condition for uniqueness in the space domain, in words, is: the sequence has no axis of symmetry or antisymmetry and it cannot be the result of a convolution with such a subsequence.

When $\phi(\omega)$ is not known for all ω, it is still possible to restore a sequence from the phase of its M-point discrete Fourier transform (DFT) provided that $M \geqslant 2N_s - 1$, where N_s is the length of the finite sequence $\{f(x)\}$. These important results, as well as several others related to uniqueness, are discussed by Hayes in [15].

The standard iterative signal restoration algorithm is a Gerchberg–Saxton type (such as used by Hayes in [15]). It is a specialization of the algorithm discussed in Section 8.2.B and we shall call it the Hayes–Lim–Oppenheim (HLO) method. To improve convergence, relaxation parameters λ_n can be introduced and the basic algorithm given by Eq. (8.2-1) is modified to

$$f_{n+1} = f_n + \lambda_n [G(f_n) - f_n]. \tag{8.4-4}$$

The operator G in Eq. (8.4-4) implements the HLO method in four steps.
Step 1: (go to frequency domain)

$$F_n(\omega) \triangleq |F_n(\omega)| e^{j\theta_n(\omega)} = \mathcal{F}[f_n(x)]$$

($\mathcal{F}$ is the Fourier transform operator).
Step 2: (correct the phase in frequency domain)

$$\tilde{F}_n(\omega) = F_n(\omega)\, e^{j[\phi(\omega) - \theta_n(\omega)]}, \qquad \phi(\omega) \triangleq \arg[F(\omega)].$$

Step 3: (return to space domain)

$$\tilde{f}_n(x) = \mathcal{F}^{-1}[\tilde{F}_n(\omega)].$$

Step 4: (implement space-domain constraint)

$$f_{n+1}(x) = \tilde{f}_n(x)\,\mathrm{rect}(x/2a).$$

For more details on the HLO method as well as how the $\{\lambda_n\}$ are chosen, see [7] and [15].

The choice of initial signal f_0 or the initial Fourier magnitude F_0 is arbitrary but should not be the zero function, i.e., $f_0 = 0$ or $F_0 = 0$. In general, convergence of $f_n(x)$ to $f(x)$ cannot be assured by using Eq. (8.4-4). In fact, $f_n(x)$ may converge to an object that depends on $f_0(x)$ (a property of the algorithm) even if the phase uniquely determines magnitude (a property of the signal). Under certain conditions, however, one can make more definitive statements about the convergence of f_n [15].

Another point to be made is the following: it is possible to use an optimization criterion for determining the λ's that can cause the iterates generated by Eq. (8.4-4) to diverge even though the f_n's from Eq. (8.2-1) converge. An example of this behavior using the optimization criterion chosen in [15] will be given later in this chapter.

B. *The POCS Algorithm for Restoration from Phase (RFP)*

Our Hilbert space $\mathscr{H}$ consists of the L_2 space of square-integrable functions over $-\infty < x < \infty$ with standard L_2 inner product ($g \in \mathscr{H}$, $h \in \mathscr{H}$)

$$(g, h) \triangleq \int_{-\infty}^{\infty} g(x)h^*(x)\,dx \tag{8.4-5}$$

and norm

$$\|g\| \triangleq (g, g)^{1/2}. \tag{8.4-6}$$

In the case of functions $f(x, y)$ of two variables x, y, the Hilbert space is the $L_{2\times 2}$ space of square-integrable functions over $-\infty < x < \infty$, $-\infty < y < \infty$ with

$$(g, h) \triangleq \int_{-\infty}^{\infty} g(x, y)h^*(x, y)\,dx\,dy \tag{8.4-7}$$

and the same norm as in Eq. (8.4-6). If we choose to restrict ourselves only to sequences, we could choose $\mathscr{H}$ to be the l_2-space of square-summable sequences $\{f(n)\}$ with inner product

$$(g, h) \triangleq \sum_n g(n)h^*(n) \tag{8.4-8}$$

and norm

$$(g, g)^{1/2} \triangleq \left[\sum_n |g(n)|^2\right]^{1/2}. \tag{8.4-9}$$

However, we shall not stress this approach.

The two sets of principal interest in the RFP problem are

$$\mathscr{C}_1 = \{h(x) : h(x) = 0, |x| > a\} \tag{8.4-10}$$

and

$$\mathscr{C}_2 = \{h(x) : \arg[H(\omega)] = \phi(\omega)\}. \tag{8.4-11}$$

$\mathscr{C}_1$ is the set of space-truncated functions and $\mathscr{C}_2$ is the set of all h's with a prescribed phase. Both $\mathscr{C}_1$ and $\mathscr{C}_2$ are closed convex sets. With P_i denoting the projection operator onto $\mathscr{C}_i$, it is not difficult to show that P_1 and P_2 are realized by

$$P_1 g = \begin{cases} g(x), & |x| \leq a, \\ 0, & |x| > a, \end{cases} \tag{8.4-12}$$

and

$$P_2 g \leftrightarrow \begin{cases} |G(\omega)| \cos[\phi(\omega) - \psi(\omega)] e^{j\phi(\omega)}, & \cos[\phi(\omega) - \psi(\omega)] \geq 0, \\ 0, & \text{otherwise}, \end{cases} \tag{8.4-13}$$

where $\phi(\omega)$ is the prescribed phase, $\psi(\omega) \triangleq \arg[G(\omega)]$, and $g(x)$ is an arbitrary element of $\mathscr{H}$. Youla in [2] furnishes useful generalizations of $\mathscr{C}_2$ and P_2 as follows: let $\mathscr{C}_2$ be defined by

$$\mathscr{C}_2 = \{h(x) : \arg[H(\omega)] = \phi(\omega) \text{ over a set } \Delta\}; \tag{8.4-14}$$

then

$$P_2 g \leftrightarrow \begin{cases} |G(\omega)| \cos[\phi(\omega) - \psi(\omega)] e^{j\phi(\omega)}, & \cos[\phi(\omega) - \psi(\omega)] \geq 0,\ \omega \in \Delta, \\ 0, & \text{otherwise for } \omega \in \Delta, \\ G(\omega), & \omega \notin \Delta. \end{cases} \tag{8.4-15}$$

For the reconstruction of $f(x)$ from its Fourier transform phase $\phi(\omega)$ we can use the two-step POCS algorithm given by Eq. (8.3-10), i.e., $f_{n+1} = T_2 T_1 f_n$, with f_0 arbitrary and $T_i = 1 + \lambda_{in}(P_i - 1)$, $i = 1, 2$, $0 < \lambda_{in} < 2$. The relaxation parameters λ_{in} may be chosen by the per-step or per-cycle optimization method described in Section 8.3.B or 8.3.C, respectively.

In Eq. (8.3-10), T_1 and T_2 can be commuted without affecting the ultimate convergence, although the *rate* of convergence may depend on the ordering.

C. Per-Step Optimization of the Relaxation Parameters in Restoration from Phase by POCS

In this case, since P_1 projects onto a linear subspace, we have from Eq. (8.3-10) that

$$\lambda_{1,\mathrm{opt}} = 1. \tag{8.4-16}$$

To compute $\lambda_{2,\mathrm{opt}}$, we use Eq. (8.3-17) with P_2 defined in Eq. (8.4-13). To simplify the notation, we make the following definitions

$$y_n \triangleq P_2 f_n, \qquad Y_n(\omega) \triangleq \mathscr{F}[y_n(x)], \tag{8.4-17}$$

so that

$$Y_n(\omega) = \begin{cases} |F_n(\omega)| \cos[\phi(\omega) - \psi_n(\omega)]\, e^{j\phi(\omega)}, & \omega \in \Omega_n, \\ 0, & \omega \in \Omega_n^{\mathrm{c}}, \end{cases} \tag{8.4-18}$$

where $\psi_n(\omega) \triangleq \arg[F_n(\omega)]$, $\psi(\omega) = \arg[F(\omega)]$, and $\Omega_n \triangleq \{\omega : \cos[\phi(\omega) - \psi_n(\omega)] \geqslant 0\}$. It is shown in [3] that $\lambda_{2,\mathrm{opt}}$ is given by

$$\lambda_{2,\mathrm{opt}} = 1 - \frac{\int_{\Omega_n^{\mathrm{c}}} |F(\omega)|\,|F_n(\omega)| \cos[\phi(\omega) - \psi_n(\omega)]\, d\omega}{\|F_n\|^2 - \|Y_n\|^2}. \tag{8.4-19}$$

We establish from Eq. (8.4-19) that $\lambda_{2,\mathrm{opt}} \geqslant 1$ since the integration is over Ω_n^{c}, for which the term on the right of the minus sign is negative. Since $|F(\omega)|$ is unknown, we approximate $\lambda_{2,\mathrm{opt}}$ by replacing $|F(\omega)|$ with $|F_n(\omega)|$ wherever the former appears.

D. Per-Cycle Optimization of the Relaxation Parameters in Restoration from Phase by POCS

The experimental results in [13] indicate that per-cycle optimization of the RPs is considerably more effective in improving convergence than the per-step method. In view of these results, we shall only consider the per-cycle case. In particular, we shall consider how per-cycle optimization of the RPs is affected for the two algorithms obtained by commuting the operators T_1 and T_2. Thus we consider the optimization of: (i) $f_{n+1} = T_2 T_1 f_n$ and (ii) $f_{n+1} = T_1 T_2 f_n$. In addition, we shall find for which algorithm, i.e., (i) or (ii), it is easier to optimise the RP's.

Case (i): Finding the RPs for $f_{n+1} = T_2 T_1 f_n$. The value of λ_2 that minimizes $\|e_{n+1}\|^2$ given by

$$\|e_{n+1}\|^2 = \|[1 + \lambda_2(P_2 - 1)] T_1 f_n - f\|^2 \tag{8.4-20}$$

is, as discussed earlier, given by

$$\lambda_{2m} = \frac{\mathrm{Re}[(f - T_1 f_n, P_2 T_1 f_n - T_1 f_n)]}{\|P_2 T_1 f_n - T_1 f_n\|^2}. \tag{8.4-21}$$

However, direct use of Eq. (8.4-21) is curtailed by not knowing λ_{1m} and f. As stated in Section 8.3.D, the alternatives to minimizing Eq. (8.4-20) are to minimize some suitable upper bound on Eq. (8.4-20) or to minimize an approximation of Eq. (8.4-20) directly. In either of these alternatives λ_{1m} can be determined approximately by a simple search technique that can be done in a computationally efficient way. This estimate of λ_{1m} is then used in Eq. (8.4-21) with the latest *estimate* of $|F|$ replacing the actual $|F|$ to compute an estimate of λ_{2m}. As we shall see, this procedure can give excellent results, but the associated computation time is much longer than for case (ii), which we discuss next.

Case (ii): Finding the RPs for $f_{n+1} = T_1 T_2 f_n$. From the discussion leading up to Eq. (8.3-20), it is easy to show that $\lambda_{1m} = 1$ since $\mathscr{C}_1$ is a closed linear subspace and for any $f \in \mathscr{C}_1$ the error is minimized by projecting orthogonally onto the subspace. With this result, we can compute λ_{2m} explicitly as

$$\lambda_{2m} = 1 + \frac{\|P_2 f_n - P_1 P_2 f_n\|^2}{\|P_1 P_2 f_n - f_n\|^2} + \frac{\mathrm{Re}[(f - P_2 f_n, P_2 f_n - f_n)]}{\|P_1 P_2 f_n - f_n\|^2}. \tag{8.4-22}$$

Equation (8.4-22) is derived in [3]. It is not difficult to show that the third term on the right in Eq. (8.4-22) is nonnegative. Therefore we obtain as a lower bound

$$\lambda_{2m} \geqslant 1 + \frac{\|P_2 f_n - P_1 P_2 f_n\|^2}{\|P_1 P_2 f_n - f_n\|^2} \triangleq \lambda_{\mathrm{L}}. \tag{8.4-23}$$

We note that

$$\|e_{n+1}\|^2 = \|P_1 f_n - f + \lambda_2 P_1 (P_2 - 1) f_n\|^2 \tag{8.4-24}$$

is a quadratic function of λ_2. Therefore, there is only one minimum and it occurs at the value of λ_{2m} given by Eq. (8.4-22). However, in the range $0 < \lambda \leqslant \lambda_{2m}$, the error is a monotonically decreasing function of λ even if $\lambda \geqslant 2$. Thus we can use λ_{L} as an approximation of λ_{2m} even if $\lambda_{\mathrm{L}} \geqslant 2$ because in this way we minimize (approximately) $\|e_{n+1}\|^2$. However, at some point and thereafter in the algorithm, we should restrict $\lambda_{\mathrm{L}} < 2$ to ensure convergence.

We close this section by pointing out that instead of using Eq. (8.4-23) as an estimate of λ_{2m} we can use Eq. (8.4-22) with our best (current) estimate

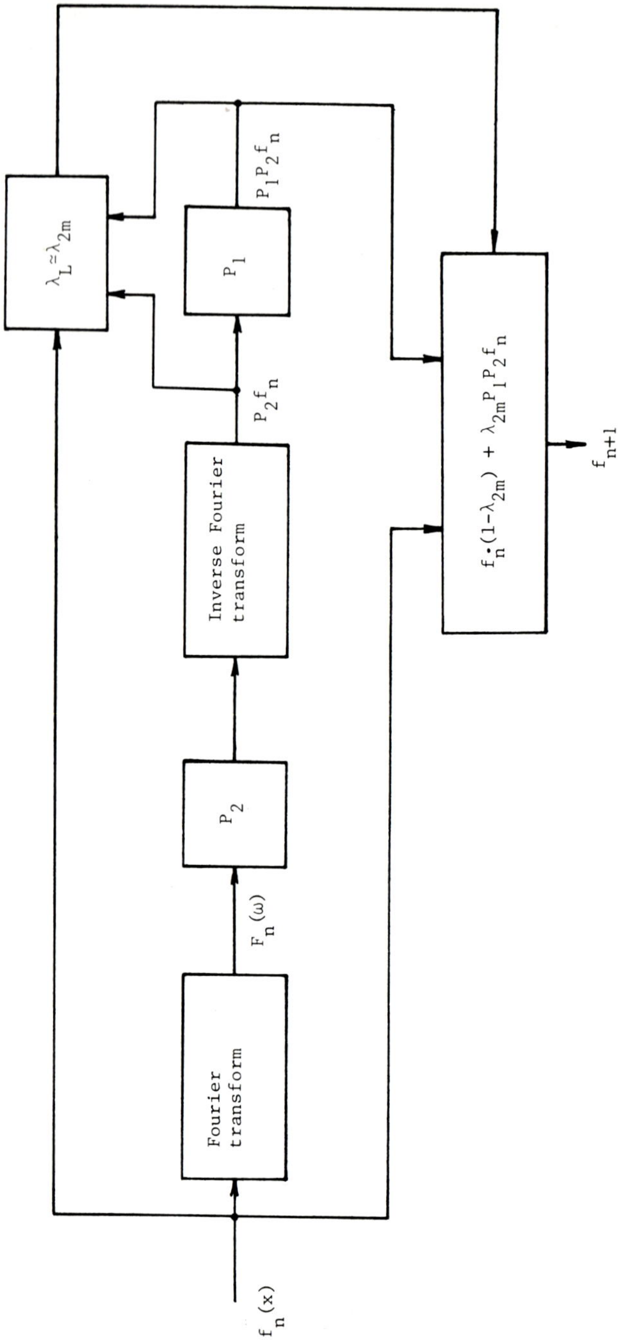

Fig. 8.4-1. Block diagram for the algorithm $f_{n+1} = T_1 T_2 f_n$ in the restoration-from-phase problem with per-cycle optimization of the RPs.

of the magnitude of $F(\omega)$ replacing the unknown $|F(\omega)|$. In the example discussed in the next section, no significant difference was observed in the convergence behaviors of $f_{n+1} = T_1 T_2 f_n$ under these two criteria.

A schematic diagram of the algorithm in this case is given in Fig. 8.4-1. The projection $P_2 f_n$ is calculated in the Fourier domain by using Eq. (8.4-13) and λ_{2m} is approximated by λ_L of Eq. (8.4-23). Since $\lambda_{1m} = 1$ we have $f_{n+1} = P_1 T_2 f_n$ and therefore $f_n \in \mathscr{C}_1$ for every $n \neq 0$. As a result we have $P_1 f_n = f_n$, which leads to

$$f_{n+1} = P_1[1 + \lambda_{2m}(P_2 - 1)]f_n = (1 - \lambda_{2m})f_n + \lambda_{2m} P_1 P_2 f_n.$$

8.5 COMPUTER SIMULATIONS OF RESTORATION FROM PHASE

A. *Examples of Restoration from Phase by POCS*

In this section two computer simulations involving the recovery of signals from the phase of their Fourier transforms are furnished. We use per-cycle optimization of the relaxation parameters and compare signal recovery by POCS with that by another method: the Hayes–Lim–Oppenheim method [7]. Additional discussion related to this simulation can be found in [13].

1. *Effect of Per-Cycle Optimization*

Consider the discrete original signal of Fig. 8.5-1, which can be defined by

$$f(x) = \begin{cases} 0.5 + 0.5\cos(\pi x/30), & x = 1, 2, \ldots, 50, \\ 0, & x = 51, 52, \ldots, 128. \end{cases} \tag{8.5-1}$$

Let F_0 denote the initial value of the Fourier transform magnitude (the input to the algorithm). The percent error at the nth iteration is defined by

$$e_n \triangleq 100 \frac{\|f_n(x) - f(x)\|}{\|f(x)\|}. \tag{8.5-2}$$

Results are obtained for the two algorithms resulting from the commutation of the operators T_1 and T_2: (1) $f_{n+1} = T_2 T_1 f_n$ and (2) $f_{n+1} = T_1 T_2 f_n$. In (1) we compute the λ's so as to minimize (approximately) Eq. (8.4-20). In (2) we set $\lambda_{1m} = 1$ and compute λ_{2m} according to Eq. (8.4-22). Since f is unknown in Eq. (8.4-22) we use the lower bound furnished by Eq. (8.4-23), i.e., $\lambda_{2m} = \lambda_L$. To guarantee convergence we require $\lambda_{2m} < 2$, but λ_L may equal or exceed 2, therefore we present the results for both $\lambda_{2m} = \lambda_L$ and

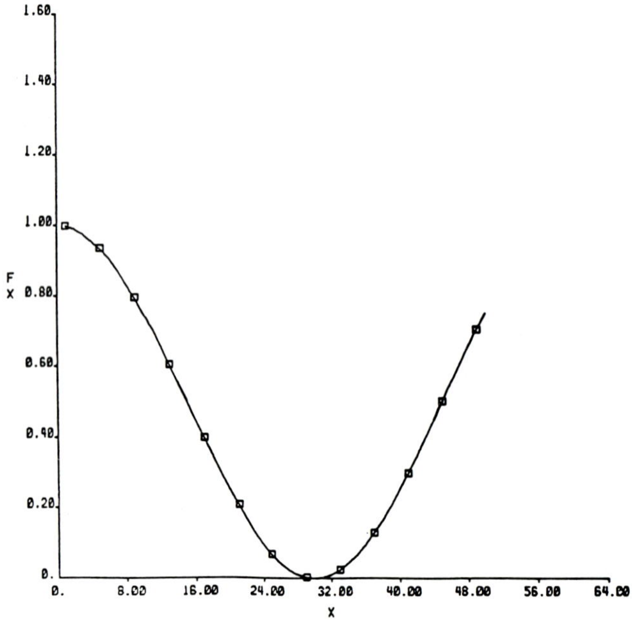

Fig. 8.5-1. Original signal to be restored from phase-only for Example 1: $f(x) = 0.5 + 0.5\cos(\pi x/30)$ for $x = 1, \ldots, 50$, and $f(x) = 0$ elsewhere.

$\lambda_{2m} = \min(\lambda_L, 1.99)$. Results were computed for 39 iterations with optimization of the RPs and with no optimization, i.e., $\lambda_1 = \lambda_2 = 1$. The results are listed in Table 8.5-1 for the two distinct initializations $F_0 = 10$ and $F_0 = 10\exp(-\omega^2/100)$. Figures 8.5-2 and 8.5-3 show the results of the restoration by $f_{n+1} = T_1 T_2 f_n$ after 39 iterations for the two initializations. The main conclusions drawn from Table 8.5-1 are:

- Per-cycle optimization significantly improves restoration over pure projections.
- The order in which the operators are applied does affect the rapidity of convergence. A "best" ordering procedure is not known at present.
- Because of the importance of good initialization, one should attempt to obtain or use *a priori* information that will get F_0 as close to $F(\omega)$ as possible.

B. Comparison of POCS with Hayes–Lim–Oppenheim Method

EXAMPLE 1

For a comparison with the POCS method we use the HLO method described before. The relaxation parameters in the HLO method are the λ_n

Table 8.5-1
Summary of Results for Restoration from Phase by POCS When Per-Cycle Optimization of Relaxation Parameters Is Used[a]

Iteration number	$F_0 = 10$: $\lambda_1 = \lambda_2 = 1.0$	$F_0 = 10$: Optimization for $f_{n+1} = T_2 T_1 f_n$	$F_0 = 10$: Optimization for $f_{n+1} = T_1 T_2 f_n$, $\lambda_{2m} = \min(\lambda_L, 1.99)$	$F_0 = 10$: Optimization for $f_{n+1} = T_1 T_2 f_n$, $\lambda_{2m} = \lambda_L$	$F_0 = 10\exp(-\omega^2/100)$: $\lambda_1 = \lambda_2 = 1.0$	$F_0 = 10\exp(-\omega^2/100)$: Optimization for $f_{n+1} = T_2 T_1 f_n$	$F_0 = 10\exp(-\omega^2/100)$: Optimization for $f_{n+1} = T_1 T_2 f_n$, $\lambda_{2m} = \min(\lambda_L, 1.99)$	$F_0 = 10\exp(-\omega^2/100)$: Optimization for $f_{n+1} = T_1 T_2 f_n$, $\lambda_{2m} \approx \lambda_L$
0	79.9	79.9	79.9	79.9	26.1	26.1	26.1	26.1
1	68.1	39.2	61.3	61.2	18.3	17.4	15.1	15.1
2	60.3	26.3	49.1	49.1	15.4	14.1	13.2	13.1
3	54.1	21.9	41.9	41.4	14.1	12.7	11.9	11.7
5	45.2	16.2	33.6	33.2	12.5	10.4	9.7	9.6
10	34.2	10.7	23.2	22.7	9.7	6.9	6.5	6.3
15	28.3	7.4	17.1	16.5	7.8	4.9	4.7	4.5
20	24.0	5.3	12.9	12.3	6.5	3.7	3.5	3.4
25	20.6	3.9	9.8	9.3	5.4	2.9	2.7	2.6
30	47.8	2.9	7.6	7.1	4.7	2.2	2.1	2.0
35	45.4	2.2	5.9	5.4	4.0	1.7	1.6	1.5
39	43.8	1.8	4.8	4.4	3.6	1.4	1.3	1.2

[a] All entries are percent errors.

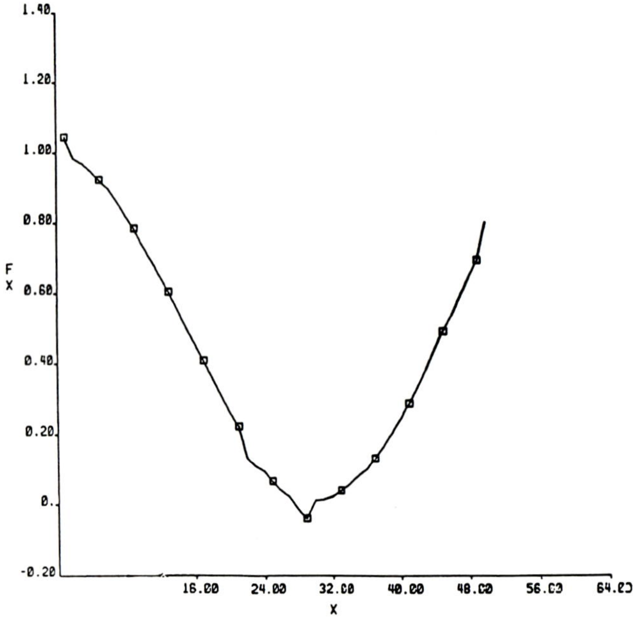

Fig. 8.5-2. Restoration of Fig. 8.5-1 using per-cycle optimization of relaxation parameters: $F_0 = 10$, 39 iterations. Algorithm is $f_{n+1} = T_1 T_2 f_n$.

in Eq. (8.4-4). The optimization criterion for this method is that described by Hayes in [15]. Table 8.5-2 summarizes the results of the comparison for the signal $f(x)$ shown in Fig. 8.5-1.

The superior convergence properties of POCS over HLO is demonstrated in the next example.

EXAMPLE 2

The signal to be restored from phase only is shown in Fig. 8.5-4 and is given by

$$f(x) = \begin{cases} [0.5 + 0.5\cos(\pi x/15)](1 - x/50), & x = 1, \ldots, 50, \\ 0, & x = 51, \ldots, 128. \end{cases} \tag{8.5-3}$$

The results are shown in Table 8.5-3 for an initialization $F_0 = 10$. As in the earlier examples, the results with unity relaxation parameters in POCS and HLO are very much alike, reducing the error after 39 iterations from 73.2% to 5.2 and 4.0%, respectively. However, when relaxation parameters are used, POCS reduces the error to 0.9% while the HLO algorithm starts to diverge at the second iteration and eventually settles at a 100% error beyond the fifth iteration. Indeed, the method of choosing λ's in HLO is such that

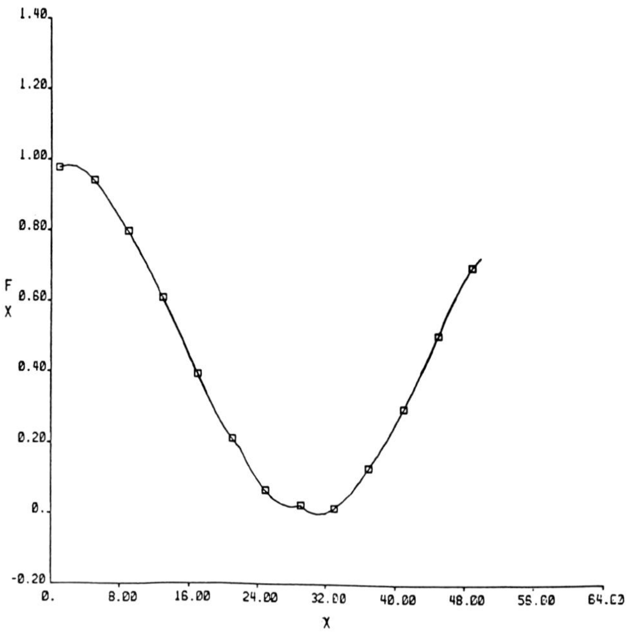

Fig. 8.5-3. Restoration of Fig. 8.5-1 using per-cycle optimization of relaxation parameters: $F_0 = 10\exp[-\omega^2/100]$, 39 iterations. Algorithm is $f_{n+1} = T_1 T_2 f_n$.

it is possible to converge to the zero signal. This cannot happen in POCS because as long as the λ's satisfy $0 < \lambda < 2$, convergence is guaranteed by the theory.

8.6 THE METHOD OF GENERALIZED PROJECTIONS

A. General

Although the POCS method can be applied to many problems of image restoration (i.e., those involving convex constraints), there are important restoration problems involving nonconvex constraints as well. For instance, digitizing a signal is not equivalent to projecting onto a convex set. Neither, for example, is constraining a signal (or its Fourier transform) to have a prescribed magnitude. The latter, of course, appears in a great many problems (see Chapters 6 and 7 of this book) where it is required to retrieve the phase when only the magnitude can be measured (the restoration from magnitude problem, which will be discussed in Section 8.7).

Table 8.5-2
Summary of Results of Comparison of POCS Algorithm with HLO Algorithm for $F_0 = 10$ and $f(x) = 0.5 + 0.5\cos(\pi x/30)$, $x = 1, \ldots, 50$ [a]

	Method of projections		Hayes–Lim–Oppenheim algorithm	
Iteration number	$\lambda_1 = \lambda_2 = 1.0$ $f_{n+1} = P_2 P_1 f_n$	Optimization of $f_{n+1} = T_2 T_1 f_n$	No optimization of λ	Method of optimizing λ as described in Ref. 15
0	79.9	79.9	79.9	79.9
1	68.1	39.2	63.3	66.1
2	60.3	26.3	52.8	59.6
3	54.1	21.9	45.7	42.8
5	45.2	16.2	37.0	35.6
10	34.2	10.7	27.2	19.1
15	28.3	7.4	21.9	14.3
20	24.0	5.3	18.3	10.7
25	20.6	3.9	15.5	8.2
30	17.8	2.9	13.3	6.3
35	15.4	2.2	11.5	4.9
39	13.8	1.8	10.3	4.1

[a] All entries are percent errors.

What can be said about the algorithm given in Eq. (8.3-5) when the projections are not onto convex sets? In general, relatively little. However, when $m = 2$ a number of interesting conclusions can be drawn.† Indeed, the investigation of the algorithm in Eq. (8.3-8) when $\mathscr{C}_1$ and $\mathscr{C}_2$ are not necessarily convex is the main consideration in this section.

For convenience, we restate here some equations and definitions.

In general, for all closed sets—not just convex ones—we call $g \triangleq P_i h$ the projection of h onto $\mathscr{C}_i$ if $g \in \mathscr{C}_i$ and

$$\|g - h\| = \min_{\text{all } y \in \mathscr{C}_i} \|y - h\|. \tag{8.6-1}$$

REMARKS

1. The projection as defined in Eq. (8.6-1) is a unique point if $\mathscr{C}_i$ is a convex set. When $\mathscr{C}_i$ is nonconvex there may be a set of points that satisfy the definition of a projection. However, in practice, we can also find a

† The constraint of $m = 2$ is not as serious a restriction as might at first be supposed because other constraint sets can be combined to form single, more complex sets for which it is still possible to derive a projection operator.

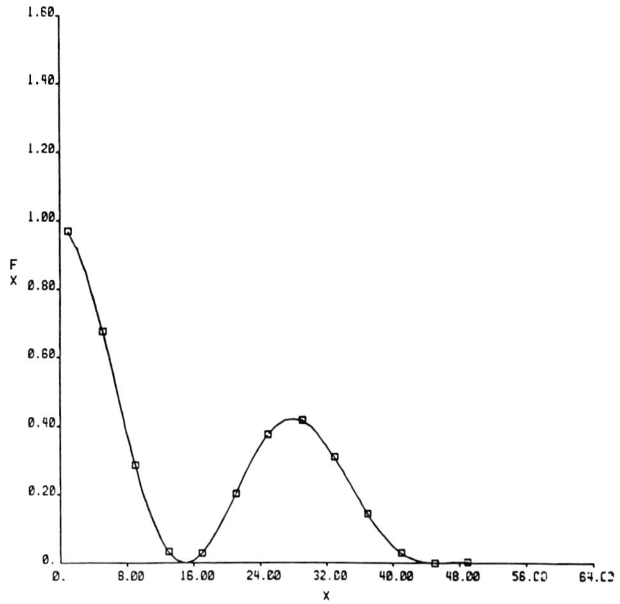

Fig. 8.5-4. Original signal to be restored by POCS and HLO algorithms for Example 2: $f(x) = [0.5 + 0.5\cos(\pi x/10)][1 - x/50]$, $x = 1, \ldots, 50$.

procedure for uniquely choosing one of these points, usually through the demand of satisfying another condition. This eliminates the ambiguity that would otherwise result from nonsingleton projection points. For instance, in the restoration from magnitude problem, in projecting onto the set of functions with prescribed Fourier magnitude, the phase of the estimate at the nth iteration uniquely defines the projection. This is explained in Section 8.6.

2. Projections always exist for convex sets. For nonconvex sets, the proof of the existence of a projection apparently does not exist, although it is difficult to conjecture a realistic situation where this would be the case. Nevertheless, to be on the safe side, we shall assume that all sets that appear in problems for which the method of generalized projections may be useful have at least one projection.

Suppose now that the image f is known to have m *a priori* properties which force f to lie in m given sets $\mathscr{C}_1, \mathscr{C}_2, \ldots, \mathscr{C}_m$. As before, a solution to the image restoration problem is a point that satisfies $f \in \mathscr{C}_0$, where

$$\mathscr{C}_0 \triangleq \bigcap_{i=1}^{m} \mathscr{C}_i. \tag{8.6-2}$$

Table 8.5-3
Summary of Results of Comparison of POCS Algorithm with HLO Algorithm for $F_0 = 10$ and $f(x) = [0.5 + 0.5\cos(\pi x/15)](1 - x/50)$, $x = 1, \ldots, 50$ [a]

	Method of projections		Hayes–Lim–Oppenheim algorithm	
Iteration number	$\lambda_1 = \lambda_2 = 1.0$ $f_{n+1} = P_2P_1f_n$	Optimization of $f_{n+1} = T_2T_1f_n$	No optimization of λ	Method of optimizing λ as described in Ref. 15
0	73.2	73.2	73.2	73.2
1	46.2	28.7	44.1	28.7
2	32.5	24.4	28.7	29.8
3	27.1	22.3	22.2	44.9
4	24.7	20.5	19.7	197.9
5	23.4	18.7	18.4	100.0
6	22.3	17.1	17.4	100.0
10	18.7	12.0	14.5	100.0
15	15.0	7.6	11.6	100.0
20	12.1	4.9	9.3	100.0
25	9.7	3.2	7.5	100.0
30	7.7	2.1	6.0	100.0
35	6.2	1.3	4.8	100.0
39	5.2	0.9	4.0	100.0

[a] All entries are percent errors.

We want to investigate the properties of the recursive algorithm given by

$$f_{n+1} = T_{1,n}T_{2,n} \cdots T_{m,n}f_n, \qquad n = 1, 2, \ldots, \tag{8.6-3}$$

where f_0 is arbitrary and

$$T_{i,n} \triangleq 1 + \lambda_{i,n}(P_i - 1), \qquad i = 1, 2, \ldots, m. \tag{8.6-4}$$

Also, 1 denotes the identity operator and $\lambda_{i,n}$ a relaxation parameter which can be changed from iteration to iteration.†

One easily verifiable result is that the Gerchberg–Saxton algorithm for restoration of an image from magnitude data does involve projection onto nonconvex sets. It is a special case of the algorithm given in Eq. (8.6-3) with $m = 2$ and $\lambda_{1,n} = \lambda_{2,n} = 1$ and can be written as

$$f_{n+1} = P_1P_2f_n, \tag{8.6-5}$$

† Note that the first subscript in $T_{i,n}$ identifies the operator while the second is associated with the iteration number and depends on it only through $\lambda_{i,n}$.

where P_1 is the projection operator onto the set $\mathscr{C}_1$ of functions which are space-limited and P_2 the projection operator onto the set $\mathscr{C}_2$ of functions which have a Fourier transform magnitude equal to some prescribed magnitude. The Gerchberg–Saxton algorithm has a "error-reduction" property (see Fienup [8]), which unfortunately does not refer to the true error, i.e.,

$$\|e_n\| = \|f_n - f\| \tag{8.6-6}$$

(f_n is the nth iterate and f is the correct image) but rather to errors vis-à-vis the original constraints on f. In this and the following sections we shall show that error reduction in the Fienup sense is equivalent to a property that we prefer to call set-distance reduction. The set-distance reduction or error-reduction property of algorithms of the type of Eq. (8.6-3) always holds for $m = 2$ but does not hold generally for $m > 2$.†

B. Restoration by Generalized Projections

As stated earlier, if one or more of the sets $\mathscr{C}_1, \mathscr{C}_2, \ldots, \mathscr{C}_m$ are nonconvex then the convergence of the algorithms given by Eq. (8.6-3) is not guaranteed. Nevertheless, the algorithm given by

$$f_{n+1} = T_1 T_2 f_n, \qquad f_0 \text{ arbitrary}, \tag{8.6-7}$$

has some properties that make it useful for image restoration.

(*Note*: From now on, for convenience only, we omit the subscript n in $\lambda_{i,n}$ and $T_{i,n}$, $i = 1, 2$ but it should be understood that λ_1, λ_2, and therefore T_1, T_2 still depend on the iteration number n.)

Since the convergence of Eq. (8.6-7) is not assured, we need some measure of performance that will allow us to gauge the performance of the algorithm during the iteration process. The performance measure is also needed as an indicator for when parameter changes are required in the algorithm in order to improve its performance. Before we define the performance measure, let us make the following definition:

Definition 1. The distance of a point h from a set‡ $\mathscr{C}$, denoted by $d(h, \mathscr{C})$, is

$$d(h, \mathscr{C}) = \min_{y \in \mathscr{C}} \|y - h\|. \tag{8.6-8}$$

From this definition and the definition of the projector P, i.e., Eq. (8.6-1),

† However, m refers to the number of sets, not constraints. More than one constraint can be associated with a single set.

‡ Unless stated otherwise, all sets are assumed to be closed.

it is clear that

$$d(h, \mathscr{C}) = \|Ph - h\|. \tag{8.6-9}$$

Now we can define the performance measure for the algorithm given by Eq. (8.6-7) as follows:

Definition 2. Let $\mathscr{C}_1$, $\mathscr{C}_2$ be any two sets with projection operations P_1, P_2, respectively, and let T_1, T_2 be defined† by Eq. (8.6-4). Let f_n be the estimate of f at the nth iteration of Eq. (8.6-7); then the performance measure at f_n, denoted by $J(f_n)$, is the sum of distances between the point f_n and the sets $\mathscr{C}_1$ and $\mathscr{C}_2$. Thus the performance is measured by a criterion which we call the summed distance error (SDE). Specifically,

$$J(f_n) \triangleq \|P_1 f_n - f_n\| + \|P_2 f_n - f_n\| \tag{8.6-10}$$

and similarly

$$J(T_2 f_n) = \|P_1 T_2 f_n - T_2 f_n\| + \|P_2 T_2 f_n - T_2 f_n\|. \tag{8.6-11}$$

Note that $J(f_n) \geqslant 0$ and $J(f_n) = 0$ if and only if $f_n \in \mathscr{C}_1 \cap \mathscr{C}_2$. The proof is obvious by examination of Eq. (8.6-10). We are now in a position to state a theorem which describes the set-distance reduction property of the recursion given by Eq. (8.6-7).

Theorem 8.6-1. The recursion given by Eq. (8.6-7) has the property

$$J(f_{n+1}) \leqslant J(T_2 f_n) \leqslant J(f_n) \tag{8.6-12}$$

for every λ_1 and λ_2 that satisfy

$$0 \leqslant \lambda_i \leqslant \frac{A_i^2 + A_i}{A_i^2 + A_i - \frac{1}{2}(A_i + B_i)}, \qquad i = 1, 2, \tag{8.6-13}$$

where

$$A_1 \triangleq \frac{\|P_1 T_2 f_n - T_2 f_n\|}{\|P_2 T_2 f_n - T_2 f_n\|}, \tag{8.6-14a}$$

$$A_2 \triangleq \frac{\|P_2 f_n - f_n\|}{\|P_1 f_n - f_n\|}, \tag{8.6-14b}$$

and

$$B_1 \triangleq \frac{(P_2 T_2 f_n - T_2 f_n, P_1 T_2 f_n - T_2 f_n)}{\|P_2 T_2 f_n - T_2 f_n\|^2}, \tag{8.6-15a}$$

$$B_2 \triangleq \frac{(P_1 f_n - f_n, P_2 f_n - f_n)}{\|P_1 f_n - f_n\|^2}. \tag{8.6-15b}$$

† Recall the dependence on iteration number n is submerged.

The proof of this theorem is somewhat tedious and will not be given here. It can be found in other places, i.e., [3] and [13].

REMARKS

1. The properties of the performance measure $J(f_n)$ make $J(f_n)$ a good quantity for monitoring the performance and controlling the parameters of the algorithm given by Eq. (8.6-7).

2. It can easily be verified, by using Cauchy's inequality, that $A_i \geqslant B_i$, $i = 1, 2$, and hence

$$A_i \geqslant \tfrac{1}{2}(A_i + B_i) \geqslant 0, \qquad i = 1, 2. \tag{8.6-16}$$

Therefore by using Eq. (8.6-16) in Eq. (8.6-13) we conclude that the value of unity will always be included in the range of λ_i, $i = 1, 2$, as given in Eq. (8.6-13). This means that the algorithm given by

$$f_{n+1} = P_1 P_2 f_n, \qquad f_0 \text{ arbitrary}, \tag{8.6-17}$$

will always have the property of set distance reduction.†

3. Theorem 8.6-1 is also true if $A_i \to \infty$, in which case the range of λ_i as given in Eq. (8.6-13) becomes $0 \leqslant \lambda_i \leqslant 1$, $i = 1, 2$.‡

4. The set-distance reduction property holds for *at least* those values of λ_i given by Eq. (8.6-13). In fact, when we examine some approximations made in the proof of the theorem we can see that this property holds for a more extended range of λ_i than that given by (8.6-13).

The set-distance reduction property cannot be generalized to more than two sets. For two sets the assertion $J(f_{n+1}) \leqslant J(f_n)$ follows *a fortiori* from Eq. (8.6-12). For more than two sets we cannot assert this [for any m, $J(f_n)$ is defined as the sum of distances from f_n to the sets $\mathscr{C}_1, \mathscr{C}_2, \ldots, \mathscr{C}_m$]. To demonstrate our inability to assert $J(f_{n+1}) \leqslant J(f_n)$ for $m > 2$ we give the following simple counterexample: consider the three sets described in Fig. 8.6-1. $\mathscr{C}_1$ and $\mathscr{C}_3$ are the circumferences of two circles and $\mathscr{C}_2$ is a line. Note that $\mathscr{C}_1$ and $\mathscr{C}_3$ are nonconvex. Suppose we want to find the point f in the intersection of $\mathscr{C}_1$, $\mathscr{C}_2$, $\mathscr{C}_3$ by the recursion

$$f_{n+1} = P_1 P_2 P_3 f_n, \qquad f_0 \text{ arbitrary}. \tag{8.6-18}$$

In Fig. 8.6-1 we indicate the initial point f_0 and the points after applying the projections $P_3, P_2, P_1, P_3, \ldots$ successively. From f_0 the iterates move in the direction of the arrows to points $f_1, f_2, f_3, \ldots$. The main conclusions

† However, that does not make the algorithm in Eq. (8.6-17) optimal.

‡ For example, $\|P_1 f_n - f_n\| \to 0$ implies $A_2 \to \infty$.

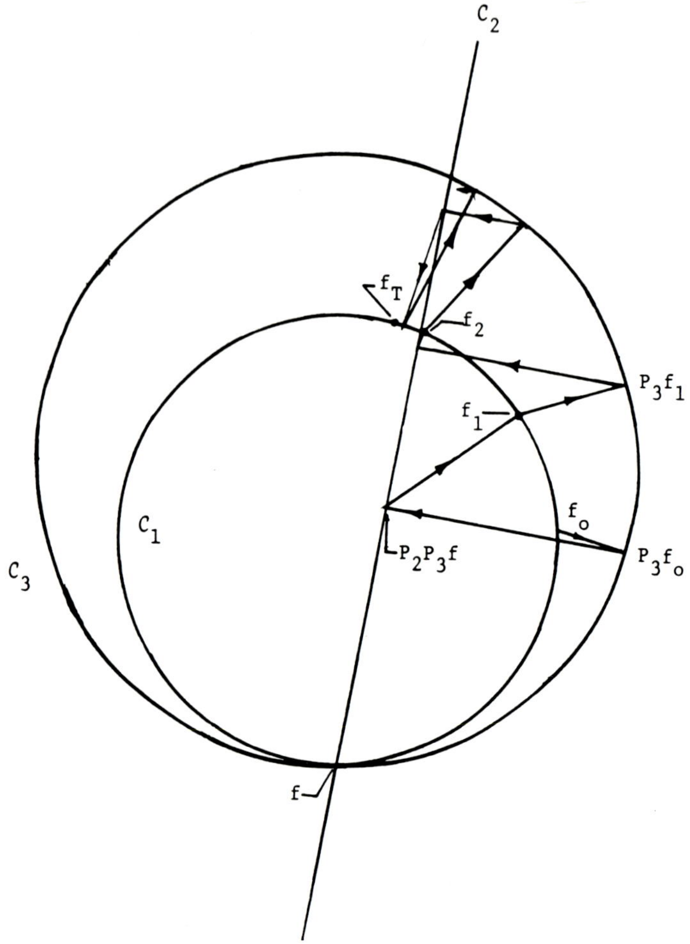

Fig. 8.6-1. Demonstration that the algorithm $f_{n+1} = P_1 P_2 P_3 f_n$ may not possess the set-distance reduction property.

for this simple example are:

1. the algorithm diverges, i.e.,

$$\|e_0\| \leqslant \|e_1\| \leqslant \|e_2\| \leqslant \cdots,$$

where $\|e_n\|$ is defined by Eq. (8.6-6);

2. for $n \geqslant 2 \quad J(f_{n+1}) \geqslant J(f_n)$;

3. there exists in this case a *trap point* f_T, i.e., a fixed point of the operator $P_1 P_2 P_3$ with $f_n \to f_T$. The point f_T satisfies $P_1 P_2 P_3 f_T = f_T$. In the general

case, therefore, the set-distance reduction property does not hold. The existence of trap points will be discussed in greater detail later.

Despite the fact that the theorem is not valid for $m > 2$, the algorithm given by Eq. (8.6-7) is not so restrictive in practice. As we have already stated, the theorem does not restrict the complexity of the sets and therefore $\mathscr{C}_1$ and $\mathscr{C}_2$ can indeed include signals with multiple constraints. However, as the sets become more complex, so do the projection operators. As we shall see, however, in image processing we can usually combine several properties of the signal f and associate them with only one set.

As a rule we can combine those properties of the signal which are easily expressed in the space domain to one set $\mathscr{C}_1$ whose associated projector P_1 can be calculated without too much effort. Similarly, the properties of the signal, which are easily expressed in the transform domain, can be combined into a second set $\mathscr{C}_2$ and the corresponding projector P_2 can again be calculated without too much effort. An example is given below.

Example

Let $\mathscr{C}_1$ be the set of functions $\{g(x)\}$ of compact support which are zero for all points $x \notin S_1$ and have a space domain magnitude which is equal to some prescribed positive function $P(x)$ in a set S_2 with $S_2 \subset S_1$ (i.e., the set S_2 is included in S_1). Thus,

$$\mathscr{C}_1 = \{g(x) : g(x) = 0, \text{ for all } x \in S_1^c \text{ and } |g(x)| = P(x), \text{ for all } x \in S_2\}, \tag{8.6-19}$$

where S_i^c is the complement of S_i. It is easily shown that for any arbitrary $f(x)$

$$y \triangleq P_1 f = \begin{cases} 0, & x \in S_1^c, \\ f(x), & x \in S_1 \cap S_2^c, \\ P(x), & x \in S_2, f(x) > 0, \\ -P(x), & x \in S_2, f(x) \leq 0. \end{cases} \tag{8.6-20}$$

Equation (8.6-20) follows from minimizing the equation

$$\begin{aligned} \|f - g\|^2 = & \int_{x \in S_1^c} |f(x) - g(x)|^2 \, dx + \int_{x \in S_1 \cap S_2^c} |f(x) - g(x)|^2 \, dx \\ & + \int_{x \in S_2, f(x) > 0} |f(x) - g(x)|^2 \, dx + \int_{x \in S_2, f(x) \leq 0} |f(x) - g(x)|^2 \, dx, \end{aligned} \tag{8.6-21}$$

where we used the condition $S_2 \subset S_1$. Since minimizing $\|f - g\|^2$ with respect to all functions $g \in \mathscr{C}_1$ is equivalent to minimizing each integral on the right-hand side of Eq. (8.6-21) independently of the others, we immediately obtain Eq. (8.6-20) for the function $y \in \mathscr{C}_1$ that minimizes $\|f - g\|^2$.

C. Optimization of the Relaxation Parameters

First we note that the method of optimization for the RPs described in Section 8.3.B for the two-step POCS algorithm cannot be applied when nonconvex sets are involved because this method strongly depends on a property of convex sets [3]. Since the summed distance error $J(f_n)$ is always nonnegative and is zero if and only if a solution is achieved, it seems reasonable to use the SDE as a criterion for the optimization of the RPs. The method of optimizing the RPs by using the SDE as a criterion will be described in detail in Section 8.7 in connection with the restoration from magnitude problem, but many of the results applied there can be used with the more general algorithm of Eq. (8.6-7), where nonconvex constraints can be of any type.

Another point that should be mentioned is that the method of optimization described here (i.e., minimization of the SDE) can be used as well in the POCS algorithm as an alternative to the method described in Section 8.3.B, which uses the restoration error as a criterion.

D. Traps and Tunnels

We define a *trap* as a fixed point† of the composition operator $T_1 T_2 \cdots T_m$ which is not a fixed point of every individual T_i, $i = 1, \ldots, m$, i.e., a point which fails to satisfy one or more of the *a priori* constraints yet satisfies

$$f_{n+1} = T_2 T_2 \cdots T_m f_n = f_n. \tag{8.6-22}$$

We say that a point f_n is in a *tunnel* if Eq. (8.6-22) is almost satisfied, which means that the change in f_n from one iteration to the next is negligible. Traps can never occur when only convex sets are involved, but tunnels (i.e., extremely slow convergence) may occur. Traps and tunnels are illustrated in Fig. 8.6-2. In general, when at least one nonconvex set is involved, traps may exist, as demonstrated in Figs. 8.6-1 and 8.6-2.

In Section 8.7 we discuss the properties of traps and tunnels in connection with the restoration from magnitude problem, but some of the results obtained there are applicable to more general restoration problems than restoration from magnitude only. We will show in Section 8.7 that when the algorithm

$$f_{n+1} = P_1 T_2 f_n, \qquad f_0 \text{ arbitrary}, \tag{8.6-23}$$

is used with P_1 linear the summed distance error $J(f_n)$ alone can be used

† The fixed point is an n-dimensional signal, e.g., a waveform ($n = 1$) or an image ($n = 2$).

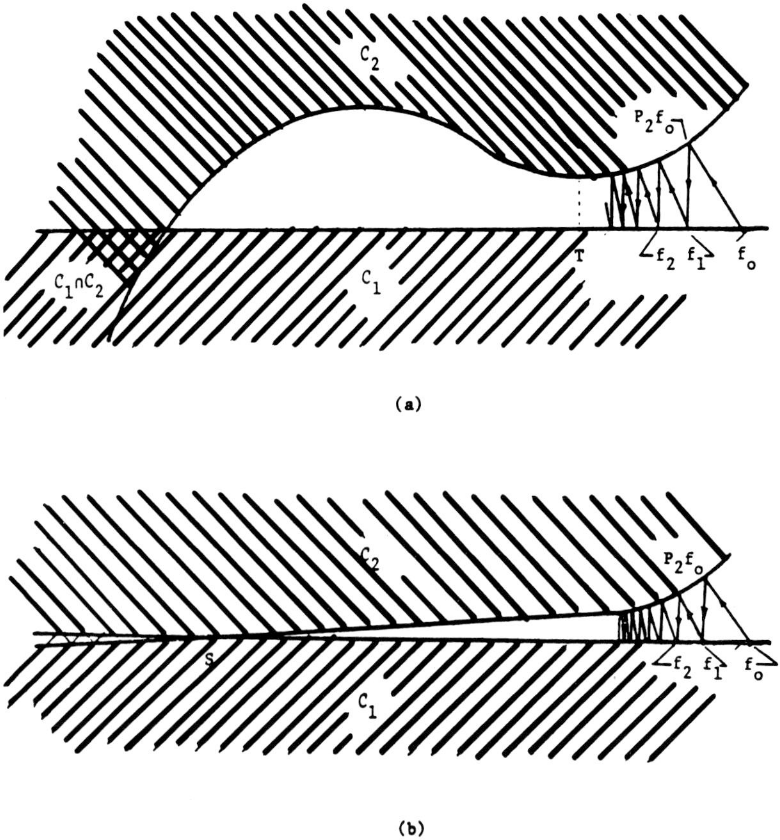

Fig. 8.6-2. Illustration of a trap and tunnel for an algorithm of the form $f_{n+1} = P_1 P_2 f_n$: (a) starting at the point f_0 the sequence $\{f_n\}$ converges to a trap point T while the true solution must belong to $\mathscr{C}_1 \cap \mathscr{C}_2$; (b) starting at the point f_0 the algorithm enters into a long tunnel toward the solution at the point S.

for detecting traps and tunnels. If P_1 is not linear, additional tests are required. We will also show that when P_1 is linear and the algorithm in Eq. (8.6-23) enters a trap, the correct solution must lie in a hyperplane orthogonal to the direction of the vector $P_2 f_n - f_n$. When in a trap and P_1 is not linear, or when in a tunnel (with P_1 either linear or not linear), a reasonable strategy is to look for a better estimate of the correct solution in a direction orthogonal to $P_2 f_n - f_n$.

8.7 SIGNAL RECOVERY FROM MAGNITUDE BY GENERALIZED PROJECTIONS

A. General

The problem of restoring a signal from magnitude information only is of great practical importance. The problem is variously called restoration from magnitude (RFM) or phase retrieval. It appears in several fields [4,16,17], such as astronomy, optics, and x-ray crystallography, where the available information about an image comes from intensity measurements in the space or spatial-frequency domain or both and from some *a priori* knowledge such as the fact that the image is space-limited and the image function is nonnegative. The RFM problem in astronomy is discussed in Chapter 7.

The related uniqueness problem refers to the question of whether a function can be uniquely defined by its Fourier transform magnitude. This problem is discussed in detail by Hayes [15] for the case of space-limited multidimensional sequences (discrete case). When dealing with uniqueness in the RFM problem it must be noted that all the functions $f(x, y)$, $-f(x, y)$, $f(x - x_0, y - y_0)$, and $f(-x, -y)$ have the same Fourier transform magnitude function and therefore uniqueness in this case is up to a sign, shift in coordinates, or coordinate reversal. It was found by Bruck and Sodin [18] that a space-limited sequence is uniquely defined by its Fourier transform magnitude if its z-transform includes at most one irreducible nonsymmetric factor. Hayes [15] also found conditions for uniqueness when the magnitude of the sequence is not known at all frequencies. Thus, assume a discrete signal of support length N with only the magnitude of an M-point discrete Fourier transform available. Then $f(\mathbf{n})$ can still be uniquely determined if its z-transform includes at most one irreducible nonsymmetric factor and $M \geqslant 2N - 1$. It is not an easy task to check a two-dimensional polynomial for irreducibility. However, it is widely believed [15,16,19,20] (based partly on experimental results) that two-dimensional images generally satisfy the uniqueness conditions. Moreover, it is known that within the set of all polynomials in more than one variable, the subset of reduçible polynomials is a nondense set of measure zero [15]. Many of these points are discussed in greater detail in Chapter 6 of this volume.

B. Generalized Projection Algorithm for the Restoration from Magnitude Problem

We illustrate the method of restoration by generalized projections by restoring a function f that is known to be space-limited and has a given prescribed magnitude function. To simplify notation and expressions we use only functions of one variable, but the extension to functions of two variables (i.e., images) is straightforward.

The two sets involved in the restoration from magnitude problem are $\mathscr{C}_1$ the set of space-limited functions (any two level amplitude constraints, i.e., $a \leqslant f \leqslant b$, can easily be added), and $\mathscr{C}_2$, the set of all functions which have a Fourier transform magnitude equal to some real positive prescribed function $M(\omega)$. Thus

$$\mathscr{C}_1 = \{g(x): g(x) = 0 \text{ for } |x| > a\}, \tag{8.7-1}$$

$$\mathscr{C}_2 = \{g(x) \leftrightarrow G(\omega): |G(\omega)| = M(\omega) \text{ for all } \omega\}. \tag{8.7-2}$$

It can easily be verified that $\mathscr{C}_1$ is convex and that $\mathscr{C}_2$ is nonconvex. The projections P_1 and P_2 onto $\mathscr{C}_1$ and $\mathscr{C}_2$, respectively are given by

$$P_1 g = \begin{cases} g(x), & |x| < a, \\ 0, & |x| \geqslant a, \end{cases} \tag{8.7-3}$$

and

$$P_2 g \leftrightarrow M(\omega)\, e^{j\phi(\omega)}, \tag{8.7-4}$$

where $\phi(\omega)$ is the phase of $G(\omega)$. $P_2 g$ is uniquely defined by Eq. (8.7-4) although $\mathscr{C}_2$ is nonconvex. The general restoration algorithm is given by $f_{n+1} = T_1 T_2 f_n$ with the T_i, $i = 1, 2$, defined by $T_i = 1 + \lambda_i(P_i - 1)$. This algorithm has the property of set-distance reduction, or the property that $\{J(f_n)\}_{n=0}^{\infty}$ is a nonincreasing sequence, for (at least) those values of λ_1, λ_2 that satisfy the inequality (8.6-13). Another form of the algorithm is obtained by commuting the transformations T_1 and T_2, i.e.,

$$f_{n+1} = T_2 T_1 f_n, \qquad f_0 \text{ arbitrary}. \tag{8.7-5}$$

These two forms of the set-distance reduction algorithm may give different results when optimization of the relaxation parameters (discussed in the next section) is used.

Fundamental Remark: When $\lambda_1 = \lambda_2 = 1$ then Eq. (8.6-7), with P_1 and P_2 as defined in Eqs. (8.7-3) and (8.7-4), reduces to the Gerchberg–Saxton algorithm, and the property that $\{J(f_n)\}_{n=0}^{\infty}$ is a nonincreasing sequence becomes equivalent to the nonincreasing error property described by Fienup [8]. Also, it is readily shown that Fienup's [8] output–output algorithm is equivalent to a special form of the set-distance reduction algorithm (for a proper choice of λ_1) given by

$$f_{n+1} = P_2 T_1 f_n, \qquad f_0 \text{ arbitrary}. \tag{8.7-6}$$

C. Optimization of the Relaxation Parameters λ_i, $i = 1, 2$

The summed distance error $J(f_n)$ is, in our view, the best performance measure that we have for this problem. Therefore, it is natural to try to choose the RPs in order to maximize the rate of reduction of the sequence $\{J(f_n)\}$. Two possible criteria for this optimization problem are the following:

1. *Per-step optimization.* Here we try to minimize $J(T_2 f_n)$ with respect to λ_2 for a given f_n, following which we minimize $J(f_{n+1}) = J(T_1 T_2 f_n)$ with respect to λ_1, where λ_2 and therefore $T_2 f_n$ are now known.
2. *Per-cycle optimization.* Here we try to minimize $J(f_{n+1})$ with respect to λ_1 and λ_2 simultaneously for a given f_n.

Since the dependence of the performance measure J on λ_1 and λ_2 in the per-step optimization is very complicated [e.g., see Eq. (8.6-11), where the λ's are buried in the T's], we cannot in general find a closed-form solution for the optimal values $\lambda_{1,\text{opt}}$ and $\lambda_{2,\text{opt}}$. Nevertheless, the following general result is shown in [3]. In the per-step optimization the optimal values for λ_1 and λ_2 satisfy

$$\lambda_{i,\text{opt}} \geqslant 1, \qquad i = 1, 2. \tag{8.7-7}$$

If P_1 is a linear operator† then $\lambda_{1,\text{opt}} = 1$. The exact calculation of the optimal values for λ_1 and λ_2 in per-step optimization needs a search through the range of λ_1 and λ_2, as is the case in the calculation for the per-cycle optimization. Since we expect the per-cycle optimization to work better (because it considers the optimization of the whole iteration), we will concern ourselves in what follows only with per-cycle optimization. As already stated for an exact calculation of λ_{1m} and λ_{2m}, which stand for the optimal values of λ_1 and λ_2 in the per-cycle case, we need a two-dimensional search. To avoid such a search we make the approximation

$$\lambda_{1m} \simeq 1 \tag{8.7-8}$$

for the algorithm $f_{n+1} = T_1 T_2 f_n$ [Eq. (8.6-7)]. This approximation becomes exact if P_1 is linear. The justification for this approximation is as follows. First we observe that no matter what λ_2 we use to compute $T_2 f_n$, the optimal value for λ_1 must be chosen to satisfy the per-step criterion. For the per-step criterion $\lambda_{i,\text{opt}} \geqslant 1$; however, since the per-step criterion generally does not yield much improvement over pure projection (in which case $\lambda = 1$) we might as well replace $\lambda_{1,\text{opt}} = 1 = \lambda_{1m}$. By using similar arguments, we can

† This is the case for P_1 as defined in Eq. (8.7-3), but if we add a positivity constraint to $\mathscr{C}_1$ then P_1 is no longer linear.

take $\lambda_{2m} \simeq 1$ when we use the algorithm:

$$f_{n+1} = T_2 T_1 f_n, \qquad f_0 \text{ arbitrary}. \tag{8.7-9}$$

In the following we discuss only the algorithm $f_{n+1} = T_1 T_2 f_n$ with $\lambda_1 = 1$, but similar arguments apply to the algorithm given in Eq. (8.7-9). With this approximation we have

$$f_{n+1} = P_1 T_2 f_n, \qquad f_0 \text{ arbitrary}. \tag{8.7-10}$$

Equation (8.7-10) implies that $f_n \in \mathscr{C}_1$ for every n except possibly for $n = 0$; therefore $\|P_1 f_{n+1} - f_{n+1}\| = 0$. Now, by using Eqs. (8.6-10) and (8.7-10) we obtain for $J(f_{n+1})$ the following relation:

$$J(f_{n+1}) = \|P_2 P_1 T_2 f_n - P_1 T_2 f_n\| = \|P_2 f_{n+1} - f_{n+1}\|. \tag{8.7-11}$$

1. *Calculation of* λ_{2m} *by Search*

Let us denote $f_{n+1}(x) \leftrightarrow |F_{n+1}|\, e^{j\phi_{n+1}(\omega)}$ a Fourier transform pair. From the definition of P_2 given by Eq. (8.7-4), we obtain $P_2 f_{n+1} = M(\omega)\, e^{j\phi_{n+1}(\omega)}$ and therefore by Eq. (8.7-11) and Parseval's relation $\|f - g\| = \|F - G\|$ we obtain

$$J^2(f_{n+1}) = \int_{-\infty}^{\infty} [M(\omega) - |F_{n+1}(\omega)|]^2 \, d\omega/2\pi, \tag{8.7-12}$$

where

$$F_{n+1}(\omega) = \mathscr{F}\{P_1[f_n + \lambda_2[P_2 f_n - f_n(x)]]\} \tag{8.7-13}$$

and $\mathscr{F}\{\cdot\}$ denotes the Fourier transform operator. To find λ_{2m} we have to minimize $J(f_{n+1})$ in Eq. (8.7-12), where $F_{n+1}(\omega)$ is given in Eq. (8.7-13). In general this can be done by a search through a range of values of λ_2. This search may be relatively fast if P_1 is linear, since in this case Eq. (8.7-13) reduces to

$$F_{n+1}(\omega) = (1 - \lambda_2) F_n(\omega) + \lambda_2 F\{P_1 P_2 f_n\}, \tag{8.7-14}$$

where we used the fact that $P_1 f_n = f_n$. This means that when P_1 is linear the search process does not involve the Fourier transform operation. When P_1 is not linear we have to use the Fourier transform operation once for each tested value of λ_2,† and therefore the search for λ_{2m} will be much slower. Nevertheless, by using search we could make several important observations regarding the use of relaxation parameters for the restoration from magnitude problem. These results are discussed in Section 8.8.

† To see this refer to Eq. (8.7-13).

D. Traps and Tunnels in Restoration from Magnitude

Traps and tunnels were defined in Section 8.6.D. The existence of traps, which occur only when nonconvex sets are involved, is postulated to be the reason for the often observed failure to converge to a valid solution.† Tunnels may occur with convex as well as nonconvex sets and they are responsible for the often observed very slow convergence toward a valid solution. In the special case of restoration from magnitude it appears that the existence of traps cannot be easily demonstrated theoretically.‡ However, the existence of traps is supported by the difficulties one encounters in restoring some signals or images from their magnitudes. The mathematical conditions that must be satisfied for traps to exist can be explored in the restoration from magnitude problem by considering the Gerchberg–Saxton algorithm

$$f_{n+1} = P_1 P_2 f_n, \qquad f_0 \text{ arbitrary.} \tag{8.7-15}$$

The reader will recall that this is the simplest form of the more general algorithm $f_{n+1} = T_1 T_2 f_n$ and corresponds to the unrelaxed case (i.e., when $\lambda_1 = \lambda_2 = 1$) involving pure projections. In Eq. (8.7-15) P_1 and P_2 and their associated sets $\mathscr{C}_1$ and $\mathscr{C}_2$ are given by Eqs. (8.7-3) and (8.7-4) and Eqs. (8.7-1) and (8.7-2), respectively. By using the notation $f_n(x) \leftrightarrow F_n(\omega) \triangleq A_n(\omega)\, e^{j\phi_n(\omega)}$, where $A_n(\omega) \triangleq |F_n(\omega)|$, we can write $P_2 f_n \leftrightarrow M(\omega)\, e^{j\phi_n(\omega)}$. Assuming f to be a correct solution, the indication of a trap in this case is $f_n \neq f$ and f_n, the nth estimation of f, is a fixed point of $P_1 P_2$, i.e.,

$$P_1 P_2 f_n = f_n. \tag{8.7-16}$$

We can replace this condition by

$$[P_2 f_n(x)] \operatorname{rect}(x/2a) = f_n(x), \tag{8.7-17}$$

where $\operatorname{rect}(x/2a)$ is 1 for $|x| \leqslant a$ and 0 elsewhere. We can write Eq. (8.7-17) in the frequency domain by taking the Fourier transform of both sides as

$$\int_{-\infty}^{\infty} M(\alpha)\, e^{j\phi_n(\alpha)} \frac{2 \sin[(\omega - \alpha)a]\, d\alpha}{\omega - \alpha} = A_n(\omega)\, e^{j\phi_n(\omega)}. \tag{8.7-18}$$

Both invalid (traps) and valid solutions to the RFM must be solutions of the form $A_n(\omega)\, e^{j\phi_n(\omega)}$ of this integral equation. The correct solution $f \leftrightarrow A_n(\omega) \exp[f\phi_n(\omega)]$ satisfies $A_n(\omega) = M(\omega)$. For traps a much weaker restriction applies: we must only have $A_n(\omega) \neq M(\omega)$ and $A_n(\omega) \geqslant 0$.

† Recall that a valid solution f is one that satisfies all the *a priori* constraints on the image.

‡ The authors are not aware of any theoretical results along these lines.

Next we discuss two results of great practical importance regarding traps and tunnels:

(i) The SDE ($J(f_n)$) can be used to detect traps. By this we mean that a trap can be detected when we observe no change in $J(f_n) > 0$ from iteration to iteration. To determine the existence of a trap from observations on the SDE we must show that $J^*(f_{n+1}) = J(f_n)$ implies that $f_{n+1} = f_n$ and vice versa; the asterisk denotes the minimum of $J(f_{n+1})$ with respect to λ_1 and λ_2. In [13] such a demonstration is furnished for some restricted cases. One can infer a trap when $J^*(f_{n+1}) > 0$ and $\|f_{n+1} - f_n\| = 0$.†

(ii) If P_1 is linear and $P_1 T_2 f_n = f_n$ then the correct solution f lies in a hyperplane orthogonal to the vector $P_2 f_n - f_n$.

Proof. First we note that from the linearity of P_1 we easily obtain $P_1 P_2 f_n = f_n \Leftrightarrow P_1 T_2 f_n = f_n$ for every λ_2. ($\Leftrightarrow$ denotes bidirectional implication.) Now, $f - f_n$ is orthogonal to $P_2 f_n - f_n$ (when $P_1 T_2 f_n = f_n$) since $(f - f_n, P_2 f_n - f_n) = (P_1(f - f_n), P_2 f_n - f_n) = (f - f_n, P_1 P_2 f_n - f_n) = 0$. The next-to-last equality makes use of the self-adjoint property of P_1.

This simple but important observation tells us that when we are in a trap or a tunnel we have to look for a solution along a direction orthogonal to the vector $P_2 f_n - f_n$. More research is needed in order to use this fact to improve the algorithm and avoid traps and tunnels. If P_1 is not linear, then the solution does not necessarily lie on an orthogonal direction to $P_2 f_n - f_n$, but we can look at this result as approximately true.‡ It may be a good or bad approximation depending on P_1, P_2, and f_n.

8.8 COMPUTER SIMULATIONS OF RESTORATION FROM MAGNITUDE

In this section we present the results of restorations from magnitude for two synthetic images: IMAGE1 and IMAGE2 given in Figs. 8.8-1 and 8.8-2, respectively. The nonzero portion of each image is confined to a region of 30×30 pixels in the center of a total field of 64×64 pixels. IMAGE1 is composed of six gray levels with minimum level 0 and maximum level 1. IMAGE2 is a two-level binary image of 0's and 1's only. For the restoration

† On a computer, however, it is not always easy to distinguish between traps and tunnels. Finite word length and/or numerical errors may mask the fact that $f_{n+1} = f_n$ (a trap) or even that $f_{n+1} \neq f_n$ (a tunnel).

‡ Unless the morphologies of the nonconvex sets are bizarre.

Fig. 8.8-1. Original image IMAGE1.

Fig. 8.8-2. Original image IMAGE2.

experiments we used the four algorithms described below:

1. The Gerchberg–Saxton algorithm given by $f_{n+1} = P_1 P_2 f_n$.
2. The algorithm $f_{n+1} = P_{1L} P_2 f_n$, where P_{1L} is the projector onto the set $\mathscr{C}_{1L}$ given by

$$\mathscr{C}_{1L} = \{g(x) : g(x) = 0 \text{ for } |x| > a \text{ and } 0 \leq g(x) \leq 1 \text{ for } |x| \leq a\}, \qquad (8.8\text{-}1)$$

i.e., $\mathscr{C}_{1L}$ is a subset of $\mathscr{C}_1$ that includes a two-level amplitude constraint.

3. The relaxed algorithm given by $f_{n+1} = P_1 T_2 f_n$.
4. The relaxed algorithm given by $f_{n+1} = P_{1L} T_2 f_n$.

In the relaxed algorithms 3 and 4, a search for the optimal value, λ_{2m}, of λ_2 was made as follows: when P_1 was used a simple search through the range $0 < \lambda_2 < 2.8$ (this range was based on previous experience) was made. Since P_1 is linear the search is relatively rapid. When P_{1L} was used the dependence of $J(f_{n+1})$ on λ_2 was modeled as being of polynomial character of degree 2. Previous experience with relaxation parameters suggests that the variation of the summed distance error, $J(f_{n+1})$, with λ is relatively smooth, hence the choice of a polynomial model. A three-points calculation was used to find the parameters of the polynomial and then to find the optimal λ_2 that minimizes $J(f_{n+1})$. In Table 8.8-1 the results of the calculations are given for the restoration of IMAGE1 with and without the addition of the two-level constraint and with initial point $f_1 = 0$ (i.e., zero magnitude and zero phase). In Tables 8.8-1 to 8.8-3 $J(f_n)$ is normalized by the square root of the energy of the unknown image, i.e., by

$$\|f\| = \left[(1/2\pi)\int_{-\infty}^{\infty} M^2(\omega)\, d\omega\right]^{1/2}. \tag{8.8-2}$$

This normalization made comparisons easier. Figure 8.8-3 gives the restored images after 30 iterations for the four algorithms described in Table 8.8-1. Figs 8.8-3a and 8.8-3b are the restorations from algorithms 1 and 3 respectively (i.e., without using the two-level constraint) and give poor but recog-

Table 8.8-1
Restoration of IMAGE1: Normalized SDE $[j(f_n)]$ as a Function of Iteration Number n for the Four Algorithms Used

	Without two-level constraint			With two-level constraint		
		Algorithm $P_1T_2f_n = f_{n+1}$			Algorithm $P_{1L}T_2f_n = f_{n+1}$	
Iteration	Algorithm $P_1P_2f_n = f_{n+1}$ $J(f_n)$	$J(f_n)$	Approx. λ_{2m}	Algorithm $P_{1L}P_2f_n = f_{n+1}$ $J(f_n)$	$J(f_n)$	Approx. λ_{2m}
0	1.0	1.0	—	1.0	1.0	—
1	0.948	0.880	2.8	0.937	0.827	3.0
2	0.725	0.709	1.45	0.382	0.391	0.944[a]
3	0.643	0.607	2.2	0.287	0.260	2.01
5	0.438	0.187	1.6	0.204	0.137	1.85
10	0.146	0.119	1.75	0.101	0.061	1.96
15	0.115	0.087	2.65	0.067	0.038	2.05
20	0.088	0.060	1.45	0.045	0.029	2.23
25	0.066	0.053	2.8	0.037	0.007	1.88
30	0.057	0.050	1.75	0.0325	0.001	1.78
35	0.053	0.045	1.9	0.028		
40	0.051	0.041	1.75	0.019		
45	0.049	0.039	1.9	0.007		
50	0.048	0.038	1.75	0.003		

[a] This value of λ_2 is clearly not optimal but it is the result of the approximate method we used to calculate λ_{2m}.

Table 8.8-2
Restoration of IMAGE2: Normalized SDE $[J(f_n)]$ as a Function of Iteration Number n for the Four Algorithms Used

Iteration	Without two-level constraint: Algorithm $P_1P_2f_n = f_{n+1}$ $J(f_n)$	Without two-level constraint: Algorithm $P_1T_2f_n = f_{n+1}$ $J(f_n)$	Without two-level constraint: Algorithm $P_1T_2f_n = f_{n+1}$ Approx. λ_{2m}	With two-level constraint: Algorithm $P_{1L}P_2f_n = f_{n+1}$ $J(f_n)$	With two-level constraint: Algorithm $P_{1L}T_2f_n = f_{n+1}$ $J(f_n)$	With two-level constraint: Algorithm $P_{1L}T_2f_n = f_{n+1}$ Approx. λ_{2m}
0	1.0	1.0	—	1.0	1.0	—
1	0.946	0.865	2.8	0.940	0.828	3.0
2	0.486	0.464	1.3	0.473	0.424	1.57
3	0.318	0.259	1.9	0.368	0.319	2.38
5	0.228	0.197	2.35	0.294	0.268	2.17
10	0.185	0.165	2.2	0.255	0.241	2.52
15	0.169	0.147	2.05	0.241	0.210	2.15
20	0.157	0.132	2.35	0.2324	0.197	2.43
25	0.144	0.107	1.9	0.2204	0.1946	1.88
30	0.134	0.098	2.05	0.2136	0.1929	2.89
40	0.1067	0.0902	2.2	0.2020	0.1901	2.8
50	0.0999	0.0859	2.5	0.1983	0.1874	2.81
70	0.0943	0.0786	2.8	0.1959	0.1848	2.63
100	0.0886	0.074	2.8	0.1942	0.1734	3.0

nizable images. The nonzero background in Figs. 8.8-3a and 8.8-3b is due to the negative portions in the restored images, which make it necessary for the sake of display to shift the level of the images upward. Figures 8.8-3c and 8.8-3d result from algorithms 2 and 4 respectively (i.e., with the two-level constraint) and show much better restorations.† Indeed, algorithm 4 yields a result indistinguishable from the original. Note the coordinate reversal of the restored images: if $f(x, y)$ is the original image then $f(-x, -y)$ is the coordinate-reversed image and has the same magnitude function. This coordinate reversal can be seen in the subsequent results as well. Table 8.8-1 and Fig. 8.8-3 show the significant improvement when relaxation parameters are used. The restorations in Figs. 8.8-3c and 8.8-3d also show the importance of the two-level constraints for good restorations with a relatively small number of iterations. Table 8.8-2 summarizes the results for the restorations of IMAGE2 for the same four algorithms as before and for the same initial point. In Fig. 8.8-4 are shown the restored images for the four algorithms after 100 iterations. The very poor restorations together

† In Chapter 11 we find that the two-level constraint has virtually no impact in restoring tomographic imagery via convex projections. There, all sets were convex.

Table 8.8-3
Normalized SDE $[J(f_n)]$ and λ_{2m} as a Function of Iteration Number n for Restoration of IMAGE1 with Two-Level Constraint and for Initial Point $f_0(x, y)$ Given by $f_0(x, y) = 0.72$ for 20×20 Points in the Center of S_f and $f_0(x, y) = 0.36$ for the Other Points of S_f, Where S_f is the Known Support of $f(x, y)$

	With two-level constraint		
	Algorithm $P_{1L}P_2f_n = f_{n+1}$	Algorithm $P_{1L}T_2f_n = f_{n+1}$	
Iteration	$J(f_n)$	$J(f_n)$	Approx. λ_{2m}
0	0.706	0.706	—
1	0.429	0.443	0.86[a]
2	0.378	0.364	2.11
3	0.350	0.320	2.44
5	0.311	0.258	2.53
10	0.242	0.200	2.24
15	0.215	0.167	2.31
20	0.193	0.149	2.32
25	0.178		
30	0.168		
35	0.159		
40	0.150		

[a] See note on Table 8.8-1.

with the very small changes in $J(f_n)$ after a certain number of iterations are symptomatic of the condition whereby the algorithm is either near a trap or approaching the correct solution through a tunnel. Table 8.8-2 and Fig. 8.8-4 also demonstrate the important result that the true error, $\|e_{100}\| = \|f - f_{100}\|$, may be very large while $J(f_n)$ may be relatively small.

Assuming then that the poor results displayed in Fig. 8.8-4 are due to traps (or tunnels), it is natural to consider a different solution path† whose "geometry" avoids the trap (or tunnel). The easiest way to implement a different solution path is to change the starting point f_0.‡ Table 8.8-3 summarizes the results of the restoration of IMAGE2 with a starting point $f_0(x, y)$ defined as follows: let S_f be the support of $f(x, y)$, i.e., 30×30 points in the center of the total field of 64×64 points. $f_0(x, y) = 0.72$ for a region of 20×20 points in the center of S_f and $f_{01}(x, y) = 0.36$ for the remaining

† For example, for the algorithm $f_{n+1} = T_1T_2f_n$ a solution path for a fixed starting point f_0 would be $f_0, T_2f_0, f_1, T_2f_1, f_2, \ldots$.

‡ Of course, it is not known *a priori* that a different f_0 will avoid traps or tunnels.

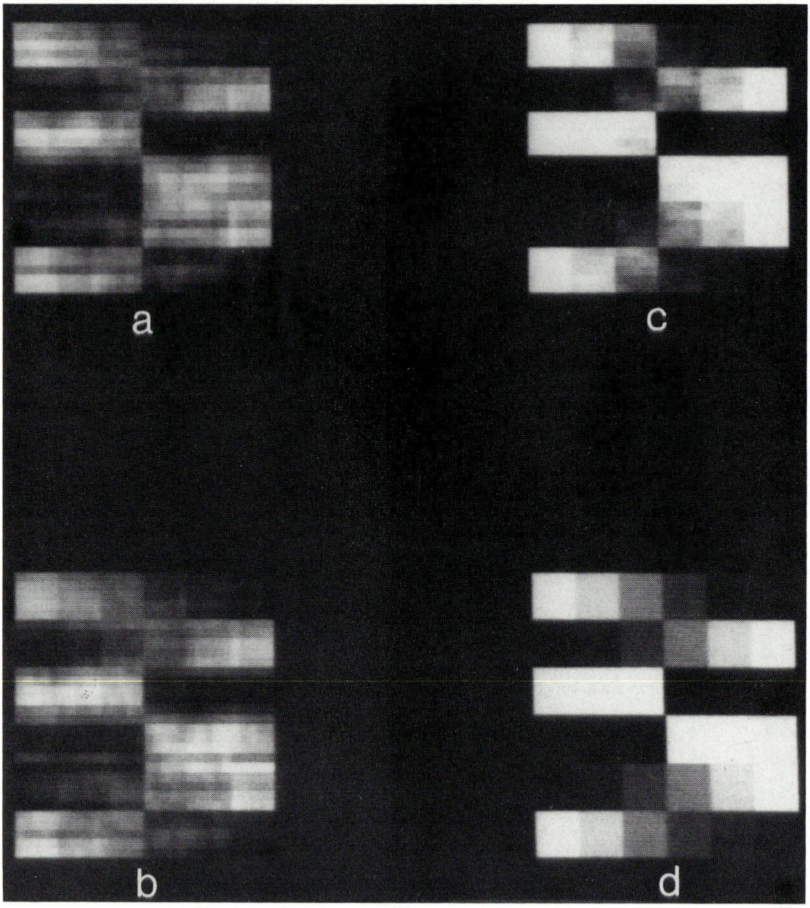

Fig. 8.8-3. Restoration of IMAGE1 after 30 iterations with initial point $f_0 = 0$: (a) by $f_{n+1} = P_1P_2f_n$; (b) by $f_{n+1} = P_1T_2f_n$; (c) by $f_{n+1} = P_{1L}P_2f_n$; (d) by $f_{n+1} = P_{1L}T_2f_n$.

points of S_f. Figure 8.8-5 shows the restored images for this case without (a) and with (b) the use of relaxation parameters. Here we clearly see the importance of the starting point for a rapid restoration. With this new $f_0(x, y)$ we get a recognizable image with pure projections and a very good image with the use of relaxation parameters after only 40 iterations. However, other than making the choice of f_0 with serendipity, we do not know at this stage of the research how to choose a starting point that will surely avoid traps and tunnels.

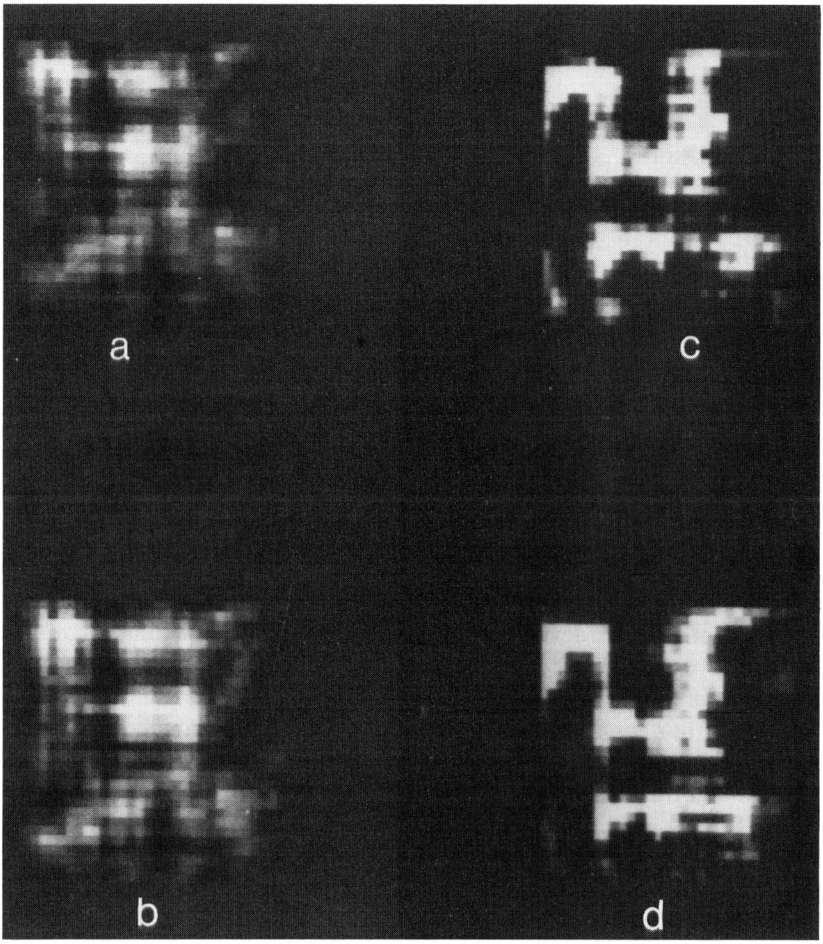

Fig. 8.8-4. Restoration of IMAGE2 after 100 iterations with $f_0 = 0$: (a) by $f_{n+1} = P_1 P_2 f_n$; (b) by $f_{n+1} = P_1 T_2 f_n$; (c) by $f_{n+1} = P_{1L} P_2 f_n$; (d) by $f_{n+1} = P_{1L} T_2 f_n$.

The main conclusions and observations drawn from these experiments are as follows:

- By using relaxation parameters we can save at least 30% over the method of pure projections in the number of iterations required to restore an image; sometimes we can save considerably more than that.
- The variation of the summed distance error with λ_2 showed the following behavior: (1) For a small number of iterations, the minimum is

8. J. R. Fienup, Phase retrieval algorithms: A comparison. *Appl. Opt.* **21**, 2758–2769 (1982).
9. D. C. Youla, Generalized image restoration by the method of alternating orthogonal projections. *IEEE Trans. Circuits and Systems* **CAS-25**, 694–702 (1978).
10. V. T. Tom, T. F. Quatieri, M. H. Hayes, and J. H. McClellan, Convergence of iterative nonexpansive signal reconstruction algorithms. *IEEE Trans. Acoust. Speech Signal Process.* **ASSP-29**, 1052–1058 (1981).
11. L. M. Bregman, The method of successive projections for finding a common point of convex sets. *Dokl. Akad. Nauk SSSR* **162**(3), 487–490 (1965).
12. L. G. Gubin, B. T. Polyak, and E. V. Raik, The method of projections for finding the common point of convex sets. *USSR Computational Mathematics and Mathematical Physics* **7**(6), 1–24 (1967).
13. A. Levi, Image Restoration by the Method of Projections with Application to the Phase and Magnitude Retrieval Problems. Ph.D. Thesis, Rensselaer Polytechnic Institute, Dept. of ECSE, Dec. 1983.
14. A. Levi and H. Stark, Signal restoration from phase by projections onto convex sets. *J. Opt. Soc. Am.* **73**, 810–822 (1983).
15. M. H. Hayes, The reconstruction of a multidimensional sequence from the phase or magnitude of its Fourier transform. *IEEE Trans. Acoust. Speech and Signal Process.* **ASSP-30**, 140–154 (1982).
16. H. A. Ferwerda, The phase problem in object reconstruction and interferometry. *Technical Digest of the Opt. Soc. Am.* **ThA1-1** (1983).
17. J. R. Fienup, *Phase retrieval in astronomy. Technical Digest of the Opt. Soc. Am.* **ThA8-1** (1983).
18. Y. M. Bruck and L. G. Sodin, On the ambiguity of the image reconstruction problem. *Opt. Commun.* **30**, 304–308 (1979).
19. R. H. T. Bates, Fourier phase problems are uniquely solvable in more than one dimension. I: Underlying theory. *Optik* **61**, 247–262 (1982).
20. J. R. Fienup, Reconstruction of an object from the modulus of its Fourier transform. *Opt. Letters* **3**, 27–29 (1978).

CHAPTER 9

Image Reconstruction from Limited Data: Theory and Applications in Computerized Tomography

Barry P. Medoff

ESL Incorporated
A Subsidiary of TRW
Sunnyvale, California, 94088-3510

9.1 INTRODUCTION

The problem of reconstructing a two-dimensional image from its line integrals occurs in many fields, including radio astronomy, electron microscopy, phase-contrast microscopy, x-ray crystallography, geophysics, synthetic aperture radar, and medical imaging, particularly cross-sectional imaging with x-ray computerized tomography (CT). The mathematical theory of image reconstruction from line integrals goes back to papers of Radon in 1917 [1] and Cramér and Wold in 1936 [2], but these early theoretical results were not known to other researchers in the various fields of application. Consequently, the basic theory has been independently rediscovered many times.

For our purposes it is important to make a distinction between two types of image reconstruction problem, which we will refer to as the *full data* problem and the *limited data* problem. In the full data problem one assumes that line integrals are known along all paths through a plane of interest. This assumption is valid in practice when one measures a large number of line integrals that are regularly spaced in the plane. Computationally efficient full data image reconstruction algorithms that reconstruct high-quality images are well known [3–5].

In the limited data problem one assumes that only an incomplete set of estimated line integrals is known; line integrals along some paths are inaccurately measured or not measured at all. In practice, for example, a limited data problem can arise in CT when the plane of interest contains a metallic structure such as a surgical clip or an artificial joint. Line integral measurements through these nearly x-ray opaque structures contain large errors caused by beam hardening and subtle motion. When conventional algorithms are applied in these limited data situations, the reconstructed images are distorted by severe streak artifact [6–8].

This chapter considers the general problem of image reconstruction from limited data. While our focus will be on applications to CT, much of our theoretical development applies to a wide class of signal recovery problems. Our main assumptions are that measured data are related to an underlying unknown signal by a bounded linear measurement operator, and that any additional *a priori* information can be represented by certain types of constraints. The novelty of our approach lies in our formulation of the problem in terms of infinite-dimensional linear operators and general inner products. With this formulation we obtain both closed-form operator pseudoinverse solutions, and iterative algorithms based on the methods of successive approximations, steepest descent, and conjugate gradients. These iterative algorithms are expressed in terms of the measurement operator, its adjoint, and the associated inner products. They are guaranteed to converge to the least-squares solution.

This chapter is based on work presented in [9], portions of which appeared in [10–14]. In the time since [9] was completed the field of signal recovery from incomplete information has been active, and many papers have been published on the subject. Perhaps closest to ours in general approach is the work of Sanz and Huang [15] and that of Musicus [16]. The other chapters of this book contain additional related material and recent results. We make extensive use of functional analysis and Hilbert space methods, more details of which can be found in Chapter 1 by Rushforth. A related approach to the limited data problem in CT, based on convex projections, is given by Sezan and Stark in Chapter 11. An approach based on Bayesian techniques is given by Hanson in Chapter 3. All this work is in a sense complementary, illustrating many points of view, a variety of analytical techniques, and several different types of application-driven constraints and objectives.

A. *Computerized Tomography and the Limited Data Problem*

The basic operation underlying CT is the measurement of line integrals. To see this, let us consider a typical CT scanner, as shown in Fig. 9.1-1.

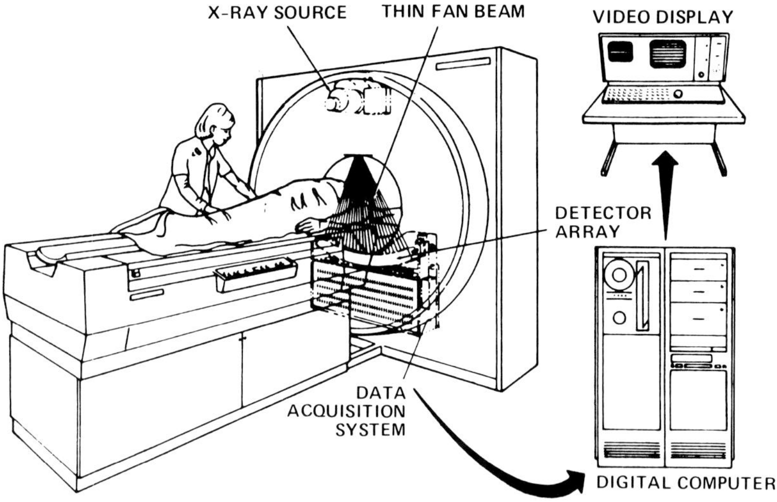

Fig. 9.1-1. Typical fan-beam x-ray CT scanner. The patient is positioned so that the plane of interest is intersected by the thin fan beam of x-rays.

This scanner embodies the fan beam geometry, which is by far the most common design. The detector array is shown in more detail in Fig. 9.1-2. Under standard simplifying assumptions, the x-ray photon intensity measured by each detector is converted by a simple preprocessing step into an accurate estimate of the line integral of the attenuation coefficient on a line from the x-ray source to that detector [4]. In a typical scanner, the detector array contains several hundred detectors, and therefore several hundred line integrals are measured at each source position. This set of line integrals measured at a single source position is referred to as a *fan-beam projection.* During each scan, fan-beam projections are measured with the source in several hundred positions equally spaced in angle around the circle, thereby generating several hundred thousand individual measurements.

The set of line integrals that is measured in a single CT scan is referred to in the literature as a *full set of line integrals*, a *complete set of line integrals*, a *shadow*, and sometimes as a *sinogram.* We will use the term shadow, and we will refer to the measurement operator that transforms an image into a set of line integrals as the *shadow transform operator.*

A *limited data* problem occurs when it is not possible to obtain a full set of regularly spaced line integral measurements. In these situations, the standard image reconstruction algorithm—convolution-backprojection—generates images that are distorted by severe streak artifact. For example, Fig. 9.1-3 shows a typical high-resolution cross-sectional image

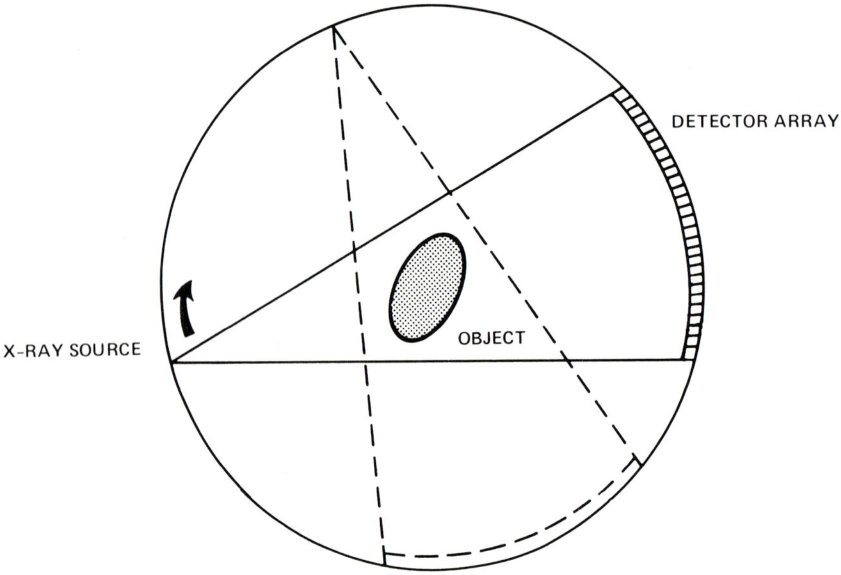

Fig. 9.1-2. Fan-beam detector array geometry in more detail. The detector array consists of several hundred discrete x-ray detectors. Each detector measures the line integral of the attenuation coefficient on a line from the x-ray source to that detector.

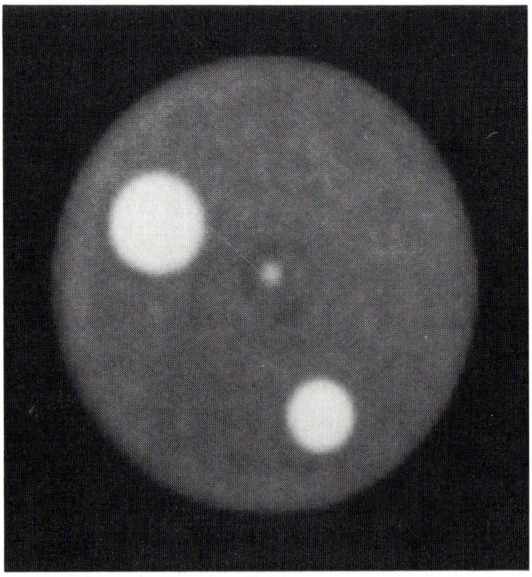

Fig. 9.1-3. Typical high-resolution cross-sectional image reconstructed from a complete set of regularly spaced line integrals by convolution-backprojection.

reconstructed from a complete set of regularly spaced line integrals by convolution-backprojection. The object is clearly seen to consist of a uniform background, within which are three smaller structures. Figure 9.1-4 is a reconstruction from an incomplete set of line integrals of the same object. In this case, fewer than 1% of the line integrals are missing. This image was also generated by convolution-backprojection. Comparing Figs. 9.1-3 and 9.1-4, we see that the absence of even a small percentage of the total set of line integrals results in the reconstructed image being distorted by severe streak artifact.

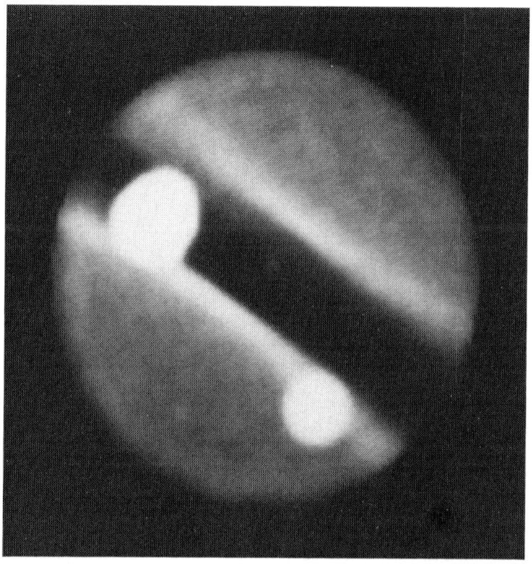

Fig. 9.1-4. Reconstruction by convolution-backprojection from an incomplete set of line integrals, showing severe streak artifact.

An example of the limited data problem is the so-called *limited angle* problem, which is illustrated by the proposed tomographic suitcase scanner [17] shown in Fig. 9.1-5a. This scanner consists of a fixed x-ray source and a fixed detector array. A suitcase sits on a moving conveyor belt, and the plane of interest moves through the thin fan beam of x-rays. Figure 9.1-5b shows the set of line integrals that is measured through a fixed point in the plane as the point moves across the fan beam. The angle of the fan beam defines a limited angular range through which line integrals are measured; thus, for example, there can be no line integral measurements along paths that are parallel to the conveyor belt. The limited angle problem is to reconstruct an image from line integral measurements in this limited angular range.

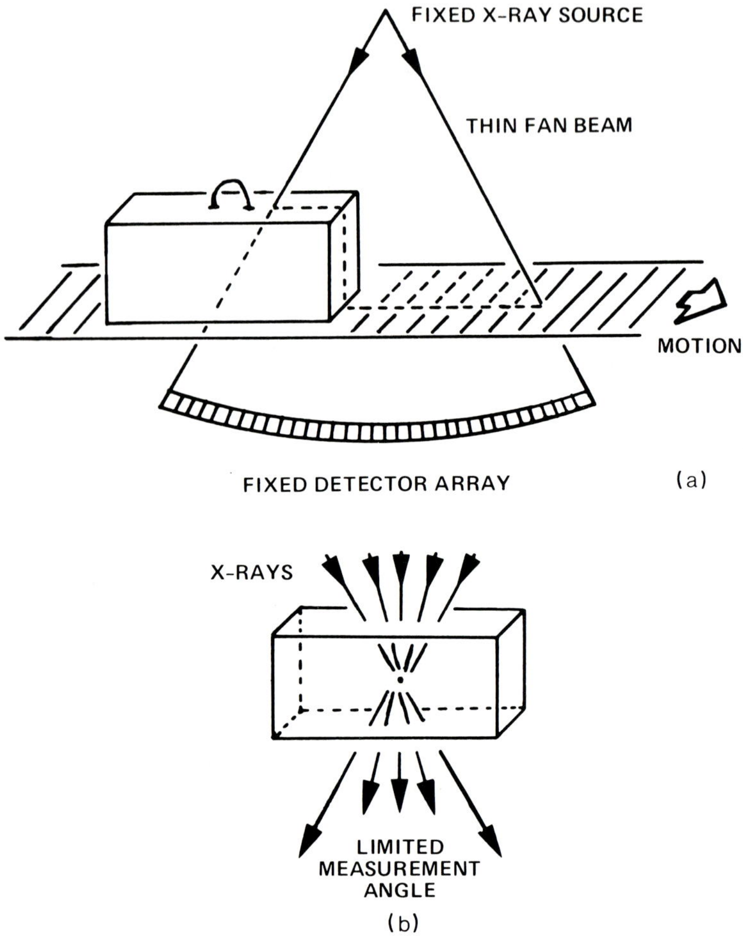

Fig. 9.1-5. Example of the limited angle problem: the tomographic suitcase scanner. (a) Diagram of the scanner. A suitcase sits on a moving conveyor belt and the plane of interest translates through the thin fan beam of x-rays. (b) Illustration of the set of line integrals that are measured through a fixed point as the point moves across the fan beam of x-rays. The limited measurement angle is equal to the angular width of the fan beam.

A second example of a limited data problem is the so-called *bagel problem*, in which the plane of interest contains a region that is opaque to x-rays. The bagel problem is illustrated in Fig. 9.1-6. In practice, a bagel problem occurs when the plane of interest passes through a metallic structure such as a surgical clip [6] or an artificial hip joint [8]; a similar problem occurs when the plane of interest passes through a bony region such as the spine.

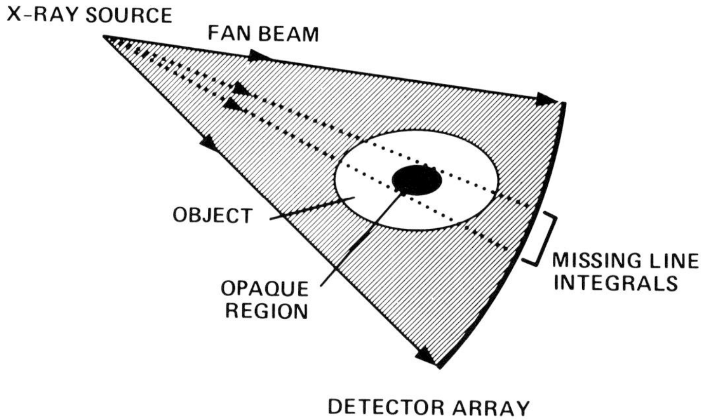

Fig. 9.1-6. Example of the bagel problem: the object of interest contains an x-ray opaque region that blocks x-rays. In each fan-beam projection there will be a set of missing line integrals corresponding to paths that intersect the opaque region. (From Medoff *et al.* [13].)

In each fan-beam projection there will be a set of measured line integrals corresponding to measurements along paths that pass through the object exterior to the opaque region; there will also be a set of missing line integrals corresponding to paths that intersect the opaque region. The x-ray opaque region, in effect, creates a "hole" in the interior of an unknown bagel-shaped region of the object. The bagel problem is to reconstruct an image in the bagel from line integrals along paths that do not pass through the hole.

B. Previous Work

We can divide the previous work on image reconstruction from limited data into two categories. The first category consists of methods that are derived from a continuous formulation of the problem. Exact solutions are known for two cases: the bagel problem [18,19] and the limited angle problem [19,20]. The limited angle problem is closely related to the problem of recovering a spatially limited object from partial knowledge of its Fourier transform [21,22], and numerous frequency domain extrapolation methods have been applied to CT [23–28].

The second category consists of reconstruction algorithms that are derived from a discrete formulation of the problem. Most of the methods in this category represent applications of standard deterministic or stochastic estimation techniques to CT. The stochastic methods include maximum likelihood [29], linear minimum variance [30–32], and maximum entropy [33,34].

The deterministic methods include matrix pseudoinverse [35–37] and matrix iterative techniques [38,39].

It has recently been recognized that the fundamental problem in image reconstruction from limited data is missing information: a limited set of line integral measurements does not determine a unique underlying image [9–13,40–42]. To arrive at a satisfactory algorithm for image reconstruction from limited data, one must first determine additional *a priori* information and then incorporate this information in a computationally efficient algorithm.

9.2 REVIEW OF IMAGE RECONSTRUCTION

This section concisely reviews the mathematical theory of image reconstruction from a complete set of line integrals.

A. Basic Functions and Operators

In our analysis of image reconstruction, we will be concerned with images, Fourier transforms of images, and shadow transforms of images. Throughout this chapter, an image is denoted by a lowercase italic letter, its Fourier transform by the same letter in uppercase italics, and its shadow transform by lowercase italics with an underbar. Thus, g denotes an image, G denotes the Fourier transform of g, and $\underline{g}$ denotes the shadow transform of g. Integral operators are denoted by bold roman letters. For example, the shadow transform operator is denoted by $\mathbf{S}$.

An *image* $g(x, y)$ is a function of two variables x and y, which are the usual Cartesian coordinates. Images usually represent physical quantities, like the distribution of attenuation coefficient in a plane. Such images must be real valued, and they might have to satisfy certain additional requirements—for example, positivity. Our mathematical formulation, however, will allow for arbitrary complex-valued images.

A *shadow* $\underline{g}(R, \theta)$ is function of two variables R and θ, where $R \in (-\infty, \infty)$ and $\theta \in [0, \pi)$. We emphasize the fact that R and θ are not polar coordinates—they are simply two independent variables.

Certain shadows are related to images by the *shadow transform*, defined by

$$\mathbf{S}g(x, y) \triangleq \int_{L(R,\theta)} g(x, y)\, dl, \qquad (9.2\text{-}1)$$

where **S** represents the *shadow transform operator* and dl is the differential arc length along the line $L(R, \theta)$ defined by

$$L(R, \theta) \triangleq \{(x, y): x \cos \theta + y \sin \theta = R\}. \qquad (9.2\text{-}2)$$

It is sometimes convenient to regard the shadow $\underset{\sim}{g}(R, \theta)$ as a one-dimensional function of R for a fixed value of θ. This will be denoted by $\underset{\sim}{g}_\theta(R)$ and referred to as the *line integral projection* of g at angle θ. The shadow transform is illustrated in Fig. 9.2-1.

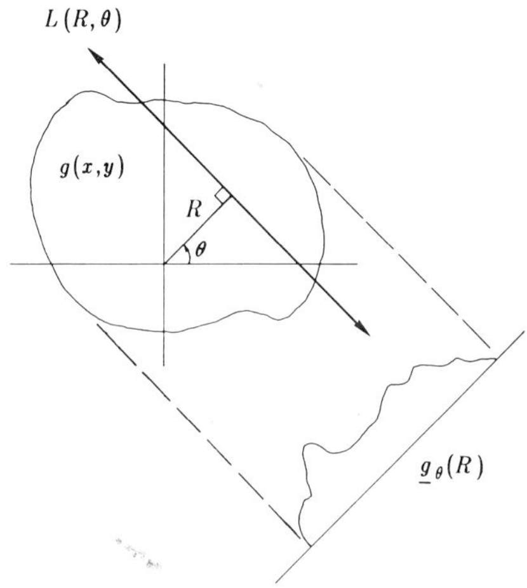

Fig. 9.2-1. Illustration of the shadow transform. The value of the shadow $\underset{\sim}{g}_\theta(R)$ for any R and θ is equal to the line integral of the image $g(x, y)$ along the line $L(R, \theta)$.

Our analysis will make extensive use of the Fourier transform. The one-dimensional Fourier transform of a function $g(x)$ is defined by

$$\mathbf{F}g(x) \triangleq \int_{-\infty}^{\infty} g(x)\, e^{-iux}\, dx, \qquad (9.2\text{-}3)$$

where **F** is the one-dimensional Fourier transform operator. The inverse transform is defined by

$$\mathbf{F}^{-1}G(u) \triangleq \frac{1}{2\pi} \int_{-\infty}^{\infty} G(u)\, e^{iux}\, du. \qquad (9.2\text{-}4)$$

The two-dimensional Fourier transform of a function $g(x, y)$ is defined by

$$\mathbf{F}_2 g(x, y) \triangleq \iint_{-\infty}^{\infty} g(x, y)\, e^{-i(ux+vy)}\, dx\, dy, \qquad (9.2\text{-}5)$$

where $\mathbf{F}_2$ is the two-dimensional Fourier transform operator. The Fourier transform $G(u, v) = \mathbf{F}_2 g(x, y)$ is a function of the radian frequency variables u and v. The inverse Fourier transform of $G(u, v)$ is defined by

$$\mathbf{F}_2^{-1} G(u, v) \triangleq \frac{1}{4\pi^2} \iint_{-\infty}^{\infty} G(u, v)\, e^{i(ux+vy)}\, du\, dv. \tag{9.2-6}$$

It is sometimes convenient to write the Fourier transform relationships in polar coordinates. With image domain polar coordinates (r, θ) defined by $x = r \cos \theta$ and $y = r \sin \theta$, and Fourier transform domain polar coordinates (ρ, ϕ) defined by $u = \rho \cos \phi$ and $v = \rho \sin \phi$, the Fourier transform relationships Eqs. (9.2-5) and (9.2-6) become

$$\mathbf{F}_2 g(r, \theta) = \int_0^{2\pi} \int_0^{\infty} g(r, \theta)\, e^{-ir\rho \cos(\theta - \phi)} r\, dr\, d\theta \tag{9.2-7}$$

and

$$\mathbf{F}_2^{-1} G(\rho, \phi) = \frac{1}{4\pi^2} \int_0^{2\pi} \int_0^{\infty} G(\rho, \phi)\, e^{ir\rho \cos(\theta - \phi)} \rho\, d\rho\, d\phi. \tag{9.2-8}$$

B. The Projection-Slice Theorem

The most fundamental result in the theory of image reconstruction is the projection-slice theorem. This theorem, proved by Bracewell in 1956 [43], expresses the relationship between an image, its shadow transform, and its two-dimensional Fourier transform in a way that is both mathematically precise and intuitively clear. It will be an important tool throughout our analysis.

Let g be an image with shadow $\underline{g}$ and two-dimensional Fourier transform G, and recall that $\underline{g}_\theta(R)$ denotes the line integral projection of g at angle θ. Then the projection-slice theorem states that the one-dimensional Fourier transform of $\underline{g}_\theta(R)$ is equal to the two-dimensional Fourier transform of the image g evaluated along a line through the origin of the two-dimensional frequency plane at an angle θ with respect to the u axis. That is,

$$G(\rho, \theta) = \mathbf{F} \underline{g}_\theta(R) = \int_{-\infty}^{\infty} \underline{g}_\theta(R)\, e^{-i\rho R}\, dR. \tag{9.2-9}$$

C. The Convolution-Backprojection Algorithm

The convolution-backprojection algorithm is a two-step procedure for recovering an image $f(x, y)$ from its shadow $\underline{f}_\phi(R)$. The algorithm was discovered by Bracewell and Riddle in 1967 [44] and later by Ramachandran and Lakshminarayanan [45]. Shepp and Logan [46] made important contributions to the discrete form of the algorithm. Lakshminarayanan [47] developed a version for the fan-beam geometry. Herman and Naparstek [48] later obtained the same fan-beam algorithm by a different derivation. The convolution-backprojection algorithm, in one form or another, is now implemented on virtually every existing CT scanner.†

We will define two operators to represent the computations in the convolution-backprojection algorithm. First, we introduce the *rho filter operator*, denoted by $\boldsymbol{\rho}$ and defined by

$$\boldsymbol{\rho}\underline{f}_{\phi}(R) \triangleq \frac{1}{4\pi^2}\int_{-\infty}^{\infty}\left[\int_{-\infty}^{\infty}\underline{f}_{\phi}(R)\, e^{-i\rho R}\, dR\right]|\rho|\, e^{i\rho R'}\, d\rho. \qquad (9.2\text{-}10)$$

This definition can also be written in terms of the one-dimensional Fourier transform operator as

$$\boldsymbol{\rho}\underline{f}_{\phi}(R) \triangleq \mathbf{F}^{-1}\{(2\pi)^{-1}|\rho|\mathbf{F}\underline{f}_{\phi}(R)\}. \qquad (9.2\text{-}11)$$

The rho filter is a linear shift-invariant operator, and therefore it can be implemented as a convolution. If $\underline{f}_{\phi}(R)$ is a line integral projection at angle ϕ, then the function $\underline{g}_{\phi}(R)$ given by

$$\underline{g}_{\phi}(R) = \boldsymbol{\rho}\underline{f}_{\phi}(R) \qquad (9.2\text{-}12)$$

is referred to as the *rho-filtered projection at angle* ϕ. We will adopt the convention that when the rho filter is applied to a two-dimensional shadow function, e.g., $\boldsymbol{\rho}\underline{f}(R, \theta)$, the operator acts on $\underline{f}$ as a one-dimensional function of R for each value of θ.

The second operator that we need is the *backprojection operator*, denoted by $\mathbf{B}$ and defined by

$$\mathbf{B}\underline{g}_{\phi}(R) \triangleq \int_{0}^{\pi} \underline{g}_{\phi}(x\cos\phi + y\sin\phi)\, d\phi. \qquad (9.2\text{-}13)$$

The backprojection operator forms an image by spreading a shadow $\underline{g}_{\phi}(R)$ back uniformly across the plane at angle ϕ and summing the result for all values of ϕ.

The convolution-backprojection algorithm is based on the fact that the backprojection and rho filter operators together form a left inverse for the shadow transform operator; that is, $\mathbf{B}\boldsymbol{\rho}\mathbf{S} = \mathbf{I}$, where $\mathbf{I}$ is the identity operator [4]. Thus, we arrive at the following two-step algorithm.

† Methods based on recovering the image by inverse transforming Eq. (9.2-9) are now in use as well. See [5].

THE CONVOLUTION-BACKPROJECTION ALGORITHM

Given a shadow $\underline{f}(R, \phi)$, reconstruct an image $f(x, y)$ as follows:

1. Form the rho-filtered projections,

$$\underline{g}_{\phi}(R) = \boldsymbol{\rho}\underline{f}_{\phi}(R). \tag{9.2-14}$$

2. Form an image by backprojecting the rho-filtered projections,

$$f(x, y) = \mathbf{B}\underline{g}_{\phi}(R). \tag{9.2-15}$$

9.3 AN INNER PRODUCT FRAMEWORK FOR IMAGE RECONSTRUCTION

The previous section reviewed the theory of image reconstruction by direct inversion of the line integral transform. In full data problems, one is able to accurately measure a large number of line integrals along paths that are regularly spaced in the plane, and algorithms that are based on direct inversion, such as convolution-backprojection, give satisfactory results. But in limited data problems, line integral measurements are inaccurate or irregularly spaced, and algorithms based on direct inversion generate reconstructed images with severe streak artifact. The difficulty is that the limited data problem is not simply one of inversion, but one of approximation and optimization. The measured data do not exactly or uniquely define a single reconstructed image. One must, therefore, precisely define what constitutes the "best" reconstruction from a given set of data and then find an algorithm for computing it. This requires the introduction of a measure of distance, which we will define mathematically in terms of a norm or inner product.

A. *Properties of Inner Products*

The general definition of inner product extends the familiar geometric concepts of length, angle, and orthogonality to arbitrary vector spaces. For our purposes, it generalizes the simple dot product in three important ways. First, inner products can be defined with arbitrary positive definite weighting functions that give relatively more or less emphasis to different elements in the vector space. Second, inner products can be defined for infinite-dimensional vector spaces, e.g., vector spaces of functions. This allows us to apply vector space methods of optimization to image reconstruction without sacrificing a continuous formulation of the problem. Third, the use of inner products permits a "coordinate-free" or "basis-independent" analysis, thereby separating features that are inherent in the problem from features that are specific to a particular choice of basis. The discussion below is necessarily brief; more details can be found in several books, such as [49,50].

An *inner product* is a mapping which associates a scalar with every pair of elements in a vector space Ω. The inner product of two vectors $f, g \in \Omega$ is represented by $\langle f, g\rangle$ and satisfies the following properties:

(1) Symmetry: $\langle f, g\rangle = \overline{\langle g, f\rangle}$

(where the bar denotes complex conjugate).

(2) Linearity: $\langle f + g, h\rangle = \langle f, h\rangle + \langle g, h\rangle$,

$$\langle \alpha f, g\rangle = \alpha\langle f, g\rangle$$

(where f, g, and h are vectors and α is a scalar).

(3) Positive definiteness: $\langle f, f\rangle > 0$ for all $f \neq 0$.

Any mapping which satisfies these properties is a valid inner product. It is clear that on any given vector space there can be many different inner products.

An inner product induces a geometry on a vector space through the definitions of length and orthogonality. The length or *norm* of a vector f is represented by $\|f\|$ and defined by $\|f\| = (\langle f, f\rangle)^{1/2}$. Two vectors f and g are said to be *orthogonal* when $\langle f, g\rangle = 0$. Orthogonality is denoted by $f \perp g$.

In the discussion above, we have defined the norm in terms of the inner product. We could just as well have defined the inner product in terms of the norm according to the so-called *polarization identity*,

$$\langle f, g\rangle = \tfrac{1}{4}[(\|f + g\|^2 - \|f - g\|^2) + i(\|f + ig\|^2 - \|f - ig\|^2)]. \quad (9.3\text{-}1)$$

The point is that the geometry of an inner product space is completely determined either by specifying the norm or by specifying the inner product; one implies the other.

B. Direct Inner Products

Now let us consider some specific choices of inner products for image reconstruction. We will view the image domain, the Fourier transform domain, and the shadow domain as three different vector spaces, and therefore we are free to choose an inner product in each domain independently. Having done this, we will find it illuminating to see how inner products in different domains are related. Note that the domain to which a particular function belongs can be determined from our notational convention: an image is denoted by a lowercase italic letter, its Fourier transform by the same letter in uppercase italics, and its shadow transform by lowercase italics with an underbar.

Starting with the image domain, the simplest inner product is the sum of squares definition. If f and g are two images, we define the *direct image domain inner product* by

$$\langle f, g \rangle \triangleq \iint_{-\infty}^{\infty} f(x, y)\overline{g(x, y)}\, dx\, dy. \tag{9.3-2}$$

Similarly, if F and G are two functions in the frequency domain, we define the *direct frequency domain inner product* by

$$\langle F, G \rangle \triangleq \frac{1}{4\pi^2} \iint_{-\infty}^{\infty} F(u, v)\overline{G(u, v)}\, du\, dv. \tag{9.3-3}$$

To relate these inner products, we recall Parseval's theorem for the two-dimensional Fourier transform,

$$\iint_{-\infty}^{\infty} f(x, y)\overline{g(x, y)}\, dx\, dy = \frac{1}{4\pi^2} \iint_{-\infty}^{\infty} F(u, v)\overline{G(u, v)}\, du\, dv. \tag{9.3-4}$$

We can succinctly restate Parseval's theorem with inner products as

$$\langle f, g \rangle = \langle F, G \rangle. \tag{9.3-5}$$

This shows that the two-dimensional Fourier transform preserves direct inner products, and therefore it preserves orthogonality and length. The inner product preserving property of the Fourier transform is crucial in the design and analysis of signal processing algorithms because it guarantees that an error which has a small energy (the square of the norm) in the frequency domain also has the same small energy in the image domain. This property also guarantees that orthogonal images have orthogonal Fourier transforms, so that Fourier transforming an orthogonal basis for the image domain generates an orthogonal basis for the frequency domain.

Now we can turn to the definition of an inner product for shadows, say $\underline{f}(R, \theta)$ and $\underline{g}(R, \theta)$. First of all, we emphasize the fact that the variables R and θ are not polar coordinates. There is nothing "natural" about defining an inner product as a double integral with respect to $R\, dR\, d\theta$, as a polar coordinate interpretation might mistakenly suggest. Instead, following the sum of squares formulation introduced above for images and Fourier transforms, we define the *direct shadow domain inner product* by

$$\langle \underline{f}, \underline{g} \rangle \triangleq \int_0^{\pi} \int_{-\infty}^{\infty} \underline{f}(R, \theta)\overline{\underline{g}(R, \theta)}\, dR\, d\theta. \tag{9.3-6}$$

Paralleling our discussion of the Fourier transform, let us determine how the direct inner product of two shadows is related to the direct inner product of two images. Let f and g be images with Fourier transforms F and G and shadow transforms $\underline{f}$ and $\underline{g}$. We first apply the projection-slice theorem and Parseval's theorem in one dimension with respect to the variable R to transform Eq. (9.3-6) into the frequency domain. Then we apply Parseval's theorem in two dimensions to go back to the image domain. We obtain

$$\langle \underline{f}, \underline{g} \rangle = \langle f, \boldsymbol{\rho}_{\mathrm{I}}^{-1} g \rangle \tag{9.3-7}$$

where $\boldsymbol{\rho}_{\mathrm{I}}^{-1}$ is the image domain inverse rho filter, defined by

$$\boldsymbol{\rho}_{\mathrm{I}}^{-1} f \triangleq \mathbf{F}_2^{-1}\{2\pi|\rho|^{-1}\mathbf{F}_2 f\}. \tag{9.3-8}$$

Therefore, the direct inner product of two shadows is not equal to the direct inner product of the corresponding two images. Unlike the Fourier transform operator, the shadow transform operator does not preserve direct inner products.

With the direct inner product for both images and shadows, the shadow transform operator is characterized by the following properties:

1. There is no Parseval relation.
2. The energy of an image is not equal to the energy of its shadow.
3. Orthogonal images do not have orthogonal shadows.

C. *The Rho Inner Product*

Let us now determine inner products that are preserved by the shadow transform operator. To this end, we first apply Parseval's theorem in two dimensions to transform the direct image domain inner product into the frequency domain. Then we write the frequency domain integral in polar coordinates and apply Parseval's theorem in one dimension with respect to the radial variable to obtain an expression in the shadow domain. We have

$$\begin{aligned}\langle f, g \rangle &= \frac{1}{4\pi^2}\int_0^{\pi}\int_{-\infty}^{\infty} F(\rho, \phi)\overline{G(\rho, \phi)}|\rho|\, d\rho\, d\phi \\ &= \int_0^{\pi}\int_{-\infty}^{\infty} \underline{f}(R, \phi)\overline{[\boldsymbol{\rho}\underline{g}(R, \phi)]}\, dR\, d\phi \\ &= \langle \underline{f}, \boldsymbol{\rho}\underline{g} \rangle. \end{aligned} \tag{9.3-9}$$

It is understood that $\boldsymbol{\rho}\underline{g}(R, \phi)$ denotes the one-dimensional rho filter acting on the variable R for every value of ϕ.

From Eq. (9.3-9) we reach the important conclusion that the direct inner product between two images can be computed in the shadow domain by the following three steps: (1) compute the shadow of each image, (2) rho filter one of the shadows, and (3) compute the direct shadow domain inner product. By this procedure, any image domain operation that is expressible as a direct inner product can be implemented as a shadow domain computation.

This result suggests the definition of a new shadow domain inner product. We define the *rho inner product* of two shadows $\underline{f}$ and $\underline{g}$ by

$$\langle \underline{f}, \underline{g} \rangle_{\boldsymbol{\rho}} \triangleq \langle \boldsymbol{\rho} \underline{f}, \underline{g} \rangle = \langle \underline{f}, \boldsymbol{\rho} \underline{g} \rangle. \tag{9.3-10}$$

Two shadows are said to be *rho orthogonal* if $\langle \underline{f}, \underline{g} \rangle_{\boldsymbol{\rho}} = 0$. The *rho norm* of a shadow $\underline{f}$ is defined by

$$\|\underline{f}\|_{\boldsymbol{\rho}} \triangleq (\langle \underline{f}, \underline{f} \rangle_{\boldsymbol{\rho}})^{1/2}. \tag{9.3-11}$$

By combining Eq. (9.3-9) with the definition of the rho inner product Eq. (9.3-10), we obtain the following new theorem for the shadow transform.

Theorem 1. The direct inner product of two images is equal to the rho inner product of their shadows. That is, if f and g are images with shadows $\underline{f}$ and $\underline{g}$, then

$$\langle f, g \rangle = \langle \underline{f}, \underline{g} \rangle_{\boldsymbol{\rho}}. \tag{9.3-12}$$

Theorem 1 expresses a property of the shadow transform that is analogous to Parseval's theorem for the Fourier transform.

With the direct inner product in the image domain and the rho inner product in the shadow domain, the following properties characterize the shadow transform operator:

1. The norm of an image is equal to the rho norm of its shadow.
2. Two images are orthogonal if and only if their shadows are rho orthogonal.
3. The inner product of two images is equal to the rho inner product of their shadows.

D. The Least-Squares Problem and the Pseudoinverse Solution

We will formulate the image reconstruction problem in terms of a linear operator **S**, which may be taken to be the shadow transform operator or any continuous or discrete approximation to it. In fact, we will only require that **S** be a bounded linear operator from a Hilbert space Ω with inner product $\langle \cdot, \cdot \rangle_{\Omega}$ to a Hilbert space Φ with inner product $\langle \cdot, \cdot \rangle_{\Phi}$, and that the measurements are represented by some vector $\underline{f} \in \Phi$. Thus we model the physical problem by the linear operator equation $\mathbf{S}f = \underline{f}$, and our first task is to define what constitutes a "solution" to this equation.

We say that an image $g \in \Omega$ is a *least-squares solution* of $\mathbf{S}f = \underline{f}$ if

$$\|\mathbf{S}g - \underline{f}\|_\Phi \leq \|\mathbf{S}f' - \underline{f}\|_\Phi \qquad \text{for all } f' \in \Omega. \tag{9.3-13}$$

An image $h \in \Omega$ is said to be a *minimum norm least-squares solution* of $\mathbf{S}f = \underline{f}$ if h is a least-squares solution and

$$\|h\|_\Omega \leq \|g\|_\Omega \tag{9.3-14}$$

for all least-squares solutions g. The minimum norm least-squares solution is the unique least-squares solution that is orthogonal to the null space of $\mathbf{S}$. Thus, in a mathematical sense, it has no component in the null space.

We define the general image reconstruction problem to be the problem of finding a minimum norm least-squares solution of $\mathbf{S}f = \underline{f}$. One must be aware of the subtle mathematical difficulty that when $\mathbf{S}$ is allowed to be an infinite-dimensional linear operator, there are situations in which $\mathbf{S}f = \underline{f}$ has no least-squares solution. This problem occurs when $R(\mathbf{S})$, the range of the bounded linear operator $\mathbf{S}$, is not closed. In general, the set of least-squares solutions will not be empty if and only if the orthogonal projection of the data $\underline{f}$ onto the closure of the range of $\mathbf{S}$ is also in the range [51,52]. We will henceforth assume that this condition holds. That is, we assume that $\underline{f} \in R(\mathbf{S}) \oplus R(\mathbf{S})^\perp$, where $\oplus$ denotes direct sum and $R(\mathbf{S})^\perp$ denotes the orthogonal complement of $R(\mathbf{S})$. In this case, the set of least-squares solutions is closed and convex, and therefore it contains a unique element of minimum norm.

If $\mathbf{S}$ is a bounded linear operator from Ω to Φ, then the pseudoinverse of $\mathbf{S}$, denoted by $\mathbf{S}^+$, is defined to be the linear operator from Φ to Ω that maps a shadow $\underline{f}$ into the minimum norm least-squares solution of $\mathbf{S}f = \underline{f}$. By definition, the pseudoinverse operator depends on the choice of inner product in both Ω and Φ.

The superscript $+$ is used to denote the pseudoinverse of a linear operator and to distinguish the pseudoinverse from the ordinary inverse, which we denote by a superscript -1. The ordinary inverse is defined only for operators that are nonsingular with range equal to the entire space Φ. For such operators, the pseudoinverse and the ordinary inverse coincide. The pseudoinverse, however, is also well defined for operators that are singular and for operators that do not have all of Φ in their range. Further discussion of the pseudoinverse operator can be found in [49–54].

We will give both closed-form and iterative methods for computing the minimum norm least-squares solution. To state these methods, we must define one additional operator. The *adjoint* of $\mathbf{S}$ is a linear operator from

Φ to Ω denoted by $\mathbf{S}^*$ and defined by the relation $\langle \mathbf{S}f, g\rangle_\Phi = \langle f, \mathbf{S}^* g\rangle_\Omega$ for all $f \in \Omega$ and $g \in \Phi$. One must realize that the adjoint is defined with respect to a particular choice of inner products; the same operator has different adjoints under different inner products. Unfortunately, the standard notation $\mathbf{S}^*$ does not explicitly indicate the inner products involved.

Some insight into the basic structure of the least-squares problem is obtained from a geometric representation of the operators $\mathbf{S}$ and $\mathbf{S}^*$. The geometric representation is based on the following well-known decomposition of the spaces Ω and Φ as direct sums of orthogonal subspaces [49]:

$$\Omega = N(\mathbf{S}) \oplus N(\mathbf{S})^\perp = N(\mathbf{S}) \oplus \overline{R(\mathbf{S}^*)}, \tag{9.3-15}$$

$$\Phi = \overline{R(\mathbf{S})} \oplus R(\mathbf{S})^\perp = \overline{R(\mathbf{S})} \oplus N(\mathbf{S}^*), \tag{9.3-16}$$

where an overbar denotes closure. By identifying the orthogonal subspaces with orthogonal lines in the plane, we obtain the picture shown in Fig. 9.3-1. We see that Ω and Φ are decomposed into four subspaces. Two of

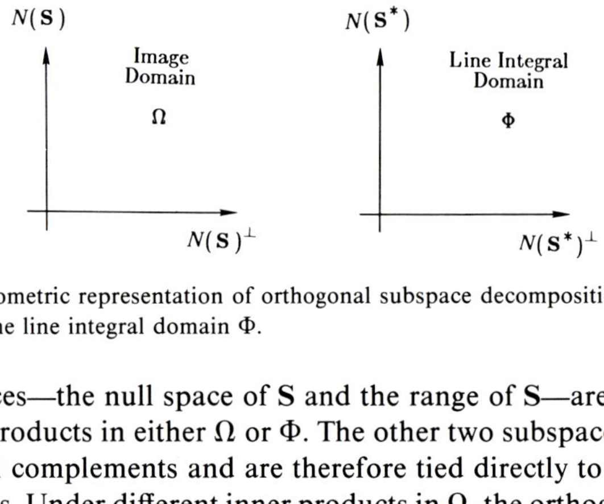

Fig. 9.3-1. Geometric representation of orthogonal subspace decompositions of the image domain Ω and the line integral domain Φ.

these subspaces—the null space of $\mathbf{S}$ and the range of $\mathbf{S}$—are independent of the inner products in either Ω or Φ. The other two subspaces are defined as orthogonal complements and are therefore tied directly to the choice of inner products. Under different inner products in Ω, the orthogonal complement $N(\mathbf{S})^\perp$ contains different elements. Similarly, under different inner products in Φ, the orthogonal complement $R(\mathbf{S})^\perp$ contains different elements.

Figures 9.3-2 and 9.3-3 depict the operators $\mathbf{S}$ and $\mathbf{S}^*$. The set Λ in Fig. 9.3-2 represents a set of least-squares solutions. The element of Λ on the $N(\mathbf{S})^\perp$ axis is the minimum norm solution. The set Γ in Fig. 9.3-3 represents a set of data vectors that have the same least-squares solutions. The element of Γ on the $N(\mathbf{S}^*)^\perp$ axis has an exact solution.

We have defined the pseudoinverse operator in terms of the solution of a minimum norm least-squares problem. The definition, however, does not provide a means for implementing the pseudoinverse operator or computing

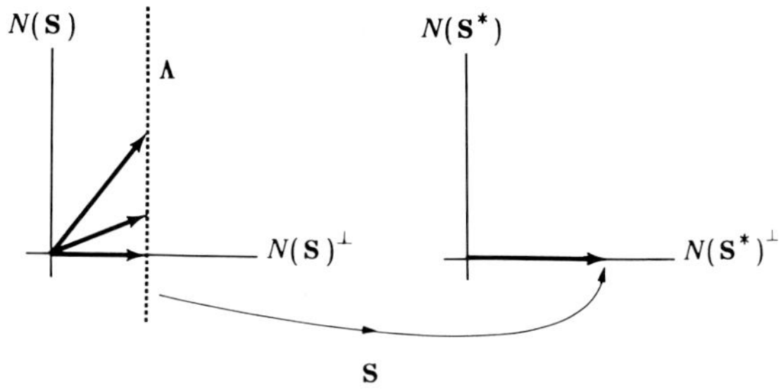

Fig. 9.3-2. Geometric representation of the linear operator S.

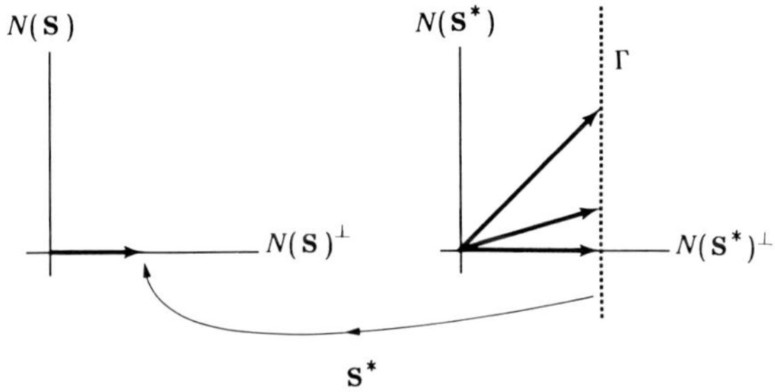

Fig. 9.3-3. Geometric representation of the adjoint operator $\mathbf{S}^*$.

the minimum norm least-squares solution. To this end, we recall two representations of the pseudoinverse [49, p. 165],

$$\mathbf{S}^+ = (\mathbf{S}^*\mathbf{S})^+\mathbf{S}^* = \mathbf{S}^*(\mathbf{S}\mathbf{S}^*)^+. \tag{9.3-17}$$

These representations express the pseudoinverse of an operator $\mathbf{S}$ in terms of the pseudoinverses of the operators $\mathbf{S}^*\mathbf{S}$ and $\mathbf{S}\mathbf{S}^*$; they are identities which can be derived based on the definition of the pseudoinverse. Their practical significance lies in the fact that one of the operators $\mathbf{S}^*\mathbf{S}$ or $\mathbf{S}\mathbf{S}^*$ can have an ordinary inverse even when the operator $\mathbf{S}$ does not. The implementation of the pseudoinverse then reduces to the implementation of an ordinary inverse and an application of the adjoint operator.

The two representations of the pseudoinverse operator in Eq. (9.3-17) can be viewed as two different computational procedures. In the first representation, the adjoint operator transforms the data into an intermediate image, and then the image domain operator $(\mathbf{S}^*\mathbf{S})^+$ transforms the intermediate image into the minimum norm least-squares solution. In the second representation, the shadow domain operator $(\mathbf{S}\mathbf{S}^*)^+$ transforms the shadow data into an intermediate shadow, and then the adjoint operator transforms the intermediate shadow into the minimum norm least-squares solution.

For applications in which neither representation of the pseudoinverse operator is easily implementable, one can instead use iterative algorithms that are guaranteed to converge to the pseudoinverse solution. Iterative algorithms based on the methods of successive approximations, steepest descent, and conjugate gradients will be given later.

E. *Incorporating a Priori Information by Constraints*

In many practical situations there is additional information about the underlying image that we would like to include in the reconstruction process. Important examples include knowledge of an outer boundary and knowledge that the image takes on specified values in certain regions of the plane. *A priori* information of this type can be formalized as the constraint that the reconstructed image must lie in a particular affine subspace of images. We recall that if S is a subspace of another space Ω, and f is any element in Ω, then the *affine subspace* $S' = f + S$ is given by

$$S' = \{s' : s' = f + s, s \in S\}.$$

The affine subspace S' is said to be a *translation* of S.

An affine subspace of images can be implicitly specified as the set of images which satisfy a *constraint equation* of the form

$$\mathbf{C}f = g, \tag{9.3-18}$$

where we will assume that the *constraint operator* $\mathbf{C}$ is a bounded linear operator on Ω with closed range and that g is in the range of $\mathbf{C}$. As an example of an affine subspace constraint, suppose that the underlying image is known to have certain values in some regions of the plane and that in other regions the image takes on unknown values. Then the constraint operator $\mathbf{C}$ sets an image to zero in the unknown regions, and g is an image which is zero in the unknown regions and equal to the known values elsewhere.

We can state the constrained least-squares minimum norm image reconstruction problem in a form that is exactly analogous to the statement of the unconstrained problem. The only modification needed is that instead of allowing f to be any element in Ω, it is restricted to lie in the affine subspace of images defined by the constraints. Some algebra [9] shows that the minimum norm least-squares solution of $\mathbf{S}f = \underline{f}$ subject to the constraint $\mathbf{C}f = g$ is given by

$$h = f_0 + (\mathbf{SP})^{+}[\underline{f} - \mathbf{S}f_0], \tag{9.3-19}$$

where $f_0 = \mathbf{C}^{+}g$ and $\mathbf{P}$ is the orthogonal projector onto $N(\mathbf{C})$ given by

$$\mathbf{P} = \mathbf{I} - \mathbf{C}^{+}\mathbf{C} = \mathbf{I} - \mathbf{C}^{*}(\mathbf{CC}^{*})^{+}\mathbf{C}. \tag{9.3-20}$$

F. Iterative Image Reconstruction Algorithms

In practice, it may not be possible to implement the closed-form representations for a particular pseudoinverse image reconstruction operator. However, by formulating the image reconstruction problem as a general least-squares problem, we can employ standard techniques to derive iterative image reconstruction algorithms that converge to the pseudoinverse solution. We will give image reconstruction algorithms based on the methods of successive approximations, steepest descent, and conjugate gradients. The algorithms are stated entirely in terms of an operator **S**, which can be any bounded linear operator, and spaces Ω and Φ with inner products $\langle\cdot,\cdot\rangle_\Omega$ and $\langle\cdot,\cdot\rangle_\Phi$.

An important feature of this formulation is that it allows for infinite-dimensional spaces, which means that the problem need not be discretized before it is solved. By choosing Ω and Φ to be infinite-dimensional spaces of continuous images, we obtain exact algorithms for solving a continuous image reconstruction problem. This is in sharp contrast to previous attempts to derive iterative least-squares image reconstruction algorithms in which the problem is first discretized and then solved by matrix techniques. By choosing Ω and Φ to be finite-dimensional spaces, we can of course obtain these matrix algorithms as a special case.

The iterative algorithms are based on standard techniques for which convergence proofs are known; see, for example, [52, p. 379] and also [54]. When the space Ω is finite-dimensional, and also when it is infinite-dimensional and the range of **S** is closed in Φ, the iterative algorithms are guaranteed to converge at a geometric rate. When the range of **S** is not closed, the algorithms are still guaranteed to converge, but the rate of convergence will only be asymptotically linear.

When **S** has a nontrivial null space, the solution toward which the algorithms converge depends on the choice of an initial estimate $\hat{f}_0 \in \Omega$. When $\hat{f}_0 = 0$ the algorithms converge to the minimum norm least-squares solution $h = \mathbf{S}^+\underline{f}$. For arbitrary $\hat{f}_0$, convergence is toward the least-squares solution

$$\hat{f} = \mathbf{S}^+\underline{f} + (\mathbf{I} - \mathbf{P}_\Omega)\hat{f}_0, \tag{9.3-21}$$

where $\mathbf{P}_\Omega$ is the orthogonal projector onto $N(\mathbf{S})^\perp$ and $\mathbf{I} - \mathbf{P}_\Omega$ the orthogonal projector onto $N(\mathbf{S})$. Note that orthogonality is now with respect to the inner product $\langle\cdot,\cdot\rangle_\Omega$.

The iterative algorithms all have computationally efficient implementations. The recursions only require applying the operator **S** and the adjoint operator $\mathbf{S}^*$ and calculating the image and shadow domain norms $\|\cdot\|_\Omega$ and $\|\cdot\|_\Phi$. The adjoint operator $\mathbf{S}^*$ is a "transposed" and possibly weighted version of the operator **S**, and therefore the computational complexity of implementing $\mathbf{S}^*$ is roughly the same as that of implementing **S**. Similar hardware or software might be used to implement both.

The simplest iterative algorithm is based on the method of successive approximations.

Successive Approximations Image Reconstruction Algorithm

Initialize:

$$\hat{f}_0 \in \Omega, \qquad 0 < \beta < 2\|\mathbf{S}\|^{-2}.$$

Iterate: for $i = 0, 1, 2, \ldots$

$$\begin{aligned} d_i &= \mathbf{S}^*[\underline{f} - \mathbf{S}\hat{f}_i], \\ \hat{f}_{i+1} &= \hat{f}_i + \beta d_i. \end{aligned} \tag{9.3-22}$$

Each iteration of the algorithm requires one transformation from image to line integrals by **S**, one transformation from line integrals to an image by the adjoint operator $\mathbf{S}^*$, one linear combination of two images, and one linear combination of two shadows.

One disadvantage of the successive approximations algorithm is that convergence is only guaranteed if β satisfies the condition $0 < \beta < 2\|\mathbf{S}\|^{-2}$, and this requires knowing the norm of **S**. This norm will usually be difficult or impossible to evaluate analytically, though it could be estimated experimentally. For practical purposes, simply choosing a sufficiently small value of β on the basis of experimental performance will usually prove to be acceptable.

Faster convergence can potentially be obtained by recognizing that in the method of successive approximations, the step length β is arbitrarily taken to have the same fixed value at each iteration. In fact, there is an optimum value at each step which is determined by a simple one-dimensional minimization, and we arrive at the following algorithm [49].

Steepest Descent Image Reconstruction Algorithm

Initialize:

$$\hat{f}_0 \in \Omega.$$

Iterate: for $i = 0, 1, 2, \ldots$

$$\begin{aligned} d_i &= \mathbf{S}^*[\underline{f} - \mathbf{S}\hat{f}_i], \\ \hat{f}_{i+1} &= \hat{f}_i + \beta_i d_i \end{aligned}$$

where

$$\beta_i = \|d_i\|_\Omega^2 / \|\mathbf{S}d_i\|_\Phi^2. \tag{9.3-23}$$

The computational requirements of steepest descent compare quite favorably with those of successive approximations. The only additional computation required to compute the optimal step size β_i is one application of $\mathbf{S}$ (line integral calculation) and evaluation of two norms. Additional storage is needed for the direction image d_i and for the line integrals $\mathbf{S}d_i$. The steepest-descent algorithm has the theoretical advantage that convergence is guaranteed without having to know the norm of the operator $\mathbf{S}$. There are no arbitrarily chosen parameters that could cause divergence of the algorithm.

The direction vectors d_i in the method of steepest descent are locally optimal in the sense of the gradient. In addition, a simple calculation shows that two successive directions d_i and d_{i+1} satisfy the orthogonality relation

$$\langle \mathbf{S}d_i, \mathbf{S}d_{i+1}\rangle_\Phi = 0. \tag{9.3-24}$$

Unfortunately, the set of direction vectors $\{d_i\}$ is not characterized by any orthogonality relation. Although the search directions generated in the method of steepest descent are locally optimal, they are not globally optimal.

Improved convergence may be obtained by going to a *conjugate direction method* which generates a set of globally optimal direction vectors $\{p_i\}$ that satisfy the condition

$$\langle \mathbf{S}p_i, \mathbf{S}p_j\rangle_\Phi = 0, \qquad i \neq j. \tag{9.3-25}$$

Direction vectors satisfying Eq. (9.3-25) are said to be *conjugate* with respect to S. For finite-dimensional problems, conjugate direction methods are guaranteed to converge in a finite number of steps.

The *conjugate gradient algorithm* [55,56] (see also [49,57] for a discussion of the operator form of the algorithm) is a particular conjugate direction method which provides simple recursions for calculating the conjugate directions p_i and the optimal step sizes β_i. The recursions can be put in many forms, one of which is the following.

Conjugate Gradient Image Reconstruction Algorithm

Initialize:

$$\hat{f}_0 \in \Omega, \qquad r_0 = p_0 = \mathbf{S}^*[\underline{f} - \mathbf{S}\hat{f}_0].$$

Iterate: for $i = 0, 1, 2, \ldots$

$$\hat{f}_{i+1} = \hat{f}_i + \beta_i p_i,$$

$$r_{i+1} = r_i - \beta_i \mathbf{S}^*\mathbf{S}p_i = \mathbf{S}^*[\underline{f} - \mathbf{S}\hat{f}_{i+1}],$$

$$p_{i+1} = r_{i+1} + \alpha_i p_i$$

where

$$\beta_i = \frac{\|r_i\|^2_\Omega}{\|\mathbf{S}p_i\|^2_\Phi}, \qquad \alpha_i = \frac{\|r_{i+1}\|^2_\Omega}{\|r_i\|^2_\Omega}. \tag{9.3-26}$$

There is a striking similarity between the recursions for the conjugate gradient algorithm Eq. (9.3-26), and those for the steepest descent algorithm Eq. (9.3-23). Both algorithms use directions that depend on the local gradient and step sizes that have the form of energy ratios. The essential difference is in the choice of the direction to move in at each step: the steepest descent algorithm moves along a line in the direction of the local gradient; the conjugate gradient algorithm moves along a line that is determined by a linear combination of the local gradient and the previous search direction. Thus, the conjugate gradient search direction is a linear combination of the local gradient and all previous gradients.

The computational requirements of the conjugate gradient and steepest descent algorithms are also surprisingly similar. In both algorithms, two norms must be evaluated at each iteration. The conjugate gradient algorithm requires additional storage for one image and additional computation to form a linear combination of two images. In exchange for this modest increase in storage and computation, the conjugate gradient algorithm generates a globally optimal set of directions.

All three iterative algorithms given above can be used to solve the constrained least-squares problem in which *a priori* information about the image is represented by an affine subspace constraint. The reconstructed image is required to lie in the affine subspace of images f that satisfy the constraint equation $\mathbf{C}f = g$, where $\mathbf{C}$ is a bounded linear operator with closed range and g is in the range of $\mathbf{C}$. From Eq. (9.3-19), the constrained least-squares solution is given by

$$h = f_0 + (\mathbf{SP})^+[\underline{f} - \mathbf{S}f_0], \tag{9.3-27}$$

where $f_0 = \mathbf{C}^+ g$ is the minimum norm element of the constraint space and $\mathbf{P}$ the orthogonal projector onto $N(\mathbf{C})$. To obtain iterative algorithms that converge to the constrained least-squares solution, the following substitutions are made in Eqs. (9.3-22), (9.3-23), and (9.3-26): $\underline{f}$ is replaced by $\underline{f} - \mathbf{S}f_0$; $\mathbf{S}$ is replaced by $\mathbf{SP}$; and $\mathbf{S}^*$ is replaced by $\mathbf{PS}^*$.

9.4 APPLICATIONS OF THE INNER PRODUCT FRAMEWORK

In the previous section we formulated and solved a generalized least-squares image reconstruction problem. By casting the analysis in terms of a linear measurement operator and its adjoint, we obtained results that apply to any choice of image and shadow domain inner products. This section presents some applications of the general theory. We show that the least-squares framework unifies certain known image reconstruction algorithms by showing them to be optimal under different norms. In particular, we show that convolution-backprojection is optimal under the rho norm. This is an important dividend of the inner product framework. Although convolution-backprojection is the standard CT reconstruction algorithm, it has never been understood as reconstructing an image that is optimal in a least-squares sense.

Several additional applications of the inner product framework can be found in [9]. These include a method for image classification from line integrals and a new analysis of the conditioning of the image reconstruction problem.

A. Convolution-Backprojection: The Pseudoinverse under the Rho Inner Product

Convolution-backprojection was derived as an algorithm for recovering a continuous function from a complete set of exact line integrals. In practice, measured line integral values will never be exact, and in fact large errors are often caused by motion or by the presence of metal objects. In such cases, there may be no image that has line integrals that are equal to the measured values. Nonetheless, convolution-backprojection will reconstruct some image from the erroneous data. Unfortunately, the theory on which convolution-backprojection is based is not sufficient to tell us what this reconstructed image represents or how its line integrals compare to the erroneous measured values.

Based on the theory of the previous section, we will now prove that from any set of line integral measurements, convolution-backprojection generates a pseudoinverse reconstruction with respect to the rho norm. The rho norm of the difference between the line integrals of this image and the measured data is minimized. Convolution-backprojection is therefore shown to be an optimal algorithm in a generalized least-squares sense.

To obtain these results, we set up the problem as follows. Let Ω be the space of square-integrable images which may be nonzero anywhere in the plane. We take the inner products to be the direct inner product in the image domain and the rho inner product in the shadow domain. Let M be some finite region in the plane, and let $\mathbf{M}$ be the masking operator which sets an image to zero outside M. The region M is chosen so that it includes all areas where the original object is nonzero.

The least-squares problem is then stated as follows. Given any measured shadow $\underline{f}$, find the image confined to the region M having a shadow that is closest to $\underline{f}$ in rho norm. Let this image be denoted by $\hat{f}$. Thus we have

$$\hat{f} = (\mathbf{SM})^{+}\underline{f}. \tag{9.4-1}$$

We will now determine an explicit representation for the pseudoinverse operator $(\mathbf{SM})^{+}$. From the general representation $\mathbf{S}^{+} = (\mathbf{S}^{*}\mathbf{S})^{+}\mathbf{S}^{*}$ with $\mathbf{S}$ replaced by $\mathbf{SM}$, we have

$$(\mathbf{SM})^{+} = [(\mathbf{SM})^{*}(\mathbf{SM})]^{+}(\mathbf{SM})^{*}. \tag{9.4-2}$$

To simplify this expression, let us determine the form of the adjoint operator $(\mathbf{SM})^{*}$. A direct calculation shows that the adjoint of $\mathbf{S}$ under the rho inner product is $\mathbf{B}\boldsymbol{\rho}$. Thus, for all $f \in \Omega$ and $\underline{g} \in \Phi$ we have

$$\langle \mathbf{SM}f, \underline{g}\rangle_{\rho} = \langle \mathbf{M}f, \mathbf{B}\boldsymbol{\rho}\underline{g}\rangle = \langle f, \mathbf{MB}\boldsymbol{\rho}\underline{g}\rangle, \tag{9.4-3}$$

and therefore the adjoint of $\mathbf{SM}$ is given by

$$(\mathbf{SM})^{*} = \mathbf{MB}\boldsymbol{\rho}. \tag{9.4-4}$$

Also, we recall that $\mathbf{B}\boldsymbol{\rho}$ is the operator form of the convolution-backprojection algorithm, and therefore $\mathbf{B}\boldsymbol{\rho}\mathbf{S} = \mathbf{I}$. The pseudoinverse of $\mathbf{SM}$ can now be written as

$$\begin{aligned}(\mathbf{SM})^{+} &= [\mathbf{MB}\boldsymbol{\rho}\mathbf{SM}]^{+}\mathbf{MB}\boldsymbol{\rho} \\ &= \mathbf{MB}\boldsymbol{\rho}.\end{aligned} \tag{9.4-5}$$

The pseudoinverse reconstruction operator Eq. (9.4-5) is equal to convolution-backprojection, followed by setting the reconstructed image to zero outside the region M. We reach the important conclusion that the image reconstructed by convolution-backprojection in any finite region of the plane, from any shadow, is the pseudoinverse reconstruction under the rho norm.

Based on the general theory of Section 9.3, we can precisely characterize the reconstructed image $\hat{f}$ and its shadow. Let the *shadow domain error* $\underline{\varepsilon}$ be defined by

$$\underline{\varepsilon} \triangleq \underline{f} - \mathbf{S}\hat{f}. \tag{9.4-6}$$

Then we have the following results:

1. The reconstructed image $\hat{f}$ is the unique image confined to the region M that minimizes the cost function

$$J(f) = \|\underline{f} - \mathbf{SM}f\|_{\rho}^{2}. \tag{9.4-7}$$

2. The shadow of the reconstructed image $\hat{f}$ is the rho-orthogonal projection of $\underline{f}$ onto the range of $\mathbf{SM}$.

3. The shadow domain error $\underline{\varepsilon}$ is rho orthogonal to the range of **SM**. In particular, it is rho orthogonal to the shadow $\mathbf{S}\hat{f}$.

4. The shadow domain error $\underline{\varepsilon}$ is in the null space of the adjoint operator $\mathbf{MB}\rho$.

B. Pseudoinverse Image Reconstruction under Direct Inner Products

With a direct inner product in both the image and the shadow domains, the image reconstruction problem becomes a standard unweighted least-squares problem. We will give an expression for the pseudoinverse reconstruction operator and then determine properties of the reconstructed image from the general properties of the pseudoinverse. It turns out that the pseudoinverse operator under direct inner products is not suitable for direct implementation; instead, we use the results of Section 9.3 to obtain iterative algorithms that converge to the pseudoinverse solution. The discrete form of the algorithm based on successive approximations is identical to the so-called simultaneous iterative reconstruction technique (SIRT) [58].

We take the space of images Ω to be those square-integrable images which may be nonzero anywhere in the plane. Let M be a finite region in the plane that is chosen to include all areas where the original object is nonzero, and let **M** be the masking operator which sets an image to zero outside M. The inner products will now be taken to be the direct inner product in both the image and shadow domains.

Given any shadow $\underline{f}$, let the image confined to the region M with shadow closest to $\underline{f}$ in the sum of squares norm be denoted by $\hat{f}$. This image is written in terms of the pseudoinverse of the operator **SM** as

$$\hat{f} = (\mathbf{SM})^{+}\underline{f}. \tag{9.4-8}$$

Let us now determine an explicit representation for the pseudoinverse operator $(\mathbf{SM})^{+}$. From the general representation $\mathbf{S}^{+} = (\mathbf{S}^{*}\mathbf{S})^{+}\mathbf{S}^{*}$ with **S** replaced by **SM**, we have

$$(\mathbf{SM})^{+} = [(\mathbf{SM})^{*}(\mathbf{SM})]^{+}(\mathbf{SM})^{*}. \tag{9.4-9}$$

To simplify this expression we must determine the form of the adjoint operator $(\mathbf{SM})^{*}$. With direct inner products for images and shadows, the masking operator **M** is self-adjoint, and the adjoint of the shadow transform operator is easily shown to be the backprojection operator **B**. Thus, for all $f \in \Omega$ and all $\underline{g} \in \Phi$ we have

$$\langle \mathbf{SM}f, \underline{g} \rangle = \langle \mathbf{M}f, \mathbf{B}\underline{g} \rangle = \langle f, \mathbf{MB}\underline{g} \rangle, \tag{9.4-10}$$

and therefore the adjoint of **SM** is given by

$$(\mathbf{SM})^* = \mathbf{MB}. \tag{9.4-11}$$

The pseudoinverse of **SM** can now be written as

$$(\mathbf{SM})^+ = (\mathbf{MBSM})^+\mathbf{MB}. \tag{9.4-12}$$

It is not possible to further simplify Eq. (9.4-12) because of the nature of the operator **MBSM**. Consider what happens when this operator acts on an image that is confined to the region *M*. Clearly the first part of this operator—the rightmost **M**—has no effect. The operator **BS** has the effect of convolving the image with the circularly symmetric point spread function $|r|^{-1}$ [4]. In this case, it filters an image that is confined to the region *M* and generates an image that is infinite in extent. The second **M** then truncates the image. This truncation makes it impossible to obtain a simple expression for the pseudoinverse of **MBSM** which appears in Eq. (9.4-12).

Since it is not possible to obtain a simple closed-form representation of the pseudoinverse operator $(\mathbf{SM})^+$, we will instead give iterative algorithms that converge to the pseudoinverse solution. We recall that the iterative algorithms of the previous section were given in terms of a general line integral operator **S** and its adjoint **S*** and the norms $\|\cdot\|_\Omega$ and $\|\cdot\|_\Phi$. In the problem that we are now concerned with, we replace **S** by **SM** and **S*** by **MB** and take the norms to be those defined by the direct inner products. With these identifications, we can give algorithms based on the methods of successive approximations, steepest descent or conjugate gradients. We will only state the successive approximations algorithm here.

Successive Approximations Image Reconstruction Algorithm

Initialize:

$$\hat{f}_0 \in \Omega, \qquad 0 < \beta < 2\|\mathbf{SM}\|^{-2}.$$

Iterate: for $i = 0, 1, 2, \ldots$

$$d_i = \mathbf{MB}[\underline{f} - \mathbf{SM}\hat{f}_i],$$
$$\hat{f}_{i+1} = \hat{f}_i + \beta d_i. \tag{9.4-13}$$

As before, we can use the general properties of the pseudoinverse to precisely characterize the solution $\hat{f} = (\mathbf{SM})^+\underline{f}$. With the *shadow domain error* $\underline{\varepsilon}$ defined by

$$\underline{\varepsilon} \triangleq \underline{f} - \mathbf{S}\hat{f} \tag{9.4-14}$$

we have the following results:

1. The reconstructed image $\hat{f} = (\mathbf{SM})^{+}\underline{f}$ is the unique image confined to the region M that minimizes the cost function

$$J(f) = \|\underline{f} - \mathbf{SM}f\|^2. \tag{9.4-15}$$

2. The shadow of the reconstructed image $\hat{f}$ is the orthogonal projection of $\underline{f}$ onto the range of **SM**, where orthogonality is now with respect to the usual sum of squares inner product.

3. The shadow domain error $\underline{\varepsilon}$ is orthogonal to the range of **SM**. In particular, it is orthogonal to the shadow $\mathbf{S}\hat{f}$.

4. The shadow domain error $\underline{\varepsilon}$ is in the null space of the adjoint operator **MB**.

If we approximate the shadow transform operator by a matrix $\hat{\mathbf{S}}$ and approximate the backprojection operator by the transpose of $\hat{\mathbf{S}}$, then Eq. (9.4-13) is identical to the SIRT algorithm [58]. We have thus derived both the SIRT algorithm and the convolution-backprojection algorithm from the same least-squares framework. SIRT converges to the least-squares solution under the sum of squares norm, while convolution-backprojection directly generates the least-squares solution under the rho norm.

In previous work on image reconstruction, algorithms have been divided into two general categories: transform-type algorithms (such as convolution-backprojection) and iterative algorithms (such as SIRT). These two categories have been viewed as two completely different approaches to image reconstruction. Our results demonstrate that both types of algorithms can be derived from the same least-squares framework simply by making an appropriate choice of inner products.

C. Image Reconstruction by Orthogonal Functions

In Section 9.3 we proved that the direct inner product of two images is equal to the rho inner product of their shadows. We will now use this theorem to obtain a new orthogonal function image reconstruction algorithm. The algorithm is valid for any image domain basis; it is therefore considerably more powerful than previously known orthogonal function image reconstruction algorithms, all of which are tied to particular sets of basis functions.

Our method is based on representing an image as a sum of orthogonal basis images, where we allow the basis images to be orthogonal with respect to a weighted inner product. The weighted inner product $\langle f, g\rangle_{\mathrm{W}}$ between two images f and g is defined in terms of the sum of squares inner product

$\langle \cdot, \cdot \rangle$ and a positive definite self-adjoint weighting operator $\mathbf{W}$ by

$$\langle f, g \rangle_{\mathbf{W}} \triangleq \langle f, \mathbf{W} g \rangle = \langle \mathbf{W} f, g \rangle. \tag{9.4-16}$$

Let $\{\phi_i(x, y)\}_{i=0}^{\infty}$ be a complete set of images that are orthogonal with respect to the weighted inner product $\langle \cdot, \cdot \rangle_{\mathbf{W}}$. We will refer to the functions $\mathbf{W}\phi_i(x, y)$ which are obtained by operating on the basis functions ϕ_i with the weighting operator $\mathbf{W}$ as *weighted basis functions.*

An image $f(x, y)$ has the series representation

$$f(x, y) = \sum_{i=0}^{\infty} a_i \phi_i(x, y),$$

where

$$a_i = \frac{1}{\langle \phi_i, \phi_i \rangle_{\mathbf{W}}} \langle f, \phi_i \rangle_{\mathbf{W}}. \tag{9.4-17}$$

We will obtain an expression for the coefficient a_i in terms of the shadow of f. To this end, we write the weighted inner product $\langle f, \phi_i \rangle_{\mathbf{W}}$ first as a direct inner product in the image domain and then as a rho inner product in the shadow domain. We have

$$\begin{aligned} a_i &= \frac{1}{\langle \phi_i, \phi_i \rangle_{\mathbf{W}}} \langle f, \phi_i \rangle_{\mathbf{W}} \\ &= (\|\phi_i\|_{\mathbf{W}})^{-2} \langle \mathbf{S} f, \mathbf{S} \mathbf{W} \phi_i \rangle_{\rho}. \end{aligned} \tag{9.4-18}$$

According to Eq. (9.4-18), the original image f is reconstructed from its shadow $\underline{f} = \mathbf{S} f$ by

$$f(x, y) = \sum_{i=0}^{\infty} (\|\phi_i\|_{\mathbf{W}})^{-2} \langle \underline{f}, \boldsymbol{\rho} \mathbf{S} \mathbf{W} \phi_i \rangle \phi_i(x, y). \tag{9.4-19}$$

This general scheme has many practical applications. For example, suppose the underlying image is known to be a linear combination of a small number of basis images. Rather than reconstruct an entire image, we can precompute the shadow of each weighted basis function, rho filter, and normalize. Then the coefficient of each basis function in the image can be determined by a simple direct shadow domain inner product. In a real-time application, this would result in substantial computational savings. This procedure will be especially valuable when there are many sets of line integrals that correspond to different linear combinations of the same small set of basis functions.

9.5 IMAGE RECONSTRUCTION FROM INCOMPLETE DATA

Many practical limited-data image reconstruction problems fall into the category of incomplete data problems. In an incomplete data problem, the full set of line integrals is divided into two subsets: a subset of measured line integrals, and a complementary subset of missing line integrals. The measured line integrals are known with great accuracy, while the missing line integrals are entirely unknown. This section applies the inner product framework of Section 9.3 to the problem of image reconstruction from incomplete data.

A. Shadow Transform and Reconstruction Operators for Incomplete Shadows

In an incomplete data image reconstruction problem, the shadow $\underline{f}$ is partitioned into two subsets: the measured line integrals $\underline{f}_y$ and the missing line integrals $\underline{f}_x$. This will be represented as a vector partitioned into two components, $\underline{f} = \binom{\underline{f}_x}{\underline{f}_y}$. We define the *partial shadow transform operators* $\mathbf{S}^y$ and $\mathbf{S}^x$ to be operators that map an image into the measured and missing parts of the shadow, so that if $\underline{f} = \mathbf{S}f$ then $\underline{f}_x = \mathbf{S}^x f$ and $\underline{f}_y = \mathbf{S}^y f$. We also define the *full data reconstruction operator* $\mathbf{R}$ as the left inverse of $\mathbf{S}$, so that $\mathbf{RS} = \mathbf{I}$. The operator $\mathbf{R}$ can be represented as convolution-backprojection, $\mathbf{R} = \mathbf{B}\boldsymbol{\rho}$.

We define the *partial reconstruction operators* $\mathbf{R}^x$ and $\mathbf{R}^y$ by

$$\mathbf{R}^x\underline{f}_x \triangleq \mathbf{R}\begin{pmatrix}\underline{f}_x\\ 0\end{pmatrix} \quad \text{and} \quad \mathbf{R}^y\underline{f}_y \triangleq \mathbf{R}\begin{pmatrix}0\\ \underline{f}_y\end{pmatrix}. \tag{9.5-1}$$

The partial reconstruction operators themselves are represented in vector form as

$$\mathbf{R} = [\mathbf{R}^x \quad \mathbf{R}^y]. \tag{9.5-2}$$

An image is therefore related to the two components of its shadow by

$$f = \mathbf{R}\underline{f} = [\mathbf{R}^x \quad \mathbf{R}^y]\begin{pmatrix}\underline{f}_x\\ \underline{f}_y\end{pmatrix} = \mathbf{R}^x\underline{f}_x + \mathbf{R}^y\underline{f}_y. \tag{9.5-3}$$

The partial shadow transform and reconstruction operators can be efficiently computed by any existing implementation of the operators $\mathbf{S}$ and $\mathbf{R}$. For example, the image $\mathbf{R}^y\underline{f}_y$ is computed by applying the reconstruction operator $\mathbf{R}$ to a shadow in which the unknown region is set to zero and the known region is set to the values determined by the measured line integrals $\underline{f}_y$.

B. Effect of Missing Information in Incomplete Shadows: Ghost Images

Let us now consider the fundamental question of uniqueness in image reconstruction from incomplete data. We will show that the known theoretical conditions under which a finite-extent image is uniquely determined by an incomplete shadow are not sufficient to determine a unique image in practice. There are finite-extent images, which we call *ghost images,* that are invisible to an incomplete set of line integrals. We demonstrate the existence of ghost images by constructing examples for several incomplete shadow problems of practical interest.

Given a partial shadow transform operator $\mathbf{S}^y$, we define a *ghost image* to be a nonzero image g such that $\mathbf{S}^y g = 0$ and $\mathbf{S}^x g \neq 0$. Thus, a ghost image g is an image that is in the null space of the partial shadow transform operator $\mathbf{S}^y$. The line integrals of a ghost image are zero along paths where data are measured; the line integrals are nonzero only along the paths where data are missing. Since line integrals through a nonzero image can only be zero if the image takes on both positive and negative values, ghost images must be bipolar. Figure 9.5-1 shows ghost images for line integral measurements in one and two directions.

Because ghost images are bipolar, it might appear that they can be eliminated by requiring the reconstructed image to be positive—but this is not the case. A ghost image by itself does not represent a candidate reconstructed image; a ghost represents the difference between images that are indistinguishable from an incomplete set of line integrals. The indistin-

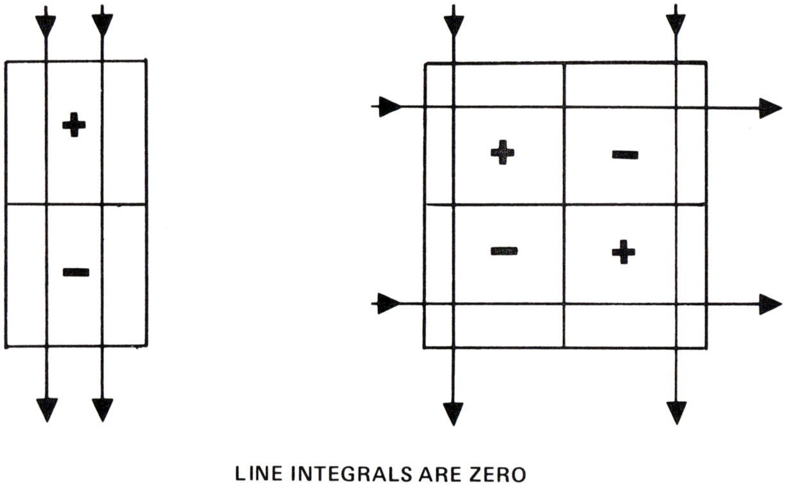

Fig. 9.5-1. Ghost images for line integral measurements in one direction (left) and two directions (right).

guishable images can be positive. Their difference, however, will be a ghost image having both positive and negative values.

The set of all ghost images is given by $N(\mathbf{S}^y)$, the null space of $\mathbf{S}^y$. From our linear operator point of view, the space of images Ω can be represented as a direct sum of a subspace of ghost images and its orthogonal complement,

$$\Omega = N(\mathbf{S}^y) \oplus N(\mathbf{S}^y)^{\perp}. \tag{9.5-4}$$

The incomplete set of measured line integrals gives no information about components of the original image that are in the null space $N(\mathbf{S}^y)$. Note that the structure of the null space $N(\mathbf{S}^y)$ is a property of the measurement operator $\mathbf{S}^y$ that is independent of the choice of inner products. The null space is a physical characteristic of the measurement process. The orthogonal complement $N(\mathbf{S}^y)^{\perp}$, on the other hand, is a function of our mathematical model that depends on $N(\mathbf{S}^y)$ and on the choice of image domain inner product.

It is important to realize that, in general, one cannot obtain an artifact-free reconstruction just by requiring the reconstructed image to lie in $N(\mathbf{S}^y)^{\perp}$. Although an image in $N(\mathbf{S}^y)^{\perp}$ has no component in the null space $N(\mathbf{S}^y)$, and therefore in a mathematical sense it can be said to have no ghost component, it will not be a correct reconstruction unless the original image itself has no component in $N(\mathbf{S}^y)$. In general, the original image will have a component in $N(\mathbf{S}^y)$. If the reconstructed image is forced to lie in $N(\mathbf{S}^y)^{\perp}$, the missing null space component (a ghost) will manifest itself in the form of streak artifact.

In theory, there are incomplete data problems—the bagel problem and the limited angle problem—in which the space of images Ω contains no ghost images. In the limited angle problem, if Ω contains only images that are finite in extent, then a unique image is determined by line integral projections in any continuous range of angles [19,59]. In the bagel problem, if Ω contains only images that are confined to a bagel-shaped region, then the set of line integrals along paths that pass through the bagel without passing through the hole determines a unique image [18,19].

In practice, algorithms that attempt to determine the theoretically guaranteed unique reconstruction fail to produce acceptable images. Mathematically, the difficulty has been traced to the poor conditioning of the incomplete data reconstruction problem [42,60–63]. The poor conditioning means that there are images in the restricted space Ω having line integrals that take on only very small values along the paths in the incomplete set of measured line integrals; the total energy in the incomplete set of measured line integrals is small compared to the energy in the image. Although such an image is not exactly in the null space of the partial shadow transform operator $\mathbf{S}^y$, it is so "close" to the null space that it will be invisible in practice. Therefore, we are justified in referring to such an image as a ghost

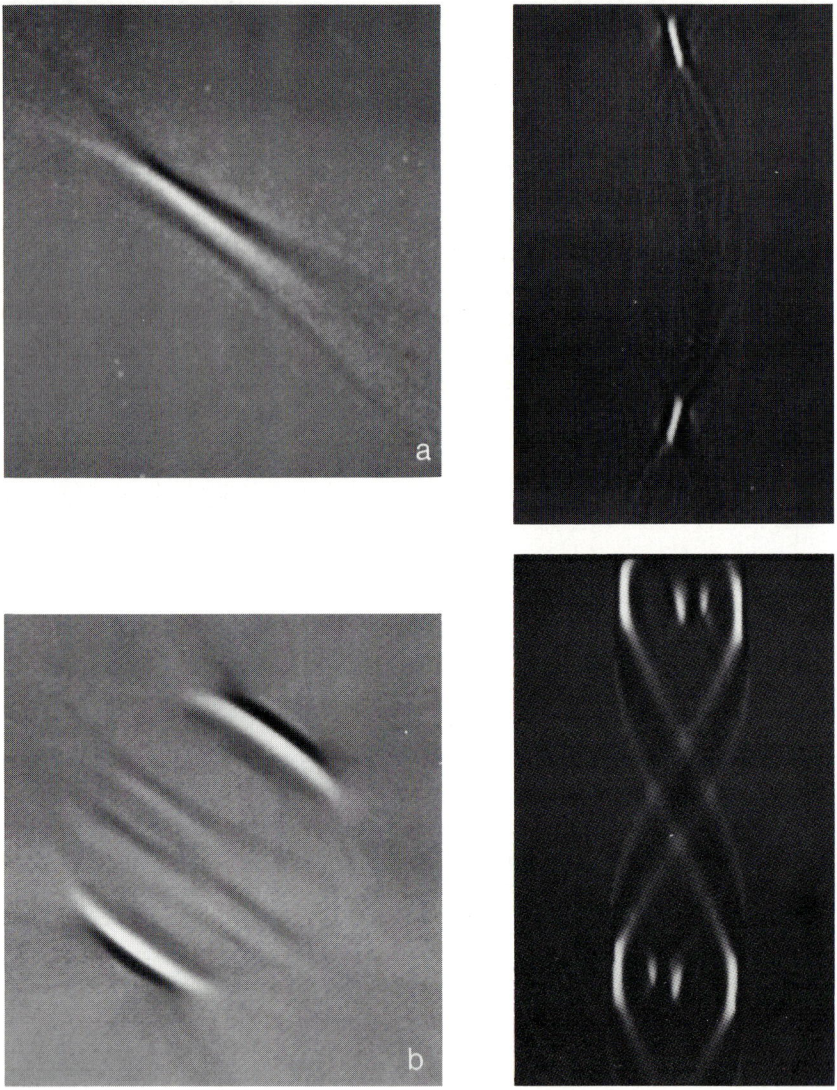

Fig. 9.5-2. Ghost images (left) and their shadows (right). (a) Ghost caused by motion. (b) Ghost caused by limited measurement angle. (c) Ghost caused by an x-ray opaque structure.

image (one might be more precise and refer to it as an approximate ghost, but for our purposes this distinction is not necessary).

Figure 9.5-2 shows three examples of ghost images. These ghost images correspond to three common types of limited-data problem: motion, limited

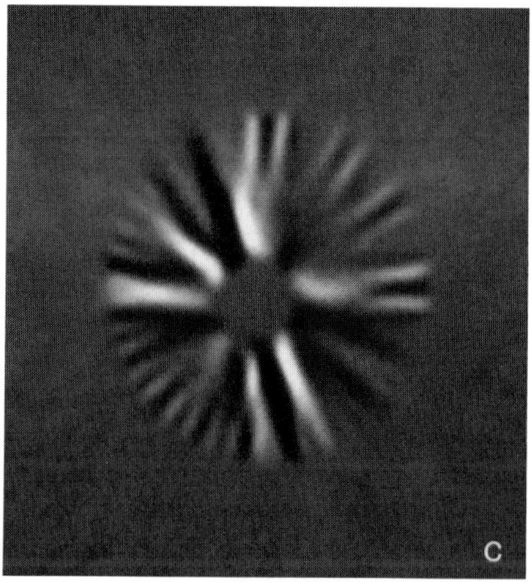

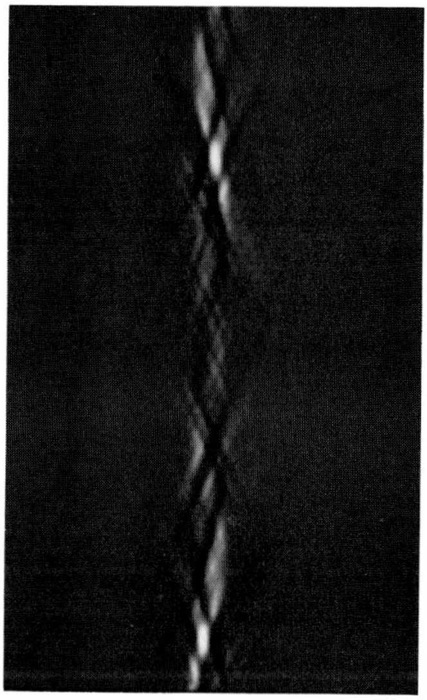

Fig. 9.5-2 (*continued*)

scanning angle, and an x-ray opaque structure in the field of view. In each example we show the ghost image and a sinogram representation of the line integrals of the image. In the sinogram, a line integral path is specified by the source location (vertical position in the sinogram) and the location of the line within the detector array (horizontal position in the sinogram); the intensity represents the line integral value, with white areas representing large positive amplitudes. The images and line integrals were calculated according to the fan-beam geometry of a Varian CT body scanner described in Section 9.1. The images represent a region that is 11 cm in diameter and spans approximately 76 detectors in each of 360 fan-beam projections spaced at 1° intervals. Black regions of an image represent negative values, and white regions represent positive values.

The essential characteristic of a ghost image is that the nonzero line integrals of significant amplitude are found only in a small region of the shadow which is not determined by the available incomplete set of line integrals. The line integrals of the ghost image in the known region of the shadow are smaller than the threshold of accuracy that is available in practice. From another point of view, any image with the characteristic that its line integrals are "bunched" in a small part of the shadow domain will be a ghost image if that part of the shadow happens to be unknown.

In spite of the importance of ghosts in practical incomplete data image reconstruction problems, surprisingly little attention has been given to determining their structure. The earliest results appear to be those of Peres [42], who constructed a ghost for the particular case of a parallel-beam geometry with complete projections missing. Examples of ghosts for arbitrary geometries with arbitrarily located missing line integrals and methods for reducing ghost artifacts are given in [9–13]. Methods for reducing ghost artifacts are also discussed in [40,41]. Several authors have examined the nonuniqueness that occurs whenever the number of projections is finite [35,64–66]. This latter work, however, is focused on determining the resolution limits and uncertainties in the full-data problem; the incomplete-data problem is not explicitly addressed.

In summary, ghost images represent a fundamental uncertainty in reconstruction from an incomplete shadow. The crucial step in obtaining acceptable reconstructed images from an incomplete shadow is to resolve this uncertainty in a way that is consistent with additional *a priori* information about the original physical object.

C. *Incorporating Additional a Priori Information*

We will represent *a priori* information in the form of constraints that must be satisfied by the reconstructed image. In this approach, constraints are implemented by constraint operators which are defined interactively by a human observer. We define $\mathbf{C}^{\mathrm{I}}$ and $\mathbf{C}^{\mathrm{S}}$ to be operators that enforce constraints on an image and shadow, respectively. For example, the constraint that an image must be nonnegative is enforced by an operator which takes as input an arbitrary image and generates as output an image with no negative values.

There is no unique way of enforcing a constraint. The nonnegativity constraint, for example, might be enforced by setting negative values to zero, or to their absolute value, or to some other positive value such as an average of surrounding positive values. Some useful constraints and possible implementations are the following.

1. Nonnegativity: set negative values to zero.
2. Known boundary: set image to zero outside the boundary.
3. Known region: set image to the known value in the specified region.
4. Known range of values: set any point above the range to the maximum value and any point below the range to the minimum value.
5. Known average value in a region: change each value in the region by the same constant so that the resulting average value satisfies the constraint.

Constraint operators can be specified interactively to take maximum advantage of a radiologist's *a priori* knowledge of anatomical structure. The procedure for specifying the constraint operators is as follows. A preliminary reconstruction using convolution-backprojection generates an image which (in spite of some streak artifact) is a recognizable image of a cross section which determines the physical structure (e.g., bone, soft tissue) represented by most pixels. From this reconstruction, a radiologist determines the outer boundary of the object and identifies parts of the object having known values. This *a priori* information is input with a cursor and defines the constraint operator.

D. Experimental Results

We have implemented an iterative algorithm for image reconstruction from incomplete data on a minicomputer that is part of a Varian fan-beam CT scanner. Here we present experimental results that demonstrate the performance of the algorithm in two different incomplete-data problems. Both are problems to which existing techniques have not been successfully applied. In both cases, our algorithm reconstructs images that show reduced streak artifact compared with images generated by conventional methods. Our second example shows how additional *a priori* information can be incorporated to improve the quality of the reconstructed image. All of the experimental results were obtained from measured x-ray data. In this respect the results are distinguished from previous studies, virtually all of which have been conducted only with synthetic computer-generated data.

In [9] we develop two classes of iterative algorithms: (1) iterative algorithms for direct reconstruction from an incomplete set of measured data and (2) iterative algorithms for estimating missing line integrals. At each iteration, the former require calculation of line integrals in the measured data region of the shadow, while the latter require calculation of line integrals in the missing-data region. Here we will be concerned with incomplete-data problems in which the missing-data region is significantly smaller than the measured-data region. For these problems, algorithms for estimating missing line integrals require fewer line integral calculations, and therefore we will focus on this class of algorithms.

The algorithm that we have implemented is based on the method of successive approximations. Let $\mathbf{C}^{\mathrm{I}}$ be an image domain constraint operator that determines an affine subspace of images by the equation $\mathbf{C}^{\mathrm{I}}f = g$. We have

$$\mathbf{C}^{\mathrm{I}}f = f_0 + \mathbf{P}f, \tag{9.5-5}$$

where f_0 is the minimum norm element of the constraint space and $\mathbf{P}$ the orthogonal projector onto $N(\mathbf{C}^{\mathrm{I}})$, the linear subspace parallel to the constraint space. In addition, we allow for constraints on the estimates of the missing line integrals which are enforced by the operator $\mathbf{C}^x$. Let Ψ denote the set of estimates that satisfy this constraint. For notational convenience, we denote the estimate of the missing line integrals at step i by $\hat{x}_i$. The successive approximations image reconstruction algorithm now takes the following form.

Estimation of Missing Line Integrals by Successive Approximations

Initialize:

$$\hat{x}_0 \in \Psi, \qquad 0 < \beta < 2\|\mathbf{I} - \mathbf{S}^x\mathbf{PR}^x\|.$$

Iterate: for $i = 0, 1, 2, \ldots$

$$\hat{x}_{i+1} = (1-\beta)\hat{x}_i + \beta\mathbf{C}^x\mathbf{S}^x\mathbf{C}^\mathrm{I}[\mathbf{R}^x\hat{x}_i + \mathbf{R}^y \underline{f}_y]. \tag{9.5-6}$$

Figure 9.5-3 shows the algorithm in schematic form.

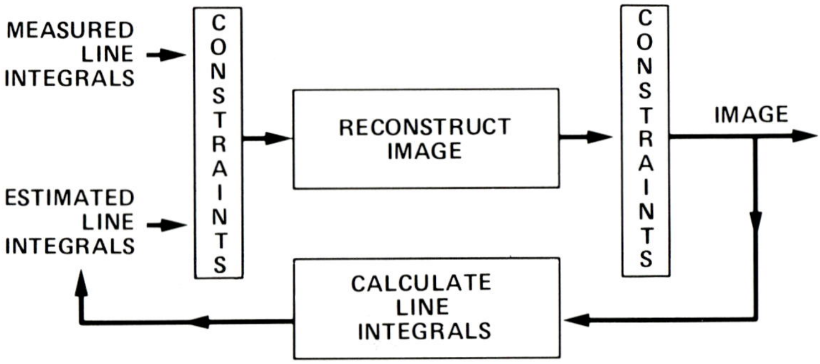

Fig. 9.5-3. Schematic diagram of the iterative algorithm for image reconstruction from incomplete data. (From Medoff *et al.* [13].)

In the first experiment, we apply the iterative algorithm to reconstruct an image when the object of interest contains an x-ray opaque structure that blocks x-rays. This is the bagel problem which was described in Section 9.1. In each projection, there will be a set of measured line integrals corresponding to rays that pass through the object exterior to the opaque region, and a set of missing line integrals corresponding to rays that intersect the opaque region. The bagel problem is to reconstruct an image in the bagel-shaped area exterior to the x-ray opaque region (a "hole" in the bagel), using only line integrals along ray paths that do not pass through the hole.

Note that in the bagel problem the measured line integrals contain no information about the hole. Therefore, we can assign an arbitrary attenuation value to the hole and then attempt to estimate the line integral values that would have been measured through an object having the actual attenuation values in the bagel and this assigned value in the hole. We assume that the outer boundary of the bagel is known. The constraints are that the reconstructed image be contained within this outer boundary and that it agree with the assigned value in the hole. There is also the obvious nonnegativity constraint on both the image and the line integrals. The complete algorithm is shown in Fig. 9.5-4. The algorithm repeats until a steady state is reached. This algorithm has been implemented on a third-generation Varian CT body scanner.

Figure 9.5-5 shows a full data reconstruction of a slice through an object that contains within it three smaller rods. Figure 9.5-6 illustrates the type of streak artifacts generated by convolution-backprojection when the central

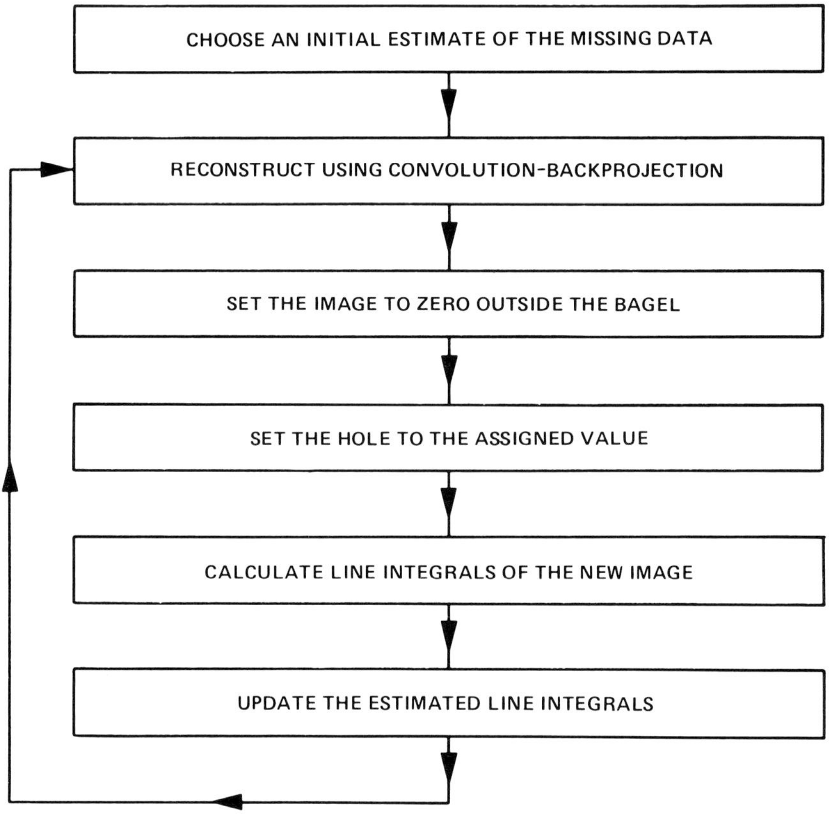

Fig. 9.5-4. The bagel algorithm. The algorithm repeats until a steady state is reached. (From Medoff *et al.* [13].)

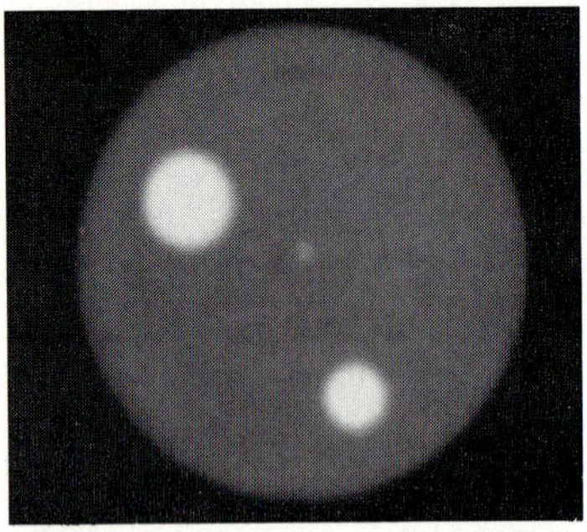

Fig. 9.5-5. Full data reconstruction from measured x-ray data showing a slice through a test object that contains within it three smaller rods.

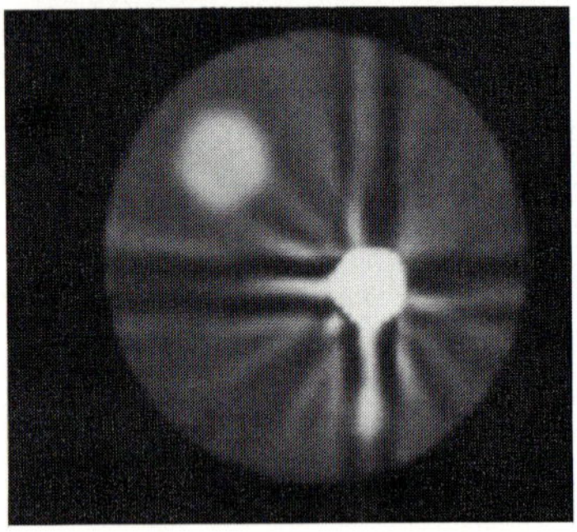

Fig. 9.5-6. Reconstruction by convolution-backprojection illustrating the type of streak artifacts generated when the central region contains an x-ray opaque structure.

region of the object contains a nearly x-ray opaque structure. Our goal is to produce an improved reconstruction in the bagel exterior to the opaque hole. In this example, approximately 20% of the nonzero line integrals in the full set of data pass through the hole. These line integrals must be estimated by the iterative algorithm. We should emphasize that this fraction of missing line integrals (20%) is an order of magnitude larger than the percentage that would be missing because of a small metal pin. This example, therefore, is a severe test of the method.

The algorithm begins by setting all the missing line integrals to an initial value based on an average of nearby measured line integrals. The full data set is then reconstructed by convolution-backprojection to produce the image shown in Fig. 9.5-7a. There is severe streak artifact, as expected, since the initial line integral estimates are not correct. This image is set to zero outside the outer boundary of the bagel. The hole is set to the assigned value, which in this case was chosen to be approximately equal to the background value in the object. The resulting image is shown in Fig. 9.5-7b. Line integrals of this image are calculated in the missing data region, and

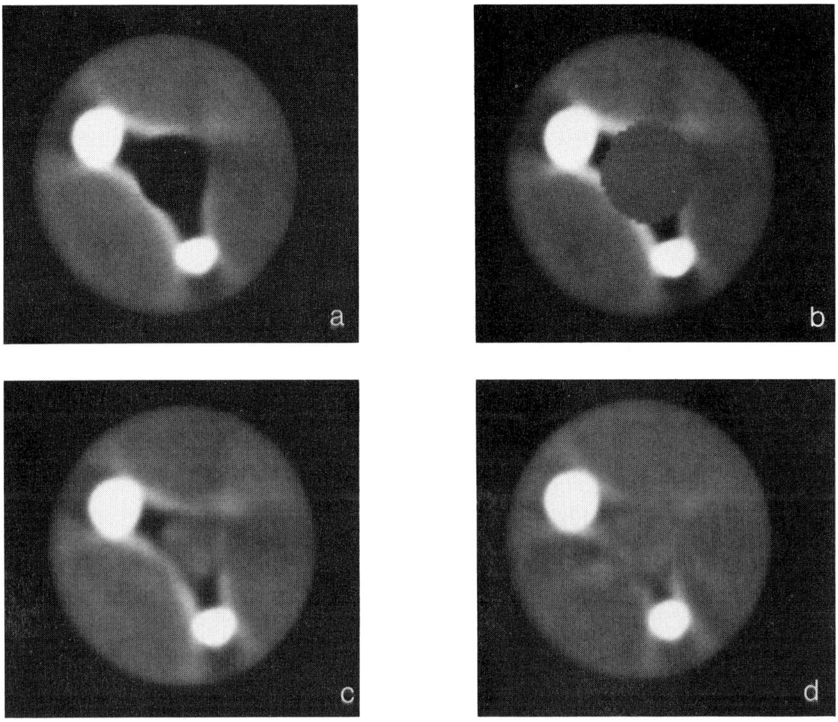

Fig. 9.5-7. First experiment. (a) Reconstruction produced by convolution-backprojection using an initial estimate of the missing line integrals. Since the initial estimates are not correct, the reconstructed image has severe artifacts. (b) Result of setting the image shown in (a) to zero outside the bagel and setting the hole to an assigned constant value. (c) Reconstruction produced after the first iteration of the bagel algorithm. (d) Steady-state reconstruction produced after 20 iterations of the bagel algorithm. Compared with the incomplete data reconstruction produced by convolution-backprojection, the background is more uniform and the streaks are greatly reduced in amplitude. (From Medoff *et al.* [13].)

the calculated values are used as new estimates of the missing data. The new full set of data is reconstructed to produce the image shown in Fig. 9.5-7c. The streaks are reduced in amplitude and the reconstructed values in the hole agree more closely with the assigned value. This image is set to zero outside the bagel, the hole is reset, and new line integrals are calculated. This procedure is repeated until, after 20 iterations, the steady-state image shown in Fig. 9.5-7d is obtained. Comparing this image with the image generated by conventional techniques, shown in Fig. 9.5-6, we see that the background is more uniform and the streaks are greatly reduced in amplitude. The reconstructed values in the hole agree closely with the value assigned to the hole by the iterative process.

A better understanding of the steady-state reconstruction is obtained by calculating the difference between the reconstructed values in the bagel produced by the iterative algorithm and the actual values of the object in the bagel. This residual error, which has the structure shown in Fig. 9.5-2c, is a ghost image that consists of alternating positive and negative regions (shown as white and black areas of the picture) arranged like spokes of a wheel. Line integrals along paths that pass through both the positive and the negative regions are zero. Only line integrals along paths through the central part of the image (corresponding to the hole in the bagel) are nonzero.

The residual error is a ghost image that is confined to the bagel and effectively invisible to measurements that do not pass through the hole. Although it can be shown that the ideal continuous bagel problem has a unique solution [18,19], the practical problem of reconstructing an image from a finite number of inexact measurements, using inexact finite precision arithmetic, suffers from a fundamental indeterminacy.

The existence of ghost images represents a fundamental uncertainty in image reconstruction from incomplete data. Sufficient information to correctly identify the ghost components is not contained in the measured data. An algorithm for image reconstruction from incomplete data that implicitly adjusts these ghost components based on some convenient mathematical criterion—for example, minimizing the variance of the reconstructed image or maximizing the entropy—will produce a unique result, but the contributions assigned to the ghost components will not match those of the underlying physical object. These components can only be correctly determined on the basis of additional *a priori* information. This additional *a priori* information must be used to determine stronger constraints that render the ghosts inconsistent.

This motivates our second experiment, in which the *a priori* knowledge of the density in a region of the underlying object will be used to reduce ghost artifacts. In this experiment, the central 20% of the nonzero line integrals are missing in 10% of the fan-beam projections that are located in a continuous angular range. This type of incomplete data problem might arise in practice due to motion in a local region of the object. Figure 9.5-8a shows the streak artifact generated when convolution-backprojection is used to reconstruct the incomplete set of data.

First, a reconstruction is obtained by using the iterative algorithm with positivity and known outer boundary constraints. The resulting steady-state image is shown in Fig. 9.5-8b. Although this image is an improvement over the image obtained by using convolution-backprojection directly on the incomplete set of data, there is still significant streak artifact. As in the

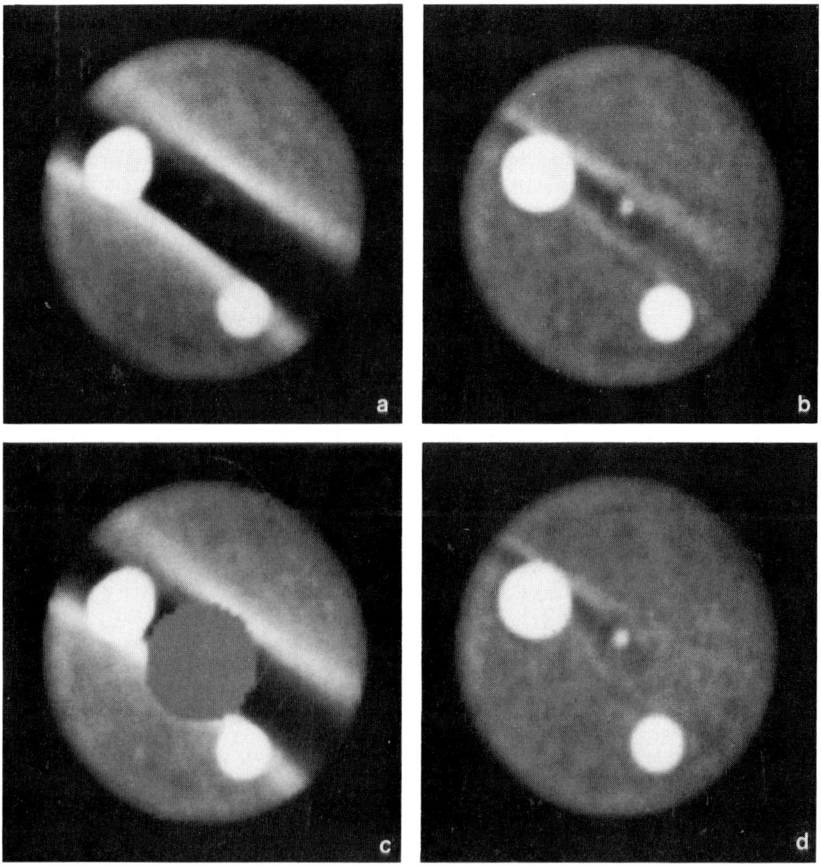

Fig. 9.5-8. Second experiment. (a) Incomplete data reconstruction with severe streak artifact produced by convolution-backprojection. In 10% of the projections, the central 20% of the nonzero line integrals are missing. (b) Steady-state image produced by the iterative algorithm using known extent and positivity constraints. There is significant residual streak artifact. (c) *A priori* knowledge of the background density in a region of the object is incorporated by a constraint operator. This operator resets the central region of the image to the known value at each step of the iteration, as shown. (d) Steady-state image produced by the iterative algorithm, using the known background constraint in addition to known extent and positivity. Compared with using only finite extent and positivity, there is significant reduction in the residual streak artifact. Note that the small structure in the central part of the object is clearly resolved. (From Medoff *et al.* [13].)

previous example, the nonzero line integrals of the residual error image are concentrated in the missing data region. This error represents a ghost that cannot be eliminated using only the positivity and known extent constraints.

In practice, it is often possible to identify streaks based on *a priori* knowledge of the structure of the object. For example, suppose that the central part of the object is known to have a certain uniform value of attenuation coefficient. This *a priori* information can be represented by a constraint operator that resets the central part of the image to the known value, as shown in Fig. 9.5-8c. The image produced by the iterative algorithm using positivity, known extent, and this additional known background information is shown in Fig. 9.5-8d. Compared with the steady-state image shown in Fig. 9.5-8b, there is substantial reduction in streak artifact. Ghost components that are not consistent with the additional constraint have been eliminated.

This experiment clearly illustrates the robustness of the iterative limited data reconstruction algorithm. Although the background constraint removes the small circular rod in the center of the image, this structure is clearly resolved in the steady-state reconstruction. Information about this structure is not retained in the estimated line integrals, but is retained in the measured data and restored at each step of the iteration. The steady-state reconstruction is a compromise between the measured data and the *a priori* information imposed by the constraints.

9.6 CONCLUSIONS

In this chapter we have presented an inner product framework for solving the problem of image reconstruction from limited data. The inner product framework enables us to formulate and solve the limited data problem by generalized least-squares techniques. The minimum norm least-squares solution is given by pseudoinverse operators and by iterative image reconstruction algorithms that are guaranteed to converge.

An important feature of our framework is that it is entirely in terms of infinite-dimensional linear operators rather than matrices, which means that the problem need not be discretized before it is solved. This allows us to retain all of the mathematical structure that is inherent in the underlying continuous problem and enables us to obtain image reconstruction algorithms that have efficient implementations.

While our focus has been on applications of the inner product framework to limited data problems in CT, the development applies to a broad class of signal recovery problems. The only requirements are that the measurements are related to an underlying unknown signal by a bounded linear measurement operator and that *a priori* information is represented in the form of an affine subspace constraint.

Two key observations that we have applied to CT also have potentially wider applicability. The first observation is that the inner product is a mathematical tool over which we have considerable freedom of choice. The inner product can be varied to obtain a family of different but clearly related reconstruction algorithms, some of which might have superior computational or numerical properties.

The second observation is that limited data problems can exhibit an inherent indeterminacy due to the presence of ghosts that are invisible to the measured data. Conditions that appear to guarantee the absence of ghosts in theory can fail to hold in practical implementations. This fundamental limitation can only be overcome by obtaining additional *a priori* information that determines a unique reconstruction, and then finding a method for incorporating this information in a reconstruction algorithm. We have introduced one such method based on affine subspace constraints and have shown that this method has considerable potential for reducing ghost artifacts in CT. It is likely that future research will provide even more powerful methods for incorporating *a priori* information in limited data problems in CT and in other fields of application.

REFERENCES

1. J. Radon, Uber die Bestimmung von Funktionen durch ihre Integralwerte Längs gewisser Mannigfaltigkeiten. *Berichte Saechsische Akademie der Wissenschaften* **69**, 262–277 (1917).
2. H. Cramér and H. Wold, Some theorems on distribution functions. *J. London Math. Soc.*, **11**, 290–294 (1936).
3. *Proc. IEEE*, **71**(3) (1983) (Special Issue on Computerized Tomography).
4. A. Macovski, "Medical Imaging Systems." New Jersey: Prentice-Hall, Inc., 1983.
5. H. Stark, J. W. Woods, I. Paul, and R. Hingorani, Direct Fourier reconstruction in computer tomography. *IEEE Trans. Acoustics, Speech, and Signal Proc.* **ASSP-29**(2), 237–245 (1981).
6. G. H. Glover and N. J. Pelc, An algorithm for the reduction of metal clip artifacts in CT reconstructions, *Med. Phys.*, **8**(6), 799–807 (1981).
7. G. S. Harell, D. F. Guthaner, R. S. Breiman, C. C. Morehouse, E. J. Seppi, W. H. Marshall, and L. Wexler, Stop-action cardiac computed tomography. *Radiology*, **123**, 515–517 (1977).
8. T. Hinderling, P. Rüegsegger, M. Anliker, and C. Dietschi, Computed tomography reconstruction from hollow projections: An application to *in vivo* evaluation of artificial hip joints. *J. Comp. Assist. Tomog.*, **3**(1), 52–57 (1979).
9. B. P. Medoff, Image Reconstruction from Limited Data, Ph.D. dissertation, Dept. of Electrical Engineering, Stanford University, Stanford, California, 1983.
10. B. P. Medoff, W. R. Brody, and A. Macovski, Image reconstruction from limited data, *Proc. International Workshop on Physics and Engineering in Medical Imaging*, Pacific Grove, California: IEEE Computer Society, March 15–18, 1982, pp. 188–192 (IEEE Cat. No. 82CH1751-7).
11. B. P. Medoff, W. R. Brody, and A. Macovski, The use of *a priori* information in image reconstruction from limited data. *Topical Meeting on Signal Recovery and Synthesis With Incomplete Information and Partial Constraints*, Incline Village, Nevada, Jan. 12–14, 1983 (Paper FA7).

12. B. P. Medoff, W. R. Brody, and A. Macovski, The use of *a priori* information in image reconstruction from limited data. *Proc. IEEE Int. Conf. on Acoust., Speech, Signal Processing,* Boston, Massachusetts, April 14–16, 1983, pp. 131–134.
13. B. P. Medoff, W. R. Brody, M. Nassi, and A. Macovski, Iterative convolution backprojection algorithms for image reconstruction from limited data. *J. Opt. Soc. Am.*, **73**(11), 1493–1500 (1983) (Special Issue on Signal Recovery).
14. B. P. Medoff, An inner product framework for image reconstruction. *Proc. IEEE Int. Conf. on Acoust., Speech, Signal Processing,* Tampa, Florida, March 26–29, 1985, pp. 1073–1076.
15. J. L. C. Sanz and T. S. Huang, Unified Hilbert space approach to iterative least-squares linear signal restoration. *J. Opt. Soc. Am.*, **73**(11), 1455–1465 (1983).
16. B. R. Musicus, Iterative algorithms for optimal signal reconstruction and parameter identification given noisy and incomplete data. Ph.D. dissertation, Dept. of Electrical Engineering and Computer Science, Massachusetts Institute of Technology, Cambridge, Massachusetts, Sept. 1982 (*RLE Tech. Rept. No. 496*).
17. W. H. Rowan *et al.*, Algorithms for limited-angle computed tomography. *Proc. International Workshop on Physics and Engineering in Medical Imaging,* Pacific Grove, California: IEEE Computer Society, March 15–18, 1982, pp. 169–177 (IEEE Cat. No. 82CH1751-7).
18. A. M. Cormack, Representation of a function by its line integrals, with some radiological applications. *J. Appl. Phys.*, **34**(9), 2722–2727 (1963).
19. M. Ein-Gal, The shadow transform—an approach to cross sectional imaging. Ph.D. dissertation, Dept. of Electrical Engineering, Stanford University, Stanford, California, 1974.
20. T. Inouye, Image reconstruction with limited angle projection data. *IEEE Trans. Nucl. Sci.*, **NS-26**(2), 2666–2669 (1979).
21. R. W. Gerchberg, Super-resolution through error energy reduction. *Optica Acta,* **21**(9), 709–720 (1974).
22. A. Papoulis, A new algorithm in spectral analysis and band-limited extrapolation. *IEEE Trans. Circuits Syst.* **CAS-22**(9), 735–742 (1975).
23. A. Lent and H. Tuy, An iterative method for the extrapolation of band-limited functions. *J. Math. Anal. Appl.*, **83**(2), 554–565 (1981).
24. M. I. Sezan and H. Stark, Image restoration by the method of convex projections: Part 2—Applications and numerical results. *IEEE Trans. Med. Imaging,* **MI-1**(2), 95–101 (1982).
25. H. Stark, D. Cahana, and H.Webb, Restoration of arbitrary finite-energy optical objects from limited spatial and spectral information. *J. Opt. Soc. Am.*, **71**(6), 635–642 (1981).
26. H. Stark and S. Cruze, Restoration of optical objects subject to nonnegative spatial or spectral constraints. *J. Opt. Soc. Am.* **72**(8), 993–1000 (1982).
27. D. C. Youla, Generalized image restoration by the method of alternating orthogonal projections. *IEEE Trans. Circuits Syst.*, **CAS-25**(9), 694–702 (1978).
28. D. C. Youla and H. Webb, Image restoration by the method of convex projections: Part 1—Theory. *IEEE Trans. Med. Imaging,* **MI-1**(2), 81–94 (1982).
29. A. J. Rockmore and A. Macovski, A maximum likelihood approach to transmission image reconstruction from projections. *IEEE Trans. Nucl. Sci.* **NS-24**, 1929–1935 (1977).
30. M. H. Buonocore, Fast minimum variance estimators for limited angle computed tomography image reconstruction. Ph.D. dissertation, Dept. of Electrical Engineering, Stanford University, Stanford, California, 1981.
31. G. T. Herman and A. Lent, Quadratic optimization for image reconstruction. I. *Comp. Graph. Im. Proc.* **5**, 319–332 (1976).

32. S. L. Wood, A. Macovski, and M. Morf, Reconstructions with limited data using estimation theory. *In* "Computer Aided Tomography and Ultrasonics in Medicine" (J. Raviv *et al.*, eds.), pp. 219–233, Amsterdam, The Netherlands: North-Holland, 1979.
33. G. Minerbo, MENT: A maximum entropy algorithm for reconstructing a source from projection data. *Comp. Graph. Im. Proc.* **10**, 48–68 (1979).
34. S. J. Wernecke and L. R. D'Addario, Maximum entropy image reconstruction. *IEEE Trans. Comp.* **C-26**, 351–364 (1977).
35. B. F. Logan and L. A. Shepp, Optimal reconstruction of a function from its projections. *Duke Math. J.* **42**(4), 645–659 (1975).
36. R. P. Tewarson, Solution of linear equations in remote sensing and picture reconstruction. *Computing* **10**, 221–230 (1972).
37. Y. S. Shim and Z. H. Cho, SVD pseudoinversion image reconstruction. *IEEE Trans. Acoust., Speech, Signal Processing* **ASSP-29**(4), 904–909 (1981).
38. R. L. Kashyap and M. C. Mittal, Picture reconstruction from projections. *IEEE Trans. Comp.* **C-24**, 915–923 (1975).
39. A. V. Lakshminarayanan and A. Lent, Methods of least squares and SIRT in reconstruction. *J. Theor. Biol.* **76**, 267–295 (1979).
40. K. M. Hanson, Limited angle CT reconstruction using *a priori* information. *Proc. IEEE Comp. Soc. Int. Symp.*, Berlin, Oct. 1982.
41. K. M. Hanson and G. W. Wecksung, Bayesian approach to limited-angle reconstruction in computed tomography. *J. Opt. Soc. Am.* **73**(11), 1501–1509 (1983).
42. A. Peres, Tomographic reconstruction from limited angular data. *J. Comp. Assisted Tomog.* **3**(6), 800–803 (1979).
43. R. N. Bracewell, Strip integration in radio astronomy. *Aust. J. Phys.* **9**, 198–217 (1956).
44. R. N. Bracewell and A. C. Riddle, Inversion of fan-beam scans in radio astronomy. *Astrophys. J.* **150**(2), 427–434 (1967).
45. G. N. Ramachandran and A. V. Lakshminarayanan, Three-dimensional reconstruction from radiographs and electron micrographs: Application of convolutions instead of Fourier transforms. *Proc. Nat. Acad. Sci. U.S.* **68**(9), 2236–2240 (1971).
46. L. A. Shepp amd B. F. Logan, The Fourier reconstruction of a head section. *IEEE Trans. Nucl. Sci.* **NS-21**, 21–43 (1974).
47. A. V. Lakshminarayanan, Reconstruction from divergent ray data. Technical Report Number 92, Dept. of Computer Science, State University of New York, Buffalo, New York, 1975.
48. G. T. Herman and A. Naparstek, Fast image reconstruction based on a Radon inversion formula appropriate for rapidly collected data. *SIAM J. Appl. Math.* **33**(3), 511–533 (1977).
49. D. G. Luenberger, "Optimization by Vector Space Methods." New York: John Wiley & Sons, Inc., 1969.
50. A. E. Taylor and D. C. Lay, "Introduction to Functional Analysis." New York: John Wiley and Sons, 1980.
51. C. W. Groetsch, "Generalized Inverses of Linear Operators." New York: Marcel Dekker, Inc., 1977.
52. M. Z. Nashed, ed., "Generalized Inverses and Applications." New York: Academic Press, 1976.
53. M. Z. Nashed and G. F. Votruba, A unified operator theory of generalized inverses. *In* "Generalized Inverses and Applications" (M. Z. Nashed, ed.), pp. 1-109, New York: Academic Press, 1976.
54. M. Z. Nashed, Generalized inverses, normal solvability, and iteration for singular operator equations. *In* "Nonlinear Functional Analysis and Applications" (L. B. Rall, ed.), pp. 311–359, New York: Academic Press, 1971.

55. F. S. Beckman, The solution of linear equations by the conjugate gradient method. *In* "Mathematical Methods for Digital Computers" (A. Ralston and H. S. Wilf, eds.), New York: John Wiley and Sons, 1960.
56. M. Hestenes and E. Stiefel, Method of conjugate gradients for solving linear systems. *J. Res. Nat. Bur. Stand. Sect. B* **49**, 409-436, 1952 (Nat. Bur. Stand. Report 1959).
57. J. W. Daniel, The conjugate gradient method for linear and nonlinear operator equations. *SIAM J. Numer. Anal.* **4**(1), 10-26 (1967). (Correction, vol. 7, no. 2, 1970, p. 277).
58. P. Gilbert, Iterative methods for the three-dimensional reconstruction of an object from projections. *J. Theor. Biol.* **36**, 105-117 (1972).
59. D. Ludwig, The Radon transform on Euclidean space. *Comm. Pure Appl. Math.* **XIX**, 49-81 (1966).
60. A. M. Cormack, Representation of a function by its line integrals, with some radiological applications. II. *J. Appl. Phys.* **35**(10), 2908-2913 (1964).
61. R. M. Lewitt and R. H. T. Bates, Image reconstruction from projections: I: General theoretical considerations. *Optik* **50**(1), 19-33 (1978).
62. R. M. Lewitt and R. H. T. Bates, Image reconstruction from projections: III: Projection completion methods (theory). *Optik.* **50**(3), 189-204 (1978).
63. R. M. Lewitt and R. H. T. Bates, Image reconstruction from projections: IV: Projection completion methods (computational examples). *Optik,* **50**(4), 269-278 (1978).
64. C. Hammaker and D. Solmon, The angle between null spaces of X-rays. *J. Math. Anal. Appl.* **62**, 1-23 (1978).
65. A. K. Louis, Ghosts in tomography—the null space of the Radon transform. *Math. Meth. Appl. Sci.* **3**, 1-10 (1981).
66. K. T. Smith, D. C. Solmon, and S. L. Wagner, Practical and mathematical aspects of the problem of reconstructing objects from radiographs. *Bull. Amer. Math. Soc.* **83**, 1227-1270 (1977).

CHAPTER **10**

Computer-Assisted Diffraction Tomography

M. Kaveh

Department of Electrical Engineering
University of Minnesota
Minneapolis, Minnesota 55455

M. Soumekh

Department of Electrical Engineering
SUNY Buffalo
Amherst, New York 14226

10.1 INTRODUCTION

In this Chapter we introduce the concept of diffraction tomography as an inverse problem of the Helmholtz equation with a complex unknown spatially varying coefficient. This type of equation governs the propagation of waves in a variety of inhomogeneous media. Using appropriate sources of excitation and a sufficient number of measurements of the scattered waves, external to the medium under study, a diffraction tomogram quantifies (images) the unknown coefficient.

The three ingredients—excitation, medium-interaction model, and external measurements—form the basis of all tomographic imaging techniques. The interaction model and measurements for the nondiffracting problems, such as those discussed elsewhere in this book, often result in Radon transforms [1] of the inhomogeneity along straight paths, leading to algorithmically efficient and accurate tomograms. In contrast, tomography based on diffraction data usually involves two distinct stages in its development. The first stage is the simplification or approximation of the model equation so that it may be inverted with a limited set of measurements in a computationally efficient manner. The second stage is concerned with signal processing issues and in a sense is a generalization of some of the algorithms for nondiffracting (straight-path) tomography.

Diffraction tomography finds its potential applications in situations where geometrical optic assumptions concerning the propagation of waves within the medium under study are not valid. Examples arise in quantitative imaging of soft tissue with ultrasound, certain seismic inversions, as well as optical and microwave imaging of objects characterized by inhomogeneous indices of refraction. The experimental examples presented in this chapter are exclusively of an ultrasonic nature. The mathematical derivations, however, are applicable to any situation for which a Helmholtz-type equation is a valid source–medium interaction model.

Figure 10.1-1 depicts the most common measurement geometry in two-dimensional transmission diffraction tomography. The source of excitation is assumed to be a plane wave and the array of receiver sensors measures the wave which consists of the primary and scattered components. The data base for inversion is obtained from coherent measurements of the scattered waves for a number of angular orientations of the object.

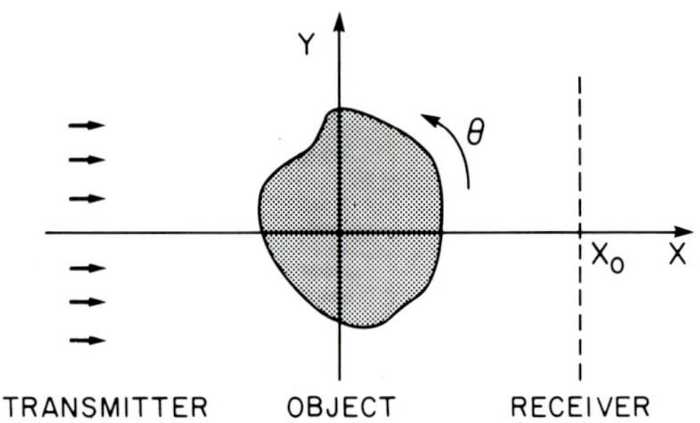

Fig. 10.1-1. Two-dimensional measurement geometry.

As alluded to earlier, the key to diffraction tomography is simplification or approximation of the wave equation to make the desired parameters of the inhomogeneity accessible to recovery. The approximations are usually based on the assumption that the wavenumber within the inhomogeneity differs only slightly from that of the known homogeneous background. In solving the direct problem of the wave equation, this assumption has been the basis for several perturbation solutions. Two approximations have been especially popular: the Born and Rytov approximations [2]–[5]. These references as well as many others in the optics literature discuss the conditions under which these approximations are valid for the direct problem. Later in this chapter we will examine the same issue for the inverse problem.

Principles of diffraction tomography using the Born approximation were first outlined by Wolf in [6]. Iwata and Nagata extended Wolf's formulation to the case of Rytov approximation [7]. The first practical formulation of the problem, including Fourier-domain reconstruction algorithms, simulation, and experimental results in ultrasonic tomography, and the coining of the term "diffraction tomography" are due to Mueller, Kaveh, and colleagues [8–11]. The generalization of the convolution-backprojection algorithm to diffraction tomography was given by Devaney [12]. Various extensions, comparisons of reconstruction methods, and applications to different media can be found in [13–20].

In the following sections we present a discussion of the basic issues encountered in diffraction tomography. We begin by exploring methods of approximating the wave equation so that it can be efficiently inverted. The signal processing aspects of the problem are presented next, followed by a treatment of diffraction tomography from limited data. The chapter closes with a comparison of some experimental diffraction tomograms based on the Born and Rytov methods. The reader should be aware that diffraction tomography is still very much in its infancy. It will be evident in the ensuing presentation that many practical issues, ranging from appropriate modeling of the physics of the problem to the design of experimental systems and algorithms, remain unsolved. We hope that this exposition will serve to stimulate further research and development on this subject.

10.2 TRANSFORMATIONS OF THE WAVE EQUATION

In this section we develop a general approach for approximately solving the inverse problem of the wave equation. This approach is motivated by the relative simplicity with which system identification is carried out in the Fourier domain for linear constant-coefficient systems. Other formulations of this problem leading to some of the same solutions are possible. These formulations are based on perturbation approximation solutions of the forward problem of the wave equation [6, 7].

We begin by examining the scalar Helmholtz equation of interest with a space-varying coefficient. As mentioned earlier, we carry out the derivation for the two-dimensional model and comment on the three-dimensional implications at the close of the chapter. Accordingly, the complex amplitude of the scalar field $\psi(\mathbf{r})$ satisfies

$$[\nabla^2 + k^2(\mathbf{r})]\psi(\mathbf{r}) = 0, \tag{10.2-1}$$

where $\mathbf{r} = [x, y]$ denotes the position vector in two-dimensional Cartesian coordinates and ∇^2 is the Laplacian operator. The function $k(\mathbf{r})$ is the spatially varying wavenumber, which in general is complex, with its real part reflecting variations in the velocity of propagation and its imaginary part signifying loss in the medium under study. In transmission diffraction tomography, $k^2(\mathbf{r})$ is expressed as

$$k^2(\mathbf{r}) = k_0^2[1 + f(\mathbf{r})],$$

where k_0 is the known real constant wavenumber of the homogeneous medium, such as water in ultrasonic tomography, surrounding the object under study and $f(\mathbf{r})$ signifies a small perturbation, $|f(\mathbf{r})| \ll 1$, to the homogeneous medium due to the presence of the object. The aim of solving the inverse problem is, once again, to approximately determine (image) $f(\mathbf{r})$, denoted the object function, based on measurements of $\psi(\mathbf{r})$ outside the object.

We now express Eq. (10.2-1) in the form of an inhomogeneous, constant-coefficient differential equation with a source term containing $f(\mathbf{r})$. This can be accomplished by reformulating the equation in terms of a general transformation on $\psi(\mathbf{r})$. Thus, let $\psi_0(\mathbf{r}) = e^{jk_0x}$ be the complex amplitude of the plane incident wave propagating in the x direction. Define a scalar differentiable function of $\alpha(\mathbf{r}) = \psi(\mathbf{r})/\psi_0(\mathbf{r})$ as $U(\mathbf{r}) = g[\alpha(\mathbf{r})]$. We have

$$\psi_0(\mathbf{r})\,\nabla^2\alpha(\mathbf{r}) = \nabla^2\psi(\mathbf{r}) - 2jk_0\hat{x}\cdot\nabla\psi(\mathbf{r}) - k_0^2\psi(\mathbf{r}), \qquad (10.2\text{-}2)$$

where ∇ is the gradient operator, $\hat{x}$ the unit vector along the x axis, and $\cdot$ denotes the dot product of two vectors. Solving Eq. (10.2-2) for $\nabla^2\psi(\mathbf{r})$ results in

$$\nabla^2\psi(\mathbf{r}) = \psi_0(\mathbf{r})\,\nabla^2\alpha(\mathbf{r}) + 2jk_0\hat{x}\cdot\nabla\psi(\mathbf{r}) + k_0^2\psi(\mathbf{r}). \qquad (10.2\text{-}3)$$

The gradient of $\alpha(\mathbf{r})$ is given by

$$\psi_0(\mathbf{r})\,\nabla\alpha(\boldsymbol{r}) = \nabla\psi(\mathbf{r}) - jk_0\hat{x}\cdot\psi(\mathbf{r}). \qquad (10.2\text{-}4)$$

Substituting for $\nabla\psi(\mathbf{r})$ in Eq. (10.2-4) and the resulting expression for $\nabla^2\psi(\mathbf{r})$ in Eq. (10.2-1) and dividing by $\psi_0(\mathbf{r})$, we obtain

$$\nabla^2\alpha(\mathbf{r}) + jk_0\hat{x}\cdot\nabla\alpha(\mathbf{r}) + k_0^2 f(\mathbf{r})\alpha(\mathbf{r}) = 0. \qquad (10.2\text{-}5)$$

Let g' and g'' denote the first and second derivatives of $g[\alpha(\mathbf{r})]$ with respect to $\alpha(\mathbf{r})$. We have

$$\nabla\alpha(\mathbf{r}) = \nabla U(\mathbf{r})/g'$$

and

$$\nabla^2\alpha(\mathbf{r}) = \nabla^2 U(\mathbf{r})/g' - [\nabla U(\mathbf{r})]^2 g''/g'^3,$$

where we have defined $[\nabla U(\mathbf{r})]^2 \triangleq \nabla U(\mathbf{r})\cdot\nabla U(\mathbf{r})$. Substituting the above two expressions in Eq. (10.2-5) results in

$$\nabla^2 U(\mathbf{r}) + 2jk_0\hat{x}\cdot\nabla U(\mathbf{r}) = -k_0^2 f(\mathbf{r})\alpha(\mathbf{r})g' + [\nabla U(\mathbf{r})/g']^2 g''. \qquad (10.2\text{-}6)$$

Equation (10.2-6) is one of the desired forms of the wave equation in terms of the transformed function $U(\mathbf{r})$. The left-hand side of the equation is a linear constant-coefficient operation on $U(\mathbf{r})$. The right-hand side explicitly contains $f(\mathbf{r})$ as a source term as well as an additional term containing $[\nabla U(\mathbf{r})]^2$. The operator on the left-hand side of Eq. (10.2-6) is invariant under the transformation $g(\cdot)$. Therefore, the effectiveness of the inversion process, i.e., the approximate recovery of $f(\mathbf{r})$, is determined by the nature of the terms other than $f(\mathbf{r})$ on the right-hand side. Before discussing this issue any further, we obtain another form of the wave equation which is especially popular.

Define a new complex function $U_1(\mathbf{r})$ as $U_1(\mathbf{r}) \triangleq \psi_0(\mathbf{r})U(\mathbf{r})$. It follows that

$$\nabla U_1(\mathbf{r}) = [jk_0\hat{x}U(\mathbf{r}) + \nabla U(\mathbf{r})]\psi_0(\mathbf{r}) \qquad (10.2\text{-}7)$$

and

$$\begin{aligned}\nabla^2 U_1(\mathbf{r}) = {} & [-k_0^2 U(\mathbf{r}) + jk_0\hat{x}\cdot\nabla U(\mathbf{r})]\psi_0(\mathbf{r}) \\ & + [jk_0\hat{x}\cdot\nabla U(\mathbf{r}) + \nabla^2 U(\mathbf{r})]\psi_0(r).\end{aligned} \qquad (10.2\text{-}8)$$

Substituting $\nabla^2 U(\mathbf{r})$ and $U(\mathbf{r})$ from Eqs. (10.2-7) and (10.2-8) into Eq. (10.2-6) gives

$$(\nabla^2 + k_0^2)U_1(\mathbf{r}) = -k_0^2\psi_0(\mathbf{r})f(\mathbf{r})\alpha(\mathbf{r})g' + \psi_0(\mathbf{r})[\nabla U(\mathbf{r})/g']^2 g''. \quad (10.2\text{-}9)$$

The operator on the left-hand side of Eq. (10.2-9) is the familiar one for the Helmholtz equation describing plane wave propagation in a homogeneous medium with wavenumber k_0. Either Eq. (10.2-6) or Eq. (10.2-9) can be the starting point for an approximate reconstruction of $f(\mathbf{r})$. Before proceeding to the formulation of the reconstruction algorithm, however, we examine the right-hand side of Eq. (10.2-6) or alternatively that of Eq. (10.2-9) more carefully. This is done to obtain a better understanding of the role played by the function $g(\cdot)$ in the final approximate recovery of $f(\mathbf{r})$.

Denote the right-hand side of Eq. (10.2-6) by $-h(\mathbf{r})$, i.e.,

$$h(\mathbf{r}) \triangleq k_0^2 f(\mathbf{r})\alpha(\mathbf{r})g' - [\nabla U(\mathbf{r})/g']^2 g''. \qquad (10.2\text{-}10)$$

$h(\mathbf{r})$ may be considered as the unknown source term of the constant-coefficient differential equation (10.2-6) with known coefficients. The success of recovering $f(\mathbf{r})$ based on Eq. (10.2-6) then clearly rests on the fidelity with which $h(\mathbf{r})$ approximates $f(\mathbf{r})$ in a given situation. Specifically, if $h(\mathbf{r})$ were to be replaced with $f(\mathbf{r})$, the resulting Eq. (10.2-6) would become an inhomogeneous linear constant-coefficient differential equation with a source term $f(\mathbf{r})$, which in principle would be recoverable from a sufficient number of measurements of $U(\mathbf{r})$. In the following we consider three specific forms for $g(\cdot)$ and discuss their implications in the above approximation issue.

A. The Rytov Transformation

The discussion above and an examination of the form of $h(\mathbf{r})$ in Eq. (10.2-10) point to a natural choice for g, i.e., one for which $\alpha(\mathbf{r})g'$ is a constant, for example, unity. Thus,

$$\frac{\partial g[\alpha(\mathbf{r})]}{\partial[\alpha(\mathbf{r})]}\alpha(\mathbf{r}) = 1,$$

resulting in

$$g[\alpha(\mathbf{r})] = \ln[\alpha(\mathbf{r})], \tag{10.2-11}$$

where $\ln(\cdot)$ denotes real or complex natural logarithm, depending on the nature of its argument. This transformation is interesting in that it is a natural choice from the general form of $h(\mathbf{r})$. It is also, coincidentally, the same as the transformation employed in the method of smooth perturbation [2, 3] in the solution of the direct problem of the wave equation, better known as the Rytov approximation. Consequently, we call $g[\alpha(\mathbf{r})] = \ln[\alpha(\mathbf{r})]$ the Rytov transformation and the resulting source function $h(\mathbf{r})$ the Rytov source. This function is now given by

$$h(\mathbf{r}) = k_0^2 f(\mathbf{r}) + [\nabla U(\mathbf{r})]^2 \tag{10.2-12}$$

Thus, $h(\mathbf{r})$ comprises of the object function $f(\mathbf{r})$ and an additive error term that depends on the variations of a logarithmically smoothed function of the normalized propagating wave. If the variations of this wave over distances of the order of a wavelength are small, $h(\mathbf{r})$ will be a good representation of $f(\mathbf{r})$ [3]. We will return to this point in Section 10.7 to assess the effects of the error term on the reconstructed image. In the meantime, we consider two more specific transformations.

B. The Born Transformation

An especially simple transformation that may be used in conjunction with Eq. (10.2-6) is a linear one given by $U(\mathbf{r}) = \alpha(\mathbf{r}) - 1$. The resulting source function is, from Eq. (10.2-10),

$$h(\mathbf{r}) = k_0^2 f(\mathbf{r}) + k_0^2 f(\mathbf{r}) U(\mathbf{r}). \tag{10.2-13}$$

Since Eq. (10.2-6) with the above $U(\mathbf{r})$, coupled with neglecting the second term on the right-hand side of Eq. (10.2-13), is used in solving the forward problem of the wave equation under the Born approximation [2, 3], we call the above function the Born transformation. It is clear from Eq. (10.2-13) that the difference between $h(\mathbf{r})$ and $f(\mathbf{r})$ in this case is proportional to the object function itself. So the requirement that $h(\mathbf{r})$ should be a close approximation to $f(\mathbf{r})$ necessitates that $|[\psi(\mathbf{r}) - \psi_0(\mathbf{r})]/\psi_0(\mathbf{r})| \ll 1$, a very severe limitation indeed. As we will see in Section 10.7, this restriction is violated very quickly in practice, resulting in an image which is a highly distorted approximation of $f(\mathbf{r})$.

The fact that the Rytov transformation has its utility mainly where rapid fluctuations of the phase function of the scattered field are small, whereas the Born transformation is valid for small overall phase changes, motivates an interesting intermediate transformation due to Z. Q. Lu [21]. This is presented next.

C. An Intermediate Transformation

In this case we consider a transformation of the form

$$g[\alpha(\mathbf{r})] = n[\alpha^{1/n}(\mathbf{r}) - 1]$$

where n is a positive integer.

This transformation is interesting in that for $n = 1$ it gives the Born transformation and for $n \to \infty$ it approaches the Rytov transformation. The resulting source function is now given by

$$h(\mathbf{r}) = k_0^2\alpha^{1/n}(\mathbf{r})f(\mathbf{r}) + \left(1 - \frac{1}{n}\right)\alpha^{1/n}(\mathbf{r})\left[\frac{\nabla\alpha(\mathbf{r})}{\alpha(\mathbf{r})}\right]^2. \qquad (10.2\text{-}14)$$

Let us expand $\alpha^{1/n}(\mathbf{r})$ in terms $\ln\alpha(\mathbf{r})$

$$\alpha^{1/n}(\mathbf{r}) = 1 + \frac{\ln\alpha(\mathbf{r})}{n} + \frac{1}{2!}\left[\frac{\ln\alpha(\mathbf{r})}{n}\right]^2 + \cdots. \qquad (10.2\text{-}15)$$

Thus,

$$h(\mathbf{r}) = k_0^2 f(\mathbf{r}) + k_0^2\left\{\frac{\ln\alpha(\mathbf{r})}{n} + \frac{1}{2!}\left[\frac{\ln\alpha(\mathbf{r})}{n}\right]^2 + \cdots\right\}f(\mathbf{r})$$
$$+ \left(1 - \frac{1}{n}\right)\alpha^{1/n}(\mathbf{r})\left[\frac{\nabla\alpha(\mathbf{r})}{\alpha(\mathbf{r})}\right]^2. \qquad (10.2\text{-}16)$$

What is revealing in Eq. (10.2-16) is that the phase of $\alpha(\mathbf{r})$ plays a dominant role in the nature of the difference between $h(\mathbf{r})$ and $f(\mathbf{r})$. For example, for a given n we may have $\ln|\alpha(\mathbf{r})|/n \ll 1$ but phase$[\alpha(\mathbf{r})]/n$ may very well be large. In this case, the error term in Eq. (10.2-16) is a complex modulation of $f(\mathbf{r})$ similar to the resulting source term from the Born transformation. Consequently, to make this error term small n should be chosen large enough that the resulting transformation approaches the Rytov one. This situation, for which low-order transformations of the type discussed in this subsection (Born being an extreme case) are clearly not very useful, is encountered in many practical applications. In these applications $f(\mathbf{r})$ has its support over dimensions of many wavelengths, resulting in large overall phase values for $\alpha(\mathbf{r})$ even though $|f(\mathbf{r})| \ll 1$.

We now summarize the important points of this section. The homogeneous Helmholtz equation with a spatially varying coefficient was recast in terms of a general transformation of the normalized propagating wave. This made it possible for us to express the equation in the form of a linear constant-coefficient operator on the transformed wave, which was independent of the transformation, with a source term that explicitly included the object function as well as a term reflecting the nonlinear nature of the problem. Accurate inversion requires the source function to be a good approximation to the object function. The Rytov transformation was introduced as a natural one for this purpose, and its relation to the Born transformation was demonstrated through an intermediate parametric function. We close this section by emphasing that the formulations given in Eqs. (10.2-6) and (10.2-9) offer the opportunity for creative design of the functions of the normalized field beyond the three explored in this section. A general approach to this design, however, is not clear. We, therefore, limit our performance comparisons to the Born and Rytov transformations. In the next section the fundamental Fourier-domain relationships that make the inversion of Eqs. (10.2-6) and (10.2-9) possible are developed.

10.3 FOURIER SLICE IN DIFFRACTION TOMOGRAPHY

In this section, the diffraction theory counterpart of the Fourier-slice property of straight-path tomography is presented. First, we give some of the notations and mathematical relations that are used in the remainder this chapter. Let $\tilde{}$ indicate the Fourier transform of the function over which it appears, and a function with subscript Θ refer to that function for the object that has been rotated by an angle θ. Define a rotation operator by

$$\mathbf{\Theta} = \begin{bmatrix} \cos\theta & \sin\theta \\ -\sin\theta & \cos\theta \end{bmatrix}$$

Let $\mathbf{\Lambda} \triangleq [\mu, \lambda]$ be the two-dimensional spatial-frequency vector with

Cartesian coordinates μ and λ. The first property of interest relates the Fourier transforms of a function and its rotated version. We have

$$\begin{aligned}\tilde{f}_\Theta(\mathbf{\Lambda}) &= \int_{-\infty}^{\infty}\int_{-\infty}^{\infty} f_\Theta(\mathbf{r})\, e^{-j\mathbf{\Lambda}\cdot\mathbf{r}}\, dx\, dy \\ &= \int_{-\infty}^{\infty}\int_{-\infty}^{\infty} f(\mathbf{r}\mathbf{\Theta})\, e^{-j\mathbf{\Lambda}\cdot\mathbf{r}}\, dx\, dy \\ &= \int_{-\infty}^{\infty}\int_{-\infty}^{\infty} f(\mathbf{r})\, e^{-j(\mathbf{\Lambda}\mathbf{\Theta})\cdot\mathbf{r}}\, dx\, dy \\ &= \tilde{f}(\mathbf{\Lambda}\mathbf{\Theta}). \end{aligned} \tag{10.3-1}$$

The second property of the Fourier transform that is needed relates the partial transform with respect to, say, the y coordinate of a two-dimensional function $g(x, y)$ to its two-dimensional transform $\tilde{g}(\mu, \lambda)$. Let $G(x_0, \nu)$ denote the partial transform defined by

$$G(x_0, \nu) = \int_{-\infty}^{\infty} g(x_0, y)\, e^{-j\nu y}\, dy. \tag{10.3-2}$$

Substituting in Eq. (10.3-2) for $g(x_0, y)$ in terms of $\tilde{g}(\mu, \lambda)$, one obtains

$$\begin{aligned} G(x_0, \nu) &= \frac{1}{4\pi^2}\int_{-\infty}^{\infty}\int_{-\infty}^{\infty} \tilde{g}(\mu, \lambda)\, e^{j\mu x_0} \int_{-\infty}^{\infty} e^{j(\lambda-\nu)y}\, dy\, d\nu'\, d\mu \\ &= \frac{1}{2\pi}\int_{-\infty}^{\infty}\int_{-\infty}^{\infty} \tilde{g}(\mu, \lambda)\, e^{j\mu x_0}\delta(\lambda - \nu)\, d\lambda\, d\mu \\ &= \frac{1}{2\pi}\int_{-\infty}^{\infty} \tilde{g}(\mu, \nu)\, e^{j\mu x_0}\, d\mu. \end{aligned} \tag{10.3-3}$$

Let us now refer to the measurement geometry of Fig. 10.1-1 with the object under study under rotation angle θ. $\psi_\Theta(\mathbf{r})$ is measured and transformed to generate $U_\Theta(\mathbf{r})$. Equation (10.2-6) can be written as

$$\nabla^2 U_\Theta(\mathbf{r}) + 2jk_0\hat{x}\cdot\nabla U_\Theta(\mathbf{r}) = -h_\Theta(\mathbf{r}). \tag{10.3-4}$$

Examining the expression for $h(\mathbf{r})$, it becomes obvious that in general $h_\Theta(\mathbf{r}) \neq h(\mathbf{r}\mathbf{\Theta})$. However, under the assumption that $h(\mathbf{r})$ is a good approximation to $f(\mathbf{r})$, we have the approximate relation

$$h_\Theta(\mathbf{r}) \simeq f_\Theta(\mathbf{r}) = f(\mathbf{r}\mathbf{\Theta}).$$

Substituting the approximatiion $f_\Theta(\mathbf{r})$ in the right-hand side of Eq. (10.3-4), Fourier transforming the result, and using the relation in Eq. (10.3-1), we obtain

$$(\mu^2 + \lambda^2 + 2k_0\mu)\tilde{U}_\Theta(\mathbf{\Lambda}) \simeq k_0^2\tilde{f}(\mathbf{\Lambda}\mathbf{\Theta}). \tag{10.3-5}$$

Using the relation in Eq. (10.3-3) in conjunction with Eq. (10.3-5) results in

$$W_{\Theta}(x_0, \nu) \simeq \frac{k_0^2}{2\pi} \int_{-\infty}^{\infty} \frac{\tilde{f}(\Lambda'\Theta)\, e^{j\mu x_0}}{(\mu - \mu_1)(\mu - \mu_2)}\, d\mu, \tag{10.3-6}$$

where $W_{\Theta}(x_0, \nu)$ is the one-dimensional Fourier transform of $U_{\Theta}(x_0, y)$ along the y coordinate and $\Lambda' = [\mu, \nu]$. The integral in Eq. (10.3-6) can be evaluated by contour integration in the complex ω plane, where $\omega = \mu + j\rho$.

To carry this operation out we resort to introducing an arbitrarily small imaginary component in the wavenumber of the medium surrounding the object. Thus, let this wavenumber now be given by $k_0(1 + j\varepsilon)$, where $\varepsilon \geqslant 0$. The isolated singularities in the integrand of Eq. (10.3-6) are given by

$$\mu_{1(2)} = -k_0(1 + j\varepsilon)(\pm)\sqrt{k_0^2(1 + j\varepsilon)^2 - \nu^2}, \qquad |\nu| \leqslant k_0. \tag{10.3-7}$$

The integration contour and the poles μ_1 and μ_2 are shown in Fig. 10.3-1.

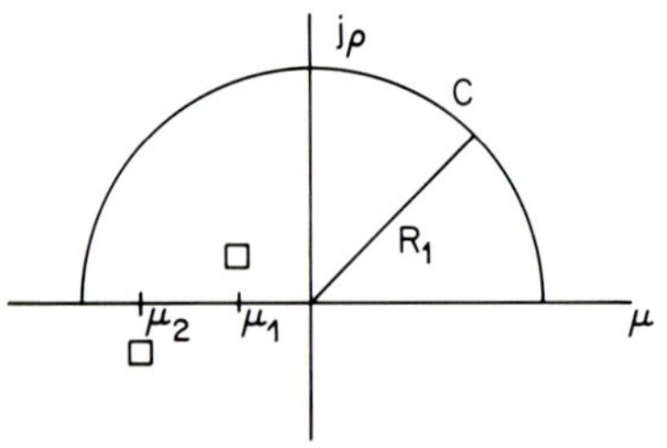

Fig. 10.3-1. Integration contour in the ω-plane.

Note that for ε sufficiently small, the transmission pole μ_1 has a positive imaginary component, whereas the reflection pole μ_2 has a negative imaginary part. Thus the contour of integration only includes μ_1 satisfying the requirement that only the transmitted wave is measured and used in the inversion process. $\tilde{f}(\Lambda'\Theta)$ is analytic inside the contour [8], and we have

$$W_{\Theta}(x_0, \nu) \simeq \lim_{\varepsilon \to 0} \lim_{R_1 \to \infty} \left\{ -\frac{k_0^2}{2\pi} \int \frac{\tilde{f}(\Lambda''(\Theta)\, e^{j\omega x_0}}{(\omega - \mu_1)(\omega - \mu_2)}\, d\omega \right\}, \tag{10.3-8}$$

where $\Lambda'' = [\omega, \nu]$. The equality in Eq. (10.3-8) follows since the integrand on C approaches zero as $R_1 \to \infty$. Evaluating Eq. (10.3-8) by the method of residues, it follows that

$$W_{\Theta}(x_0, \nu) \simeq \frac{j\tilde{f}(\Lambda_1\Theta)\, e^{j\mu_1 x_0} k_0^2}{\mu_1 - \mu_2}, \qquad |\nu| \leqslant k_0, \tag{10.3-9}$$

where $\boldsymbol{\Lambda}_1 = [\mu_1, \nu]$ and $\mu_{1(2)} = -k_0(\pm)\sqrt{k_0^2 - \nu^2}$. Alternatively,

$$\tilde{f}(\boldsymbol{\Lambda}_1\boldsymbol{\Theta}) \simeq \frac{2j}{k_0^2}\sqrt{k_0^2 - \nu^2}\, e^{-j\mu_1 x_0} W_\Theta(x_0, \nu), \qquad |\nu| \leq k_0. \quad (10.3\text{-}10)$$

Equation (10.3-10) embodies the Fourier-domain relationship between the one-dimensional Fourier transform of the transformed measured field and the two-dimensional Fourier transform of $f(\mathbf{r})$ which is to be reconstructed. As ν is varied, the vector $\boldsymbol{\Lambda}_1\boldsymbol{\Theta}$, $|\nu| \leq k_0$, describes a semicircular contour in the $[\mu, \lambda]$ plane, over which $\tilde{f}(\cdot)$ is obtained from $W_\Theta(x_0, \nu)$. Figure 10.3-2 shows the contour for a given angle Θ and the process by

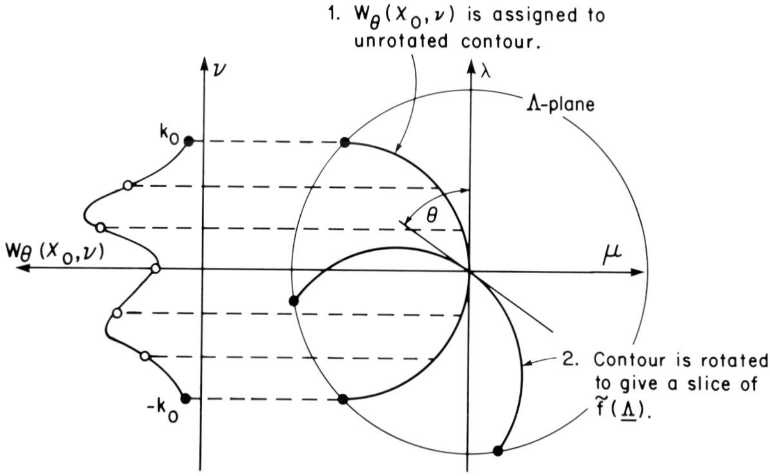

Fig. 10.3-2. Generation of a diffraction Fourier slice.

which values of $W_\Theta(x_0, \nu)$ are mapped onto this contour to approximately generate a semicircular "slice" of the function $\tilde{f}(\boldsymbol{\Lambda})$. The reconstruction is accomplished by obtaining a dense coverage of the frequency plane with these diffraction Fourier slices by varying Θ. We denote the collection of these contours the *reconstruction contours* (RCONs). In the next section we will discuss the practical issues concerning digital approaches to such a reconstruction problem.

We close this section by commenting on the relationship between tomography based on diffractionless phenomena, referred to here as straight-path tomography, and diffraction tomography. The straight-path tomography formulation can be obtained from the previously discussed situation by considering the limit as the source wavelength approaches zero. Under this assumption $k_0 \to \infty$, and the RCON in Fig. 10.3-2 over the object frequency range is well approximated by a straight-line tangent to the circle at the origin. The resulting mapping represents the well-known Fourier-slice property of straight-path tomography.

10.4 RECONSTRUCTION ALGORITHMS

In the previous section we set the basis for approximately recovering $f(\mathbf{r})$ from a set of measurements of the transmitted fields for different angular orientations of the object. The data base was generated on rotated semicircles denoted as RCONs in the spatial frequency domain. Figure 10.4-1 shows

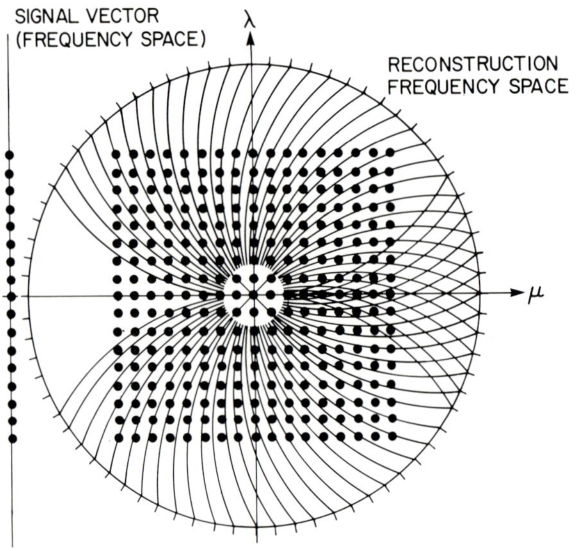

Fig. 10.4-1. Fourier domain reconstruction contours and a regular sampled-data grid.

a set of such RCONs superimposed on a uniform grid in the two-dimensional Fourier domain. Computational concerns naturally lead one to an interpolation of the data onto the uniform grid and use of the fast Fourier transform (FFT) to obtain the object function. In fact, this type of interpolation/reconstruction is common in other types of tomographic imaging as well. It therefore behooves us to develop a general formalism for this type of reconstruction that is equally applicable to a variety of RCONs. We first introduce the general formulation of the problem. This is followed by two types of reconstruction techniques: spatial domain and Fourier domain. Throughout, the more familiar straight-path tomography problem is included for comparison with reconstruction in diffraction tomography.

Consider a two-dimensional function $f(\mathbf{r})$ as given, for example, in Eq. (10.2-9). We assume that the measured (computed) information about $f(\mathbf{r})$ is given in the Fourier domain in terms of an observed function $O(\alpha, \beta)$, where (α, β) are related to the Cartesian coordinates (μ', λ') via a known transformation matrix $\mathbf{T}$. That is

$$\tilde{f}(\mu', \lambda') = O(\alpha, \beta), \tag{10.4-1}$$

with

$$\begin{bmatrix} \mu' \\ \lambda' \end{bmatrix} = \mathbf{T}(\alpha, \beta) = \begin{bmatrix} T_1(\alpha, \beta) \\ T_2(\alpha, \beta) \end{bmatrix}. \tag{10.4-2}$$

T_1 and T_2 are assumed to be continuous. The loci defined by $\mathbf{T}$ are the RCONs. Examples of (α, β) are

1. The polar coordinates found in conjunction with the Fourier-slice identity of straight-path tomography, with

$$\mathbf{T}(\alpha, \beta) = \begin{bmatrix} \cos\alpha & -\sin\alpha \\ \sin\alpha & \cos\alpha \end{bmatrix} \begin{bmatrix} 0 \\ \beta \end{bmatrix} \qquad \text{for } 0 \leqslant \alpha < \pi, |\beta| < \infty. \tag{10.4-3}$$

2. The semicircles given in Eq. (10.3-10) with radius k_0 associated with diffraction tomography. In this case $\mathbf{T}$ is given by

$$\mathbf{T}(\alpha, \beta) = \begin{bmatrix} \cos\alpha & -\sin\alpha \\ \sin\alpha & \cos\alpha \end{bmatrix} \begin{bmatrix} -k_0 + \sqrt{k_0^2 - \beta^2} \\ \beta \end{bmatrix}$$
$$\text{for } 0 \leqslant \alpha \leqslant 2\pi, |\beta| \leqslant k_0. \qquad (10.4\text{-}4).$$

As mentioned previously, $O(\alpha, \beta)$ is usually only available over a finite number of variables (α_n, β_m). The objective of the reconstruction algorithms is to obtain an estimate of $f(\mathbf{r})$ based on the set of available $O(\alpha_n, \beta_m)$. It is at this point that the question of interpolation has traditionally arisen. Those who have favored Fourier-domain reconstruction have interpolated $O(\alpha_n, \beta_m)$ onto a uniform grid, using one of the traditional polynomial interpolation functions, to obtain an estimate of $f(\mu_i, \lambda_j)$ [22]. This approach ignores the specific and intimate relation that exists between (μ', λ') and (α, β) and, therefore, is suboptimum. A notable exception to this approach, for angularly band-limited data in straight-path tomography, has been proposed by Stark and co-workers [23] in the form of exact Fourier-domain interpolation in radial and angular coordinates. In the following, we discuss a technique that is based on the relation mentioned above and operates on the exact integral relations between $f(x, y)$, $O(\alpha, \beta)$, and $\tilde{f}(\mu', \lambda')$. We note in passing that the Fourier-slice relation and its diffraction theory counterpart are based on Fourier integrals, leading one to continue using the integral relations in formulating the interpolations as well.

Consider the inverse Fourier transform equation for $f(x, y)$

$$f(x, y) = \frac{1}{4\pi^2} \int_{-\infty}^{\infty} \int_{-\infty}^{\infty} \tilde{f}(\mu', \lambda') \exp[j(\mu' x + \lambda' y)] \, d\mu' \, d\lambda'. \quad (10.4\text{-}5)$$

For simplicity, let us assume that the mapping represented by **T** is one-to-one. Therefore, with the help of Eq. (10.4-1), we can make a change of variables in Eq. (10.4-5) to obtain:

$$f(x, y) = \frac{1}{4\pi^2} \int_{-\infty}^{\infty} \int_{-\infty}^{\infty} O(\alpha, \beta) \exp[j\{T_1(\alpha, \beta)x + T_2(\alpha, \beta)y\}] J(\alpha, \beta) \, d\alpha \, d\beta, \quad (10.4\text{-}6)$$

where $J(\alpha, \beta)$ is the Jacobian of the transformation, given by

$$J(\alpha, \beta) = \left| \frac{\partial(\mu', \lambda')}{\partial(\alpha, \beta)} \right|. \quad (10.4\text{-}7)$$

For data obtained at different rotation angles θ, we let $\beta = \nu$ and $\alpha = \theta$. The Jacobians in Eq. (10.4-7) for the straight-path and diffraction cases are given by

$$J(\theta, \nu) = |\nu|, \qquad \text{straight-path}$$

$$J(\theta, \nu) = \frac{k_0 |\nu|}{\sqrt{k_0^2 - \nu^2}}, \qquad |\nu| < k_0, \text{ diffraction.}$$

Equation (10.4-6) sets forth the possibility of two types of reconstruction, each implying interpolation in a different domain. The first, denoted by *backpropagation* for diffraction tomography [12] and convolution backprojection for straight-path tomography [1], performs the effective interpolation in the spatial domain. The second, which has been called unified Fourier reconstruction (UFR), is again based on Eq. (10.4-6) but does the interpolation in the Fourier space [24].

A. Spatial Domain Reconstruction

We begin by deriving the convolution backprojection algorithm for straight-path reconstruction. Notice that in this case (ν, θ) constitute the polar coordinates in the Fourier space.

Let $[x', y'] \equiv [x, y]\mathbf{\Theta}^T$ denote the Cartesian coordinates rotated by $-\theta$. Substituting Eq. (10.4-3) into Eq. (10.4-6) results in

$$f(x, y) = \frac{1}{2\pi} \int_0^{\pi} \left[\frac{1}{2\pi} \int_{-\infty}^{\infty} |\nu| O(\theta, \nu) \exp(jy'\nu) \right] d\nu \, d\theta. \qquad (10.4\text{-}8)$$

The term inside [ ] is a one-dimensional inverse Fourier transform (IFT). In practice, of course, data are only available over finite number of angles θ_i, $i = 1, \ldots, K$, and in the form of N-point discrete Fourier transforms (DFT). An $N \times N$ reconstruction then becomes

$$f(x_k, y_l) = \frac{1}{2\pi} \sum_{i=1}^{K} f'(y'_j, \theta_i), \qquad (10.4\text{-}9)$$

where

$$f'(y'_j, \theta_i) = \text{IDFT}[|\nu_j| O(\theta, \nu_j)], \qquad (10.4\text{-}10)$$

and (x_k, y_l), for an angle θ_i and rotated coordinate y'_j, are taken as the nearest grid points satisfying

$$x \sin \theta_i + y \cos \theta_i = y'_j, \qquad j = 1, \ldots, N. \qquad (10.4\text{-}11)$$

The computational burden of the above process using the one-dimensional FFT consists of K N-point FFTs followed by N^2 assignments based on Eq. (10.4-11). Due to the nature of the assignment of the values of the pixels we consider this approach to fall in the category of interpolation in the spatial domain.

We next turn our attention to the diffraction counterpart of the convolution backprojection method. This method was originally proposed by Devaney and denoted the *backpropagation algorithm.* The algorithm can, once again, be developed from the general formulation in Eq. (10.4-6). Using our earlier notation, it follows that

$$f(x, y) = \frac{1}{2\pi} \int_0^{2\pi} \left\{ \frac{1}{2\pi} \int_{-k_0}^{k_0} \frac{k_0 |\nu|}{\sqrt{k_0^2 - \nu^2}} O(\theta, \nu) \times \exp[j(\gamma(\nu)x' + \nu y')] \, d\nu \right\} d\theta, \qquad (10.4\text{-}12)$$

where

$$\gamma(\nu) = -k_0 + \sqrt{k_0^2 - \nu^2}.$$

Again, for a discrete set of parameters and sampled image reconstruction we have

$$f(x_k, y_l) = \frac{1}{2\pi} \sum_{i=1}^{K} f'(y'_j, x'_m, \theta_i), \qquad (10.4\text{-}13)$$

where

$$f'(y'_j, x'_m, \theta_i) = \text{IDFT}\left[\frac{k_0|\nu_j|}{\sqrt{k_0^2 - \nu_j^2}} O(\theta_i, \nu_j)\, e^{\gamma(\nu_j)x'_m}\right]. \qquad (10.4\text{-}14)$$

Equation (10.4-14) differs from Eq. (10.4-10) in one major respect. In the present case the partial reconstruction $f'(y'_j, x'_m, \theta_i)$, is a function of x'_m. Thus for an $N \times N$ reconstruction it must be evaluated for N values of x'_m. Then (x_k, y_l) are taken as the nearst grid points satisfying

$$[x \; y]\Theta^{-1} = [x'_m \; y'_j], \qquad j, m = 1, \ldots, N. \qquad (10.4\text{-}15)$$

The computational requirement for this algorithm is on the order of (KN) N-point FFTs followed by N^2 assignments in the spatial domain. In summary, the backpropagation reconstruction technique contains the same theoretical elegance for diffraction tomography that backprojection offers for straight-path tomography. However, the backprojection method is substantially less computation-intensive than the backpropagation method.

To evaluate the performance of the various methods in reconstruction, we use the Shepp–Logan phantom with its Fourier transform evaluated on the diffraction RCONs [25]. Figure 10.4-2 shows the reconstructed phantom

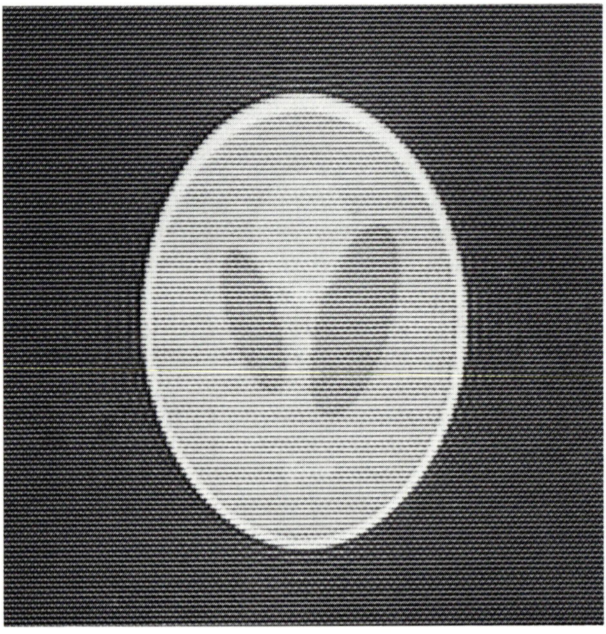

Fig. 10.4-2. The Shepp–Logan phantom.

as the (128×128) point IDFT of the exact phantom Fourier transform evaluated on a (128×128) square grid. This image is used as the ideal reconstruction, i.e., one for which no interpolation has been performed. Figures 10.4-3a and 10.4-3b show (128×128) point backpropagation reconstructions of the phantom based on 64 and 32 profiles, respectively. The sum of squared errors in reconstruction within the largest ellipse, denoted as error, is also given in the figure captions. We now turn our attention to a Fourier domain reconstruction technique that is closely related to the backprojection and backpropagation methods.

B. Frequency Domain Reconstruction

As mentioned earlier, one may perform the reconstruction efficiently in the Fourier domain by interpolating the values of O from the RCONs onto a uniform two-dimensional grid that is suitable for an FFT operation. Earlier attempts in this regard for both the straight-path and diffraction cases centered around some form of simple polynomial interpolation. An example of such interpolation known as the bilinear interpolation method [22, 25] is illustrated in Fig. 10.4-4. Denoting the distances between $[\mu_i, \lambda_k]$ and its four closest data points by a, b, c and d, the bilineär interpoaltion method is described by

$$\tilde{f}(\mu_i, \lambda_k) = \frac{O(\alpha_n, \beta_{m-1})/a + O(\alpha_n, \beta_m)/b + O(\alpha_{n+1}, \beta_m)/c + O(\alpha_{n+1}, \beta_{n-1})/d}{1/a + 1/b + 1/c + 1/d}, \tag{10.4-16}$$

where the arguments of $O(,)$ indicate the four nearest neighbors of the grid point at (μ_i, λ_k). We classify this type of interpolation as grid-driven. That is, one assigns a value to a grid point based on some of the data that are closest to it. Figure 10.4-5 shows reconstructed images based on 64 and 32 profiles using the bilinear method. Note that the results are both visually and numerically inferior to those generated by the backpropagation method. In fact, Fourier domain methods such as this have generally been considered inferior in quality of reconstruction to the previously discussed spatial domain methods.† A common characteristic of the *ad hoc* Fourier techniques is that they ignore the exact integral relation in Eq. (10.4-15) between $O(\alpha, \beta)$ and $f(\mathbf{r})$. In the following we discuss a method called unified Fourier reconstruction (UFR) method, which derives its interpolation

† Fourier-domain reconstructions of comparable quality with the backprojection reconstructions in straight-path tomography have been reported in [26].

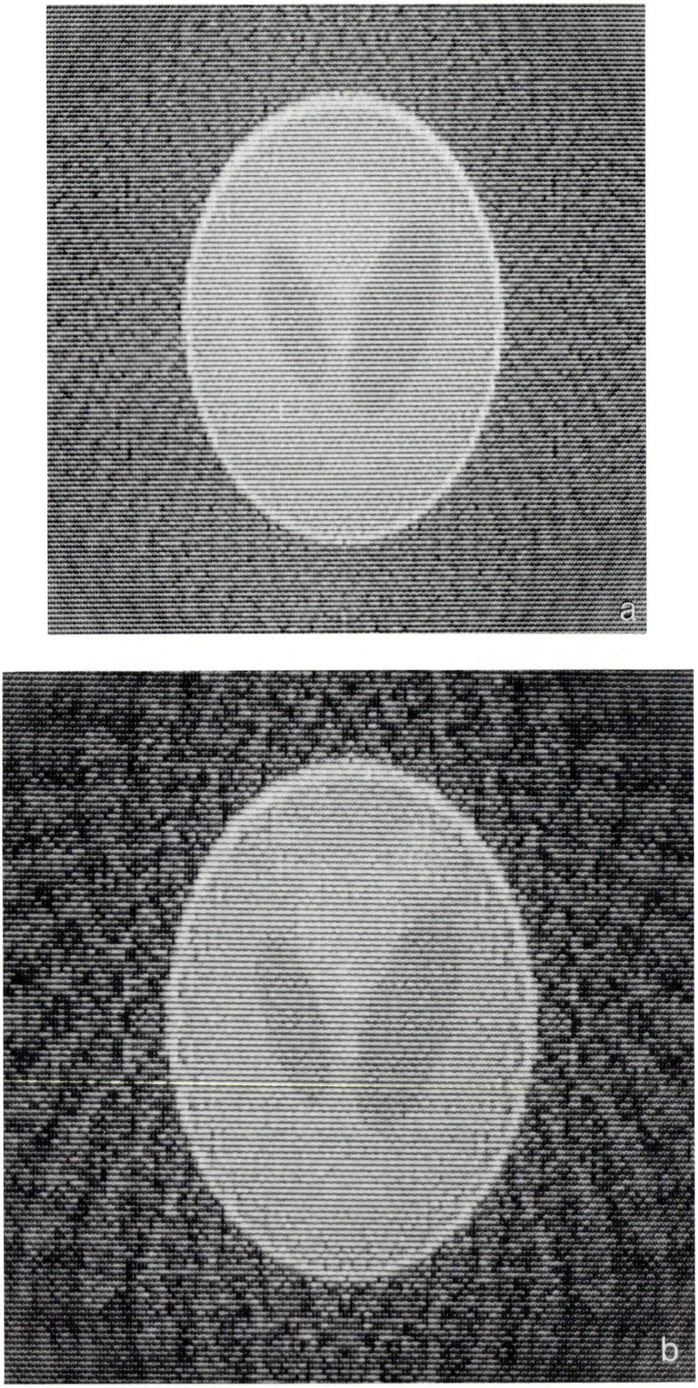

Fig. 10.4-3. Backpropagation reconstructions: (a) 64 profiles, error = 30.7; (b) 32 profiles, error = 51.2.

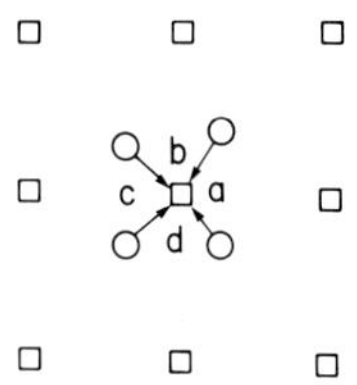

Fig. 10.4-4. Bilinear interpolation procedure.

scheme from Eq. (10.4-6). To do so, we temporarily maintain the continuous-space formalism of Eq. (10.4-6).

Let $I(x, y)$ be the indicator function for the support of the region of the image D that we desire to reconstruct. That is, $I(x, y) = 1$, $(x, y) \in D$ and $I(x, y) = 0$ otherwise. Usually D is taken as a circle of radius R that is large enough to accommodate the expected support $f(x, y)$. It can, however, be smaller to focus attention on a certain portion of the image. In any event, we define the spatially windowed object function by q. Then

$$q(x, y) = I(x, y)f(x, y) \tag{10.4-17}$$

and

$$\tilde{q}(\mu, \lambda) = \tilde{I}(\mu, \lambda)^*\tilde{f}(\mu, \lambda), \tag{10.4-18}$$

where * indicates the two-dimensional convolution integral. Note that when I includes the support of f then $q = f$. Then, using Eq. (10.4-6) in Eq. (10.4-18), we obtain the exact relation at any Cartesian grid point (μ_i, λ_k) in the frequency domain as

$$\tilde{q}(\mu_i, \lambda_k) = \frac{1}{4\pi^2}\int_{-\infty}^{\infty}\int_{-\infty}^{\infty} \tilde{I}[\mu_i - T_1(\alpha, \beta),$$
$$\lambda_k - T_2(\alpha, \beta)]O(\alpha, \beta)J(\alpha, \beta)\, d\alpha\, d\beta. \tag{10.4-19}$$

Equation (10.4-19) is an exact consequence of the basic relation expressed in Eq. (10.4-6). At this point we can express the fact that $O(\alpha, \beta)$ is only available over a finite number of points (α_n, β_m) through an ideal sampling representation. That is,

$$O(\alpha, \beta) = \sum_{n=1}^{K}\sum_{m=1}^{N} O(\alpha_n, \beta_m)\, \delta(\alpha - \alpha_n, \beta - \beta_m), \tag{10.4-20}$$

where $\delta(x, y)$ is the two-dimensional Kronecker delta function. Substituting Eq. (10.4-20) into Eq. (10.4-21), the exact [within the sampling errors offered by (α_n, β_m)] reconstruction formula in the Fourier domain is obtained. We

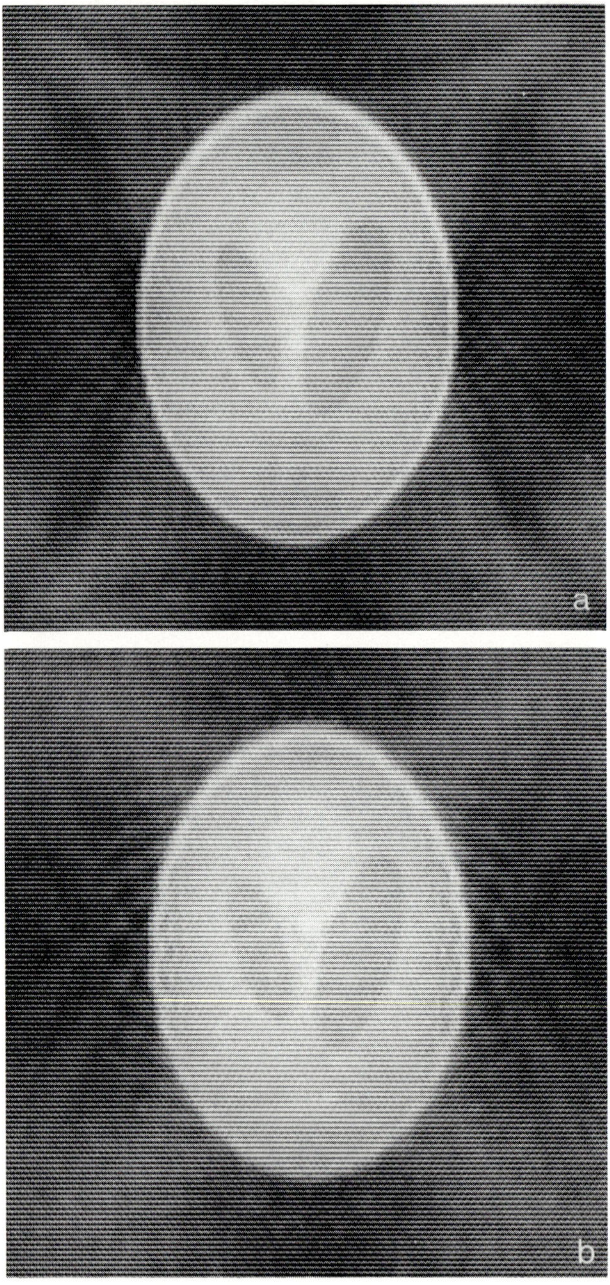

Fig. 10.4-5. Bilinear reconstruction: (a) 64 profiles, error = 93; (b) 32 profiles, error = 137.5.

have

$$\tilde{q}(\mu_i, \lambda_k) = \sum_{n=1}^{K} \sum_{m=1}^{N} J(\alpha_n, \beta_m) O(\alpha_n, \beta_m), \tilde{I}[\mu_i - T_1(\alpha_n, \beta_m), \lambda_k - T_2(\alpha_n, \beta_m)]. \quad (10.4\text{-}21)$$

The form of Eq. (10.4-21) is reminiscent of Nyquist's interpolation formula for a sampled signal. The interpolation kernel in the present case has two distinct components. The scale factor $J(\alpha_n, \beta_m)$ appropriately normalizes the density of the irregularly spaced available data and is crucial to the success of this algorithm. $\tilde{I}(\cdot, \cdot)$ is the continuous frequency-space interpolation function containing information about the support of the function to be reconstructed. An example of $I(\mathbf{r})$ is a disk of radius R. In this case,

$$\tilde{I}(\mu, \lambda) = \frac{J_1(R\sqrt{\mu^2 + \lambda^2})}{R\sqrt{\mu^2 + \lambda^2}},$$

where J_1 is a first-order Bessel function. When R is on the order of half the image dimensions, the main lobe of $\tilde{I}$ has a width on the order of the grid separation in the frequency domain. Thus the main contribution of the kernel $\tilde{I}$ in Eq. (10.4-21), for a given data point at (α_n, β_m), is to the grid points that are closest to (α_n, β_m). Consequently, for computational efficiency we approximate $\tilde{q}(\mu_i, \nu_k)$ in terms of a small number of its neighboring data points. This assignment is obviously data-driven. Thus

$$\tilde{q}(\mu_i, \lambda_k) \simeq \sum_{n} \sum_{m} J(\alpha_n, \beta_m) O(\alpha_n, \beta_m) \tilde{I}[\mu_i - T_1(\alpha_n, \beta_m), \lambda_k - T_2(\alpha_n, \beta_m)], \quad (10.4\text{-}22)$$

where indices (n, m) are such that $[\mu_i, \nu_k]$ is one of L nearest neighbors of $\mathbf{T}(\alpha_n, \beta_m)$.

The computational requirement for realizing Eq. (10.4-22) is on the order of LN^2 operations followed by a two-dimensional FFT, where L is typically 4 or 9. The assignment operation indicated in Eq. (10.4-22) is shown graphically in Fig. 10.4-6 for nine nearest neighbors. Figures 10.4-7 and 10.4-8 show reconstructions of the Shepp–Logan phantom using four and

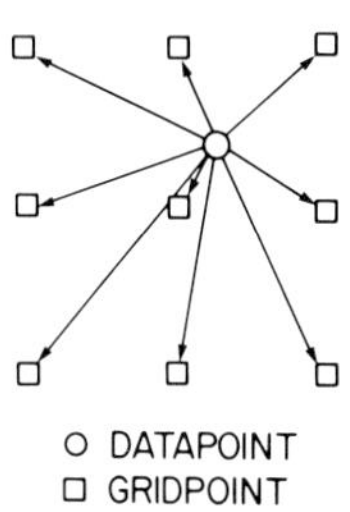

Fig. 10.4-6. Nine-neighbor UFR interpolation procedure.

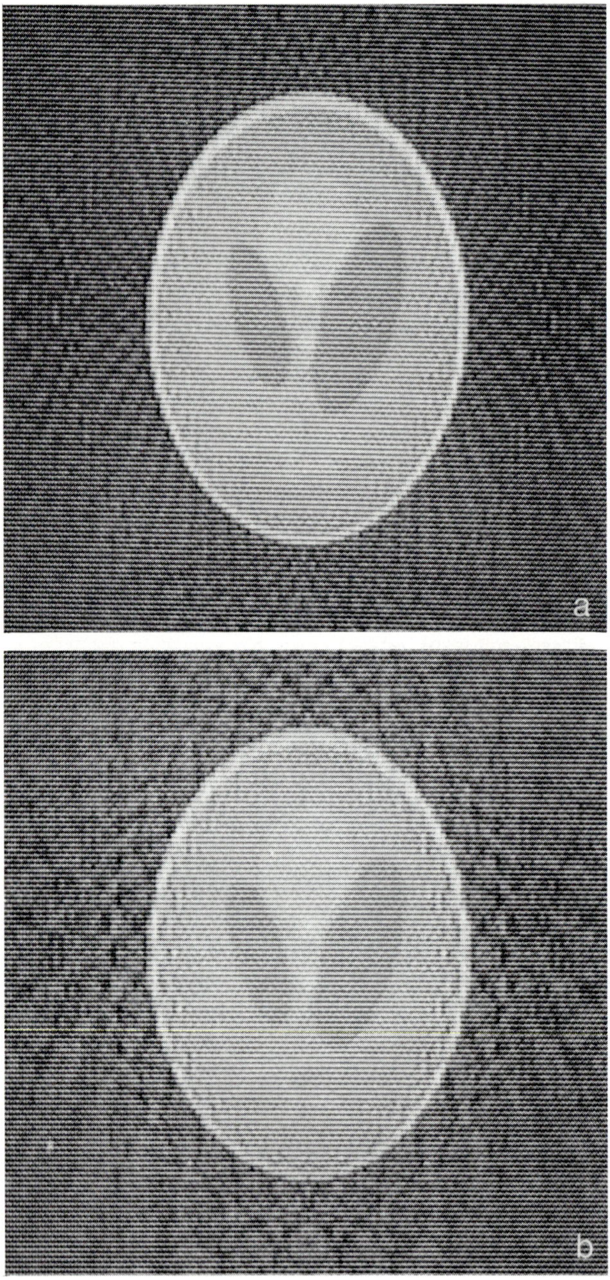

Fig. 10.4-7. Four-neighbor UFR reconstruction: (a) 64 profiles, error = 29.5; (b) 32 profiles, error = 44.5.

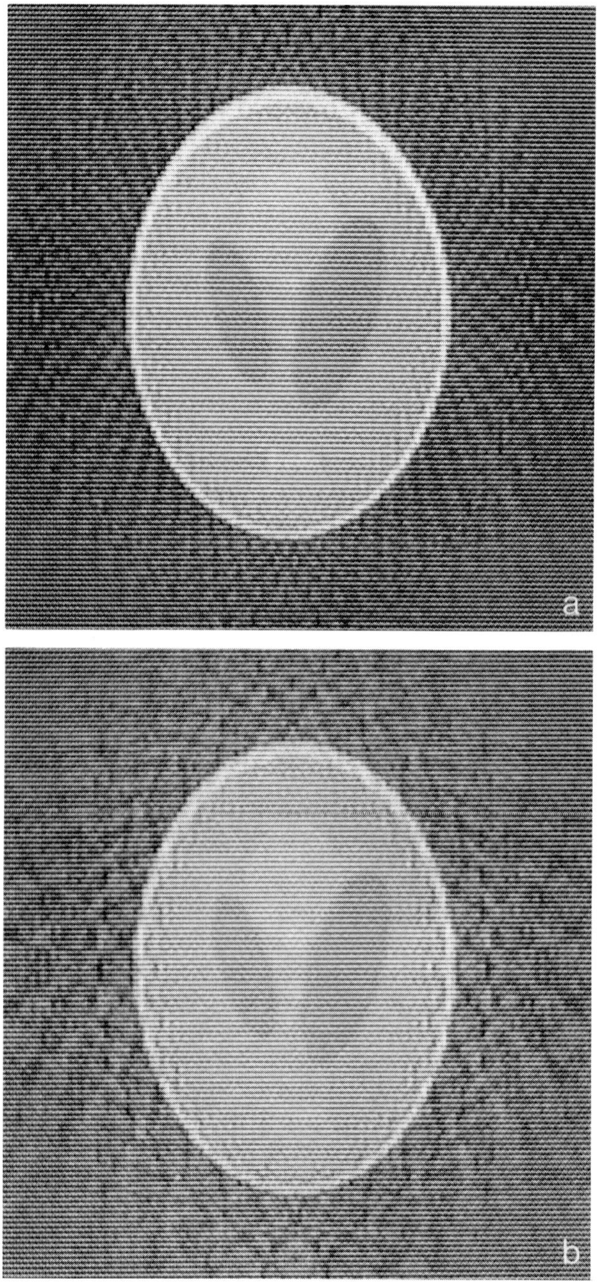

Fig. 10.4-8. Nine-neighbor UFR reconstruction: (a) 64 profiles, error = 23; (b) 32 profiles, error = 32.

nine neighbors, respectively. The computation time for the nine-neighbor case is slightly higher than that for the bilinear interpolation algorithm and about 1/20 of that for the backpropagation algorithm. The squared error of this algorithm is substantially less than that of the bilinear method and comparable to that of backpropagation. Further evidence of the effectiveness of the UFR algorithm is given in Fig. 10.4-9. In this case cuts through 64- and 32-profile (64 × 64) point reconstructions of two concentric cylinders are shown for the UFR and backpropagation reconstructions.

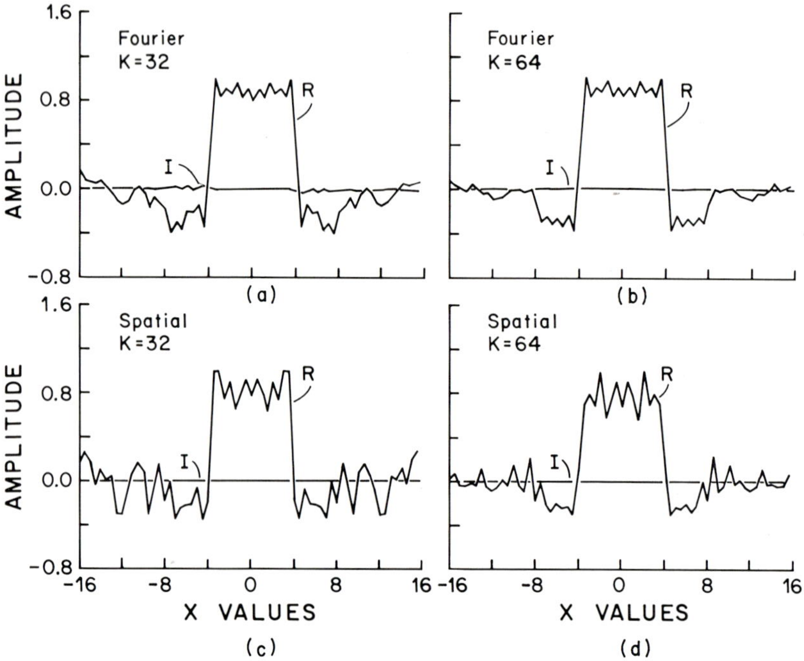

Fig. 10.4-9. Cuts through the reconstructions of two concentric cylinders using the back-propagation and UFR algorithms.

We close this section by giving the UFR formula for the approximate reconstruction of $f(\mathbf{r})$ or alternatively $\tilde{f}(\mu_i, \lambda_k)$ based on the measurement of profiles for different angular rotations θ_m of the object. Substituting the expression for $\tilde{f}(\boldsymbol{\Lambda\theta})$ in Eq. (10.3-10) for $O(\alpha_n, \beta_m)$ and the diffraction J in Eq. (10.4-31), we obtain

$$\tilde{f}(\mu_i, \lambda_k) \simeq \sum_n \sum_m \frac{2j}{k_0} e^{-j\gamma_m} |\nu_m| W_{\Theta_n}(x_0, \nu_m) \tilde{I}(\mu_i - \cos\theta_n \gamma_m + \sin\theta_n \nu_m, \lambda_k - \sin\theta_n \gamma_m - \cos\theta_n \nu_m), \qquad (10.4\text{-}32)$$

where indices (n, m) are such that $[\mu_i, \lambda_k]$ is one of the L nearest neighbors of $\mathbf{T}(\theta_n, \nu_m)$ and $\gamma_m = -k_0 + \sqrt{k_0^2 - \nu_m^2}$. We stress here that the approximate equality in Eq. (10.4-32) is now a result of both the linear approximation in Eq. (10.3-5) and interpolation errors in the reconstruction stage.

10.5 RECONSTRUCTION FROM LIMITED ANGULAR DATA

In the previous section we discussed algorithms for reconstructing the object function on a uniform grid in the spatial domain when the values of the observed function are known at equally spaced, suitably sampled points in the (θ, ν) domain. However, in many circumstances the number and/or range of available object angular profiles are limited by the physical constraints of the imaging system. Such constraints cause regions of unknown data or gaps in the spatial-frequency plane that prevent one from directly applying either the Fourier or the spatial domain reconstruction algorithm mentioned earlier. This is known as the limited angle problem [27-29].

A different case arises when segments of some of the object profiles cannot be measured due to the presence of obstacles between the source radiation and the object [30]. This is known as the incomplete object-projections-data problem. In general, the effect of the incomplete projections data is more severe than that of the limited angle data. Much has been reported concerning the limited angle and incomplete data problems in straight-path tomography systems. Many of these methods are variations of the two-dimensional Gerchberg-Papoulis iterative reconstruction algorithm [31, 32] and are discussed elsewhere in this book.

In diffraction tomography, iterative correction of the object profiles is, in practice, not possible. The difficulty lies in determining the scattered field at the receiver line (object profiles) from the estimated object function in each iteration (i.e., the solution of the forward-scattering problem). To solve the forward-scattering problem in each iteration, one is required to approximate the spatially varying differential equation (10.2-5) that relates the object profile to the object function. Obviously, the approximation errors introduced in the forward-scattering stage propagate and compound with each iteration. Thus, the iterative schemes might yield highly erroneous results.

The spatially varying nature of the wave equation is also the major source of difficulty in the incomplete data problems of diffraction tomography, i.e., when an obstacle is present. The analysis of diffraction and refraction phenomena within and between an inhomogeneous object and a hard-scattering obstacle is complex. It is nearly impossible to trace refracted rays and account for all multiple scattering effects within an object. The effects of obstacles in diffraction tomography cannot simply be interpreted as the absence of certain segments in the object projections (as is the case in straight-path tomography). Consequently, there is no numerically manageable reconstruction and/or restoration scheme that is applicable when diffraction plays an important role in the scattering within an object surrounded by opaque structures.

In this section we examine the limited angle problem in diffraction tomography. We do not intend to deal with incomplete and limited angle data problems in any general detail. These are covered for straight-path tomography in Chapter 9 of this book. Our aim is to differentiate between problems of this nature in straight-path and diffraction tomography systems and also to present an algorithm that is applicable to the partially known data problem in diffraction tomography. The algorithm is a noniterative interpolation scheme for approximately recovering missing object RCONs in the frequency domain. This clearly can also be viewed as a first-stage interpolation before application of the UFR procedure. The technique exploits the functional properties of the observed function such as its smoothness and periodicity with respect to the propagation angle, in conjunction with certain *a priori* information about the test object, to restore the unknown data. We begin our formulation by discussing the general interpolation problem for a periodic function. We then specialize the results to restoring the unavailable object projections in both diffraction and straight-path tomography systems.

A. Restoration of Unknown Profiles

The limited angle problem in diffraction tomography, when the range of the available data is 90°, is illustrated in Fig. 10.5-1. We start by introducing a convenient mapping of the observed function. Let $P(\theta', \rho')$ be the mapping of $\tilde{f}(\mu', \lambda')$ into the polar spatial frequency coordinates, i.e.,

$$P(\theta', \rho') = \tilde{f}(\mu', \lambda'), \qquad (10.5\text{-}1)$$

where

$$\theta' = \operatorname{arctg}(\lambda'/\mu'), \qquad \rho' = \sqrt{\mu'^2 + \lambda'^2},$$

and arctg is the four-quadrant inverse tangent operator. Suppose for a constant value of ρ', $P(\theta', \rho')$ is known over a discrete finite set of values of θ'. We denote this set by S and its complement by S^c. $P(\theta', \rho')$ is a periodic function of θ' with period 2π. Hence, if $\theta'_i \in S$ then the values of $P(\theta', \rho')$ are also known at $\theta' = \theta'_i + 2\pi k$, $k = 0, \pm 1, \pm 2, \ldots$.

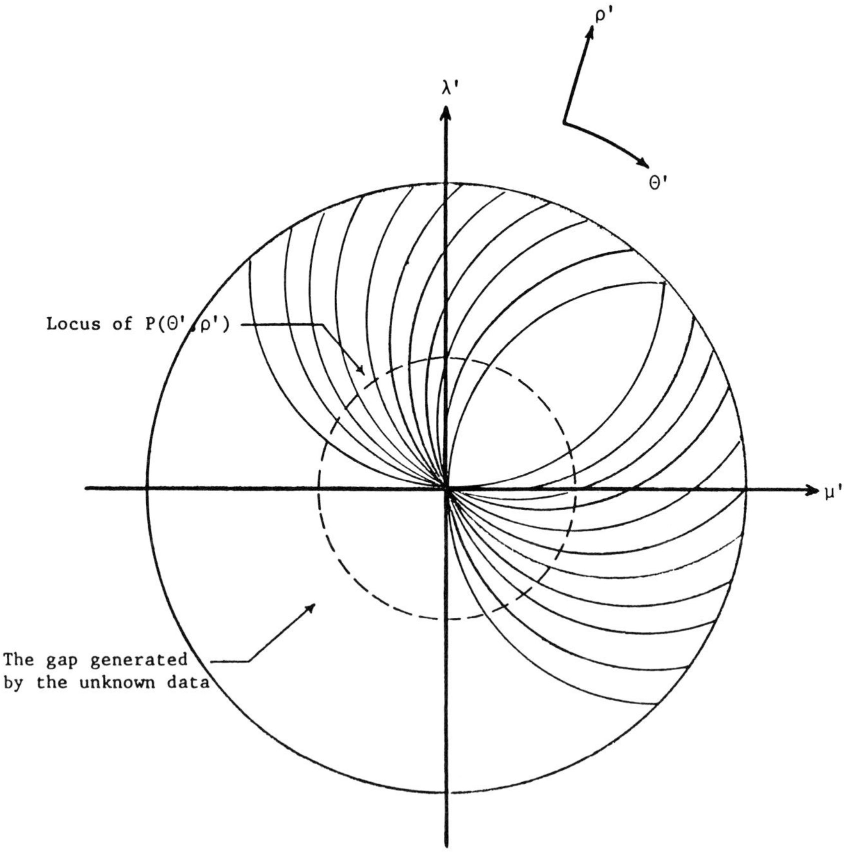

Fig. 10.5-1. Fourier domain coverage with diffraction profiles over a 90° interval.

Without loss in generality we assume that $\theta' = 0$ is an element of S and fit a periodic interpolating function over $[0, 2\pi]$ that fits $P(\theta'_j, \rho')$ for all $\theta'_j \in S$. For any $\theta'_i \in S_c$, the value of $P(\theta'_i, \rho')$ can be determined from the periodic interpolating function. In the following we present an interpolation scheme based on *trigonometric polynomials.*

We can represent $P(\theta', \rho')$ by the following Fourier series expansion:

$$P(\theta', \rho') = \sum_{n=-\infty}^{\infty} p_n(\rho')\, e^{jn\theta'}. \tag{10.5-2}$$

Assume that the object function is contained within a disk of radius R. Thus, the fast fluctuations of $P(\theta', \rho')$ in the (θ', ρ') domain correspond to the function [5]

$$P_1(\theta', \rho') = e^{j2\pi R\rho' \cos \theta'}. \tag{10.5-3}$$

The right-hand side of Eq. (10.5-3) has the following Fourier series expansion:

$$e^{j2\pi R\rho' \cos \theta'} = \sum_{n=0}^{\infty} \varepsilon_n j^n J_n(2\pi R\rho') \cos(n\theta'), \tag{10.5-4}$$

where $\varepsilon_0 = 1$, $\varepsilon_n = 2$ for $n \geqslant 1$ and J_n is a Bessel function of the first kind, nth order. The $J_n(2\pi R\rho)$ and, as a consequence, the magnitude of the components of the Fourier series in Eq. (10.5-4) become very small for $n > 2\pi R\rho'$. Hence, the following truncated Fourier series is a good approximation of the original infinite series:

$$P_1(\theta', \rho') \approx \sum_{n=0}^{M} \varepsilon_n j^n J_n(2\pi R\rho') \cos(n\theta'), \tag{10.5-5}$$

where

$$M \approx 2\pi R\rho'. \tag{10.5-6}$$

This implies that the following truncated Fourier series expansion is a good approximation of $P(\theta', \rho')$ with M chosen according to Eq. (10.5-6):

$$P(\theta', \rho') \approx \sum_{n=-M}^{M} p_n(\rho')\, e^{jn\theta'}. \tag{10.5-7}$$

Let the number of the elements in S be K_1. In practice, $K_1 \geqslant 2M + 1$. Hence, we can write the following set of K_1 linear equations:

$$P(\theta_i', \rho') \approx \sum_{n=-M}^{M} p_n(\rho')\, e^{jn\theta_i'}, \qquad \theta_i' \in S. \tag{10.5-8}$$

Using $\hat{p}_n(\rho')$ as an estimate of $p_n(\rho')$, we can write

$$\mathbf{E}\hat{\mathbf{p}} = \mathbf{P}, \tag{10.5-9}$$

where E is a $K_1 \times (2M + 1)$ matrix with elements $\{e^{jn\theta_i'}\}$, $\hat{\mathbf{p}}$ is a column vector of size $(2M + 1)$ of the estimates of the truncated Fourier series coefficients $\{p_n(\rho')\}$, and $\mathbf{P}$ is a column vector of size K_1 of the available data $\{P(\theta_i', \rho')\}$. The least-squares solution of Eq. (10.5-9) is given by

$$\hat{\mathbf{p}} = (\mathbf{E}^T\mathbf{E})^{-1}\mathbf{E}^T\mathbf{P}. \tag{10.5-10}$$

Finally, $P(\theta', \rho')$ can be estimated for any value of θ' with the help of Eqs. (10.5-7) and (10.5-10).

The above results form the basis for restoring the unknown values of the observed function in diffraction and straight-path tomography, as described below.

1. *Straight-Path Tomography*

In this case, we have

$$P(\theta', \rho') = 0(\theta, \nu), \tag{10.5-11}$$

where

$$\theta' = \theta \qquad \text{and} \qquad \rho' = \nu.$$

Thus, the scheme described above is directly applicable in straight-path tomography. It turns out that a similar periodic interpolation can be used on the object projections in the spatial domain. This method can then be utilized in the incomplete data problems.

2. *Diffraction Tomography*

For this problem

$$P(\theta', \rho') = O(\theta, \nu), \tag{10.5-12}$$

where

$$\theta' = \theta + \operatorname{arctg}(\nu/\gamma) \qquad \text{and} \qquad \rho' = \sqrt{\nu^2 + \gamma^2}.$$

These equations can easily be verified from Eq. (10.4-4) and the expression given for γ in Eq. (10.4-12). Note that for a constant value of $\pm\nu$, ρ' also assumes a constant value. Thus, for $\rho' = |\nu_m|$, the interpolating points are the $(\theta_i, \pm\gamma_m)$ pairs where the values of $O(\theta_i, \pm\nu_m)$ are known. Therefore, the accuracy of this interpolation method depends on the gap generated in the θ' domain, and not in the θ domain, by the unknown projections. This is an interesting feature in diffraction tomography, where it is required that one collect the object profiles over a 360° angular interval. For example, in acoustic microscopy based on the principles of diffraction tomography, physical constraints on scanning the specimen (test object) limit the range of available object profiles to close to 180°. This translates to a 180° gap in the θ domain. However, by formulating the interpolation problem in terms of $P(\theta', \rho')$, the gap generated by the unknown data in the θ' domain is very small at low and midband spatial frequencies. It can be shown that the maximum gap is 90°, occurring at $\rho' = k_0$.

B. Examples

The object function considered here is a superposition of 3, 5, and 10 wavelength cylinders of different gray levels. The available profiles for this object were obtained from the analytical expression for the Fourier transform of a homogeneous disk. Figure 10.5-2 shows the reconstruction results when 16 object profiles are available in a 90° angular interval. Figure 10.5-3 displays the results when eight object profiles are available in a 45° angular interval. In both figures "limited data" refers to UFR reconstruction without preliminary interpolation as described in this section. The improvement due to interpolation is obvious.

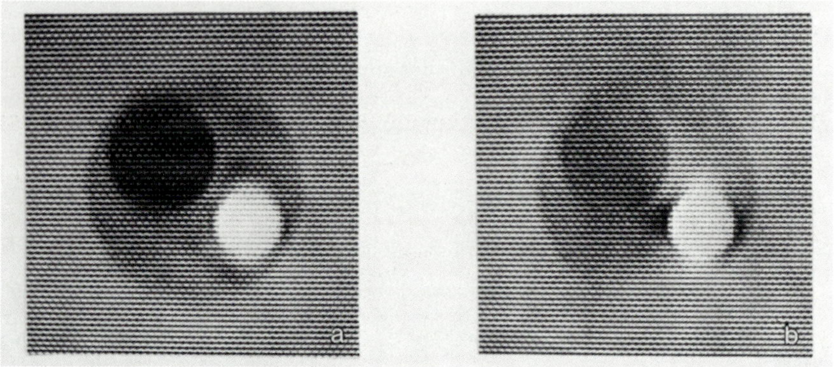

Fig. 10.5-2. Reconstructed images with sixteen profiles in a 90° angular interval. (a) interpolated, (b) limited data.

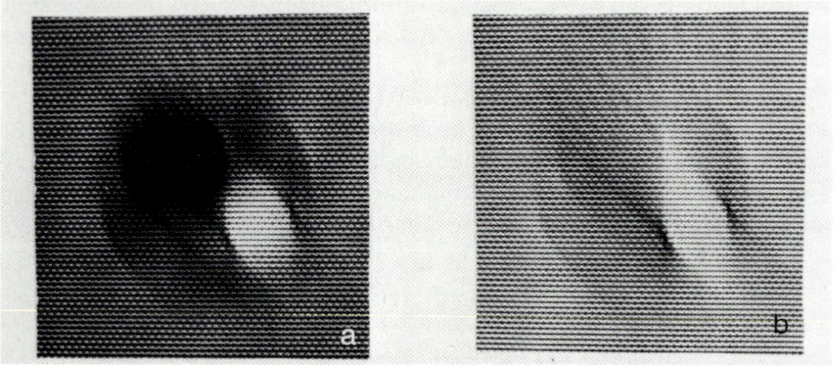

Fig. 10.5-3. Reconstructed images with eight profiles in a 45° angular interval. (a) interpolated, (b) limited data.

10.6 PHASE DETERMINATION

We mentioned in Section 10.2 that the Rytov transformation results in the preferred mode of first-order approximation to diffraction tomography.

There is a basic drawback to this technique, however, stemming from the need to obtain the complex logarithm of the normalized field. In this section we described a method which has been found to be effective in determining the imaginary part (unwrapped phase)† of the logarithm of the normalized field.

Denote the phase of $U_\Theta(x_0, y)$ by $\phi(\theta, y)$ and its principal value by $\phi_1(\theta, y)$. Then

$$\phi(\theta, y) = \phi_1(\theta, y) + 2\pi l \qquad \text{for some integer } l.$$

Ordinarily $\phi_1(\theta, y)$ is obtained through the use of an inverse tangent routine on the imaginary and real parts of $U_\Theta(x_0, y)$. In the phase unwrapping process the integer l is determined as a function of θ and y, resulting in the actual phase function, $\phi(\theta, y)$.

Traditionally, in signal processing applications, phase unwrapping has been a problem in the determination of the phase of the Fourier transform of a one-dimensional signal. As such, various constraints related to such Fourier transforms have been incorporated in the development of viable algorithms (see, e.g., [33, 34]). Unfortunately many of these constraints are not physically meaningful for the problem at hand, where the phase in the spatial domain is to be retrieved, and they have consequently proved to be ineffective in processing experimental data.

The algorithm discussed here appeals to four physical constraints on the measured set of data for diffraction tomography. The first follows from the fact that the phase of the normalized field is nearly zero outside the shadow region of the object. The second constraint assumes continuity of $\phi(\theta, y)$ as a function of y for a weakly scattering medium. This is similar to a condition often imposed in traditional phase unwrapping situations. The third constraint demands continuity on the low-frequency behavior of $\phi(\theta, y)$ as a function of θ. This constraint is a consequence of the observation that the large-scale variation in the phase is determined by the average velocity of propagation (time of flight) through the object at any given y. The fourth constraint is an auxiliary one that is intimately related to the reasons for the third one. Here one assumes an approximate knowledge, *a priori* or *a posteriori*, of the low-frequency variation of the phase at any θ. This again may be obtained from associated time-of-flight measurements, or by low-pass filtering (bandwidth approximately inverse of the diameter of the object) the in-phase and quadrature components of the measured fields prior to the computation of the phase. We denote this approximate

† Any complex function $F = |F| \exp(\psi_F)$ is uniquely specified by magnitude and phase (modulo 2π). Thus ψ_F and $\psi_F + 2\pi l$ (l an integer) specify the same function for a given $|F|$. However, the specification of log F requires knowing l. The determination of l and the principal value of ψ_F is the phase unwrapping problem.

phase function by $\phi_L(\theta, y)$ and demand that $\phi(\theta, y)$ not differ substantially from this low-frequency approximation.

We now consider the practical situation, where the principal value and the low-frequency approximation to the unwrapped phase are available over a discrete set of y and θ; i.e., $\phi(\theta_i, y_j)$ and $\phi_L(\theta_i, y_j)$, $i = 1, \ldots, k$ and $j = 1, \ldots, N$, are given. The constraints that were discussed above may be summarized as:

$$
\begin{aligned}
&\text{(i)} && \phi(\theta_i, y_1) = \phi_1(\theta_i, y_1)\\
& && \phi(\theta_i, y_N) = \phi_1(\theta_i, y_N)\\
&\text{(ii)} && |\phi_i, y_j) - \phi(\theta_i, y_{j-1})| < \pi\\
&\text{(iii)} && |\phi(\theta_i, y_j) - \phi(\theta_{i-1}, y_j)| < \pi\\
&\text{(iv)} && |\phi(\theta_i, y_i) - \phi_L(\theta_i, y_j)| < \alpha\pi,
\end{aligned}
$$

where $\alpha < 1$ is a factor that depends on the knowledge of the experiment and the size of the average velocity of propagation through the medium under study. We conjecture that in most practical situations condition (iv) can be simplified to comparing the average values of $\phi(\theta_i, y_j)$ and $\phi_L(\theta_i, y_j)$ over y. In the following, we summarize the phase unwrapping process and demonstrate the method by using ultrasonic data obtained from a breast phantom. Such a phantom is used for evaluating ultrasonic imaging techniques for the detection of breast tumors.

Consider the ith profile. Invoking constraint (i), commence the phase unwrapping for this profile by setting ϕ at one of the receiver edges, e.g., y_1 equal to its principal value:

Step (1). Set $\phi(\theta_i, y_1) = \phi_1(\theta_i, y_1)$. Further, from (ii) we can write

$$|\phi(\theta_i, y_2) - \phi(\theta_i, y_1)| < \pi.$$

Or, equivalently, we have

$$|\phi_1(\theta_i, y_2) + 2\pi 1_2 - \phi(\theta_i, y_1)| < \pi. \tag{10.6-1}$$

The values of $\phi_1(\theta_i, y_2)$ and $\phi(\theta_i, y_1)$ are known. Hence there exists a unique integer value for 1_2 which guarantees inequality (10.6-1). In like manner, the values of $1_j, j = 3, 4, \ldots, N$, can be obtained.

Step (2). Set $\phi(\theta_i, y_j) = \phi_1(\theta_i, y_j) + 2\pi 1_j$, where 1_j is an integer that ensures (ii). Repeat this step for all values of j, starting with the sample adjacent to y_1.

It should be noted that at any point, e.g., y_j, the value of 1_j is found with the help of the information available about the previous sample point. Therefore, a sample with low signal-to-noise ratio can corrupt the phase value of succeeding samples. A quick test for the possibility of occurrence of such a failure in step (2) is to make use of constraint (i) again. It is evident from (i) that the unwrapped phase at y_N should be equal to its principal value. In other words, the value of 1_N that was found in step (2) should be equal to zero. Therefore, if 1_N is equal to zero, one proceeds to unwrap the next profile. Otherwise, this profile is tagged for further processing based on other constraints to counter the erroneous effects of the noisy samples.

Step (3). Test $\phi(\theta_i, y_N)$ for constraint (i). If this constraint is not satisfied tag the profile as incorrect and proceed to the next profile. If (i) is satisfied go to the next step of further validating the accuracy of unwrapping.

The step of validating $\phi(\theta_i, y_N)$ obtained from step (3) involves a check against criterion (iv). Here, one checks to see if the resultant $\phi(\theta_i, y_j)$ is close, on the average, to $\phi_L(\theta_i, y_j)$.

Step (4). Compare the correct $\phi(\theta_i, y_j)$ from step (3) with $\phi_L(\theta_i, y_i)$ according to constraint (iv). If satisfied, tag $\phi(\theta_i, y_j)$ as valid and go to the next profile. Continue steps (1)–(4) until all the profiles are unwrapped and at least a subset is validated. The validated profiles are used to unwrap the rest.

Now assume that the $(i-1)$st profile is a valid one and the ith one is not. To unwrap the ith profile we invoke the additional constraint given in (iii). Thus,

Step (5). Set $\phi(\theta_i, y_j) = \phi_1(\theta_i, y_j) + 2\pi l_j$, where l_j is an integer that ensures $|\phi(\theta_{i-1}, y_j) - \phi(\theta_i, y_j)| < \pi$. Repeat for all values of j.

At this point, the ith profile should properly be unwrapped at all points for which the signal-to-noise ratio is high. There are situations where step (5) produces noisy samples of the profile. This can result in the appearance of small spikes in the phase of the ith profile. Further, the erroneous data may be used for the $(i+1)$st profile, causing a propagation of the noisy samples. This may be circumvented to some extent by appropriately low-pass filtering the unwrapped profile.

The above procedure is now demonstrated by using 60 profiles of 128 samples each from a breast phantom insonified by an 800-kHz ultrasonic signal. Using steps (1)–(4), only two valid phase profiles, the 10th and the 19th, were found to be valid. Figure 10.6-1 shows four profiles to indicate the various problems encountered and the appropriate solutions. In these figures $\hat{\phi}$ denotes an unwrapped phase on the basis of only constraints (i) and (ii).

Figure 10.6-1a shows the correctly unwrapped phase using steps (1)–(4). Figure 10.6-1b shows $\hat{\phi}$ when steps (1) and (2) were followed. Step (3) detected the divergence of $\hat{\phi}$ and step (5) performed the appropriate unwrapping using the 10th profile. Figure 10.6-1c shows a $\hat{\phi}$ that satisfies constraint (i) but fails the validation in step (4). The ϕ is obtained using

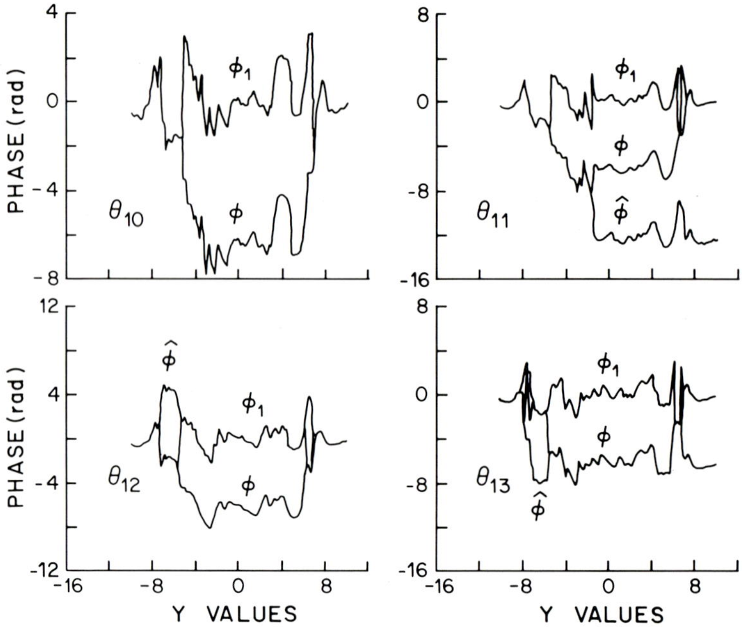

Fig. 10.6-1. Unwrapped phase profiles from a breast phantom.

step (5). Finally, Fig. 10.6-1d shows a $\hat{\phi}$ that satisfies validation in step (4) but fails the test in step (1). The phase is unwrapped on the basis of the 12th profile.

10.7 EXPERIMENTAL RESULTS AND COMPARISON OF BORN AND RYTOV METHODS

In this section we discuss the performance of the Born and Rytov methods in diffraction tomography. It is extremely difficult to give a general treatment of the errors in the reconstructed images with these techniques. Many of the nuances of the methods, however, can be demonstrated through examples generated by numerical simulation and experimental measurements of scattered fields from simple objects (phantoms). Some theoretical results are also available [33]. However, the complexity of such a treatment precludes its inclusion in this chapter. The tomograms displayed in this section were all obtained by using a nearest-neighbor or a UFR reconstruction technique. The emphasis in these examples is on the effect of approximation errors from choosing a particular transformation on the tomograms.

These "modeling" errors generally are the dominant sources of degradation in the recovered images.

A major difficulty in evaluating diffraction tomographic methods is acquisition of the appropriate data set. Since the data consist of the scattered fields from inhomogeneous structures, a simulation requires the forward solution of a generally complex boundary value problem, i.e., the solution of Eq. (10.2-1). Since this is normally a nontrivial task, existing simulation-based studies use soft cylindrical inhomogeneities as building blocks. The scattered fields from such objects can be obtained in terms of series solutions [35–37] and their extensions. Such techniques have been the sources of data generation for the examples presented here.

Experimental studies in diffraction tomography also present unique design and measurement challenges. First and foremost, experimental objects are no longer two-dimensional. Thus, care must be taken for the system and the object under study to approximately satisfy the conditions set forth in a two-dimensional formulation. This can be accomplished to a large extent by using phantoms that have cylindrical symmetry. At the time of this writing no three-dimensional reconstructions were available, largely due to the complexity of scanner design.

Figure 10.7-1 shows a diagram of an experimental ultrasonic diffraction tomography system at the University of Minnesota. The transmitting transducer generates an approximate *CW* planar wave. The scattered wave is coherently detected at a spatial sampling period of about one-half the wavelength of the insonifying field. For each rotation of the phantom, the in-phase and quadrature components of the wave are stored and constitute the sample profile $\psi_\Theta(x_0, y_j)$ for a particular angle θ. Processing of the data is then accomplished as indicated in Sections 10.3, 10.4, and 10.6.

The first set of examples are based on simulated data, scattered from centered cylindrical scatterers with six wavelength radii. Sixty-four profiles were used. Figure 10.7-2 shows slices of the reconstructed images for 4 and 8% positive deviations from the background with the Born transformation, while Fig. 10.7-3 shows the same reconstructions with the Rytov method. All plots show both the real and imaginary parts of the reconstructed images, which in this case should ideally be real and positive. The breakdown of the Born approximation even for such a weak disturbance is evident. Two points are worth noting. The largest error in the Rytov tomograms is clearly concentrated in the vicinity of the edges of the disturbance. This is to be expected since the additive error term in Eq. (10.2-13) has its largest contribution around the edges. The second point is the extremely poor performance of the Born method. Referring to Eq. (10.2-13) for the Born source term, this result is not too surprising. The error term in this case is a product of the object function and the normalized, complex, scattered

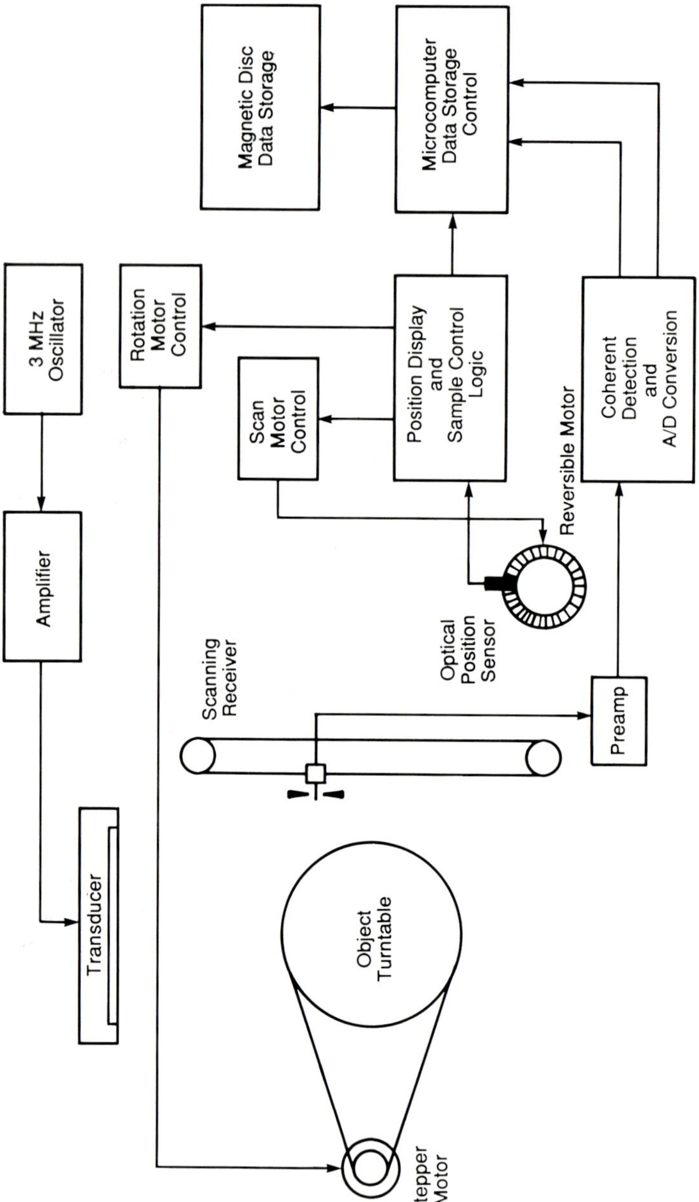

Fig. 10.7-1. An experimental ultrasonic diffraction tomography system.

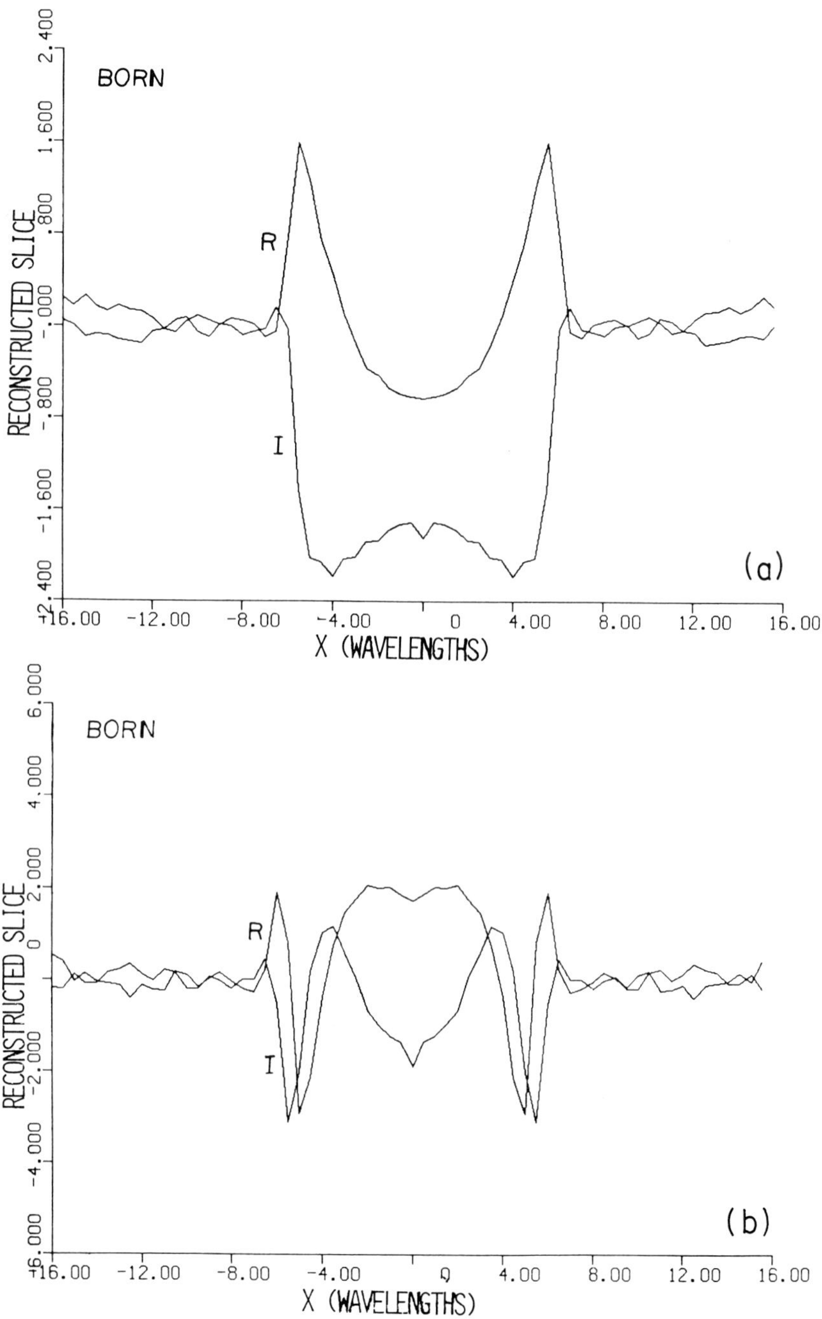

Fig. 10.7-2. Slices of Born reconstructions of cylindrical scatterers with radius R equal to six wavelengths: (a) $f(\mathbf{r}) = 0.04$, $|\mathbf{r}| \leq R$; (b) $f(\mathbf{r}) = 0.08$, $|\mathbf{r}| \leq R$.

field. As a result, the final reconstructed image is highly distorted as shown in Fig. 10.7-2, with mixing of the real and imaginary parts. It is interesting to observe, however, that the error contribution outside the support of the object function is very small. This again follows the fact that the source error term for each profile is proportional to the object function, thus zero outside its support.

Figure 10.7-4 further dramatizes the difference in performance between the Born and Rytov methods. The simulated data for this example were obtained from a complex circular object function

$$f(\mathbf{r}) = \begin{cases} 0.1 + j0.1 & \lambda_0 < |r| < 2\lambda_0, \\ 0.1 & |r| < \lambda_0, \end{cases}$$

where λ_0 is the source wavelength. The phenomena that are due to source error terms are again clearly demonstrated in these figures. In the case of the Rytov tomogram, the real and imaginary parts of the object function are distinguished and represented with reasonably good fidelity. The two components of the object function are completely mixed in the Born tomogram, rendering it useless.

Tomograms based on experimental measurements also confirm the general characteristics of the Born and Rytov methods observed above. The main source of difficulty in processing such data for the Rytov reconstruction is the previously discussed phase unwrapping stage due to the effects of receiver noise, quantization, and aperture sampling. Nevertheless, successful experimental tomograms have been obtained with two examples described below.

Figures 10.7-5 and 10.7-6, respectively, depict the Born and Rytov tomograms of a cylindrical phantom made of gelatin with a 120-wavelength diameter containing one 10-wavelength diameter hole. The phantom was immersed in water, so the hole was filled with water. The velocity of propagation for ultrasound in the gelatin was estimated as about 3% higher than in water with negligible attenuation. Thus, the tomograms are mainly real.

The Born tomogram again is devoid of any quantitative information, although the imaginary part gives the location of the hole. The real part in Fig. 10.7-5 gives a completely erroneous representation of the wavenumber inside the object. Notice that the deviation of the real part of the image from the surrounding water is only indicated at the boundary of the phantom. In contrast to the Born tomogram, that obtained with the Rytov transformation shows good fidelity of reconstruction, as demonstrated in Fig. 10.7-6.

The final experimental result presented is a Rytov diffraction togogram of a breast phantom at an ultrasound frequency of approximately 800k Hz. This is indeed a difficult phantom to image. The difficulties stem from the

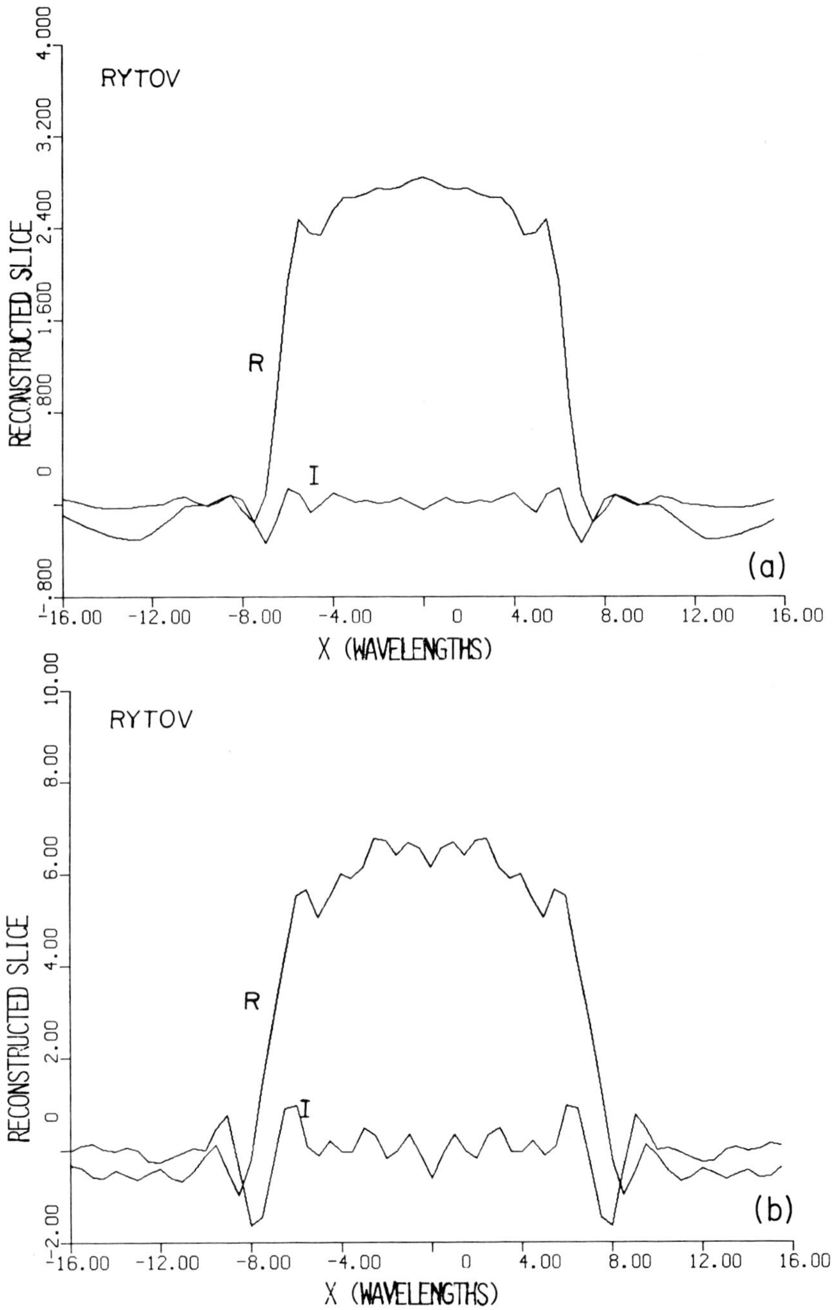

Fig. 10.7-3. Slices of Rytov reconstruction of the same cylinders as in Fig. 10.7-2.

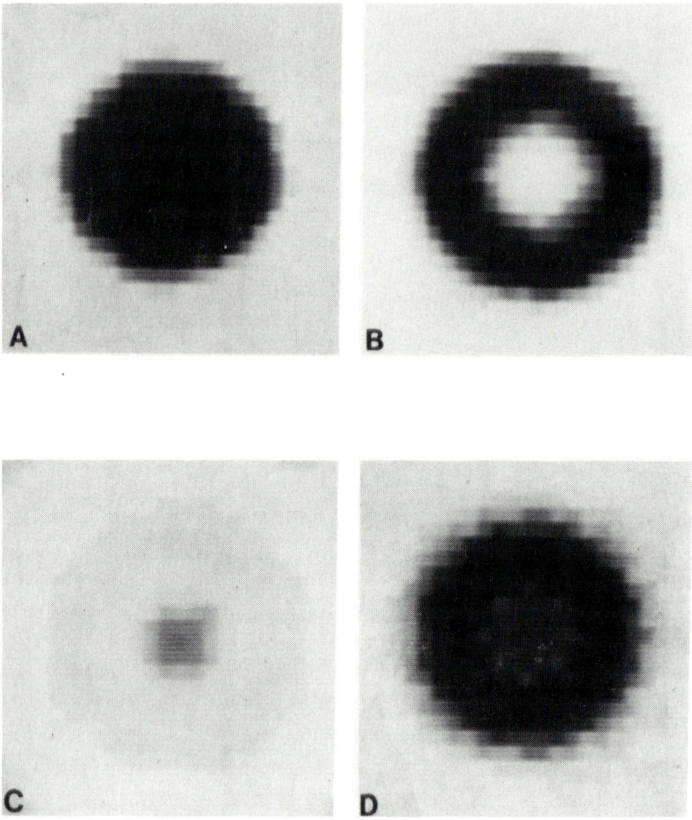

Fig. 10.7-4. Rytov and Born tomograms of a complex disturbance $f(\mathbf{r}) = 0.1 + j0.1$, $\lambda_0 < |\mathbf{r}| < 2\lambda_0$; $f(\mathbf{r}) = 0.1$, $|\mathbf{r}| < \lambda_0$: (A) Rytov, real; (B) Rytov, imaginary; (C) Born, real; (D) Born, imaginary.

complex three-dimensional and lossy nature of the object function. These attributes contribute to making the phase unwrapping stage a challenging one, as discussed in Section 10.6. Figure 10.7-7 shows the real part of the Rytov tomogram of this phantom, using the unwrapped phase profiles, some of which were shown in Fig. 10.6-1. The image shows structures similar to those obtained with x-ray and time-of-flight ultrasound tomograms. However, an exact assessment of its quantitative accuracy is thus far not available.

In summary, the results presented in this section confirm the prediction of the earlier sections of this chapter that the Born transformation in diffraction tomography is of extremely limited use. On the other hand, the

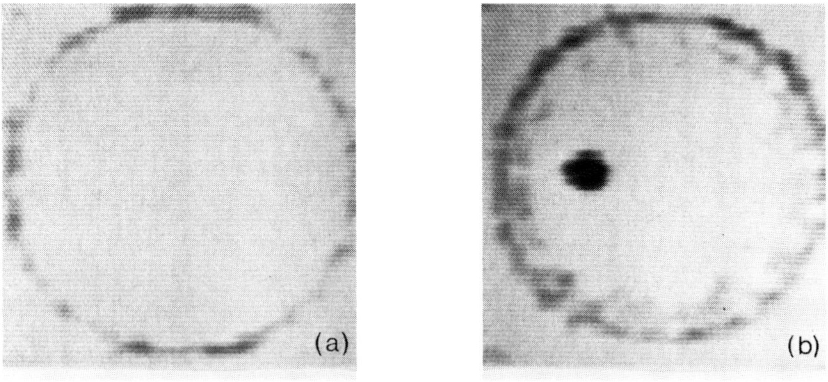

Fig. 10.7-5. Born tomogram of a gelatine cylinder with a water-filled hole: (a) real part; (b) imaginary part.

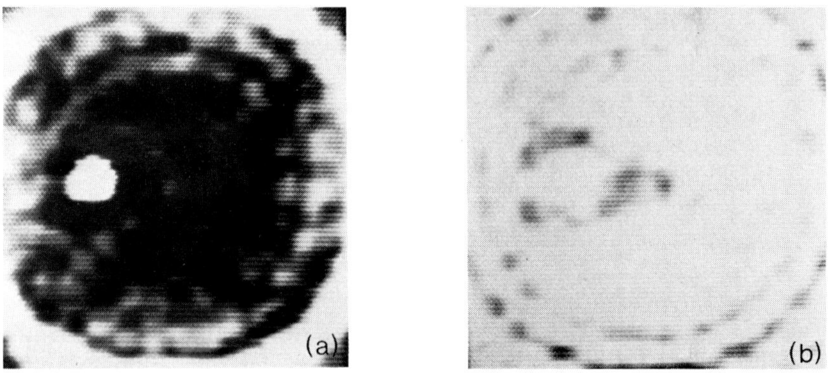

Fig. 10.7-6. Rytov tomogram of the same phantom as in Fig. 10.7-5: (a) real part; (b) imaginary part.

Rytov transformation was shown to have the potential of providing computationally efficient diffraction tomograms in certain practidal situations where phase unwrapping is possible and three-dimensional effects can be minimized or accounted for.

10.8. CONCLUDING REMARKS

Diffraction tomography is the process by which wave field data are translated into quantitative information, usually in the form of an image,

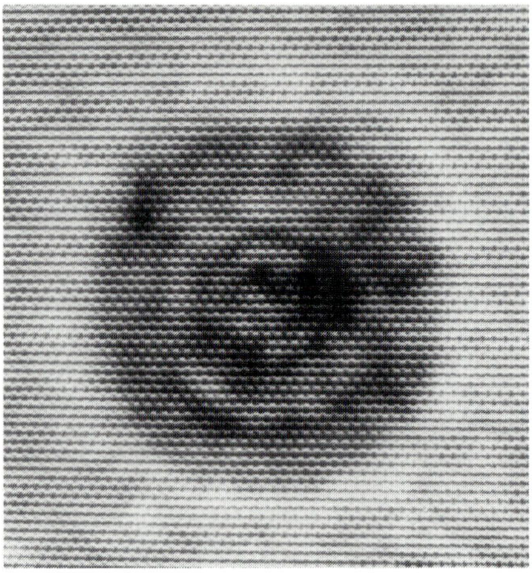

Fig. 10.7-7. Real part of the Rytov tomogram of a breast phantom.

about certain characteristics of a medium. In this chapter we presented approximate solutions to this problem when the propagation model is given by a two-dimensional Helmholtz-type equation. This equation is the only one for which, under the small-perturbation assumptions, mildly successful tomography algorithms have been developed thus far.

We close this chapter with observations about some unsolved practical issues pertaining to diffraction tomography. Even within the small-perturbation constraints alluded to earlier, the two-dimensional reconstruction algorithm suffers seriously when applied to media with complex three-dimensional structures. Three-dimensional inversion algorithms are, in principle, simple extensions of the ones discussed here and may be found in [8] and [12]. Unfortunately, experimental design, measurement, and signal processing requirements for such a formulation are very difficult. Partial solutions, intermediate between two- and three-dimensional schemes have been proposed [38]. No simulation or experimental results on any of these methods are available, however.

The key to the utility of any diffraction tomography method lies first and foremost in the accuracy of the source–medium interaction or wave equation model. Such an equation must be derived and justified for any given class of applications considered. An example is given in [39] for modeling ultrasound propagation in soft biological tissue. Results on signal processing aspects of this problem, such as phase unwrapping, reconstruction, and interpolation algorithms, are by no means conclusive; these areas and need further exploration.

Of course, the grail in diffraction tomography is the exact inversion of the wave equation under consideration. The search for this ultimate prize will continue to excite and often frustrate researchers in this field.

ACKNOWLEDGMENTS

Our work on diffraction tomography has been a collaborative effort with R. K. Mueller of the University of Minnesota. It has also benefited from the efforts of R. Iverson, T. Coulter, R. Rylander, D. Fishbaine, M. Chaplin, and S. Bedros, all formerly or presently with the University of Minnesota, and the insights of Z. Q. Lu of Nankai University. J. Greenleaf and his staff at the Biodynamics Research Unit, Mayo Clinic, have been the source of much of our recent experimental data. Our research on this subject has been generously supported by the National Science Foundation.

REFERENCES

1. A. C. Kak (1979). Computerized tomography with X-Ray, Emission and Ultrasonic Sources. *Proc. IEEE* **67**, 1245–1272.
2. L. A. Chernov (1969). "Wave Propagation in a Random Medium," McGraw-Hill, New York.
3. V. I. Tatarski (1961). "Wave Propagation in a Turbulent Medium," McGraw-Hill, New York.
4. J. W. Strohbehn (1968). Line of sight wave propagation through turbulent atmosphere. *Proc. IEEE* **56**, 1301–1318.
5. A. Ishimaru (1978). "Wave Propagation and Scattering in Random Media," Academic Press.
6. E. Wolf (1969). Three-dimensional structure determination of semi-transparent objects from holographic data. *Optics Commun.* **1**, 153–156.
7. K. Iwata and R. Nagata (1970). Calculation of three-dimensional refractive index distribution from interferograms. *J. Opt. Soc. Am.* **60**, 133–135.
8. R. F. Mueller, M. Kaveh, and G. Wade (1979). Reconstructive tomography and applications to ultrasonics. *Proc. IEEE* **67**, 567–587.
9. M. Kaveh, R. K. Mueller, and R. D. Iverson (1979). Ultrasonic tomography based on perturbation solutions of the wave equation. *Computer Graphics and Image Process.* **9**, 105–116.
10. R. K. Mueller, M. Kaveh, and R. D. Iverson (1980). A new approach to acoustic tomography using diffraction techniques. *In* "Acoustical Imaging" (A. F. Metherell ed.). Vol. 8, p. 615. Plenum, New York.
11. M. Kaveh, R. K. Mueller, R. Rylander, T. R. Coulter, and M. Soumekh (1980). Experimental results in ultrasonic diffraction tomography. *In* "Acoustical Imaging" (K. Wang ed.). Vol. 9, p. 433. Plenum, New York.
12. A. J. Devaney (1982). A filtered backpropagation algorithm for diffraction tomography. *Ultrasonic Imaging* **4**, 336–350.
13. M. L. Oristaglio (1983). A geometric approach to the filtered backpropagation algorithm. *Ultrasonic Imaging* **5**, 30–37.
14. A. J. Devaney (1984). Geophysical diffraction tomography. *IEEE Trans. Geoscience and Remote Sensing* **22**, 3–13.

15. M. F. Adams and A. P. Anderson (1982). Synthetic aperture tomographic (SAT) imaging for microwave diagnostics. *Proc. IEE part H 129*, 83–88.
16. F. Stenger and S. A. Johnson (1979). Ultrasonic transmission tomography based on the inversion of the helmholtz wave equation for plane and spherical wave insonification. *Appl. Math. Notes* **4**, 102–127.
17. M. Kaveh, M. Soumekh, and R. K. Mueller (1981). A comparison of Born and Rytov approximations in acoustic tomography. *In* "Acoustical Imaging (J. Powers ed.). Vol. 11, p. 325. Plenum, New York.
18. J. F. Greenleaf (1983). Computerized tomography with ultrasound. *Proc. IEEE* **71**, 330–337.
19. M. Slaney, A. Kak, and L. Larson (1984). Limitations of Imaging with First Order Diffraction Tomography. *IEEE Trans. Microwave Theory and Techniques* **32**, 8–15.
20. B. D. James, C. W. Yang, and W. M. Boerner (1975). Extension of scalar to vector propagation tomography—A computer numerical approach. *In* "Inverse Methods in Electromagnetic Imaging" (W. M. Boerner ed.) NATO ASI Series, Vol. 143, p. 1147. Reidel, Dordrecht.
21. M. Kaveh, M. Soumekh, Z. Q. Lu, R. K. Mueller, and J. F. Greenleaf (1982). Further results on diffraction tomography using Rytov's approximation. *In* "Acoustical Imaging" (E. A. Ash and C. R. Hill eds.). Vol. 12, p. 599. Plenum, New York.
22. R. M. Mersereau and A. V. Oppenheim (1974). Digital reconstruction of multidimensional signals from their projections. *Proc. IEEE* **62**, 210–229.
23. H. Stark, J. W. Woods, I. Paul, and R. Hingorani (1981). Direct fourier reconstruction in computer tomography. *IEEE Trans. Acoust., Speech and Signal Process.* **29**, 237–245.
24. M. Kaveh, M. Soumekh, and J. F. Greenleaf (1984). Signal processing for diffraction tomography. *IEEE Trans. Sonics and Ultrasonics* **31**, 230–238.
25. S. X. Pan and A. C. Kak (1983). A computational study of reconstruction algorithms for diffractive tomography: Interpolation vs. filtered-backpropagation. *IEEE Trans. Acoust., Speech and Signal Process.* **31**, 1262–1275.
26. H. Stark, J. W. Woods, I. Paul, and R. Hingorani (1981). An investigation of computerized tomography by direct Fourier inversion and optimum interpolation. *IEEE Trans. Biomed. Eng.* **28**, 496–505.
27. K. C. Tam and V. Peres-Mendez (1981). Tomographical imaging with limited-angle input. *J. Opt. Soc. Am.* **71**, 582–592.
28. M. Sezan and H. Stark (1982). Image restoration by the method of convex projections. *IEE Trans. Med. Imag.* **1**, 95–101.
29. F. A. Grunbaum (1980). A study of fourier space methods for limited angle image reconstruction. *Numer. Funct. Anal. and Optimiz.* **2**(1), 31–42.
30. B. Medoff, W. Brody, and A. Macovski (1983). The use of *a priori* information in image reconstruction from incomplete data. *In* "Proceedings of ICASSP", pp. 131–134.
31. R. W. Gerchberg (1974). *Super-resolution through error energy reduction. Opt. Acta* **21**, 709–720.
32. A. Papoulis (1975) A new algorithm in spectral analysis and band-limited extrapolation. *IEEE Trans. Circuits and Systems* **22**, 735–742.
33. J. Tribolet (1977). A new phase unwrapping algorithm. *IEEE Trans. Acoust. Speech and Signal Process.* **25**, 170–177.
34. K. Steiglitz and B. Dickinson (1982). Phase unwrapping by factorization. *IEEE Trans. Acoust., Speech and Signal Process.* **30**, 984–991.
35. M. Soumekh, and M. Kaveh (1986). A Theoretical Study of Model Approximation Errors in Diffraction Tomography. *IEE Trans. Ultrasonics, Ferroelectrics and Frequency Control.* **33**, 10–20.
36. P. M. Morse and K. U. Ingard (1968). "Theoretical Acoustics". McGraw-Hill, New York.

37. R. D. Iverson (1978). A simulation in the tomographic reconstruction of images by means of acoustic scattering. M. S. Thesis, University of Minnesota.
38. Z. Q. Lu, M. Kaveh and R. K. Mueller (1984). Diffraction tomography using beam waves: Z-average reconstruction. *Ultrasonic Imaging* **6**, 95–102.
39. R. K. Mueller (1980). Diffraction tomography. I: The wave equation. *Ultrasonic Imaging* **2**, 213–222.

CHAPTER **11**

Applications of Convex Projection Theory to Image Recovery in Tomography and Related Areas

*M. Ibrahim Sezan**

Henry Stark

Electrical, Computer, and Systems Engineering Department
Rensselaer Polytechnic Institute
Troy, New York 12180

11.1 INTRODUCTION

In computerized tomography (CT)—irrespective of the method of reconstruction—it is generally assumed that 360° view data are available for processing. What can one do in the way of reconstructing good quality imagery when less than 360° view data are available? This is the subject of this chapter. Assuming reconstruction based on direct Fourier transform inversion, we seek to recover the information in the missing view data by using the mathematical theory of convex projections discussed in Chapter 2. The technique is also valid for other imaging modalities such as nuclear magnetic resonance (NMR) and imaging by synthetic aperture radar.

There are several reasons why an incomplete view data situation might prevail in CT. In medical CT there may be insufficient time for full-view data acquisition of time-varying anatomical situations. In other applications,

* Present address: Research Laboratories, Eastman Kodak CO., Rochester, NY 14650.

e.g., on-site industrial inspection, incomplete view data can result when obstructions prevent full-view data collection or when imaging a moving part or a time-varying event. In NMR the length of time necessary to acquire the data may limit the number of views one can take of a moving organ.

As stated above, we combine herein the method of projections† onto convex sets (POCS), discussed by Youla in Chapter 2, with the direct Fourier method (DFM) [1] of reconstructing an image from (line-integral) projections into a single recovery/reconstruction algorithm which we call POCS–DFM. This algorithm is then used to reconstruct good-quality imagery even when not all the view data are available. Tuy [2] dealt with a similar problem and used POCS to restore missing projection data in connection with reconstruction via filtered convolution backprojection. Tuy's algorithm requires the generation of projection data at each iteration that are consistent with *a priori* constraints. In general, however, it is often easier and more natural to apply constraints to the image or its spectrum rather than its CT projections. Moreover, there are important applications, for example in NMR tomography, where the Fourier samples, but not the projections, of the image are directly measured. Therefore we are motivated to study the image recovery problem from a different viewpoint than Tuy's, one that we hope can be generalized for various applications. The POCS–DFM algorithm interpolates and/or extrapolates the missing Fourier domain information by the method of convex projections and reconstructs an image by direct Fourier inversion. As a case study in this chapter, we use POCS–DFM to reconstruct a simulated human thorax cross section from x-ray projection data that are angularly view-limited. We discuss potential applications of the POCS–DFM algorithm to related areas such as NMR imaging, ultrasound tomography, and spotlight-mode synthetic aperture radar imaging.

Why POCS–DFM? With regard to the method of recovery, POCS offers two noteworthy advantages: (i) it enables any number of *a priori* convex-type (see Chapter 2) constraints to be incorporated in the algorithm, and (ii) it guarantees convergence: weak, i.e., inner product convergence,‡ in general, strong convergence in practice. With regard to the method of reconstruction, the direct Fourier method offers several advantages as well: speed and convenience. The speed advantage results from the $O(N^2 \log N)$ operations required to reconstruct an $N \times N$ image using the FFT as opposed to

† Projection onto convex sets refers to mathematical projections while projections in CT refer to line integrals.

‡ We say that a sequence of vectors f_k in a Hilbert space $\mathscr{H}$ converges weakly to the vector f if $\lim_{k\to\infty}(f_k, h) = (f, h)$ for every $h \in \mathscr{H}$. For more on weak convergence see N. I. Akhiezer and I. M. Glazman, Theory of Linear Operators in Hilbert Space, Vol. I, Frederick Ungar, 1978.

$O(N^3)$ for filtered convolution backprojection. The convenience advantage results from the ease with which the space and Fourier domain constraints can be satisfied.

In the following section we demonstrate how POCS is used to restore a synthetic image from its partial Fourier spectrum. Next, we review the direct Fourier method in order to set the stage for a demonstration of the POCS–DFM algorithm.

11.2 DEMONSTRATION OF THE METHOD OF PROJECTIONS ONTO CONVEX SETS: SPECTRAL EXTRAPOLATION OF IMAGES

The recovery of missing Fourier components of an image (spectral extrapolation) is an important problem that occurs in many fields involving imaging. In direct Fourier transform tomography, it is a practically important operation in the generation of a high-quality image when less than full-view data is available.

Spectral extrapolation is treated by many authors in the literature [3–13]. In parallel-beam x-ray CT each view gives the projection data from which a central slice (i.e., the Fourier data on a line passing through the origin) of the discrete Fourier transform (DFT) of the image is obtained. This phenomenon, known as the projection theorem or central slice projection theorem, is discussed in Section 11.3. When projection data for all views are obtained, the Fourier plane is packed with samples of the Fourier data on a polar grid. After interpolation to a Cartesian grid, an inverse DFT is computed which yields the desired image. When viewing is limited to a certain angular range the available Fourier data will then be confined to a *data cone* with subtended angle of less than 180°† and restoration, i.e., extrapolation of all the Fourier data, is desirable before inverting the Fourier data to form an image. The spectral extrapolation problem also arises in NMR tomography, ultrasound tomography, and spotlight-mode synthetic aperture radar imaging.

We now study the restoration of a synthetic image from its partial Fourier spectrum. This study demonstrates an application of POCS and constitutes the ground work for the POCS–DFM recovery/reconstruction algorithm. Typical constraint sets and associated projection operators for the recovery of such an image are given below.

† As an example, Fig. 11.2-3 shows a simulated full-view Fourier spectrum (180° data cone) while Fig. 11.2-4 shows a partial view (90° data cone) Fourier spectrum. These figures will be discussed in greater detail later.

A. Sets and Projection Operators

Because we will deal with space-limited objects, defined by their reflectance, transmittance, or absorbance over a region Ω, the functions of interest in this study will be assumed to be members of $L_{2\times 2}(\Omega)$, the space of functions $f(x, y)$ of two variables x, y square integrable over $\Omega \subset \mathscr{R}^2$. The associated Hilbert space $\mathscr{H}$ is $L_{2\times 2}(\Omega)$ with inner product and norm defined by, respectively,

$$(g, h) \triangleq \iint_{\Omega} g(x, y)h^*(x, y)\, dx\, dy, \tag{11.2-1}$$

$$\|g\| \triangleq (g, g)^{1/2}, \tag{11.2-2}$$

where $h, g \in L_{2\times 2}(\Omega)$. The following sets will be used:

1. $\mathscr{C}_1$: The set of all functions in $\mathscr{H}$ that vanish almost everywhere† (a.e.) outside a prescribed region $\mathscr{S} \subset \Omega$. $\mathscr{C}_1$ is an extension of the set $\mathscr{T}_a$ introduced in Chapter 2 and is a closed linear manifold (CLM) without interior. Given an atbitrary $f \in \mathscr{H}$, its projection onto $\mathscr{C}_1$ is realized by

$$P_1 f = \begin{cases} f, & (x, y) \in \mathscr{S}, \\ 0, & (x, y) \notin \mathscr{S}. \end{cases} \tag{11.2-3}$$

2. $\mathscr{C}_2$: The set of all functions in $\mathscr{H}$ whose Fourier transforms assume a prescribed value G over a closed region $\mathscr{L}$ in the u–v Fourier plane. This set is essentially the "dual," i.e., the frequency domain analog, of $\mathscr{C}_a(g)$—the linear variety introduced in Chapter 2. The projection of an arbitrary $f \in \mathscr{H}$ onto $\mathscr{C}_2$ is realized by

$$P_2 f \leftrightarrow \begin{cases} G(u, v), & (u, v) \in \mathscr{L}, \\ F(u, v), & (u, v) \notin \mathscr{L}, \end{cases} \tag{11.2-4}$$

where $F(u, v) = \mathscr{F}[f(x, y)] \leftrightarrow f(x, y)$ and $\mathscr{F}$ denotes the Fourier transform operator. In the case of incomplete view data, $G(u, v)$ is known only in a data cone with subtended angle of less than 180°.

3. $\mathscr{C}_3$: The set of all real nonnegative functions in $\mathscr{H}$ that satisfy the energy constraint

$$\iint_{\Omega} |f(x, y)|^2\, dx\, dy \leq E \triangleq \rho^2, \tag{11.2-5}$$

† Except, possibly, over a point set of measure zero.

where $\rho = \sqrt{E}$. To show convexity we write

$$\|\mu f_1 + (1-\mu)f_2\| \leq \mu\|f_1\| + (1-\mu)\|f_2\| \leq \mu\rho + (1-\mu)\rho = \rho \quad (11.2\text{-}6)$$

for any f_1 and f_2 in $\mathscr{C}_3$ and $0 \leq \mu \leq 1$. To demonstrate closure we must show that given a sequence $\{f_n\}$ with limit f (written $f_n \to f$),† $\{f_n\} \subset \mathscr{C}_3$ implies $f \in \mathscr{C}_3$. Let $f \triangleq \tilde{f}_1 + j\tilde{f}_2$ where $\tilde{f}_1, \tilde{f}_2$ are real. Then

$$\|f_n - f\|^2 \to 0$$

requires that

$$\iint_\Omega (f_n - \tilde{f}_1)^2\,dx\,dy + \iint_\Omega \tilde{f}_2^2\,dx\,dy \to 0,$$

which implies that $\tilde{f}_2 = 0$ a.e. Thus, f is real. Now assume $f < 0$ over some region $\delta \subset \Omega$; then

$$\iint_{\Omega-\delta} (f_n - \tilde{f}_1)^2\,dx\,dy + \iint_\delta (f_n + |\tilde{f}_1|)^2\,dx\,dy \to 0, \quad (11.2\text{-}7)$$

which contradicts $\|f_n - f\| \to 0$. Hence, f cannot be negative. Finally assume $\|f\| = \rho + \varepsilon$, $\varepsilon > 0$. Then

$$\begin{aligned}\|f_n - f\|^2 &= \|f_n\|^2 + \|f\|^2 - 2|(f_n, f)|\\ &\geq \|f_n\|^2 + \|f\|^2 - 2\|f_n\|\,\|f\|\\ &= (\|f_n\| - \|f\|)^2 \geq \varepsilon^2.\end{aligned}$$

But this leads to a contradiction since $\|f_n - f\| \to 0$. Thus $\varepsilon = 0$ and $\mathscr{C}_3$ is closed.

Now consider the projection of an arbitrary $f \in \mathscr{H}$ onto $\mathscr{C}_3$. To describe the projection onto $\mathscr{C}_3$, we write $f \triangleq f_1 + jf_2$ where $f_1 = \mathrm{Re}\{f\}$ and $f_2 = \mathrm{Im}\{f\}$ and let f_1^+ be the rectified portion of f_1, i.e., $f_1^+ = f_1$ when $f_1 > 0$ and $f_1^+ = 0$ when $f_1 \leq 0$. Then the projection of f onto $\mathscr{C}_3$ is realized by [13]

$$P_3 f = \begin{cases} 0, & f_1 < 0,\\ f_1^+, & E_1^+ \leq E,\\ (\sqrt{E/E_1^+})f_1^+, & E_1^+ > E,\end{cases} \quad (11.2\text{-}8)$$

where E_1^+ is the energy in f_1^+, i.e.,

$$E_1^+ \triangleq \iint_\Omega (f_1^+)^2\,dx\,dy. \quad (11.2\text{-}9)$$

† $f_n \to f$ is equivalent to $\lim_{n\to\infty}\|f_n - f\| = 0$.

4. $\mathscr{C}_4$: The subset of all real nonnegative functions in $\mathscr{H}$. This set, already discussed in Chapter 2, is a closed convex cone with vertex ϕ (the zero function). The projection onto this set is

$$P_4 f = \begin{cases} f_1, & f_1 \geqslant 0 \quad (f = f_1 + jf_2), \\ 0, & \text{otherwise.} \end{cases} \tag{11.2-10}$$

5. $\mathscr{C}_5$: The set of all real functions in $\mathscr{H}$ whose amplitudes must lie in a prescribed closed intrval $[a, b]$, $a \geqslant 0$, $b > 0$, $a < b$. To show convexity of $\mathscr{C}_5$ note that

$$\mu f_1 + (1-\mu) f_2 \geqslant \mu a + (1-\mu) a = a \tag{11.2-11}$$

and

$$\mu f_1 + (1-\mu) f_2 \leqslant \mu b + (1-\mu) b = b \tag{11.2-12}$$

for any f_1 and $f_2 \in \mathscr{C}_5$ and $0 \leqslant \mu \leqslant 1$. The set is closed by definition. The projection onto $\mathscr{C}_5$ is realized by the rule

$$P_5 f = \begin{cases} a, & f(x, y) < a, \\ f(x, y), & a \leqslant f(x, y) \leqslant b, \\ b, & f(x, y) > b, \end{cases} \tag{11.2-13}$$

for f arbitrary in $\mathscr{H}$.

B. Image Description and Numerical Considerations

The image consists of three nested rectangles, the longest being 24×32 pixels, centered on a 64×64 pixel field; it is shown in Fig. 11.2-1 The image is represented by a sequence $f(m, n)$, where $f(m, n)$ represents the gray level at pixel coordinates (m, n) and $m = 1, \ldots, N$, $n = 1, \ldots, N (N = 64)$. The gray levels are confined to $0 \leqslant f(m, n) \leqslant 1$ and are 0.4, 0.8, and 1.0 from the largest to the smallest rectangle, respectively. The background level is zero.

The two-dimensional discrete Fourier transform of a sequence $\{f(m, n)\}$ is given by

$$F(k, l) = \sum_{m=0}^{N-1} \sum_{n=0}^{N-1} f(m, n) W_N^{mk} W_N^{nl}, \tag{11.2-14}$$

where

$$W_N \triangleq e^{-j(2\pi/N)}.$$

If the elements $f(m, n)$ are real then $F(k, l)$ enjoys certain Hermitian symmetries. For example, it can be demonstrated that $F(N-k, N-l) =$

Fig. 11.2-1. The image. It consists of three nested rectangles of gray levels 1.0 (center), 0.8, and 0.4. (From Sezan and Stark [13].) © 1982 *IEEE.*

$F^*(k, l)$, $F(N-k, 0) = F^*(k, 0)$, and $F(0, N-l) = F^*(0, l)$. These symmetries can be seen in Fig. 11.2-2, which is the DFT of the image in Fig. 11.2-1.

To simulate the data cone on what is essentially a Cartesian raster it is important to understand the Hermitian properties of the two-dimensional DFT so that the Hermitian symmetry of the data within the cone is preserved.

Another requirement for properly simulating the data cone in CT by DFT is to have the origin, i.e., the dc value of the Fourier plane, at the center of the 64×64 field. This facilitates superimposing the cone mask on the Fourier plane. The cone mask $\chi_c(\theta)$ is the characteristic function

$$\chi_c(\theta) = \begin{cases} 1, & \text{over all } \theta \text{ within the cone over which } F(u, v) \text{ is given,} \\ 0, & \text{otherwise.} \end{cases} \tag{11.2-15}$$

When convenient, we shall omit the argument θ in $\chi_c(\theta)$. To shift the dc value of the Fourier transform to the origin the image elements $f(m, n)$ are premultiplied by $(-1)^{m+n}$ before taking the DFT. That this shifts the DFT to the origin is easily demonstrated by the following example in one

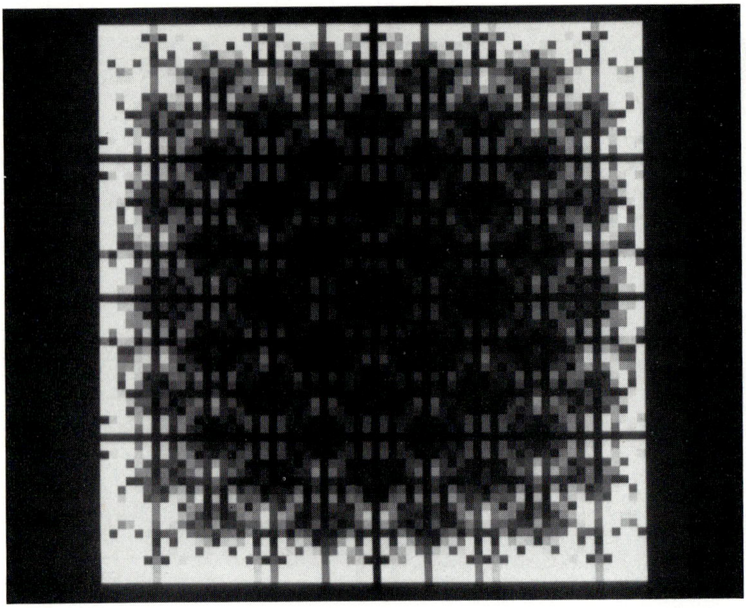

Fig. 11.2-2. Discrete Fourier transform (DFT) of the image in Fig. 11.2-1. (From Sezan and Stark [13].) © 1982 *IEEE.*

dimension: let $f'(n) \triangleq f(n)(-1)^n$. Then

$$\begin{aligned} F'(l) &\triangleq \sum_{n=0}^{N=1} f'(n)\, e^{-j(2\pi/N)nl} \\ &= \sum_{n=0}^{N-1} f(n)\, e^{-j(2\pi/N)n(l\pm N/2)} \\ &= \begin{cases} F(l-N/2), & l = N/2, \ldots, N-1, \\ F(l+N/2), & l = 0, \ldots, (N/2)-1. \end{cases} \end{aligned} \qquad (11.2\text{-}16)$$

The centered DFT and the cone-truncated data are shown, respectively, in Figs. 11.2-3 and 11.2-4. A 90° data cone is used in the experiments. The pixel values outside the data cone are set to zero.

C. Algorithms

Both the Gerchberg–Papoulis (G–P) algorithm and more general algorithms of the form†

$$f_{k+1} = T_m T_{m-1} \cdots T_1 f_k, \qquad k = 0, 1, \ldots, \qquad (11.2\text{-}17)$$

† See Eqs. (2.4-3) and (2.4-4) in Chapter 2 as well as Theorem 2.4-1.

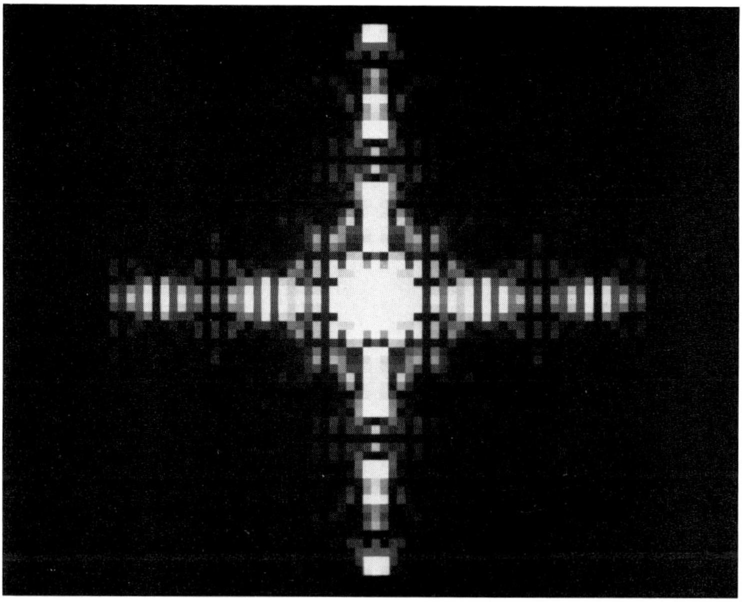

Fig. 11.2-3. The DFT of Fig. 11.2-1 shifted so that the origin is at the center of the Fourier plane. (From Sezan and Stark [13].) © 1982 *IEEE.*

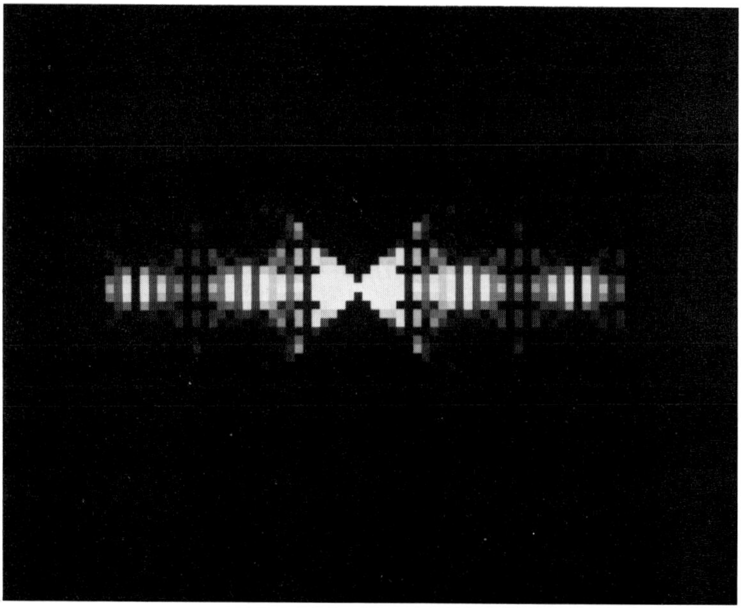

Fig. 11.2-4. Data in the data cone furnished to the algorithms. This is the function $G(u, v)$ in Eq. (11.2-4). (From Sezan and Stark [13].) © 1982 *IEEE.*

where† $T_i \triangleq I + \lambda_i(P_i - I)$ and $0 < \lambda_i < 2$, $i = 1, 2, \ldots, m$, are implemented. The convergence properties of these algorithms were discussed in Chapter 2. The three algorithms G–P, UNIRELAX, and RELAX are now briefly described.

1. Gerchberg–Papoulis (G–P)

The familiar G–P algorithm [11,12] and its offspring [3–7] are special cases of POCS. The G–P algorithm can be written in compact form, for our problem as

$$f_{k+1} = P_2 P_1 f_k, \qquad k = 0, 1, \ldots, \tag{11.2-18}$$

where P_2 and P_1 are the projectors given in Eqs. (11.2-4) and (11.2-3), respectively. The G–P algorithm is a linear algorithm in the sense that it involves projections onto convex sets which are linear subspaces or at most linear varieties (here $\mathscr{C}_1$ is a linear subspace and $\mathscr{C}_2$ is a linear variety).

2. UNIRELAX

This algorithm, in general, involves pure projections onto convex sets and realizes the iteration

$$f_{k+1} = P f_k, \qquad k = 0, 1, \cdots. \tag{11.2-19}$$

Here P is a composition operator composed of projectors P_i, $i \in \{1, 2, 3, 4, 5\}$, defined in Section 11.2.A. UNIRELAX differs from G–P in that it uses additional *a priori* known constraints. For example, for $P \triangleq P_5 P_2 P_3 P_1$, the solution must be consistent with space-limitedness, energy constraint, level constraint, and the measured data.

3. RELAX

This algorithm is a generalization of UNIRELAX and realizes the iteration

$$f_{k+1} = T f_k, \qquad k = 0, 1, \ldots, \tag{11.2-20}$$

where T is a composition operator composed of relaxed projectors $\{T_i\}$, where $T_i \triangleq I + \lambda_i(P_i - I)$ and $0 < \lambda_i < 2$, $i \in \{1, 2, 3, 4, 5\}$. RELAX has the additional freedom over UNIRELAX is that the λ_i can be adjusted for more rapid convergence.

If the λ_i are set to unity, then RELAX reduces to UNIRELAX. The λ_i, if properly chosen, can be used to offset the slow convergence which might

† I is the identity operator.

result when only the P_i (pure projections) are used. As pointed out in Chapter 2, slow convergence is likely near the points of tangency of two convex sets. In Chapter 2 Youla shows that the T_i are generally not projectors but have the same fixed points as the P_i, and every fixed point of T and T_i, $i = 1, 2, \ldots$, is a feasible solution, i.e., a point in the intersection of closed convex sets.

We now summarize the *a priori* known facts and assumptions about the image:

A Priori Constraints	Actual
1) Image support confined to rectangular region $\Delta x = 55$ pixels, $\Delta y = 57$ pixels.	1) Image support in rectangular region $\Delta x = 32$ pixels, $\Delta y = 24$ pixels.
2) Gray levels f satisfy $0 \leq f \leq 1$.	2) Gray levels f satisfy $0 \leq f \leq 1$.
3) Energy over 64×64 pixel^2 field cannot exceed $\rho^2 = 268.5$.	3) Energy in image is 267.0 over the 32×24 pixel^2 object field. No energy outside.
4) Initially known spectrum is $G(u, v) = F(u, v)\chi_c(\theta)$. [$\chi_c(\theta)$ is 90° cone].	4) Spectrum of image is $F(u, v)$.

In this example, the sets $\mathscr{C}_1, \mathscr{C}_2, \mathscr{C}_3, \mathscr{C}_5$, consisting of $L_{2\times 2}$ functions (note $\mathscr{C}_5 \subset \mathscr{C}_4$ so $\mathscr{C}_4$ is not needed), are explicitly defined as follows:

(i) $\mathscr{C}_1$ is the set of all functions that vanish (a.e.) outside a rectangular region of length 55 pixels and width 57 pixels.

(ii) $\mathscr{C}_2$ is the set of all functions whose Fourier transforms assume the measured or prescribed value $G(u, v)$ over the set of points within the Fourier-plane data cone.

(iii) $\mathscr{C}_3$ is the set of all functions whose norm (energy) is less than or equal to 268.5 pixels squared.

(iv) $\mathscr{C}_5$ is the set of all functions whose gray levels f are in the range [0, 1].

The formula used for computing enrgy is

$$E \triangleq \sum_{m=1}^{64} \sum_{n=1}^{64} f^2(m, n) = 267.0. \tag{11.2-21}$$

D. Results: Noise-Free Restoration

The basis for comparing the results of applying the different algorithms was the percent error e_k at the kth iteration defined by

$$e_k \triangleq 100\|f - f_k\| / \|f\|, \tag{11.2-22}$$

where f denotes the original image.

A summary of the results is given in Table 11.2-1. In the first (leftmost) column is the name of the algorithm; in the second and third columns are the recursion relations and relaxation parameters that were used; and in the final column is the error after the 30th iteration. The starting error e_0 is computed from Eq. (11.2-22) and f_0, the starting image, is computed from the inverse FFT of the initially known incomplete Fourier spectrum.

Table 11.2-1
Summary of the Results (Noise-Free Restoration)

Algorithm		Relaxation Parameters	Percent Error At 30th Iteration
Name	Recursion		
			(e_0=63.45)
G-P	$P_2P_1f_k=f_{k+1}$	$\lambda_1=\lambda_2=1.0$	39.435
UNIRELAX1	$P_2P_3P_1f_k=f_{k+1}$	$\lambda_1=\lambda_2=\lambda_3=1.0$	21.193
UNIRELAX2	$P_5P_2P_3P_1f_k=f_{k+1}$	$\lambda_1=\lambda_2=\lambda_3=\lambda_5=1.0$	21.110
RELAX1	$T_2T_3T_1f_k=f_{k+1}$	$\lambda_1=\lambda_3=1.9995$ $\lambda_2=1.75$	18.124
RELAX2	$T_5T_2T_3T_1f_k=f_{k+1}$	$\lambda_1=\lambda_3=1.9995$ $\lambda_2=1.925$ $\lambda_5=0.25$	17.689
RELAX3	$T_5T_2T_3T_1f_k=f_{k+1}$	$\lambda_1=\lambda_3=1.9995$ $\lambda_2=1.75$ $\lambda_5=0.25$	17.362

In this particular problem, the *a priori* knowledge of the level constraint, i.e., that $0 \leqslant f \leqslant 1$, furnished insignificant help in recovering the image. This can be seen by comparing the results of UNIRELAX1, which omitted P_5, and UNIRELAX2, which included it; no improvement in convergence was observed. Essentially the same behavior was observed when relaxed operators were used (RELAX1 versus RELAX2). This demonstrated that not all *a priori* information is equally important in a particular problem and that some care should be used in deciding what to include if we do not wish to suffer from excessive computation. On the other hand, the level constraint assumes considerable importance in the phase retrieval problem (see Chapter 8).

The possibility of improving the restoration rate by choosing relaxation parameters different from unity is suggested by comparing the results of UNIRELAX1 and UNIRELAX2 (all λ's set to unity) versus RELAX1, RELAX2, and RELAX3. The latter furnishes a 4% improvement after 30 iterations. The λ's used in RELAX3 were chosen somewhat arbitrarily and do not represent an optimal set. However, perhaps more interesting than what happens after 30 iterations is the marked decrease in error after only a few iterations when appropriate relaxation parameters are used. The error histories for G–P, UNIRELAX1, and RELAX1 are compared in Fig. 11.2-5.

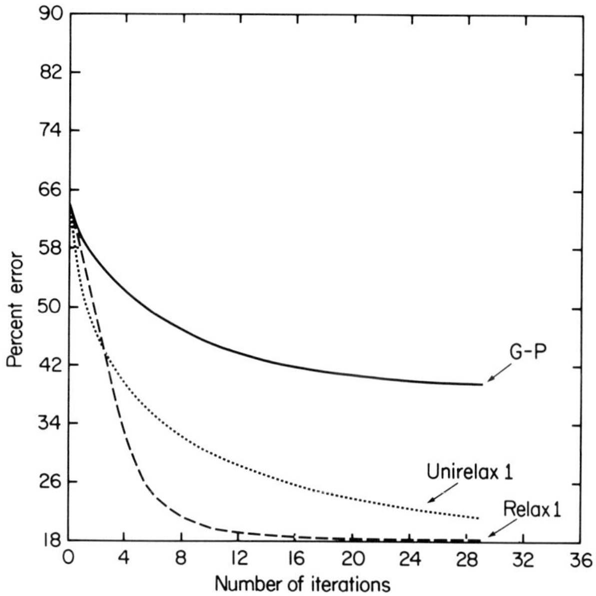

Fig. 11.2-5. Error performance of the G–P, UNIRELAX1, and RELAX1 algorithms. (From Sezan and Stark [13].) © 1982 *IEEE*.

After eight iterations RELAX1 has generated an image with an approximately 20% error (down from an initial error of 63%); the errors in the restored image when UNIRELAX1 and G–P are used are 32 and 46%, respectively. Figure 11.2-6 compares the G–P, UNIRELAX1, and RELAX1 restorations with the original image at the end of the 30th iteration.

E. Conclusions Drawn from Noise-Free restoration

Although our restorations involved the same (single) image, the results of our study suggest that incorporation of additional *a priori*, often nonlinear but convex-type, constraints improves the restoration. This can be seen by comparing the G–P restoration with UNIRELAX and RELAX restorations. The improvement is mainly due to the additional energy constraint, which also contains a built-in nonnegativity constraint. Our results also suggest

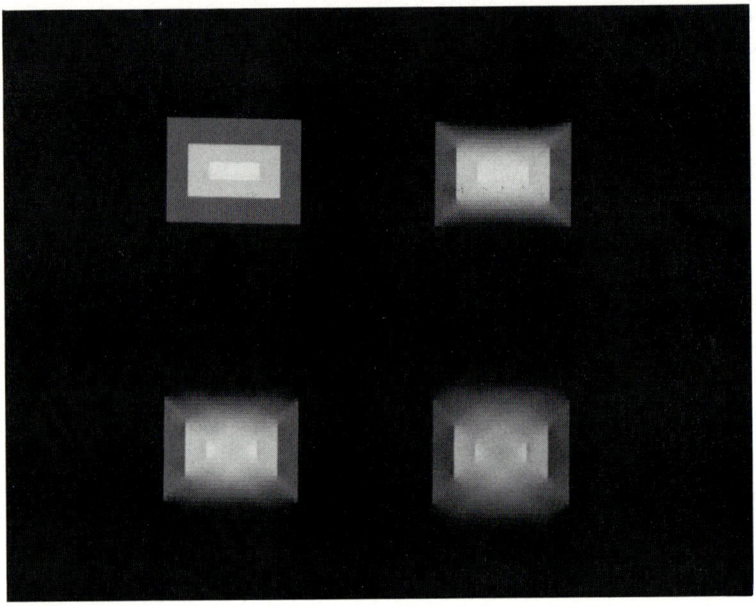

Fig. 11.2-6. Restoration of the image. Counterclockwise from upper left: original image, UNIRELAX1, G-P, RELAX1. (From Sezan and Stark [13].) © 1982 *IEEE*.

that nonunity relaxation parameters can be used to improve the restoration rate. Ideally one would want to optimize the choice of λ's at each iteration to yield best results. However, it is not possible to determine the best choice of λ's in the general case because these depend on the unknown image. Nevertheless, approximate methods such as those described by Levi and Stark in Chapter 8 can be useful. There an approximate per-step and per-cycle optimization of the relaxation parameters λ_1 and λ_2 is derived when only two convex projections are involved in the restoration. These results are successfully applied in problems involving image restoration from phase or Fourier magnitude.

Recently, using different constraints, Lent and Tuy [8] used the method of POCS to extrapolate simulated tomographic imagery. Their results also indicate the considerable versatility of this method.

F. Regularization of the Spectral Extrapolation Problem

When the given Fourier datum is noiseless, it is consistent with the space-limitedness constraint on the image. When the datum is noisy, however, it generally will be inconsistent with the space-limitedness constraint and other *a priori* constraints. For this reason, the solution set $\mathscr{C}_0$ may be empty and it is possible for the iterates to diverge. This behavior is characteristic of ill-posed (or incompletely posed) problems. The problem of restoring an image from its partial Fourier spectrum is ill-posed. Although the solution to an ill-posed problem may be unique, its dependence on the available data is discontinuous. The methods for regularization (i.e., make an ill-posed problem well-posed) of ill-posed or incompletely posed problems are discussed by several authors [14–16]. These methods involve the addition of one or more *a priori* constraints to the problem and thus restrict the set of admissible solutions. Due to its nature, POCS is very well suited for such regularizations. A very simple regularization is to introduce an upper bound for the total energy of the image. This can be achieved by projecting during each iteration onto $\mathscr{C}_3$, i.e., the set of energy-constrained functions. However, such regularization does not explicitly involve a noise parameter that is somehow descriptive of the level of noise. Another example of noise smoothing that does involve a noise parameter and still involves projecting onto a convex set is given below. It is closely related to P_{11} described by Youla in Chapter 2 except that the smoothing is done in the frequency domain.

Let $\tilde{G}(u, v)$ be the given Fourier transform data over a region $\mathscr{L}$ (i.e., a point set) of the u–v plane, and assume that it is related to the true (noiseless) data $G(u, v)$ by

$$\tilde{G}(u, v) = \begin{cases} G(u, v) + n(u, v), & (u, v) \in \mathscr{L}, \\ 0, & (u, v) \notin \mathscr{L}, \end{cases} \tag{11.2-23}$$

where u, v are Cartesian spatial frequencies and $n(u, v)$ is the noise at (u, v). We define the characteristic function $\tilde{\chi}$ by

$$\tilde{\chi} \triangleq \begin{cases} 1, & (u, v) \in \mathscr{L}, \\ 0, & (u, v) \notin \mathscr{L}. \end{cases} \tag{11.2-24}$$

Then $\tilde{G}(u, v) = \tilde{\chi}\tilde{G}(u, v)$, and if $F(u, v)$ is the spectrum of the noiseless image $f(x, y)$, then $G(u, v) = \tilde{\chi}F(u, v)$. Consider the set $\mathscr{C}_6$ defined as the set of all real-valued functions $h \in \mathscr{H}$ whose Fourier transform $H(u, v)$ satisfies

$$\|\tilde{\chi}H(u, v) - \tilde{G}(u, v)\| \leq \sigma. \tag{11.2-25}$$

In words: $\mathscr{C}_6$ is the set of all real functions in $\mathscr{H}$ whose Fourier transforms over $\mathscr{L}$ are within an L_2 distance σ from the given data $\tilde{G}(u, v)$. To show convexity: let $f_1, f_2 \in \mathscr{C}_6$ and let $0 \leq \mu \leq 1$. Then we must show that

$f_3 \triangleq \mu f_1 + (1-\mu) f_2 \in \mathscr{C}_6$. But

$$\begin{aligned}\|\tilde{\chi}F_3 - \tilde{G}\| &= \|\mu\tilde{\chi}F_1 + (1-\mu)\tilde{\chi}F_2 - \tilde{G}\| \\ &= \|\mu(\tilde{\chi}F_1 - \tilde{G}) + (1-\mu)(\tilde{\chi}F_2 - \tilde{G})\| \\ &\leq \mu\|\tilde{\chi}F_1 - \tilde{G}\| + (1-\mu)\|\tilde{\chi}F_2 - \tilde{G}\| \\ &\leq \mu\sigma + (1-\mu)\sigma = \sigma, \end{aligned} \tag{11.2-26}$$

and hence $\mathscr{C}_6$ is convex.

Let the sequence $\{f_n\} \subset \mathscr{C}_6$ converge to $f = \tilde{f}_1 + j\tilde{f}_2$, where $\tilde{f}_1 \triangleq \mathrm{Re}\{f\}$, $\tilde{f}_2 \triangleq \mathrm{Im}\{f\}$. Thus $f_n \to f$ and to prove closure we must show that $f \in \mathscr{C}_6$. From definition of convergence we have

$$\|f_n - f\|^2 = \iint_{\Omega} (\tilde{f}_1 - f_n)^2 \, dx \, dy + \iint_{\Omega} \tilde{f}_2^2 \, dx \, dy \to 0. \tag{11.2-27}$$

So we conclude that $\tilde{f}_2$ must be zero a.e., i.e., f is real. Now $f_n \to f$ implies $F_n \to F$ and since norm $(\|\cdot\|)$ is a continuous mapping, $\|\tilde{\chi}F_n - \tilde{G}\| \leq \sigma$ implies that $\|\tilde{\chi}F - \tilde{G}\| \leq \sigma$. Therefore $\mathscr{C}_6$ is a closed convex set.

Let f be an arbitrary function in $\mathscr{H}$ with $\mathrm{Re}\{f\} \triangleq f_1$ and $\mathrm{Im}\{f\} \triangleq f_2$. Also let $F_1(u, v) \triangleq \mathscr{F}[f_1(x, y)]$. Then the projection of f onto $\mathscr{C}_6$ is realized by

$$P_6 f \leftrightarrow \begin{cases} F_1, & \|\tilde{\chi}F_1 - \tilde{G}\| \leq \sigma, \\ \tilde{G} + \sigma \dfrac{\tilde{\chi}F_1 - \tilde{G}}{\|\tilde{\chi}F_1 - \tilde{G}\|} + (1-\tilde{\chi})F_1, & \|\tilde{\chi}F_1 - \tilde{G}\| > \sigma. \end{cases} \tag{11.2-28}$$

The proof of Eq. (11.2-28) is given in [17] and [18].

G. Noise Simulation and Numerical Considerations

The noisy Fourier data in the data cone are simulated in the following manner: first, zero-mean white Gaussian noise of known variance (corresponding to a certain signal-to-noise ratio) is added to the original image (including the background); second, the noisy image is Fourier transformed; and third, the cone mask is applied to the resulting transform to form $\tilde{G}(u, v)$. The characteristic function $\tilde{\chi}$ can now be defined in terms of a cone mask function defined by

$$\tilde{\chi} \triangleq \chi_c(\theta) = \begin{cases} 1, & \text{over all } \theta \text{ within the cone over which the data are given} \\ 0, & \text{otherwise.} \end{cases} \tag{11.2-29}$$

As before, a 90° data cone is used in the experiments and the pixel values outside the data cone are set to zero. The Fourier data inside the cone corresponding to noise at signal-to-noise ratios (SNR) of 20 and 5 dB are shown in Figs. 11.2-7 and 11.2-8, respectively. The quantity $\|G - \tilde{G}\|^2$ is used to estimate the parameter σ to be used in the experiments. A summary of the *a priori* known facts and the assumptions made about the image is given below:

A priori constraints	Actual
1) Image support confined to rectangular region $\Delta X = 55$ pixels, $\Delta Y = 57$ pixels.	1) Image support is in rectangular region $\Delta X = 32$ pixels, $\Delta Y = 24$ pixels.
2) Gray levels f satisfy $0 \leq f \leq 1$.	2) Gray levels f satisfy $0 \leq f \leq 1$.
3) Energy over 64×64 pixel^2 field cannot exceed $\rho^2 = 268.50$.	3) Energy in image is 267.0 over the 32×24 pixel^2 object field. No energy outside.
4) Initially known spectrum: $\tilde{G}(u, v) = G(u, v) + n(u, v)$, where $G(u, v) = F(u, v)\chi_c(\theta)$ and $n(u, v) = n(u, v) \times \chi_c(\theta)$ is the noise contamination. [$\chi_c(\theta)$ is 90° cone mask].)	4) Spectrum of the image is $F(u, v)$.
5) For SNR = 5 dB, $\sigma^2 = 1.53 \times 10^5$. For SNR = 20 dB, $\sigma^2 = 4818$.	5) For SNR = 5 dB, $\|G - \tilde{G}\|^2 = 1.52 \times 10^5$. For SNR = 20 dB, $\|G - \tilde{G}\|^2 = 4814.1$.

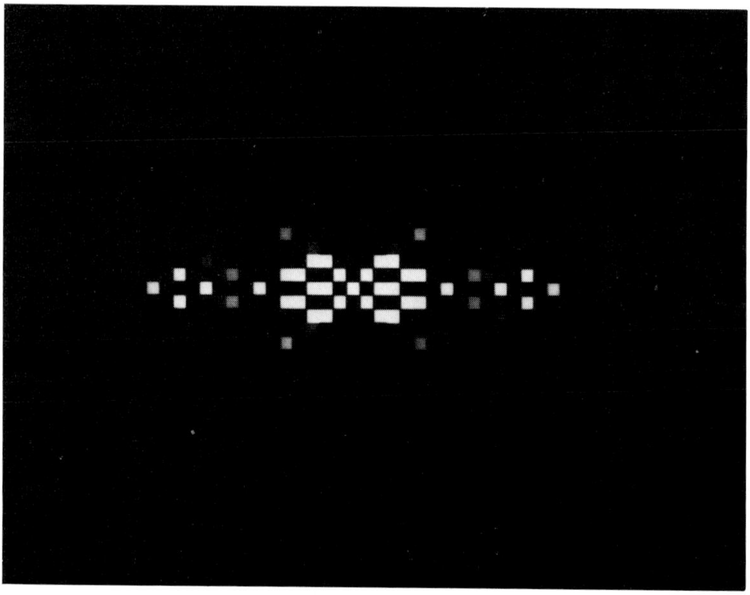

Fig. 11.2-7. Data in the data cone when the image has been corrupted by moderate noise (SNR = 20 dB). (From Sezan and Stark [17].)

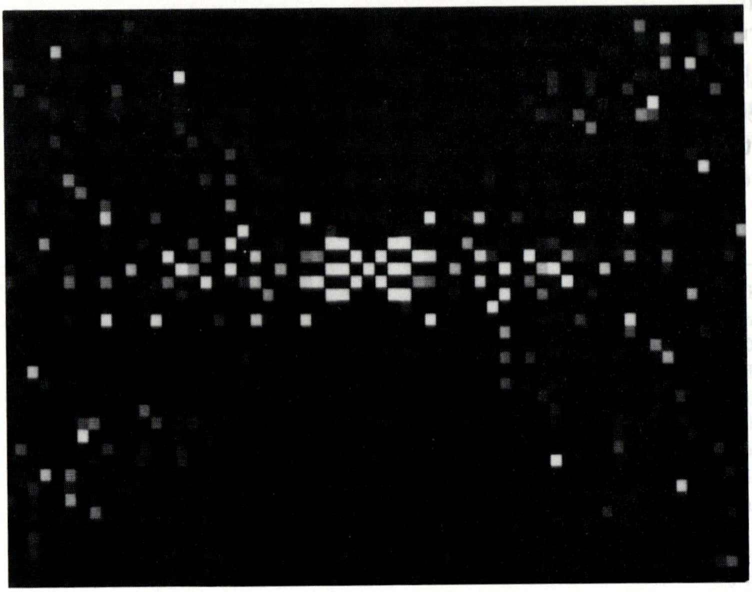

Fig. 11.2-8. Data in the data cone when the image has been corrupted by heavy noise (SNR = 5 dB). (From Sezan and Stark [17].)

H. Results: Restoration from Noisy Data

The basis for comparing results was the percent error e_k at the kth iteration defined in Section 11.2.D. A summary of the results is given in Table 11.2-2.

As before, the algorithms labeled UNIRELAX involved pure projections; those labeled RELAX involved relaxed projections (nonunity relaxation parameters). In the first (leftmost) column of Table 11.2-2 is the name of the algorithm; in the second and third columns are the recursion relations of the algorithms and the values of relaxation parameters that were used; and in the last two columns are the percentage errors at the 80th iteration for the two noise levels SNR = 5 dB and SNR = 20 dB. Whenever the P_2 operator is used it should be understood that the noisy data cone information is used as the *a priori* known spectral information.

The error histories for G-P, UNIRELAX1, and UNIRELAX2 are compared in Fig. 11.2-9a. The G-P algorithm starts to diverge at the 19th iteration and continues diverging until, at the 80th iteration, the error reaches 68%. In contrast, UNIRELAX1 and UNIRELAX2 fail to diverge, although the former does not converge monotonically. In the case of UNIRELAX1, the operator P_3, placed last in the order of projection operators, ensures that

Table 11.2-2
Summary of the Results (Restoration from Noisy Data)

Algorithm		Relaxation parameters	Percent error at 80th iteration	
Name	Recursion		SNR=5 dB	SNR=20 dB
			(e_0=73.615)	(e_0=63.800)
G-P	$P_2P_1f_k=f_{k+1}$	$\lambda_1=\lambda_2=1.0$	68.371	40.144
UNIRELAX1	$P_3P_5P_2P_1f_k=f_{k+1}$	$\lambda_1=\lambda_2=1.0$ $\lambda_3=\lambda_5=1.0$	45.222	19.204
UNIRELAX2	$P_3P_5P_6P_1f_k=f_{k+1}$	$\lambda_1=\lambda_3=1.0$ $\lambda_5=\lambda_6=1.0$	62.129	29.098
UNIRELAX3	$P_5P_2P_3P_1f_k=f_{k+1}$	$\lambda_1=\lambda_2=1.0$ $\lambda_3=\lambda_5=1.0$	46.601	19.216
RELAX1	$T_3T_5T_2T_1f_k=f_{k+1}$	$\lambda_1=\lambda_3=1.9995$ $\lambda_2=1.75$ $\lambda_5=0.25$	69.632	24.021
RELAX2	$T_3T_5T_6T_1f_k=f_{k+1}$	$\lambda_1=\lambda_3=\lambda_6=1.9995$ $\lambda_5=0.25$	57.860	28.779
RELAX3	$T_5T_2T_3T_1f_k=f_{k+1}$	$\lambda_1=\lambda_3=1.9995$ $\lambda_2=1.75$ $\lambda_5=0.25$	73.452	25.161

the iterates never exceed the energy bound of 268.50 and thus regularizes the problem. The regularization, in case of UNIRELAX2, is due to the noise smoothing operator P_6. UNIRELAX1, which uses the noisy data directly (by projecting onto $\mathscr{C}_2$), outperforms UNIRELAX2, which uses the noisy data only as the center of the noise sphere in which the noiseless (unknown) data are assumed to be. The conditions for Fig. 11.2-9b are the same as those for Fig. 11.2-9a except that the SNR = 20 dB. In this case divergence of the G–P algorithm is very slight, although it is outperformed by the UNIRELAX1 and UNIRELAX2 algorithms. Indeed, UNIRELAX1 has reduced the error from 64 to 20% in 60 iterations, after which the efficacy of further error reduction is greatly reduced.

In Section 11.2D it was shown that nonunity relaxation parameters could be used effectively in the noiseless case to increase the convergence rate. However, designs that optimize a performance criterion in the noiseless case often perform poorly in the noisy case. This proved to be the case here, and it can be seen by comparing the results of UNIRELAX1, UNIRELAX2, and UNIRELAX3 with RELAX1, RELAX2, and RELAX3, respectively. Moreover, from Fig. 11.2-10, which gives the error histories for RELAX3 in high-noise (SNR = 5 dB), moderate-noise (SNR = 20 dB), and noiseless environments, we see the following: (1) The monotonic error

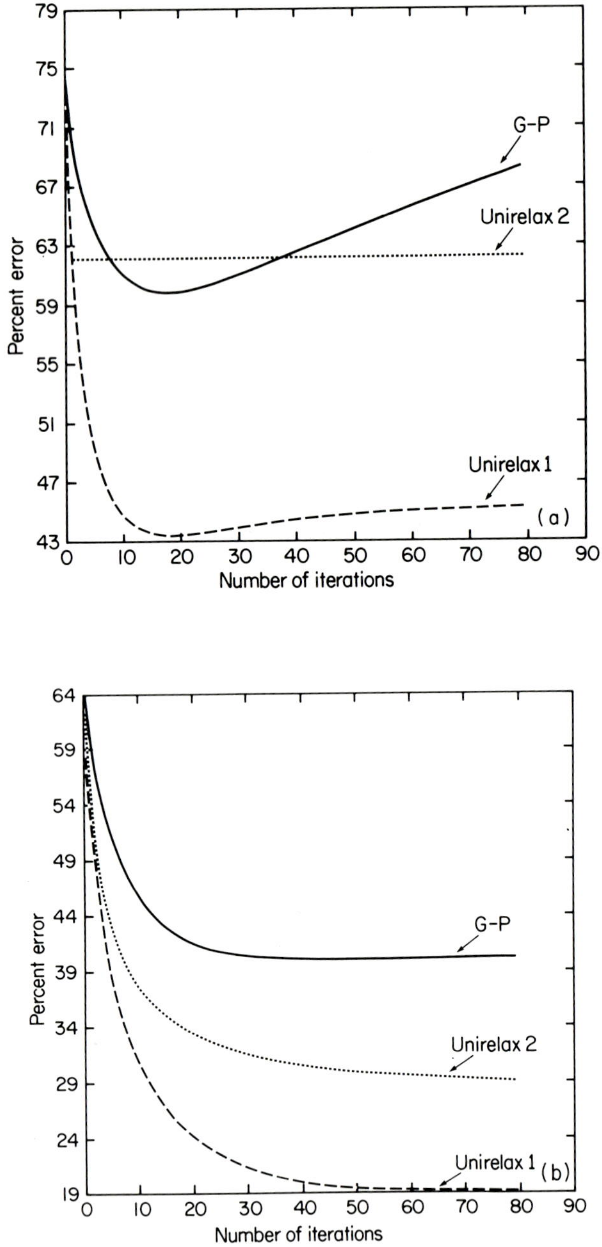

Fig. 11.2-9. Error performance of G–P, UNIRELAX1, and UNIRELAX2 in restoring the image from noisy data: (a) SNR = 5 dB; (b) SNR = 20 dB. (From Sezan and Stark [17].)

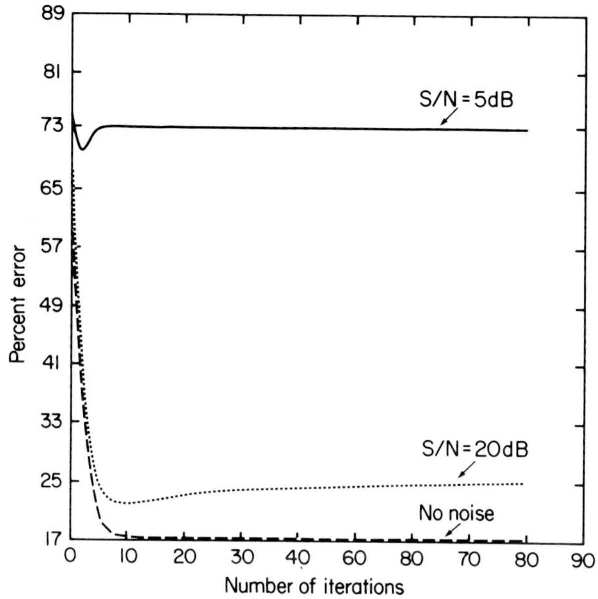

Fig. 11.2-10. Error performance of RELAX3 in restoring the image from noise-free data, moderately noisy data (SNR = 20 dB), and very noisy data (SNR = 5 dB). (From Sezan and Stark [17].)

decrease with iteration number k commonly observed in noiseless recursive image restoration is not realized in a high-noise environment. (2) When there is significant error correction, it usually occurs within a relatively small number of iterations. Beyond this number, further restoration is very slow in the noiseless case and nonexistent in the noisy case. (3) In the high-noise environment (SNR = 5 dB) attempts at restoration using the same relaxation parameters that have good results in the noiseless case fail.

In the high-noise environment (SNR = 5 dB) the restorations are of poor visual quality. Figure 11.2-11 compares the G–P and UNIRELAX1 restorations with the original and noisy original (SNR = 20 dB) at the end of 80 iterations. In this case UNIRELAX1 furnishes a reasonable restoration: all three rectangles are easily resolved and gray levels have been reproduced with reasonable fidelity. Considerable blurring in the vertical direction is the weakest feature of the restoration.

I. Conclusions Drawn from Restoration in the Presence of Noise

From the results given in the previous section we see that incorporating the energy level constraint or the noise level constraint or both in the POCS

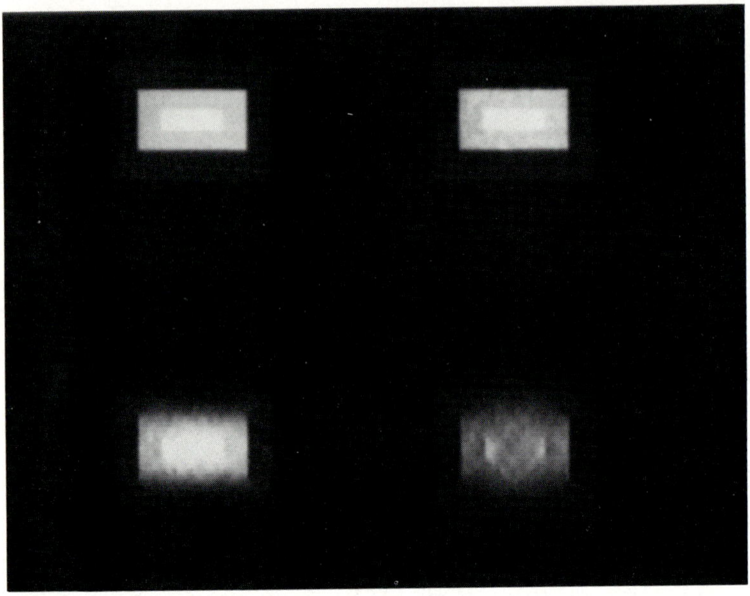

Fig. 11.2-11. Restoration of the image. Clockwise from upper left: original image, noisy image (SNR = 20 dB), G–P, UNIRELAX1. (From Sezan and Stark [17].)

algorithm prevents divergent behavior. On the other hand, the Gerchberg–Papoulis algorithm not involving such regularizers diverges in the presence of noise.

The choice of relaxation parameters is very noise-sensitive. In particular, restorations using unit relaxation parameters (pure projections) yield better restorations than those using nonunity relaxation parameters.

11.3 REVIEW OF THE DIRECT FOURIER METHOD

The central problem addressed in this chapter is image recovery in direct Fourier transform tomography with limited view-data. In this section we present a review of the direct Fourier method for reconstructing an image from line integral projections.

The direct Fourier method (henceforth DFM) is based on the central slice projection theorem [1]. Let $\mu(x, y)$ represent a function in Cartesian coordinates, whose image is desired [e.g., in CAT, $\mu(x, y)$ is the x-ray absorbtivity at position x, y]. Let $\mu_\phi(\hat{x}, \hat{y})$ represent the same distribution in a coordinate system $\hat{x}$–$\hat{y}$ rotated from x–y by an angle ϕ. The projection data at view angle ϕ are defined as

$$p_\phi(\hat{x}) \triangleq \int_L \mu_\phi(\hat{x}, \hat{y})\, d\hat{y}, \tag{11.3-1}$$

where L is the beam path. The central slice projection theorem states that

$$P_\phi(\rho) = M(\rho, \phi), \qquad \rho \geq 0, \quad 0 < \phi \leq 2\pi, \tag{11.3-2a}$$

where

$$P_\phi(\rho) = \int_{-\infty}^{\infty} p_\phi(\hat{x})\, e^{-2\pi\rho\hat{x}}\, d\hat{x} \tag{11.3-2b}$$

and $M(\rho, \phi)$ denotes the two-dimensional (2-D) Fourier transform, in polar coordinates, of the cross-sectional absorbtivity $\mu(x, y)$. In practice $p_\phi(\hat{x})$ is the data actually obtained at the discrete detector locations $\{\hat{x}_j\}$. In place of Eq. (11.3-2b) the discrete Fourier transform is used to obtain $P_\phi(\rho)$ at the discrete set of spatial frequencies $\{\rho_n\}$. Since the view angles are also a discrete set $M(\rho, \phi)$ is known at points on a polar grid $\{\rho_n, \phi_k\}$, $n = 1, 2, \ldots, N_D$, $k = 1, 2, \ldots, 2K + 2$, where N_D is the number of detectors, K the highest harmonic in the angularly band-limited image, and $\phi_k = [2\pi/(2K + 2)]k$. Here we immediately see that if we could compute the 2-D inverse Fourier transform of $M(\rho_n, \phi_k)$ we would obtain the desired image. In practice, however, the 2-D Fourier transform is realized via the fast Fourier transform (FFT), which requires knowledge of the data on a rectangular grid of Cartesian points $\{u_l, v_m\}$. The data, however, are known on a polar grid of points $\{\rho_n, \phi_k\}$ which do not coincide with the Cartesian points $\{u_l, v_m\}$. Hence interpolation is required, and specifically interpolation from points on a polar grid to points on a Cartesian grid, to use the FFT for inverting $M(\rho_n, \phi_k)$. However, inexact interpolation is a major source of error in reconstruction by direct Fourier methods even when the interpolation errors are not large (Lewitt [19] discussed this point in some detail). Attempts at implementing DFM using nearest-neighbor and linear interpolation were made by Mersereau and Oppenheim [20]. However, we found that these fail to give good results in CT. Recently, an exact interpolation procedure for K-angularly band-limited functions was derived [1,21,22] which—when combined with direct Fourier imaging—produced images equal to or surpassing in quality those produced by convolution backprojection. Moreover, these results were obtained under practical constraints, such as approximating exact interpolation with a finite-sample formula and prefiltering non-K-bandlimited signals to avoid aliasing. The interpolation formula is given in the following theorem.

Theorem. Let $\mu(x, y)$ be space-limited to $2A$. Let its Fourier transform in polar coorinates $M(\rho, \phi)$ be angularly band-limited to K.† Then $M(\rho, \phi)$ can be reconstructed from its polar samples $M(\rho_n, \phi_k) = M(n/2A, 2\pi k/N)$ from

$$M(\rho, \phi) = \sum_{n=-\infty}^{\infty} \sum_{k=0}^{N-1} M\left(\frac{n}{2A}, \frac{2\pi k}{N}\right) \operatorname{sinc}\left[2A\left(\rho - \frac{n}{2A}\right)\right] \sigma\left(\phi - \frac{2\pi k}{N}\right), \tag{11.3-3}$$

where N is an even integer satisfying $N > 2K$ and

$$\sigma(\phi) \triangleq \sin[(N-1)\phi/2]/N \sin(\tfrac{1}{2}\phi). \tag{11.3-4}$$

To compute‡ $M(u_l, v_m)$, Eq. (11.3-3) is used with

$$\rho_{lm} \triangleq \sqrt{u_l^2 + v_m^2}, \tag{11.3-5}$$

$$\phi_{lm} \triangleq \cos^{-1}(u_l/\rho_{lm}). \tag{11.3-6}$$

Thus the set $\{u_l, v_m\}$ can be computed directly from the set $\{\rho_n, \phi_k\}$. In practice, a truncated interpolation involving a finite number of cardinal functions is used. Thus, Eq. (11.3-3) is replaced by

$$M(\rho, \phi) = \sum_{n=n_\rho - L_\rho}^{n_\rho + L_\rho} \sum_{k=k_\phi - L_\phi}^{k_\phi + L_\phi} M\left(\frac{n}{2A}, \frac{2\pi k}{N}\right) \operatorname{sinc}\left[2A\left(\rho - \frac{n}{2A}\right)\right] \sigma\left(\phi - \frac{2\pi k}{N}\right), \tag{11.3-7}$$

where $[2A\rho] \triangleq n_\rho$ and $[N\phi/2\pi] \triangleq k_\phi$ are the nearest neighbors of the point about which the expansion is done. The notation $[x]$ here means the largest integer less than or equal to x. If $L_\rho = L_\phi = 0$, the above approximation reduces to the nearest-neighbor approximation. It is possible to adjust the relative proportions of radial and azimuthal interpolation by adjusting the parameters L_ρ and L_ϕ, respectively. In practice, image quality is improved when the truncation is tapered rather than abruptly truncated, the latter meaning that every point in the truncation window is given unit weight while those outside are given zero weight. The taper normally used is the Cartesian product of two identical one-dimensional triangular windows $\omega(n)$ of the form

$$\omega(n) = \max(1 - |n|/W, 0), \qquad n = 0, 1, 2, \ldots, \tag{11.3-8}$$

where W is a parameter. Then $W = \infty$ is the abrupt truncation case.

After interpolating the 2-D Fourier spectrum from a polar to a Cartesian grid, a 2-D inverse FFT is computed to obtain the desired image. The direct Fourier method is illustrated in Fig. 11.3-1.

† This means that the Fourier series of $M(\rho, \phi)$ has no more than $2K + 1$ terms, i.e., $M(\rho, \phi) = \sum_{k=-K}^{K} c_k e^{jk\phi}$.

‡ To avoid excessive notation we use the same symbol M whether it is a function of Cartesian or polar points.

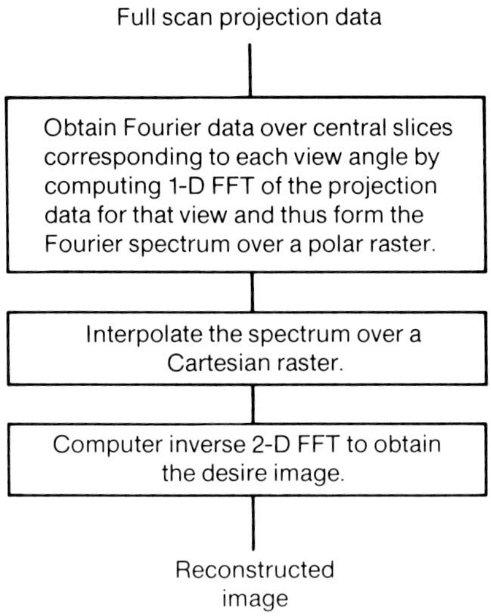

Fig. 11.3-1. Direct Fourier method (DFM).

Before closing the discussion on the direct Fourier method, several remarks are in order. First, since the object is of finite support, the projection data $p_\phi(x)$ will also have finite support and when these data are Fourier transformed aliasing errors will be introduced since the Fourier transform $P_\phi(\rho)$ cannot have finite support.† This aliasing will produce distortion in the reconstructed image. The effect of aliasing can be reduced by choosing N_D (number of detectors) large enough. Moreover, in practical situations the finite width of the x-ray beam produces smoothed projection data, which is somewhat akin to low-pass filtering of the projection data. Hence, in practice the effect of aliasing artifacts can be made tolerably smaller.

A second remark concerns the concept of an angularly band-limited function. The case $K = 0$ corresponds to an isotropic function in frequency, which implies isotropy in space. For example, a circular disk at the origin of the x–y coordinate system would correspond to $K = 0$. Its Fourier transform would also be isotropic. As K increases, one generally observes increasing departures from isotropy. In general, the functions $\mu(x, y)$ of interest will not be angularly band-limited to any finite K. The effect of artifacts in the reconstructed image due to the object having a nonfinite K can be reduced by decreasing the angular spacing between the successive views, i.e., increasing the number of views. Alternatively, one can angularly band-limit the data prior to sampling and thereby avoid the aliasing error due to nonfinite K. How this is done is explained in [1].

† An object cannot simultaneously be space-limited and band-limited.

11.4 POCS–DFM ALGORITHM

If projection data are available over a limited angular range, then the image Fourier transform $M(\rho, \phi)$ can be interpolated over a Cartesian grid within this angular range (a data cone). The pixels remaining outside this angular range are set to zero as a first approximation. The missing frequency information is then restored by POCS and the desired image is obtained by inverse transforming. The complete algorithm is called the POCS–DFM algorithm and it is illustrated in Fig. 11.4-1.

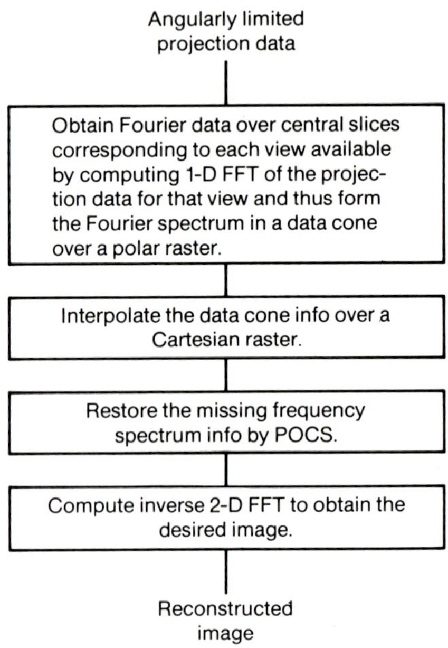

Fig. 11.4-1. POCS–DFM algorithm for reconstructing from angularly limited projection data.

11.5 A CASE STUDY: RECONSTRUCTION OF A THORAX CROSS SECTION FROM ANGULARLY LIMITED X-RAY PROJECTION DATA

A. Description of the Data

The 128 × 128 pixel human thorax phantom† is shown in Fig. 11.5-1. The simulated parallel-beam projection data assume an x-ray energy of 70 keV,

† The thorax phantom is widely used in the literature. It was designed by G. Herman and A. Lakshminarayanan.

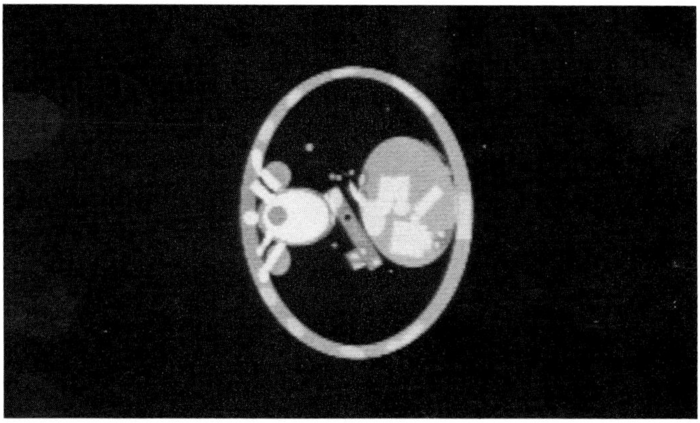

Fig. 11.5-1. Thorax phantom (through eighth vertebra). (From Sezan and Stark [23].) © 1984 *IEEE*.

128 detectors spaced apart by 0.3 cm, and 360 views over 360°. The interpolated Fourier spectrum with $L_\rho = 3$, $L_\phi = 1$, i.e., 21 interpolation points using tapering corresponding to $W = 5$, and the resulting reconstruction from complete view data using the direct Fourier method are shown in Fig. 11.5-2.

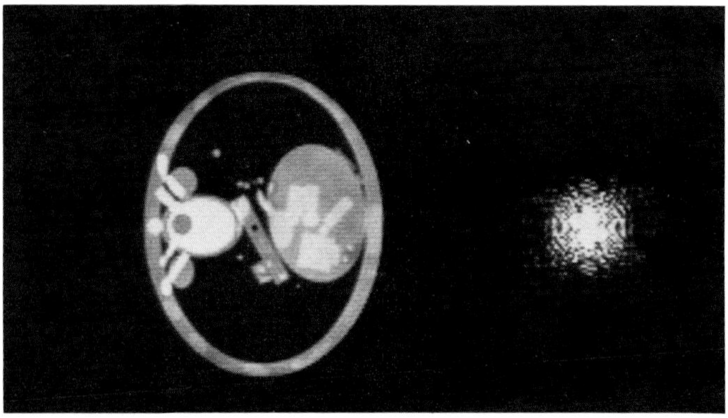

Fig. 11.5-2. DFM reconstruction (left) and interpolated spectrum. (From Sezan and Stark [23].)

Reconstruction by POCS–DFM from angularly limited projection data is attempted for the following three cases; projection data is limited to the view range of (i) $[-80°, 80°]$, (ii) $[-67°, 67°]$, (iii) $[-45°, 45°]$. The formula of Eq. (11.3-7) with $L_\rho = 3$, $L_\phi = 1$ is used to interpolate $M(\rho, \phi)$ from polar to Cartesian points. The Fourier spectra for full-view data and the above three cases are shown after interpolation in Fig. 11.5-3.

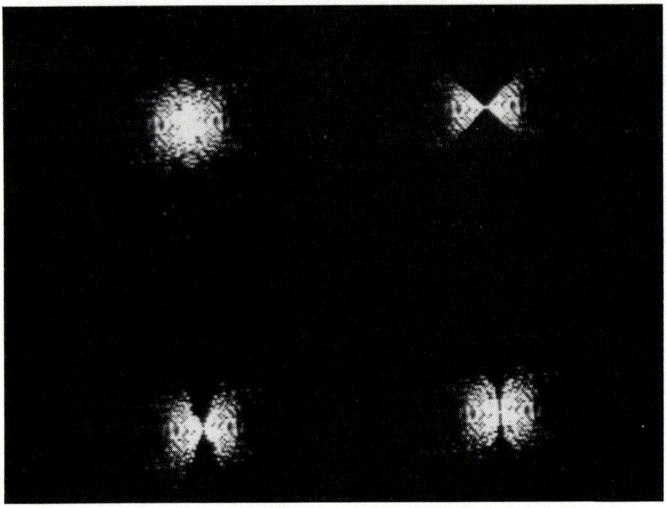

Fig. 11.5-3. Interpolated Fourier spectrum. Clockwise from upper left: full-view projection data, view limited to [−45°, 45°], view limited to [−80°, 80°], view limited to [−67°, 67°]. (From Sezan and Stark [23].) © 1984 *IEEE*.

B. Results

As before, the basis for comparing results was the percent error e_k defined by

$$e_k \triangleq 100\frac{\|f - f_k\|}{\|f\|}, \tag{11.5-1}$$

where f denotes the image reconstructed from *complete* view data.

A summary of the *a priori* known facts and assumptions made about the image are given below:

A priori Constraints	Actual
1) Image support confined to rectangular region of length 124 pixels and width 103 pixels.	1) Image support is elliptical (Fig. 11.5-1).
2) Gray levels f satisfy $0 \leq f \leq 0.4$.	2) Gray levels f satisfy $0 \leq f \leq 0.38$.
3) Energy over 128×128 pixel^2 field cannot exceed $\rho^2 = 284.0$.	3) Energy over 128×128 pixel^2 field is 282.7.

For this particular problem the sets $\mathscr{C}_1$, $\mathscr{C}_2$, $\mathscr{C}_3$, $\mathscr{C}_5$ (since $\mathscr{C}_5 \subset \mathscr{C}_4$, we do not use $\mathscr{C}_4$) are as follows:

(i) $\mathscr{C}_1$ is the set of all functions that vanish (a.e.) outside a rectangular region of length 124 pixels and width 103 pixels.

(ii) $\mathscr{C}_2$ is the set of all functions whose Fourier transforms agree (a.e.) with the values obtained by measurement in the spectral data cone. There are three cases: $[-45°, 45°]$, $[-67°, 67°]$, and $[-80°, 80°]$. The data cone of $[-45°, 45°]$ is associated with view data limited to 180°, the data cone of $[-67°, 67°]$ with view data limited to 268°, and the data cone of $[-80°, 80°]$ with view data limited to 320°.

(iii) $\mathscr{C}_3$ is the set of all functions whose norm (energy) is less than or equal to 284 pixels squared.

(iv) $\mathscr{C}_5$ is the set of all functions whose gray level f is confined to the range $0 \leqslant f \leqslant 0.4$.

The associated projection operators are P_1, P_2, P_3, and P_5. Table 11.5-1 furnishes a summary of the experimental results.

The first (leftmost) column of Table 11.5-1 gives the spectral data cone over which data are available; in the second column is the name of the

Table 11.5-1
Summary of the Results

Available View Range	Restoration Algorithm			Percent Error	
	Name	Recursion	Relaxation Parameters	e_0 (zero-order solution)	e_{30}
[-45, 45]	G-P UNIRELAXL	$P_2P_1f_k=f_{k+1}$ $P_5P_2P_3P_1f_k=f_{k+1}$	$\lambda_1=\lambda_2=1.0$ $\lambda_1=\lambda_2=1.0$ $\lambda_3=\lambda_5=1.0$	52.419	47.511 42.057
[-67, 67]	G-P UNIRELAX RELAX	$P_2P_1f_k=f_{k+1}$ $P_2P_3P_1f_k=f_{k+1}$ $T_2T_3T_1f_k=f_{k+1}$	$\lambda_1=\lambda_2=1.0$ $\lambda_1=\lambda_2=\lambda_3=1.0$ $\lambda_1=\lambda_3=1.9995$ $\lambda_2=1.0$	45.000	22.203 17.837 16.184
[-80, 80]	G-P UNIRELAX RELAX	$P_2P_1f_k=f_{k+1}$ $P_2P_3P_1f_k=f_{k+1}$ $T_2T_3T_1f_k=f_{k+1}$	$\lambda_1=\lambda_2=1.0$ $\lambda_1=\lambda_2=\lambda_3=1.0$ $\lambda_1=\lambda_3=1.9995$ $\lambda_2=1.0$	41.361	15.485 12.100 9.352

restoration algorithm; in the third and fourth columns are the recursion relations and the associated values of the relaxation parameters, respectively; the next column is the error between the full scan image and the image immediately reconstructed from the limited view data (the so-called zero-order solution); and the last column is the error at the 30th iteration for the three cases of angularly limited projection data.

1. *Results for the [−45°, 45°] Data Cone*

In this case, the reconstructed image at the end of 30 iterations is of low visual quality for all restoration algorithms. The reconstruction with UNIRELAXL (i.e., where level constraints are also applied) restoration outperforms the reconstruction with the Gerchberg–Papoulis restoration. The error performances of the reconstructions are compared in Fig. 11.5-4.

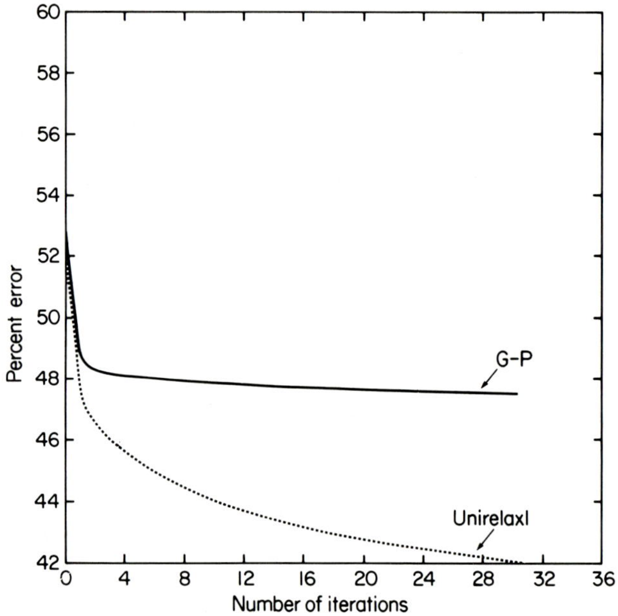

Fig. 11.5-4. Error performance of G–P and UNIRELAXL in reconstructing from projection data available over [−45°, 45°]. (From Sean and Stark [23].) © 1984 *IEEE.*

2. *Results for the [−67°, 67°] Data Cone*

In this case the three different restoration algorithms G–P, UNIRELAX, and RELAX are each, in turn, combined with the direct Fourier method of reconstruction to realize a better image than the zero-order solution. The zero-order solution is the image generated by inverse Fourier transforming the available data without extrapolation. It results in an image whose error, as computed by Eq. (11.5-1) is 45%. In Fig. 11.5-5 the zero-order solution is compared with full-view data reconstruction as well as reconstructions using G–P, UNIRELAX, and RELAX restorations. The error performances of these three reconstructions are compared in Fig. 11.5-6. With the RELAX restoration, an error less tan 17% is reached at the end of 30 iterations and the resulting reconstruction has fairly good visual quality (see, Fig. 11.5-5).

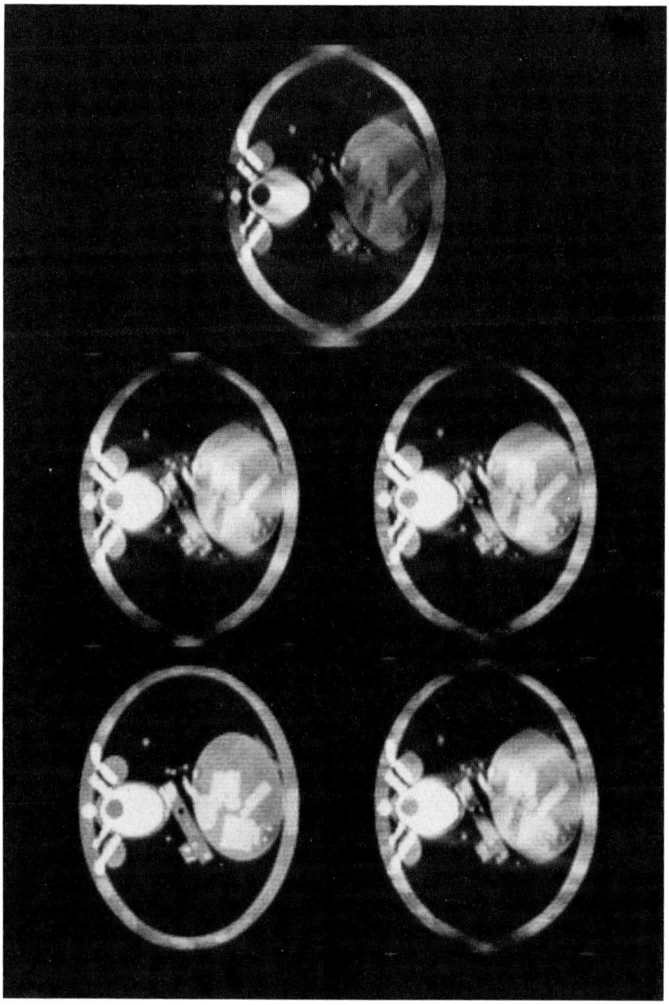

Fig. 11.5-5. Reconstructions when projection data are limited to [−67°, 67°]. (Top) Zero-orer reconstruction; (below) clockwise from upper left: G-P, RELAX, UNIRELAX, and full-view data reconstruction. (From Sezan and Stark [23].) © 1984 *IEEE*.

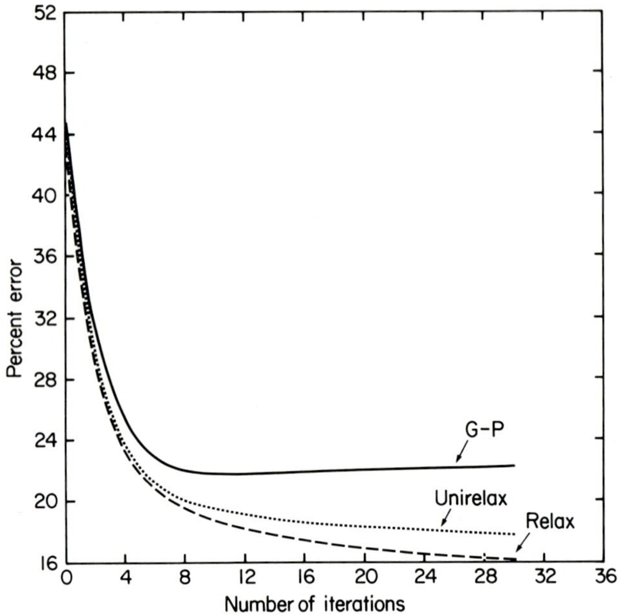

Fig. 11.5-6. Error performance of G-P, UNIRELAX, and RELAX in reconstructing from projection data available over $[-67°, 67°]$. (From Sezan and Stark [23].) © 1984 *IEEE.*

3. *Results for the $[-80°, 80°]$ Data Cone*

The zero-order image (for which the error is $e_0 = 42\%$), the full-view image, and the reconstructed images at the end of 30 iterations using RELAX, UNIRELAX, and G-P restorations are compared in Fig. 11.5-7. As in the previous case, reconstructions using *a priori* restoration are significantly better than the zero-order reconstruction. Figure 11.5-8 compares the error performance of the reconstructions using RELAX, UNIRELAX, and G-P algorithms; the errors are respectively 9.4, 12.1, and 15.5% at the end of 30 iterations. The G-P and RELAX reconstructions are compared in Fig. 11.5-9. The superior performance of RELAX is especially evident in the reconstruction of the elliptical structure (note the obliterations in the G-P case) and the reduced plumelike clouding from the tips of bright objects.

C. Conclusions

The main conclusion drawn from the results of our case study is the following: when the projection data are angularly view-limited, the images

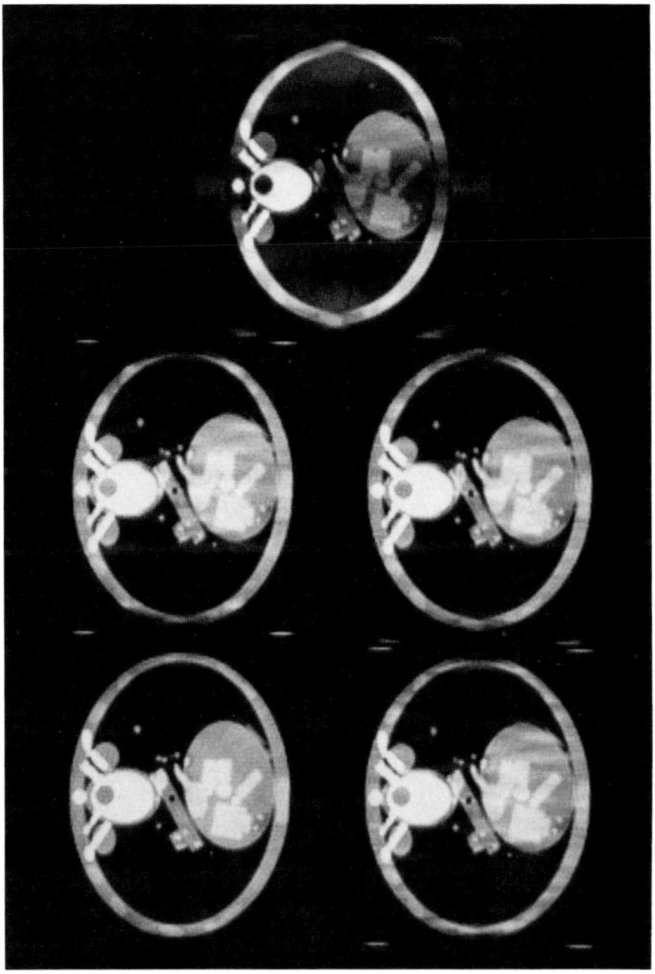

Fig. 11.5-7. Reconstructions when projection data are limited to $[-80°, 80°]$. (Top) Zero-order reconstruction; (below) clockwise from upper left: G–P, RELAX, UNIRELAX, and full-view data reconstruction. (From Sezan and Stark [23].) © 1984 *IEEE*.

obtained by direct Fourier reconstruction incorporating *a priori* POCS restoration are superior to the zero-order image obtained by direct inversion of the incomplete Fourier spectrum.

More central to the objective of this study is the comparison of performance with the G–P and more general algorithms, i.e., UNIRELAX and RELAX, which incorporate additional *a priori* constraints. Consistent with the results given in Section 11.2, the additional energy level constraint

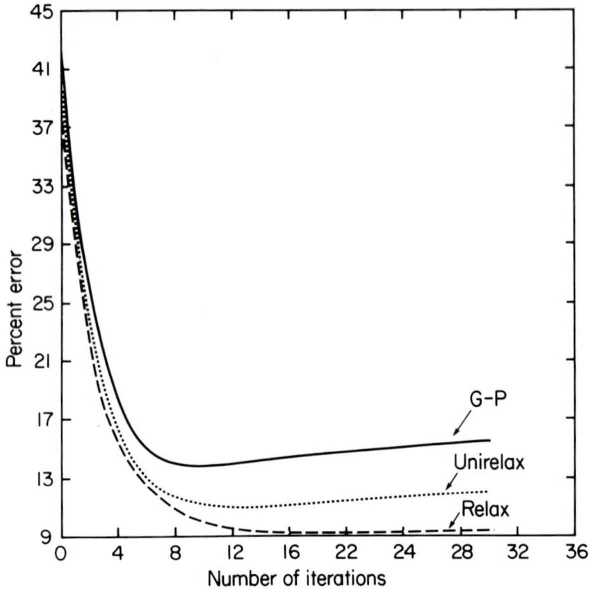

Fig. 11.5-8. Error performance of G–P, UNIRELAX, and RELAX in reconstructing from projection data available over [−80°, 80°]. (From Sezan and Stark [23].) © 1984 *IEEE.*

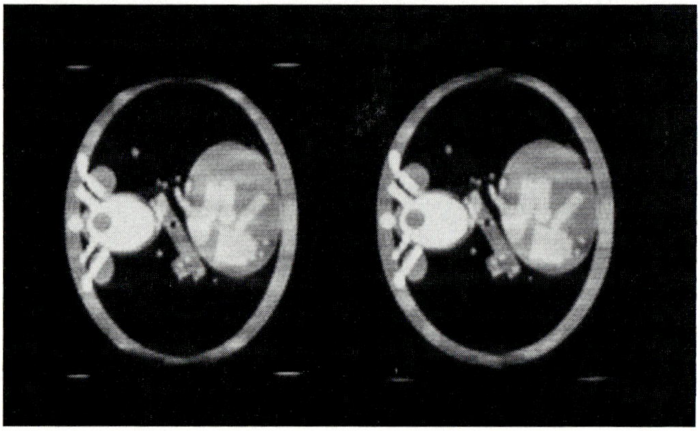

Fig. 11.5-9. (Left) G–P and (right) RELAX reconstructions at 30th iteration. (From Sezan and Stark [23].) © 1984 *IEEE.*

proved critical in obtaining better images. In the [−45°, 45°] (data cone) case, however, none of the algorithms was able to produce an image with good visual quality. The practical significance of this result is that there is a level of data insufficiency that prevents recovery of the missing data regardless of how sophisticated the recovery algorithm is. However, when this is not the case then further improvement is best obtained by using relaxed projections, in agreement with the results obtained when restoring the synthetic image in the absence of noise. But, as before, the values of the nonunity relaxation parameters are somewhat heuristically chosen and they do not represent an optimum set of values.

We discuss potential applications of POCS–DFM in areas other than x-ray CT in the following section.

11.6 APPLICATIONS TO RELATED AREAS

In this section we discuss the application of the POCS–DFM algorithm to spotlight-mode synthetic aperture radar (SAR) imaging, NMR tomography, ultrasound computerized tomography (UCT), and the synthesis of functions satisfying moment constraints.

A. *Spotlight-Mode Synthetic Aperture Radar*

The following discussion is based on the Munson *et al.* [24] tomographic formulation of spotlight-mode SAR.

At an observation angle ϕ (see Fig. 11.6-1) the radar transmits a frequency-modulated (FM) chirp pulse $\mathrm{Re}\{s(t)\}$, where

$$s(t) = \begin{cases} \exp[j(\omega_0 t + at^2)], & |t| \leqslant T/2, \\ 0, & \text{otherwise}, \end{cases} \tag{11.6-1}$$

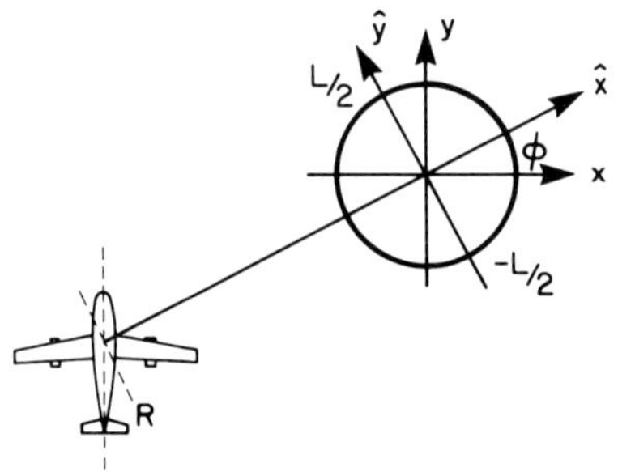

Fig. 11.6-1. Zero depression angle geometry for data collection in spotlight-mode SAR.

$2a$ is the FM rate, and ω_0 is the radio-frequency (RF) carrier frequency. It is shown in [24] that under certain assumptions the return signal $C_\phi(t)$ at an observation angle ϕ and a distance R from the target has the form

$$C_\phi(t) = (A/2)P_\phi\{(2/c)[\omega_0 + 2a(t - 2R/c)]\} \qquad (11.6\text{-}2)$$

and is available for $-(T/2) + 2(R + L)/c \leq t \leq (T/2) + 2(R - L)/c$. In Eq. (11.6-2) A is the propagation attenuation, c the speed of light, L the diameter that bounds the target, and $P_\phi(\cdot)$ the Fourier transform of the projection $p_\phi(x)$ as in Section 11.3, i.e.,

$$p_\phi(\hat{x}) \triangleq \int_{-L/2}^{L/2} \mu_\phi(\hat{x}, \hat{y})\, d\hat{y}. \qquad (11.6\text{-}3)$$

Here $\mu_\phi(\hat{x}, \hat{y})$ denotes the complex-valued reflectivity density distribution of the target defined in a coordinate system $\hat{x}$–$\hat{y}$ rotated from x–y by an angle ϕ. By the central slice projection theorem the function $P_\phi(\rho)$ with $\rho \geq 0$, $0 < \phi \leq 2\pi$, gives the value of the two-dimensional Fourier transform $M(u, v)$ of the reflectivity density distribution $\mu(x, y)$ at a polar point (ρ, ϕ) in the u–v Fourier plane. Hence, in practice, the sampled returns at discrete observation angles ϕ_k, $k = 1, 2, \ldots, K$, with $|\phi_k| \leq \Phi$ provide the samples of $M(u, v)$ over a polar raster within the annular segment Q described by

$$Q = \{(\rho, \phi): (2/c)(\omega_0 - aT) \triangleq \rho_1 \leq \rho \leq \rho_2 \triangleq (2/c)(\omega_0 + aT), \quad |\phi| \leq \Phi\}, \qquad (11.6\text{-}4)$$

where it is assumed that $\omega_0 \pm aT \gg 4aL/c$, which is a reasonable assumption for SAR. The geometry of the annulus segment Q is shown in Fig. 11.6-2.

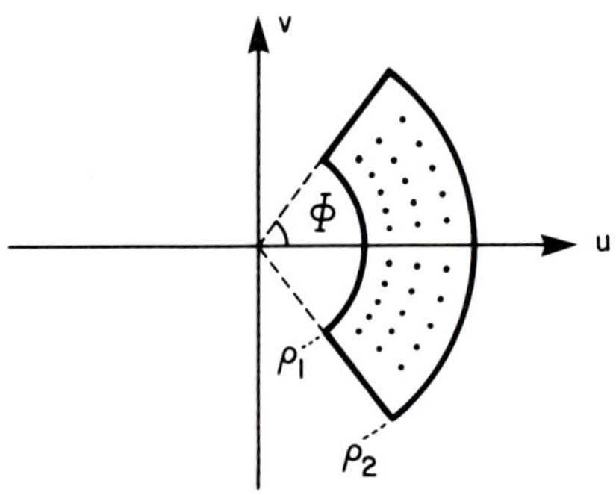

Fig. 11.6-2. Annulus segment Q where the Fourier samples are measured.

The angular extent of Q is limited by the range of observation angles; its radial extent is determined by the frequency content of the transmitted signal. Therefore Fourier samples of the reflectivity density distribution can only be determined partially from measurements. The following procedure can be used to reconstruct the image distribution (the complex magnitude of the reflectivity density distribution is imaged):

(i) Interpolate the measured Fourier data over a Cartesian grid within Q by using truncated cardinal function expansion.†

(ii) Restore the missing Fourier data by the method of POCS and fill in an $N \times N$ array of Fourier samples.

(iii) Compute the inverse FFT (IFFT) to obtain the desired image distribution.

The above procedure for obtaining $\mu(x, y)$ is indeed the POCS–DFM algorithm.

Since, unlike x-ray CT, the image distribution in SAR is complex-valued, the closed convex sets $\mathscr{C}_3$ and $\mathscr{C}_5$ defined in Section 11.2 must be modified. The modified sets $\tilde{\mathscr{C}}_3$ and $\tilde{\mathscr{C}}_5$ are described in the following.

$\tilde{\mathscr{C}}_3$: The set of all functions in $\mathscr{H}$ that satisfy the energy constraint

$$\|f\|^2 = \int_\Omega |f(x, y)|^2 \, dx \, dy \leqslant E \triangleq \rho^2, \qquad f \triangleq f_1 + jf_2. \tag{11.6-5}$$

$\tilde{\mathscr{C}}_5$: The set of all functions in $\mathscr{H}$ whose complex magnitudes satisfy the following constraint:

$$0 \leqslant |f| \leqslant b, \qquad b > 0. \tag{11.6-6}$$

The projection operators projecting onto $\tilde{\mathscr{C}}_3$ and $\tilde{\mathscr{C}}_5$ are realized by [18]

$$\tilde{P}_3 f = \begin{cases} f, & \|f\|^2 \leqslant E, \\ f\sqrt{E}/\|f\|, & \|f\|^2 > E, \end{cases} \tag{11.6-7}$$

and

$$\tilde{P}_5 f = \begin{cases} f, & 0 \leqslant |f| \leqslant b, \\ (f/|f|)b, & |f| > b, \end{cases} \tag{11.6-8}$$

respectively (f arbitrary in $\mathscr{H}$).

B. Nuclear Magnetic Resonance Tomography

Nuclear magnetic resonance imaging is intrinsically a three-dimensional imaging modality. However, by selective excitation [25–27] two-dimensional tomographic imaging of the spatial nuclear magnetization distribution within a selected slice can be obtained.

† See Section 11.3.

By using a line integral projection reconstruction (LPR) technique [26,27], Fourier samples of the image distribution are obtained over a polar raster in the frequency domain. In this technique the gradient is angularly varied and the NMR signal is recorded for a set of angular positions of the gradient field. (This is analogous to collecting angular view data in x-ray CT.)

At a particular view angle $\phi_k = 2\pi k/K$, $k = 1, 2, \ldots, K$, the imaging equation, i.e., the relation between the measured signal $s_{\phi_k}(t)$ and the nuclear magnetization distribution $M(x, y)$ within the selected slice of support Ω, has the following form [27]:

$$s_{\phi_k}(t) = C \iint_{\Omega} M(x, y) \exp[-j\gamma(G_{x(k)}x + G_{y(k)}y)t]\, dx\, dy, \tag{11.6-9}$$

where

$$G_{x(k)} \triangleq |\mathbf{G}| \cos \phi_k, \tag{11.6-10}$$

$$G_{y(k)} \triangleq |\mathbf{G}| \sin \phi_k \tag{11.6-11}$$

are the x and y components of the angularly varying gradient field vector $\mathbf{G}$, respectively, and C is a constant and γ is the gyromagnetic ratio. By substituting Eqs. (11.6-10) and (11.6-11) into Eq. (11.6-9) we get

$$s_{\phi_k}(t) = C \iint_{\Omega} M(x, y) \exp[-j\gamma|\mathbf{G}|(x \cos \phi_k + y \sin \phi_k)t]\, dx\, dy. \tag{11.6-12}$$

In practice $s_{\phi_k}(t)$ is sampled with interval Δt. Thus Eq. (11.6-12) becomes

$$s_{\phi_k}(n\, \Delta t) \triangleq s_{\phi_k}(n) = C \iint_{\Omega} M(x, y) \exp[-j\gamma|\mathbf{G}|n(x \cos \phi_k + y \sin \phi_k)\, \Delta t]\, dx\, dy,$$

$$n = 1, 2, \ldots, N. \tag{11.6-13}$$

But from Eq. (11.6-13) we see that the temporal signal $s_{\phi_k}(n)$ contains spatial frequency information, and in fact $s_{\phi_k}(n)$, $n = 1, 2, \ldots, N$, $k = 1, 2, \ldots, K$, is the two-dimensional Fourier transform of the nuclear magnetization $M(x, y)$ sampled at $(\gamma|\mathbf{G}|n\, \Delta t \cos \phi_k, \gamma|\mathbf{G}|n\, \Delta t \sin \phi_k)$, or equivalently at $(\rho_n \cos \phi_k, \rho_n \sin \phi_k)$, where $\rho_n \triangleq \gamma|\mathbf{G}|n\, \Delta t$. Therefore the Fourier spectrum of the image distribution within the selected slice can be directly measured over a polar raster in the frequency domain. If full-view spectrum information is available the direct Fourier inversion method (DFM) can be used to reconstruct the nuclear magnetization distibution. In case of incomplete-view measurements the missing spectrum can be restored via POCS prior to Fourier inversion. This restoration followed by Fourier inversion is indeed the POCS-DFM reconstruction algorithm. Incomplete-view measurements are motivated in NMR by the desire for reduction of data collection time.

C. *Ultrasound Computerized Tomography*

In ultrasound computerized tomography (UCT) the spatial distribution of the real part of the refractive index, which involves the sonic velocity, and the imaginary part of the refractive index, which involves the attenuation coefficient, can be imaged. Neglecting diffraction effects, the line integrals, i.e., projection data, of both sonic velocity and the attenuation coefficient can be estimated from measurements. However, if there are opaque obstructions such as bones or gas bubbles around the soft tissue of interest, projection data for those views yield no information and the viewing becomes essentially angularly view limited. In this case POCS-DFM reconstruction is directly applicable.

In diffraction tomography it may still be possible to use POCS-DFM when the measured data are incomplete. Kaveh and Soumekh in Chapter 10 consider diffraction tomography and solve the wave equation with certain assumptions. The Fourier transform of the spatial refractive index distribution can be obtained on circular arcs that pass through the origin of the frequency plane (see also [28,29]). The data in the Fourier transform plane can be filled in by scanning the object by the insonifying source or varying the frequency of the source. The data on circular arcs can be interpolated over a Cartesian grid by an accurate interpolation technique discussed in Chapter 10. Once this is done it seems reasonable that POCS-DFM can be used in diffraction tomography as well.

D. *Incorporation of a priori Moment Information into Signal Recovery and Synthesis Problems*

In this section we consider a procedure for the recovery of an unknown signal $f(x)$ that is consistent with a set of *a priori* known generalized moments. We utilize the method of projections onto convex sets (POCS). The procedure may have application in several settings: (1) in synthesizing a probability density function from a set of measured moments, (2) in estimating an optical field from moment-type measurements, and (3) in CT *reconstruction* by POCS as an alternative to both convolution backprojection and direct Fourier inversion. Let $\mathcal{H} \triangleq L_2(\Omega)$, $\Omega \subset \mathcal{R}$, be the space of real-valued square-integrable functions. For $y_1, y_2 \in \mathcal{H}$ the inner product and norm are defined by

$$(y_1, y_2) \triangleq \int_\Omega y_1(x) y_2(x)\, dx, \tag{11.6-14}$$

$$\|y_1\| \triangleq (y_1, y_2)^{1/2}, \tag{11.6-15}$$

respectively. We now consider the following constraints on the first $M+1$ generalized moments of the function $f(x) \in \mathcal{H}$:

$$\begin{aligned} l_0 &\leq (g_0, f) \leq u_0, \\ l_1 &\leq (g_1, f) \leq u_1, \\ &\vdots \\ l_M &\leq (g_M, f) \leq u_M, \end{aligned} \tag{11.6-16}$$

where $l_0, l_1, \ldots, l_M$ and $u_0, u_1, \ldots, u_M$ are well-defined lower and upper bounds, respectively. The weighting functions $g_0(x), g_1(x), \ldots, g_M(x)$ are well-defined functions in $\mathcal{H}$. We assume that the l_i, u_i, and $g_i(x)$, $i = 0, \ldots, M$, are known *a priori*. The constraints in Eq. (11.6-16) restrict the function $f(x)$ to lie in a family of sets $\mathcal{C}_r$, $r = 0, 1, \ldots, M$, in $\mathcal{H}$. For $r \in \{0, 1, \ldots, M\}$ $\mathcal{C}_r$ is defined as [all the $\mathcal{C}_r$ are subsets of $L_2(\Omega)$]

$$\mathcal{C}_r \triangleq \{h : l_r \leq (g_r, h) \leq u_r\}, \qquad u_r \geq l_r. \tag{11.6-17}$$

In words: $\mathcal{C}_r$ is the set of all functions $\{h(x)\}$ that satisfy the moment constraint $l_r \leq (g_r, h) \leq u_r$. For any r, $\mathcal{C}_r$ is a closed convex set [30]. Therefore *a priori* moment information can be incorporated into the method of POCS. For an arbitrary $y(x) \in \mathcal{H}$, the projection operator P_r projecting onto $\mathcal{C}_r$ is realized by [30]

$$P_r y \triangleq \begin{cases} y(x), & l_r \leq M_r \leq u_r, \\ y(x) - [(M_r - u_r)/K_r] g_r(x), & M_r > u_r, \\ y(x) + [(l_r - M_r)/K_r] g_r(x), & M_r < l_r, \end{cases} \tag{11.6-18}$$

where $M_r \triangleq (g_r, y)$ and $K_r \triangleq \|g_r\|^2$. For $u_r = l_r \triangleq m_r$, the projection is realized by

$$P_r y_r = \begin{cases} y(x), & M_r = m_r, \\ y(x) + [(m_r - M_r)/K_r] g_r(x), & M_r \neq m_r. \end{cases} \tag{11.6-19}$$

To demonstrate the use of *a priori* moment information via the method of POCS, we consider as an example the function

$$f_R(x) = \begin{cases} x \exp(-x^2/2), & x \in \Omega \triangleq [0, 5], \\ 0, & \text{otherwise}, \end{cases} \tag{11.6-20}$$

which is recognized as a truncated Rayleigh probability density function with parameter $\alpha = 1$. We let $\operatorname{supp} f_R = \Omega = [0, 5]$. For $g_r(x) \triangleq x^r$, the moments of $f_R(x)$ up to order K are given by

$$m_r = (x^r, f_R), \qquad r = 0, 1, \ldots, K. \tag{11.6-21}$$

We consider the following problem:

Problem. Find a nonnegative function $f(x)$, $x \in \Omega$, with prescribed moments m_r, $r = 0, 1, \ldots, K$ as defined in Eq. (11.6-21).

Before proceeding further with our problem, the following remarks are in order:

(i) Although the solution to this problem exists, it is not unique. In other words, there are generally nonnegative functions other than $f_R(x)$ having the prescribed moments m_r, $r = 0, 1, \ldots, K$. However, if our goal is the synthesis of such a function the uniqueness is unimportant.

(ii) The moment information restricts the function $f(x)$ to lie in the closed convex sets $\mathscr{C}_r$, $r = 0, 1, \ldots, K$. The nonnegativity constraint restricts $f(x)$ to lie in the closed convex set $\mathscr{C}_p$, defined by

$$\mathscr{C}_p \triangleq \{h(x) : h(x) \geq 0, x \in \Omega\}. \tag{11.6-22}$$

For an arbitrary $y(x) \in L_2(\Omega)$, the projection onto $\mathscr{C}_p$ is realized by

$$P_p y = \begin{cases} y(x), & y(x) \geq 0, \\ 0, & y(x) < 0. \end{cases} \tag{11.6-23}$$

Indeed, other *a priori* information and constraints can be incorporated into the problem. However, we will restrict our attention to moment information and the nonnegativity constraint. For the experiment we assume *a priori* knowledge of moments up to order 4 ($K = 4$). The moments of the function $f_R(x)$, $x \in \Omega \triangleq [0, 5]$, are given in Table 11.6-1. The following two functions

Table 11.6-1
Moments of $f_R(x)$, $x \in [0, 5]$

m_0	m_1	m_2	m_3	m_4
1.0000	1.2533	1.9999	3.7594	7.9973

are used to intialize the POCS algorithm:

$$f_{01}(x) \triangleq \exp[-(x - 1.253)^2/2], \qquad x \in [0, 5], \tag{11.6-24}$$

and

$$f_{02}(x) \triangleq 0, \qquad x \in [0, 5]. \qquad (1.6\text{-}25)$$

The moments of $f_{01}(x)$ and $f_{02}(x)$ up to order 4 are given in Table 11.6-2.

Table 11.6-2
Initialization Functions, Their Moments, and Corresponding Percent Errors e_{01} and e_{02}

$f_0(x)$, $x\epsilon[0,5]$	m_0	m_1	m_2	m_3	m_4	e_{0i} $i = 1,2$
$f_{01}(x) = \exp[-(x-1.253)^2/2]$	2.2435	3.2656	6.3304	14.4409	36.9743	103.4041
$f_{02}(x) = 0$	0.0000	0.0000	0.0000	0.0000	0.0000	100.0000

In general, the percent error e_k at the kth iteration is defined as

$$e_k \triangleq 100\|f_k - f_R\|/\|f_R\|$$

and, as a special case, the initial percent error e_{0i}, $i = 1, 2$, is defined as

$$e_{0i} \triangleq 100\|f_{0i} - f_R\|/\|f_R\|, \qquad i = 1, 2. \qquad (11.6\text{-}26)$$

The errors e_{01} and e_{02} are given in the last column of Table 11.6-2.

Table 11.6-3 summarizes the results obtained for $f_0(x) = f_{01}(x)$. The leftmost column is the recursion relation of the algorithm. The next five columns furnish the resulting moments m_0, m_1, m_2, m_3, m_4, respectiely, at the end of 100 iterations. The last column, e_{100}, is the percent error in estimating the unknown function by f_{100}. The results for $f_0(x) = f_{02}(x)$ are

Table 11.6-3
Summary of the Results for $f_0(x) = f_{01}(x)$

Algorithm	m_0	m_1	m_2	m_3	m_4	e_{100}
$P_pP_3P_pP_2P_pP_1P_pP_0\ f_k = f_{k+1}$	0.9998	1.2565	1.9860	3.7594	8.6382	16.3185
$P_pP_4P_pP_3P_pP_2P_pP_1P_pP_0\ f_k = f_{k+1}$	0.9995	1.2593	1.9853	3.6828	7.9973	16.1552
$P_pP_3P_pP_4P_pP_2P_pP_1P_pP_0\ f_k = f_{k+1}$	0.9995	1.2589	1.9954	3.7594	8.4183	16.0685
$f_R(x)$, $x\epsilon[0,5]$	1.0000	1.2533	1.9999	3.7594	7.9973	---

summarized in Table 11.6-4. In general, the good performance of POCS is evident when the resulting moments at the end of 100 iterations and the initial moments given in Table 11.6-2 are compared with the prescribed moments. As expected, the value of the fourth moment m_4 is improved when the operator P_4 is included in the composition operator T. At any iteration, provided that the final nonnegativity correction is not a major one, the moment associated with the last moment operator within T will be the one that agrees the most with the prescribed value. This is the case for m_3 with $T = P_pP_3P_pP_4P_pP_2P_pP_1P_pP_0$ and $T = P_pP_3P_pP_2P_pP_1P_pP_0$, and for m_4 with $T = P_pP_4P_3P_pP_pP_2P_pP_1P_pP_0$ when $f_0(x) = f_{01}(x)$. The moment values at each iteration of the algorithm $Tf_k = P_pP_4P_pP_3P_pP_2P_pP_1P_pP_0f_k = f_{k+1}$ are plotted in Fig. 11.6-3. For both of the initializations the resulting functions at the end of 100 iterations furnish moments in excellent agreement with the *a priori* known moments; Fig. 11.6-4 compares the resulting functions. The fact that the recovered functions, even after 100 iterations, do not match $f_R(x)$ is, in part, due to the nonuniqueness of a solution constrained by a finite set of moments.

Table 11.6-4
Summary of the Results for $f_0(x) = f_{02}(x)$

Algorithm	m_0	m_1	m_2	m_3	m_4	e_{100}
$P_pP_3P_pP_2P_pP_1P_pP_0\ f_k = f_{k+1}$	1.0040	1.2032	2.0178	3.9361	8.3900	35.3866
$P_pP_4P_pP_3P_pP_2P_pP_1P_pP_0\ f_k = f_{k+1}$	1.0040	1.2043	2.0141	3.9105	8.2860	35.0347
$P_pP_3P_pP_4P_pP_2P_pP_1P_pP_0\ f_k = f_{k+1}$	1.0040	1.2043	2.0143	3.9108	8.2864	35.0604
$f_R(x),\ x\varepsilon[0,5]$	1.0000	1.2533	1.9999	3.7594	7.9973	---

Finally, we note that the moment constraint can be used in tomographic image reconstruction when the projection data are incomplete. To illustrate this fact we let $f(x, y)$ be a 2-D distribution in the Hilbert space of square-integrable functions, i.e., $f(x, y) \in L_{2\times2}(\Omega)$, $\Omega \subset R^2$. For a well-defined function $g(x, y) \in L_{2\times2}(\Omega)$ the moment equation becomes

$$m = \iint_{\Omega} g(x, y)f(x, y)\, dx\, dy. \tag{11.6-27}$$

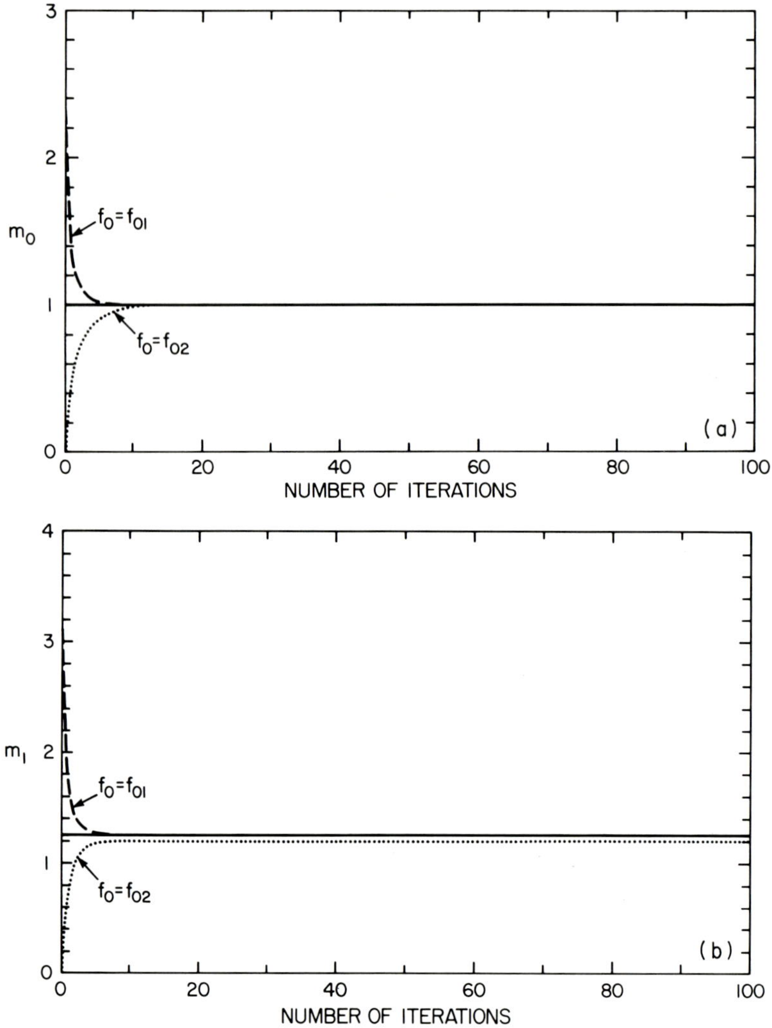

Fig. 11.6-3. Moment values versus number of iterations for $T = P_pP_4P_pP_3P_pP_2P_pP_1P_pP_0$: (a) m_0, (b) m_1, (c) m_2, (d) m_3, (e) m_4. (Solid lines indicate correct values.)

If we define $g(x, y)$ as

$$g(x, y) \triangleq \chi_L(x, y), \tag{11.6-28}$$

where $\chi_L(x, y)$ is the characteristic function of a strip L, i.e.,

$$\chi_L(x, y) = \begin{cases} 1, & (x, y) \in L, \\ 0, & \text{otherwise}, \end{cases} \tag{11.6-29}$$

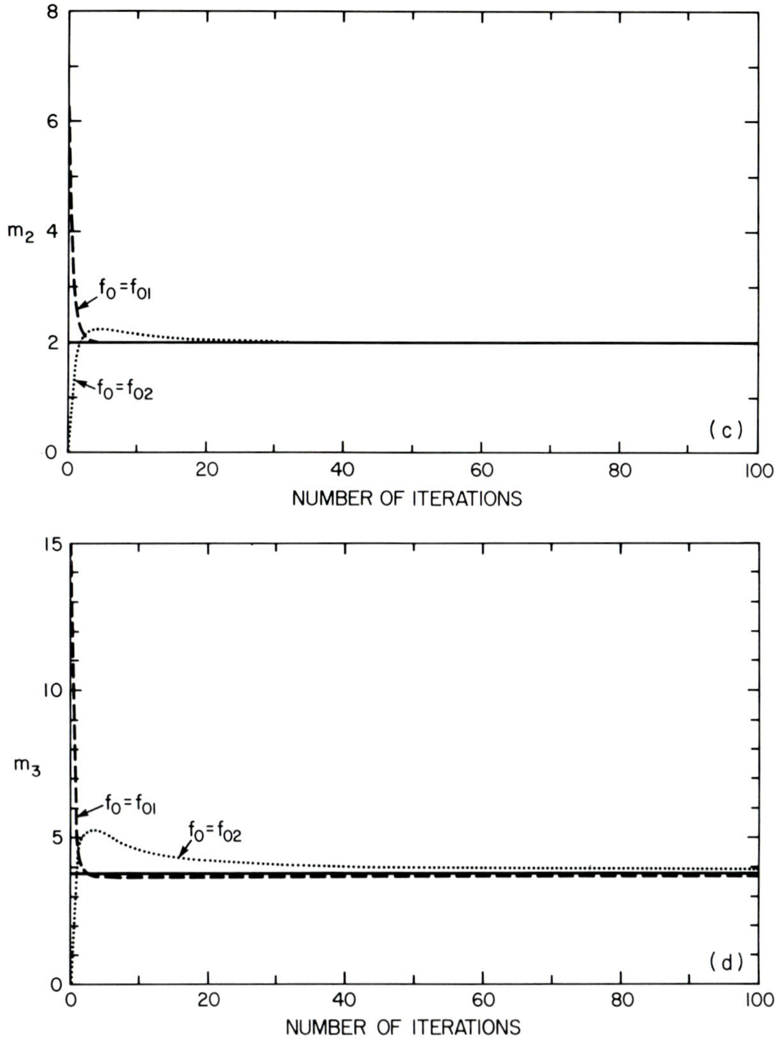

Fig. 11.6-3. (*continued*)

then Eq. (11.6-27) can be written as

$$m_L = \iint_{\Omega} \chi_L(x, y) f(x, y)\, dx\, dy. \tag{11.6-30}$$

(We assume that $L \cap \Omega \neq \varnothing$.) But Eq. (11.6-30) is the strip integral, or equivalently the projection of $f(x, y)$ along the strip L. Thus *a priori* bounds

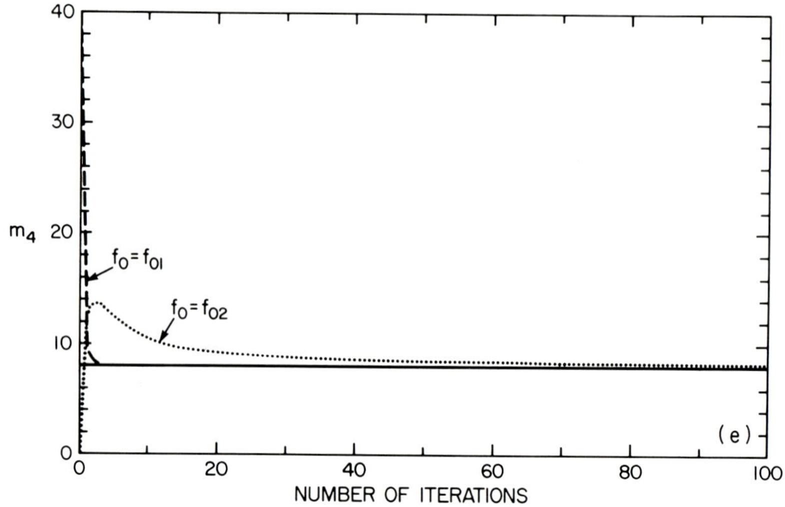

Fig. 11.6-3. (*continued*)

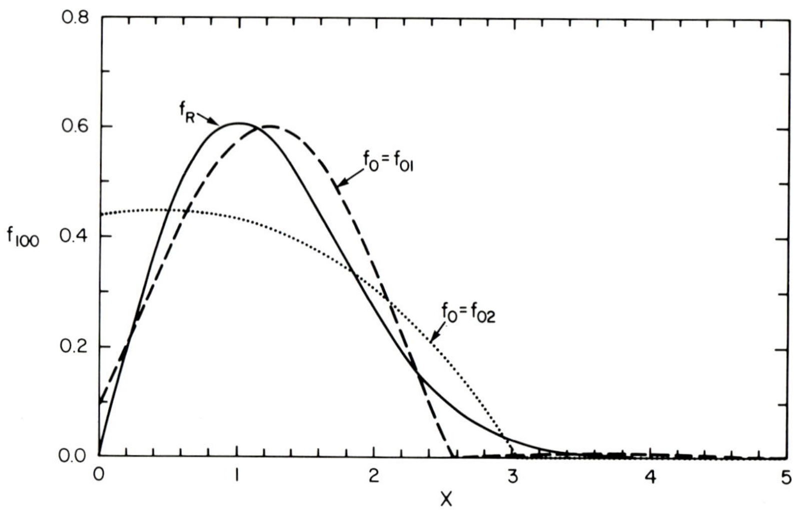

Fig. 11.6-4. $f_R(x)$ and the resulting functions at the end of 100 iterations of the algorithm $P_pP_4P_pP_3P_pP_2P_pP_1P_pP_0f_k = f_{k+1}$.

on projections are convex-type constraints. Consequently, these constraints can be incorporated into the tomographic image reconstruction algorithms discussed in this chapter and elsewhere in this book.

11.7 SUMMARY

Our aim in this chapter has been to show how the method of projection onto convex sets can be applied in image or signal recovery from incomplete data and *a priori* constraints. First, we applied POCS to spectral extrapolation of images. A synthetic image was recovered from its partial Fourier spectrum. The additional *a priori* constraints incorporated into the problem via the method of POCS improved the reconstruction in general and regularized the problem in the presence of noise. This preliminary study constituted the groundwork for a case study where the POCS–DFM algorithm, synthesized from the method of POCS and the direct Fourier method, was used to reconstruct an x-ray tomogram of a simulated human thorax. Reconstructions from full-view and limited-view projection data were performed.

POCS–DFM, being a Fourier domain algorithm, is also useful for applications where the Fourier samples of the image function are directly measured. The potential applications of the POCS–DFM algorithm to such direct Fourier imaging schemes as NMR tomography, ultrasound computerized tomography, and spotlight-mode synthetic aperture radar imaging were discussed.

Finally, we considered the incorporation of *a priori* generalized moment information in the signal synthesis problem. We used the method of POCS to synthesize a function consistent with a set of prescribed moments. Use of the moment information in the context of tomographic image reconstruction was also discussed, and it was shown that the strip integrals obtained in computerized tomography are equivalent to moments. This suggests that POCS is useful not only as a recovery algorithm but also as a technique for *image reconstruction* in projection tomography.

REFERENCES

1. H. Stark, J. W. Woods, I. Paul, and R. Hingorani (1981). *IEEE Trans. Biomed. Eng.* **BME-28**, 496–505.
2. H. K. Tuy (1983). *Technical Digest of Topical Meeting on Signal Recovery and Synthesis with Incomplete Information and Partial Constraints*, **FA1-1**, **FA1-4**.
3. K. C. Tam (1982). *IEEE Trans. Nucl. Sci.* **NS-29**, 512–515.
4. K. C. Tam and V. Perez-Mendez (1981). *J. Opt. Soc. Am.* **71**, 582–592.
5. T. Sato, S. J. Norton, M. Linzer, O. Ikeda, and M. Hirama (1981). *Appl. Opt.* **20**, 395–399.
6. H. Stark, D. Cahana, and H. Webb (1981). *J. Opt. Soc. Am.* **71**, 635–642.
7. H. Stark, D. Cahana, and G. J. Habetler (1981). *Opt. Lett.* **6**, 259–260.
8. A. Lent and H. K. Tuy (1981). *Jour. of Math. Analysis and Applications* **83**, 553–565.
9. A. K. Louis (1980). *Math. Meth. in the Appl. Sci.* **2**, 209–220.
10. D. C. Youla (1978). *IEEE Trans. Circuits and Systems* **CAS-25**, 694–702.
11. A. Papoulis (1975). *IEEE Trans. Circuits and Systems* **CAS-22**, 735–742.

12. R. W. Gerchberg (1974). *Opt. Acta* **21**, 709–720.
13. M. I. Sezan and H. Stark (1982). *IEEE Trans. on Medical Imaging* **MI-1**, 95–101.
14. A. Tikhonov (1963). *Sov. Math.* **4**, 1035–1038 and 1624–1627.
15. K. Miller (1970). *SIAM J. Math. Anal.* **1**, 52–74.
16. M. Bertero, C. D. Mol, and G. A. Viano (1978). *Optics Letters* **3**, 51–53.
17. M. I. Sezan and H. Stark (1983). *Appl. Opt.* **22**, 2781–2789.
18. M. I. Sezan (1984). Ph.D. Dissertation, Rensselaer Polytechnic Institute.
19. R. M. Lewitt (1983). *IEEE Proc.* **71**, 390–408.
20. R. M. Mersereau and A. Oppenheim (1974). *IEEE Proc.* **62**, 1319–1338.
21. H. Stark (1979). *J. Opt. Soc. Am.* **69**, 1519–1525.
22. H. Stark and M. Wengrowitz (1983). *IEEE Trans. on Acoust., Speech and Signal Proc.* **ASSP-31**, 1329–1331.
23. M. I. Sezan and H. Stark (1984). *IEEE Trans. on Medical Imaging* **MI-3**, 91–98.
24. D. C. Munson, Jr., J. D. O'Brien, and W. K. Jenkins (1983). *IEEE Proc.* **71**, 917–925.
25. P. Mansfield and P. G. Morris (1982). "NMR Imaging in Biomedicine", Academic Press, New York.
26. Z. H. Cho, H. S. Kim, H. B. Song, and J. Cumming (1982). *IEEE Proc.* **70**, 1152–1173.
27. W. S. Hinshaw and A. H. Lent (1983). *IEEE Proc.* **71**, 338–350.
28. J. F. Greenleaf (1983). *IEEE Proc.* **71**, 330–337.
29. S. K. Kenue and J. F. Greenleaf (1982). *IEEE Trans. on Sonics and Ultrasonics* **SU-29**, 213–217.
30. M. I. Sezan and H. Stark (1985). (Accepted for publication in *Jour. of Math. Analysis and Applications.*)

CHAPTER 12

Image Synthesis: Discovery Instead of Recovery

Bahaa E. A. Saleh

Department of Electrical and Computer Engineering
University of Wisconsin
Madison, Wisconsin 53706

12.1 Introduction

An established practice in microphotography is the introduction of deliberate distortions in the artwork in order to compensate for known distortions in the optical system of the microphotographic camera [1]. For example, to prevent the rounding of line ends and formation of fillers at line intersections, which occur at very high reductions, serifs are drawn at the line corners, and thicknesses of lines at line intersections are sharply reduced and brought to a point as indicated in Fig. 12.1-1. The size of the serifs depends on the angle of taper chosen, the total reduction required, and the lens aperture used [2]. This practice is also common in the printing of reticles [1]. The corrections are usually based on experience gained by a process of trial and error. Another practice is to measure geometric distortions introduced by the imaging system by imaging an accurate test grid and incorporating a counter geometric correction in the original drawing

Fig. 12.1-1. Compensation of image thickening at line intersections and rounding at corners.

[1]. This chapter reviews recently developed techniques for the systematic determination of optimum masks for these and similar problems.

The problem can be regarded as the counterpart to the usual signal/image recovery (reconstruction, or restoration) problem, which was addressed in detail elsewhere in this book. It may be called signal/image "discovery," "construction," "synthesis," or "design." An image restoration problem involves the recovery of an actual image that has been distorted by a given imaging system. Image synthesis, on the other hand, aims at "discovering" an input image which, when distorted by a given imaging system (e.g., a display device), yields a desired output pattern. The sought image is discovered (designed, constructed, or synthesized), rather than recovered, reconstructed, or restored.

Aside from the design of masks for microphotography and microlithography, the problem of image synthesis is encountered in microfilming, laser printing, and other forms of high-resolution display or pattern production. Certainly the smallest improvement of the resolving powers of microlithographic systems by means of intelligent mask manipulation would be most welcome by the microelectronics industry. Another area to which image synthesis applies is that of design of aids for the visually impaired. Here, the human visual system is the imaging system. No access is possible to the system's output. But the input (viewed) image may be preprocessed to compensate for the distortion introduced by the eye–brain system itself.

It is evident that the problems of image restoration and image synthesis are very similar. In both problems the imaging system is known; the output image is known and the input is to be found. But the motive is quite different—discovery versus recovery. One subtle difference has to do with the existence of a solution. In an image restoration problem the measured output results from an actual, albeit unknown, input. In an image synthesis problem, on the other hand, it is possible that no input image is capable of producing the desired output image. Therefore, in image synthesis the problem of utmost concern is that of *existence* of a solution. If more than one solution exists, the choice among them is a matter of convenience. In an image restoration problem one is usually concerned with *uniqueness* and recoverability. Typically, more than one input image could produce the measured output image, and a choice is usually based on *a priori* statistical information that might be available.

Whereas image restoration is a problem of signal *estimation*, image synthesis is a *control* problem. Existence of an input that produces a certain output is related to the concept of controllability in control theory. The emphasis in control theory, however, is on feedback systems, and on time signals where causality leads to difficult problems of realizability. In imaging systems the independent variable is position instead of time and causality does not apply. Feedback systems are also not common (except perhaps in adaptive optics). Such systems will not be covered in this chapter. The difficulty with images is in the generalization of one-dimensional results to the two-dimensional regime. In signal theory such generalizations are sometimes trivial; they are, however, often not possible, as we will see in this chapter.

The image synthesis problem is a problem of inversion. If the mapping that represents the imaging system has an inverse, then finding the input that produces a given output and conditions of solution existence is a straightforward problem. However, often that mapping cannot be inverted. This is where the difficulty lies. Mathematical approaches to the problem, which have been so far rather *ad hoc* in nature, depend on the nature of the imaging system and on other constraints placed on the synthesized image. This dependence sets the plan of presentation for this chapter.

We begin in Section 12.2 with a general formulation of the image synthesis problem, followed by a discussion of a number of special imaging systems of interest. In Section 12.3 the imaging system is an ideal diffraction-limited optical system. Resolution of the system as pertinent to the problems of image recovery and image synthesis is discussed. Conditions under which superresolution is achievable for both cases are examined. Section 12.4 then proceeds to a system composed of a cascade of a bandlimiter followed by a hardlimiter. This represents a diffraction-limited optical system employing an ideal high-contrast detector that yields binary output images. The one-dimensional problem is first presented and an exact solution is shown always to exist. Connection with the problem of recovering a bandlimited function from its zero crossings is established. Methods of two-dimensional image synthesis based on linear programming and on alternating orthogonal projections are then presented. Section 12.5 is a brief discussion of the synthesis problem when both the generated pattern and the input masks are restricted to be binary images. Coherent imaging systems using high-contrast detection are the subject of Section 12.6. Here the synthesized mask is allowed to be complex-valued. The method of alternating orthogonal projections is used.

12.2 GENERAL FORMULATION OF THE PROBLEM

Consider an imaging system represented by the operator S, whose input and output images $f(\mathbf{x})$ and $g(\mathbf{x})$ are related by

$$g = Sf \tag{12.2-1}$$

as illustrated schematically in Fig. 12.2-1. The system S is known. A real nonnegative image $g(\mathbf{x})$, whose values in a given domain $\mathbf{x} \in \mathscr{D}$ are prescribed, is to be generated at the output of the system.

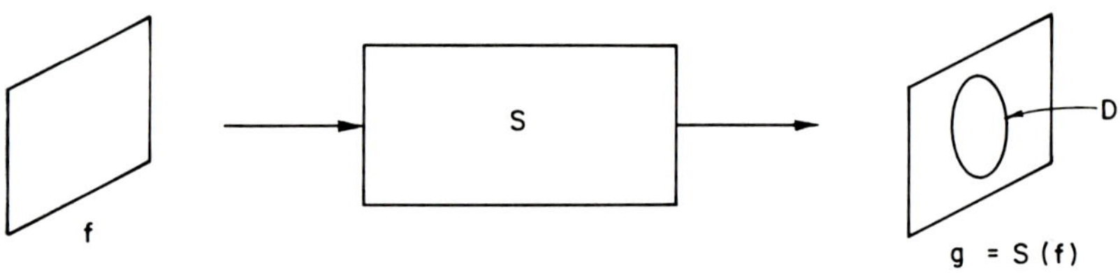

Fig. 12.2-1. Model for image formation.

The image synthesis problem addresses the following questions:

(i) Does an image $f \in \mathscr{F}$, which satisfies the given constraints represented by the set $\mathscr{F}$, exist such that the system output $Sf(\mathbf{x}) = g(\mathbf{x})$ for all $\mathbf{x} \in \mathscr{D}$?

(ii) If "yes," how may f be determined?

(iii) If "no," how may we determine an approximate input image f which produces an output $g = Sf$ that is closer to the desired g according to some criterion (metric)?

Examples of sets $\mathscr{F}$, from which the input image is to be selected, are

(i) The set of real nonnegative functions (appropriate for incoherent imaging systems).

(ii) The set of complex-valued functions (appropriate for coherent imaging systems).
(iii) The set of binary-valued functions that have the value 0 or 1 (appropriate for images recorded as binary masks).
(iv) The set may also be restricted to functions of finite extent.

An example of a possible optimization criterion for finding an approximate solution is to minimize a norm

$$\|Sf - g\| = \min.$$

Usually such minimization problems yield a solution similar to that encountered in image restoration.

It is obvious that the above questions cannot be answered under general conditions of an arbitrary system S and an arbitrary set $\mathcal{F}$. Answers have been found for only few special cases, which must be considered separately. Those will be presented in the following sections.

12.3 DIFFRACTION-LIMITED SYSTEMS

In this section we discuss the image synthesis problem for a linear shift-invariant bandlimited imaging system S assuming that the sought after image f is real-valued. This situation is applicable to incoherent imaging through a diffraction-limited isoplanatic optical system [3].

A. *Existence*

A linear shift-invariant system is characterized by its impulse response function $h(\mathbf{x})$ or its transfer function $H(\boldsymbol{\omega})$. The Fourier transforms $F(\boldsymbol{\omega})$ and $G(\boldsymbol{\omega})$ of the images $f(\mathbf{x})$ and $g(\mathbf{x})$ are related by

$$G(\boldsymbol{\omega}) = H(\boldsymbol{\omega})F(\boldsymbol{\omega}). \tag{12.3-1}$$

A bandlimited system has a transfer function $H(\boldsymbol{\omega})$ that has a compact support, say $\mathcal{B}$, i.e.,

$$H(\boldsymbol{\omega}) = 0, \qquad \omega \notin \mathcal{B}. \tag{12.3-2}$$

Frequencies not in $\mathcal{B}$ are not transmitted through the system. For example, in the one-dimensional case, the set $\mathcal{B}$ is an interval $[-B/2\pi, B/2\pi]$, where B is the cutoff frequency (in lines per millimeter) of the system.

It is obvious that if the desired image $g(\mathbf{x})$ contains frequencies outside $\mathcal{B}$, then an exact solution of the image synthesis problem cannot exist. The desired image cannot be transmitted through the system. A solution exists if and only if $H(\boldsymbol{\omega})$ is nonzero for values of $\boldsymbol{\omega}$ for which $G(\boldsymbol{\omega})$ is nonzero. The solution may then be determined from

$$F(\boldsymbol{\omega}) = G(\boldsymbol{\omega})/H(\boldsymbol{\omega}). \tag{12.3-3}$$

Examples of inverse-filter precompensation (pre-equalization) of the transfer function of diffraction-limited optical systems are reported in [4].

The above conditions of existence are helpful if the domain $\mathscr{D}$, over which the function $g(\mathbf{x})$ is prescribed, extends over the entire $\mathbf{x}$ plane. If $\mathscr{D}$ is of finite extent, i.e., the function $g(\mathbf{x})$ is known only for $\mathbf{x} \in \mathscr{D}$, then the Fourier transform $G(\boldsymbol{\omega})$ cannot be determined. The existence condition cannot be assessed. Then a solution to the synthesis problem exists if it is possible to extrapolate $g(\mathbf{x}), \mathbf{x} \in \mathscr{D}$, so that $g(\mathbf{x})$ is bandlimited to the system's bandwidth.

B. Resolution

The resolving power of an imaging system is determined by its response to two points or two thin lines. The minimum distance between the two points or lines for which the system's response has two discernible peaks is the resolution distance. According to Sparrow's criterion [5], the resolving limit is that for which the two peaks in the response barely merge into one peak with maximum flatness (vanishing second derivative). If the line response function $H(\omega)$ is rectangular with a cutoff angular frequency $2\pi B$ (ideal bandlimited system), then Sparrow's two-line resolution distance is $d = 0.66/B$.

When postprocessing of the detected image (image restoration) is allowed, in theory one can always determine from the system's response whether the input is one or two lines, even if the response is a single blurred peak, and regardless of the distance between the lines. The ability to resolve two lines is then extended beyond the classical resolution limit. This is known as superresolution [6–17]. In fact, with postprocessing, if the input image $f(\mathbf{x})$ is of finite extent, then its frequency components that have not been transmitted by the bandlimited system may still be recovered by use of analytic continuation or other sophisticated postprocessing methods [6–17]. However, it is also well known that the problem is ill-posed; i.e., small noise, which may be present in the data, results in very large errors in estimating the exact form of the input image.

Enhanced resolution can also be substantially improved by appropriate design of the image formation system itself, e.g., proper allocation of the available bandwidth in the spatial frequency domain [18–24]. The image-gathering system may also be designed specifically to enhance the performance of the postprocessing image restoration algorithm to be used [25].

Is superresolution also possible in image synthesis? That is, is it possible to produce through a bandlimited system an image composed of two discernible peaks separated by a distance smaller than the system's classical resolution distance $d = 0.66/B$? To be more precise, examine the synthesis of an image $f(\mathbf{x})$ that generates through a bandlimited system a function $g(\mathbf{x})$ whose dependence on x has the profile sketched in Fig. 12.3-1, and that is independent of y [here, $\mathbf{x} = (x, y)$]. Our prescription of the function $g(\mathbf{x})$ is limited to an area D forming a strip $x \in [a, b]$ parallel to the y axis. The behavior of g outside D is of no interest to us. Within D, the function $g(x, y)$ crosses a given level θ in only two regions forming only two peaks separated by a distance d.

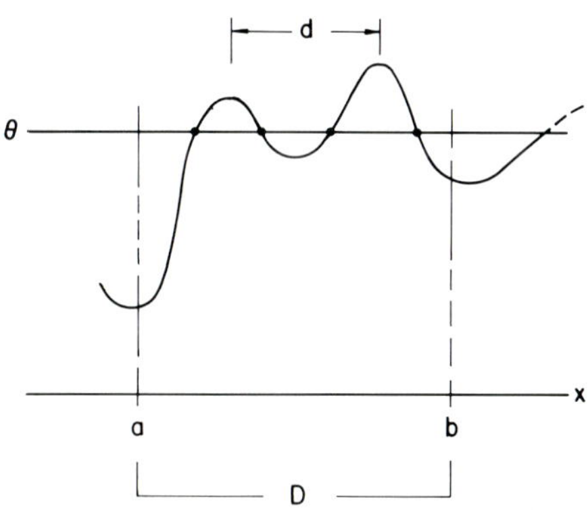

Fig. 12.3-1. A double-peaked signal defined in an interval $D = [a, b]$.

It has been shown in [26] that synthesis of this function is always possible regardless of how small the distance d is in relation to the resolution distance $0.66/B$. This is accomplished by using as an input a finite number of properly located impulses of appropriate amplitudes (instead of only two impulses). For example, to generate a double-peaked symmetric output, only five impulses are sufficient. One can equally use five bandlimited functions $\mathrm{sinc}(2Bx)$ of appropriate amplitudes and displacements.

Unlimited superresolution is therefore possible in this image synthesis problem. Note, however, that as the distance between the two peaks becomes smaller, the double-peaked function dips only very slightly, as illustrated in Fig. 12.3-2. Note also that we have limited ourselves to an area D in the output signal domain that is of finite extent. This result is similar to that in the image restoration case—a condition for achieving superresolution in image restoration is that the unknown input image is of finite extent.

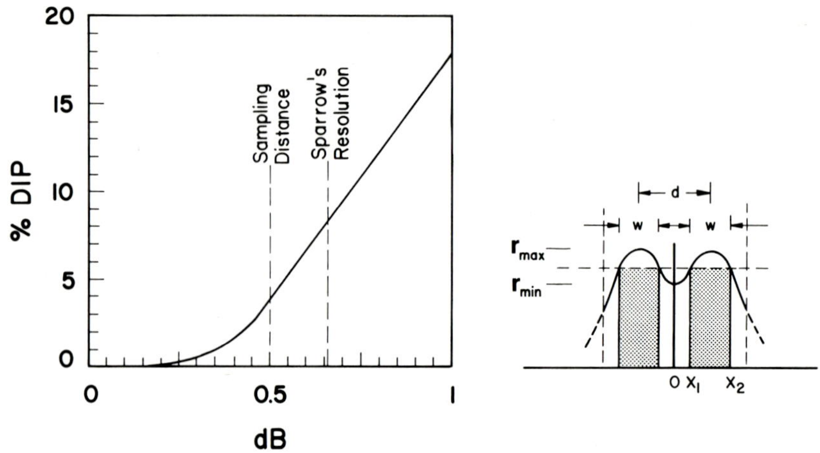

Fig. 12.3-2. Percent dip $100 \times (r_{\max} - r_{\min})/(r_{\max} + r_{\min})$ in a double-peaked bandlimited signal (used to generate an image of two lines of thickness W and separation d) as a function of the separation–bandwidth product dB. It is assumed that $d = 3W$. The signal is constructed using a superposition of five functions $\operatorname{sinc}(2Bx)$ of different amplitudes and locations.

It is not surprising that a bandlimited signal can undergo rapid variations that are locally faster than the highest frequency it contains. A signal may be "programmed" to have a "local" instantaneous frequency which is larger than its bandwidth. The key condition is that this be *confined to a local region of finite extent.* More will be said about this problem in Section 12.4.

C. Linear Imaging Systems with Nonlinear Detectors

Imaging systems usually use nonlinear detectors (photographic film, photoelectric devices, or biological cells as in vision). The nonlinearity is usually a point nonlinearity of the saturation type illustrated in Fig. 12.3-3. The overall system is a cascade of a bandlimited linear system whose output is an image $r(\mathbf{x})$ and a point nonlinearity $g = \phi(r)$. If the nonlinearity ϕ is invertible, then synthesis of a prescribed image g through this composite

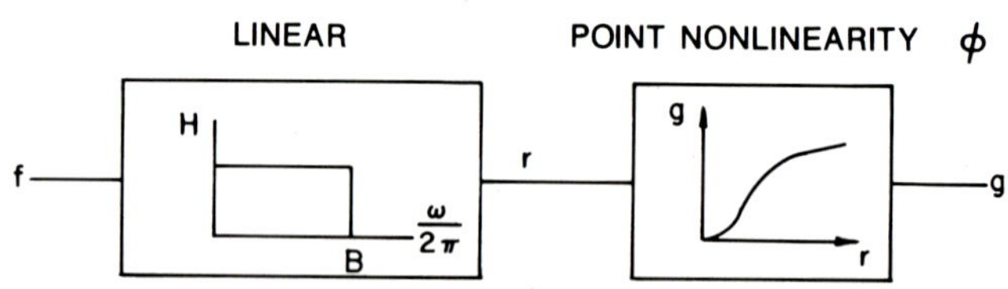

Fig. 12.3-3. Model of a linear bandlimited system followed by a point nonlinearity.

system is equivalent to synthesis of an image $r = \phi^{-1}(g)$ through the linear part of the system. The conclusions stated in Sections 12.3.A and 1.3.B then apply.

If the point nonlinearity ϕ is not invertible, we cannot start from the prescribed image g and proceed backward to determine the needed input f. A radically different approach must be taken. Instead, an appropriate approach is to formulate a set of conditions that the function $r(\mathbf{x})$ must satisfy in order to produce the correct output g and to be a possible output of the linear system. This approach has proved to be successful in the special case when the point nonlinearity is of the hardlimiting kind, as the next sections demonstrate.

12.4 BINARY IMAGES THROUGH DIFFRACTION-LIMITED SYSTEMS

A. The System

In this section, image synthesis is considered when the imaging system (Fig. 12.4-1) is composed of a cascade of a linear bandlimited system whose

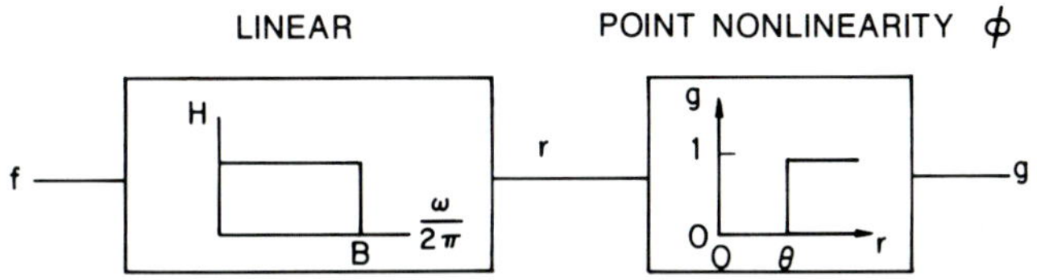

Fig. 12.4-1. Model of a linear bandlimited system followed by a hardlimiter.

transfer function has compact support $\mathscr{B}$, followed by a point nonlinearity of the hardlimiting kind,

$$
\begin{aligned}
g &= 1, \qquad r \geqslant \theta, \\
&= 0, \qquad r < \theta,
\end{aligned}
\tag{12.4-1}
$$

where θ is a constant. This severe nonlinearity models detection systems of very high contrast, such as photoresists used in microlithography. The output of such a system is a binary image of values (0, 1). Image synthesis involves the generation of a binary image, prescribed in a given area $\mathscr{D}$, through this system. In this section, the input function f to be found is allowed to be any real-valued function. In Section 12.5, f is restricted further to the class of binary-valued functions. In Section 12.6, however, the restriction is relaxed to the class of complex-valued functions.

The binary function g is completely defined by the regions (i.e., point sets) $\mathscr{D}_+$ and $\mathscr{D}_-$ within which it has a value 1 or 0,

$$\begin{aligned} g(\mathbf{x}) &= 1, \qquad \mathbf{x} \in \mathscr{D}_+, \\ &= 0, \qquad \mathbf{x} \in \mathscr{D}_-. \end{aligned} \tag{12.4-2}$$

The union of the two regions $\mathscr{D}_+$ and $\mathscr{D}_-$ constitutes the region $\mathscr{D}$.

The synthesis problem can be considered solved if a real-valued function $r(\mathbf{x})$ is found that satisfies the following conditions:

(i) It is bandlimited to the system's bandwidth.
(ii) It satisfies the inequalities

$$\begin{aligned} r(\mathbf{x}) &\geqslant \theta, \qquad \mathbf{x} \in \mathscr{D}_+, \\ &< \theta, \qquad \mathbf{x} \in \mathscr{D}_-. \end{aligned} \tag{12.4-3}$$

Obviously when r is hardlimited it produces the desired binary image g. Because r is bandlimited, it is a possible output of the linear bandlimited system. The corresponding input f can then be directly determined.

B. Connection with the Zero-Crossing Problem

The image synthesis problem for the bandlimiter–hardlimiter system is equivalent to finding a bandlimited image that has values greater or equal to threshold in a region $\mathscr{D}_+$ and below threshold elsewhere (in $\mathscr{D}_-$). There is obviously no loss of generality if the threshold θ is taken to be zero. After all, a constant can be added to the found function r without changing its bandlimited nature. The image synthesis problem is then equivalent to finding a bandlimited function that is nonnegative in a region $\mathscr{D}_+$ and negative in $\mathscr{D}_-$.

Because multiplying a bandlimited function by a minus sign does not change its bandlimited nature, we can state that the present image synthesis problem is equivalent to the problem of finding a bandlimited function whose zero crossings are specified (at the boundary C of the region $\mathscr{D}_+$, separating it from the region $\mathscr{D}_-$).

The recovery of bandlimited signals from their zero crossings has been a subject of recent interest [27–33]. Logan [28] showed that for an interesting subclass of one-dimensional bandpass functions, zero crossings of a function uniquely determine the function (within a multiplicative constant). The band must be less than one octave in width (highest frequency less than twice lowest frequency), and the function should have no zeros in common with zeros of its Hilbert transform other than real simple zeros. An algorithm for the reconstruction of such functions from their zero crossings was advanced by Voelcker and Requicha [31]. Zeevi and Rotem [32], Gavlely *et al.* [33], and Curtis *et al.* [34] generalized Logan's theorem to two-dimensional functions and developed reconstruction algorithms.

Although this is encouraging inasmuch as the image synthesis problem is concerned, it is not directly applicable. In image synthesis we are interested not in the recovery of a signal from its zero crossings but in the discovery (synthesis) of a signal with prescribed zero crossings in an area of finite extent. Logan's theorem and its generalizations are not helpful in answering the question of existence in image synthesis. Of course, if it is known that the signal exists, then those recovery techniques can be useful in the discovery problem. They can also be used to construct approximate images.

C. Exact Solution in the One-Dimensional Case

It was recently shown that a one-dimensional signal $r(x)$ of bandwidth B that has a finite number of prescribed zero crossings $\lambda_1, \lambda_2, \ldots, \lambda_K$ in an interval $D = [a, b]$ can always be found regardless of how narrow B is or how close the zero crossings are [35]. This means that a solution for the one-dimensional image synthesis problem always exists. A finite number of lines of arbitrary widths and locations can, in principle, be printed as closely as desired by using a diffraction-limited reduction camera and an ideal film of infinite contrast.

The function $r(x)$ that satisfies the above conditions is not unique. There are many possible solutions. The function has the form

$$r(x) = \prod_{i=1}^{\infty}[1 - (x/x_i)^2]. \tag{12.4-4}$$

It has zeros at $\pm x_i$, $i = 1, 2, \ldots, \infty$, selected according to the following rules:

Any finite number of zeros (say $x_1, x_2, \ldots, x_M$) are placed within the interval $[0, a]$.

(2) The next K zeros x_i, $i = M+1, M+2, \ldots, M+K$, are chosen to be the prescribed zeros λ_i, $i = 1, 2, \ldots, K$, which lie in the interval $[a, b]$.

(3) An infinite number of zeros x_i, $i = M+K+1, \ldots, \infty$ are placed in the interval $[b, \infty]$; they must form a sequence that satisfies the following conditions:

(a) The zeros are sufficiently separated that there exists a constant $c > 0$ such that $x_{i+1} - x_i \geqslant c$ for all $i = 1, 2, \ldots, \infty$.

(b) In the limit $i \to \infty$, the zeros must coincide with zeros of the function sinc($2Bx$) (a bandlimited function of bandwidth B whose zeros are $\mu_i = i/2B$, $i = 1, 2, \ldots$), i.e.,

$$\lim_{i\to\infty} \mu_i/x_i = 1. \tag{12.4-5}$$

(c) The zeros must make the function $r(x)$ square-integrable.

Because there are many ways of selecting the zeros outside the interval of interest, the solution is not unique. The simplest choice for those zeros that satisfy all conditions is to make them coincide with the zeros $\mu_i = i/2B$ of the sinc function

$$\text{sinc}(2Bx) = \prod_{i=1}^{\infty}[1 - (x/\mu_i)^2],$$

as illustrated in Fig. 12.4-2. The solution is then

$$r(x) = \text{sinc}(2Bx)\frac{\prod_{i=1}^{K}[1 - (x/\lambda_i)^2]}{\prod_{j\in J}[1 - (x/\mu_j)^2]}, \tag{12.4-6}$$

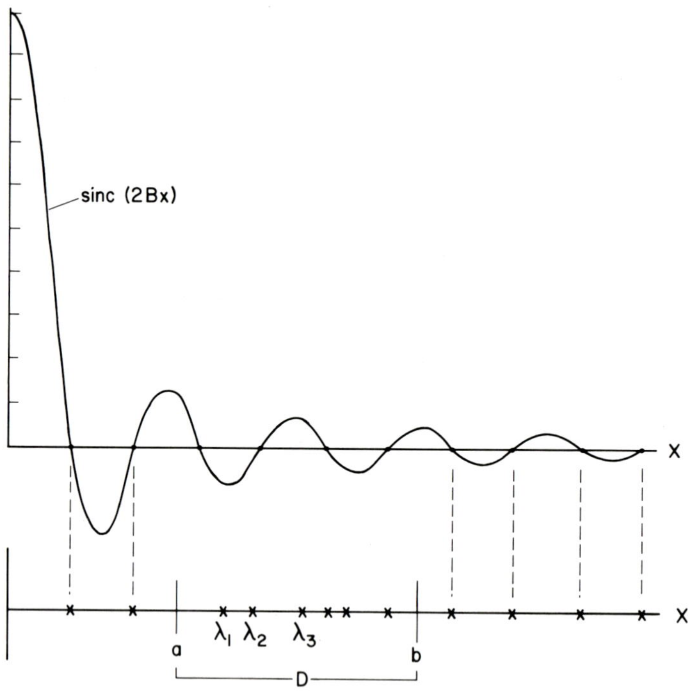

Fig. 12.4-2. Zeros of the bandlimited function of Eq. (12.4-4).

where λ_i are the prescribed zeros, $\mu_j = j/2B$ are the zeros of the sinc($2Bx$) function, and J is the set of j for which μ_j lies inside the interval $D = [a, b]$.

Figure 12.4-3 displays examples of bandlimited functions generated by use of this formula. In all cases the function $r(x)$ has the correct zeros (as it should). However, as the zeros become much closer than a sampling distance ($1/2B$), the excursion of the function above and below zero in the interval of interest D become small compared with the values that the function takes outside that interval. When superresolution is attempted, most of the signal energy lies outside D.

The price for superresolution in image synthesis is therefore energy loss (an affordable price in some situations) and a stringent demand for sharp performance of the hardlimiter. This is where detection noise also comes in.

The role played by noise is simpler to assess in the image synthesis problem than in the conventional image restoration problem. In image synthesis, if noise is present at the detector (at the output of the bandlimiter), then provided that its rms value is smaller than the excursions that the bandlimited signal undergoes above or below threshold and that the hard-limiter is ideal (infinite contrast), its effect will be minimal. However, when the usually defined resolution of the system is locally exceeded by a large margin, the variations of the signal above and below threshold within the interval of interest D are minute compared with the values that the constructed signal has outside D. Most of the signal energy then lies outside D. The system is then very sensitive to noise. The imaging system is very inefficient from an energy point of view, but very efficient in condensing a large amount of information in a given interval—this is an interesting energy/information trade-off.

The above formula can also be applied to the recovery problem. If a bandpass signal that satisfies Logan's conditions [28] is clipped, then it is uniquely determined from its zero crossings. But the bandpass signal is also a bandlimited signal; therefore it can be recovered by use of Eq. (12.4-6).

D. Existence of a Solution in the Two-Dimensional Case

In Section 12.4.C we showed that in the one-dimensional case the synthesis problem (for the bandlimiter–hardlimiter imaging system) always has a solution. This result can be extended to circularly symmetric two-dimensional patterns (e.g., rings or spoke patterns used for manufacturing reticles).

Can the result be extended to an arbitrary two-dimensional pattern? Does a solution to the two-dimensional image synthesis problem always exist regardless of system's bandwidth? If one could come up with one counter-

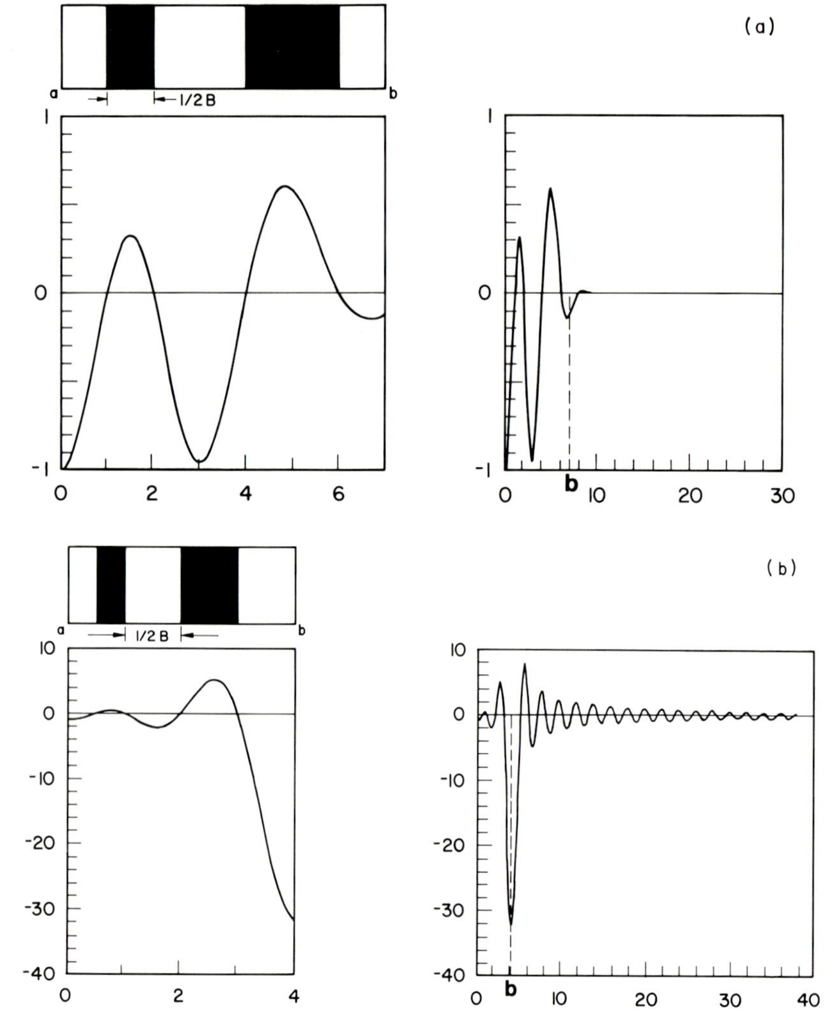

Fig. 12.4-3. Three examples of one-dimensional bandlimited functions determined by use of Eq. (12.4-4). All distances are in units of sampling distance $1/2B$. Two plots are given for each example—the left plot focuses on the interval $[a, b]$ marked by the boundaries of the image; the right plot shows the behavior of the function outside the interval of interest. In (a) the distance $b = 7/2B$; in (b) $b = 4/2B$; and in (c) $b = 2.5/2B$. (Adapted from Bucklew and Saleh [35].)

example, then it would be concluded that the answer to our question is negative. One such example is that for which the zero crossings form a discontinuous line such as the contour of a sharp corner. A bandlimited signal cannot have a discontinuous zero-crossing line.

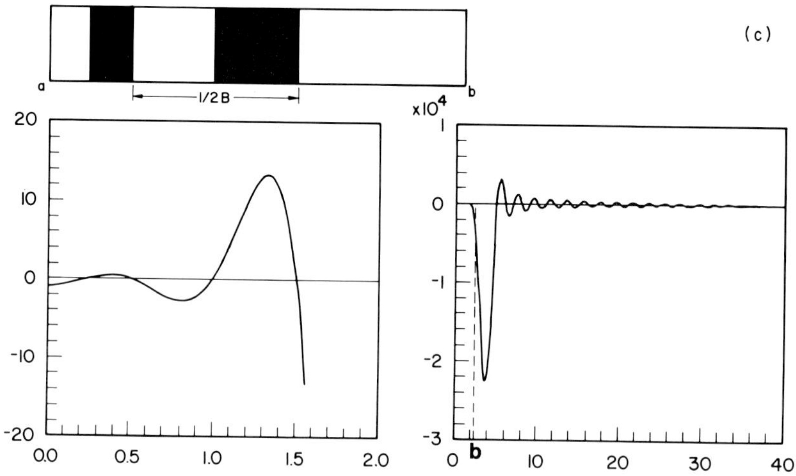

Fig. 12.4-3. *(continued)*

Since conditions of existence (or nonexistence) of a solution are not known, we can only pursue the problem further by discussing methods of finding a solution when one exists, or methods of finding approximate solutions. A number of such methods will be discussed in the following subsections.

E. Solutions Based on Linear Programming Methods

The most direct method of solving the synthesis problem in the two-dimensional case is to use numerical methods of linear programming [36]. The objective is to find a bandlimited function $r(\mathbf{x})$ in the region $\mathscr{D}$, whose Fourier transform has compact support $\mathscr{B}$ and which is positive in a specified region $\mathbf{x} \in \mathscr{D}_+$ and negative elsewhere in $\mathscr{D}$. The idea is to select a finite number N of sampling points $\mathbf{x}_i$ at which we require the inequality condition to be satisfied. There are a number of techniques for finding a bandlimited function whose values at a number of points satisfy equalities or inequalities [37–39].

We write the function $r(\mathbf{x})$ as a superposition of a set of bandlimited functions, say sampling functions centered around a number N of sampling points $\mathbf{x}_i$:

$$r(\mathbf{x}) = \sum_{i=1}^{N} K_i \gamma(\mathbf{x} - \mathbf{x}_i), \tag{12.4-7}$$

where $\gamma(\mathbf{x})$ is the inverse Fourier transform of the characteristic function

$$\begin{aligned} \Gamma(\boldsymbol{\omega}) &= 1, \qquad \boldsymbol{\omega} \in \mathscr{B}, \\ &= 0, \qquad \text{otherwise}. \end{aligned} \tag{12.4-8}$$

For example, if the support $\mathscr{B}$ is a square of width $2B$ centered around the origin, then $\gamma(\mathbf{x}) = \text{sinc}(2Bx)\,\text{sinc}(2By)$. The function $r(\mathbf{x})$ of Eq. (12.4-7) is bandlimited regardless of the values of K_i. The constants K_i must now be chosen to ensure that $r(\mathbf{x})$ has the correct sign at all sampling points

$$\begin{aligned} r(\mathbf{x}_j) &\geqslant 0, \qquad \mathbf{x}_j \in \mathscr{D}_+, \\ &< 0, \qquad \mathbf{x}_j \in \mathscr{D}_-, \qquad j = 1, 2, \ldots, N. \end{aligned} \tag{12.4-9}$$

By substituting from Eq. (12.4-9) into Eq. (12.4-7) we obtain

$$\sum_{i=1}^{N} C_{ji} K_i \geqslant 0, \qquad j = 1, 2, \ldots, N, \tag{12.4-10}$$

where

$$C_{ji} = b_j \gamma(\mathbf{x}_j - \mathbf{x}_i), \tag{12.4-11}$$

and

$$\begin{aligned} b_j &= 1, \qquad \mathbf{x}_j \in \mathscr{D}_+, \\ &= -1, \qquad \mathbf{x}_j \in \mathscr{D}_-, \qquad j = 1, 2, \ldots, N. \end{aligned} \tag{12.4-12}$$

The constants K_i can be determined by solving the set of inequalities (12.4-10). This is a standard linear programming problem.

It should be noted that if the imaging system is ideal, in the sense that its transfer function $H(\boldsymbol{\omega})$ is constant within the support $\boldsymbol{\omega} \in \mathscr{B}$, then the above solution is equivalent to using an input function $f(\mathbf{x})$ which is a set of impulses at the sampling points $\mathbf{x}_i$:

$$f(\mathbf{x}) = \sum_{i=1}^{N} K_i \delta(\mathbf{x} - \mathbf{x}_i). \tag{12.4-13}$$

The function $\gamma(\mathbf{x})$ is now the impulse response function $h(\mathbf{x})$ of the system. In fact, one can also expand $f(\mathbf{x})$ in terms of $h(\mathbf{x})$ as

$$f(\mathbf{x}) = \sum_{i=1}^{N} K_i h(\mathbf{x} - \mathbf{x}_i). \tag{12.4-14}$$

Although the expansion (12.4-7) of the function $f(\mathbf{x})$ seems to be *ad hoc* and arbitrarily chosen, it is more general than it appears, as discussed in detail in [36]. The accuracy of this method depends on the number of sampling points used. However, when that number is large, the computations are substantial. The question of existence of a solution to Eq. (12.4-10) has not been readily answered. Theoretically, methods of testing the existence of solutions to linear programming problems do exist [40].

F. Solutions Using Method of Alternating Projection

The method of alternating orthogonal projections pioneered by Gerchburg [12], Papoulis [41], Youla [42], and others [43–48], which has been very successful in image recovery problems (see Chapters 2, 8, and 11), is also applicable to image synthesis.

1. Method of Projections on Convex Sets

We first review briefly the method of projections on convex sets as described by Youla and Webb [43]. This is an iterative algorithm for finding a function r in a Hilbert space $\mathcal{H}$ that lies in the intersection C_0 of M closed convex subsets $\{C_i\}$ of $\mathcal{H}$:

$$r \in C_0 = \bigcap_{i=1}^{M} C_i. \tag{12.4-15}$$

The sets $\{C_i\}$ correspond to M constraints or properties that the function r must satisfy. To find a function r which has all the M properties, all that is needed is a projection operator P_0 that projects members of $\mathcal{H}$ onto C_0. When it is not possible to determine P_0, an alternative approach is to determine the projection operator P_i which projects on the set C_i, $i = 1, \ldots, M$, and form a composition operator

$$T = P_M \cdots P_2 P_1. \tag{12.4-16}$$

When the operator T is repeatedly applied on an arbitrary function $r_0 \in \mathcal{H}$,

$$r_n = T^n r_0, \qquad n = 1, 2, \ldots, \tag{12.4-17}$$

the sequence $\{r_n\}$ converges to a vector in C_0.

The operator T is not, in general, the projection operator onto C_0, but every point in C_0 is a fixed point of every P_i and therefore a fixed point of T. It has been shown [49] that if a nonexpansive mapping $T: \mathcal{H} \to \mathcal{H}$ is asymptotically regular and has at least one fixed point, then for any $r \in \mathcal{H}$, the sequence $\{T^n r\}$ is necessarily weakly convergent to a fixed point of T. Note also that if all C_i's are linear, the sequence $\{T^n r\}$ converges strongly to P_0 for every $r \in \mathcal{H}$. The projection operators defined by Youla and Webb [43] satisfy these conditions. They assign to $r \in \mathcal{H}$ its nearest neighbor $P_i r$

in C_i. The vector $f - P_i f$ is "normal" to the "tangent plane" to C_i at the point $P_i r$. This is why convergence is not at a geometric rate. Methods of accelerating convergence by extending the projections beyond the contours have also been developed [43,47].

2. *Application to Image Synthesis*

In this image synthesis problem, $r(\mathbf{x})$ is restricted to have two properties:

Property 1. $r(\mathbf{x})$ is bandlimited, i.e., its Fourier transform $R(\boldsymbol{\omega}) = 0$ for $\boldsymbol{\omega} \notin \mathcal{B}$.

Property 2. $r(\mathbf{x})$ must have nonnegative and negative values in the regions $\mathcal{D}_+$ and $\mathcal{D}_-$, respectively, i.e.,

$$\begin{aligned} r(\mathbf{x}) &\geq 0, \qquad \mathbf{x} \in \mathcal{D}_+, \\ r(\mathbf{x}) &< 0, \qquad \mathbf{x} \in \mathcal{D}_-. \end{aligned} \tag{12.4-18}$$

We consider C_1 and C_2 to be the subsets of functions that have properties 1 and 2, respectively. It is not difficult to show that the two sets are convex. Their corresponding projection operators P_1 and P_2 must be determined. As discussed in [43], P_1 is the bandlimiting operator defined by

$$P_1 f(\mathbf{x}) = F^{-1}[\Gamma(\boldsymbol{\omega}) R(\boldsymbol{\omega})], \tag{12.4-19}$$

where

$$\begin{aligned} \Gamma(\boldsymbol{\omega}) &= 1, \qquad \boldsymbol{\omega} \in \mathcal{B} \\ &= 0, \qquad \text{elsewhere,} \end{aligned} \tag{12.4-20}$$

in which F^{-1} denotes the operation of inverse Fourier transformation.

The projection operator P_2 may be obtained by minimizing the distance between $P_2 r \in C_2$ and r. The result is the following rule [43]: if at a given $\mathbf{x}$, $r(\mathbf{x})$ satisfies the appropriate inequality in Eq. (12.4-17), keep it unchanged; if it does not, force it to zero. Thus,

$$\begin{aligned} P_2 r(\mathbf{x}) &= r(\mathbf{x}), \qquad \mathbf{x} \in \mathcal{D}_+ \text{ and } r(\mathbf{x}) \geq 0; \text{ or} \\ &\qquad\qquad\quad\;\; \mathbf{x} \in \mathcal{D}_- \text{ and } r(\mathbf{x}) < 0. \\ &= 0, \qquad \text{otherwise.} \end{aligned} \tag{12.4-21}$$

Now that the two projection operators P_1 and P_2 have been determined, the iterative algorithm can be implemented by using

$$r_n = T^n r_0, \qquad n = 1, 2, \ldots, \tag{12.4-22}$$

where $T = P_2 P_1$. After every iteration the function r_n satisfies the inequality constraints ($r_n \in C_2$) but is not necessarily bandlimited ($r_n \notin C_1$). As n increases, r_n converges to a bandlimited function. A measure of bandlimitedness, and therefore of convergence, is the parameter

$$\alpha_n = \int_{\mathscr{B}} R_n(\boldsymbol{\omega})\, d\boldsymbol{\omega} \bigg/ \int_{-\infty}^{\infty} R_n(\boldsymbol{\omega})\, d\boldsymbol{\omega}, \qquad (12.4\text{-}23)$$

where $R_n(\boldsymbol{\omega})$ is the Fourier transform of $r_n(\mathbf{x})$. As n increases α_n should approach one. By checking α_n after each iteration, convergence is monitored. The procedure is illustrated in Fig. 12.4-4.

3. *An Alternative Method*

Instead of using the projection operator P_2 in the iterative algorithm described above, other nonexpansive operators may be attempted. One such operator is based on changing the sign of the function at positions where it does not satisfy the inequality, instead of forcing it to zero.

$$\begin{aligned} P_2' r(\mathbf{x}) &= \ \ |r(\mathbf{x})|, \qquad \mathbf{x} \in \mathscr{D}_+, \\ &= -|r(\mathbf{x})|, \qquad \mathbf{x} \in \mathscr{D}_-. \end{aligned} \qquad (12.4\text{-}24)$$

Compared to P_2, the operator P_2' generates larger discontinuities at the boundaries between the intervals $\mathscr{D}_+$ and $\mathscr{D}_-$. This was expected to increase the possibility that the bandlimited version of the function (when P_1 is applied to it) crosses threshold at these boundaries (where it is supposed to). Although the operator P_2' is not the projection operator, it is nonexpansive and asymptotically regular [49]; convergence of the algorithm is therefore ensured [43].

4. *Performance*

Tests on a number of examples lead to the following conclusions [50]. For patterns of thin lines separated by distance $d > 1/2B$, the iterative algorithm (using P_2 or P_2') always converged to a solution. About the same number of iterations were needed for both operators. They both converged to the same function. However, using P_2 usually resulted in a better first iteration. A one-dimensional and a two-dimensional example illustrating typical results are shown in Figs. 12.4-5 and 12.4-6. For separations smaller than $1/2B$, the constructed functions have the correct sign everywhere, thus offering superresolution, but their magnitudes become small, making their use impracticable. This confirms results obtained when other approaches are used.

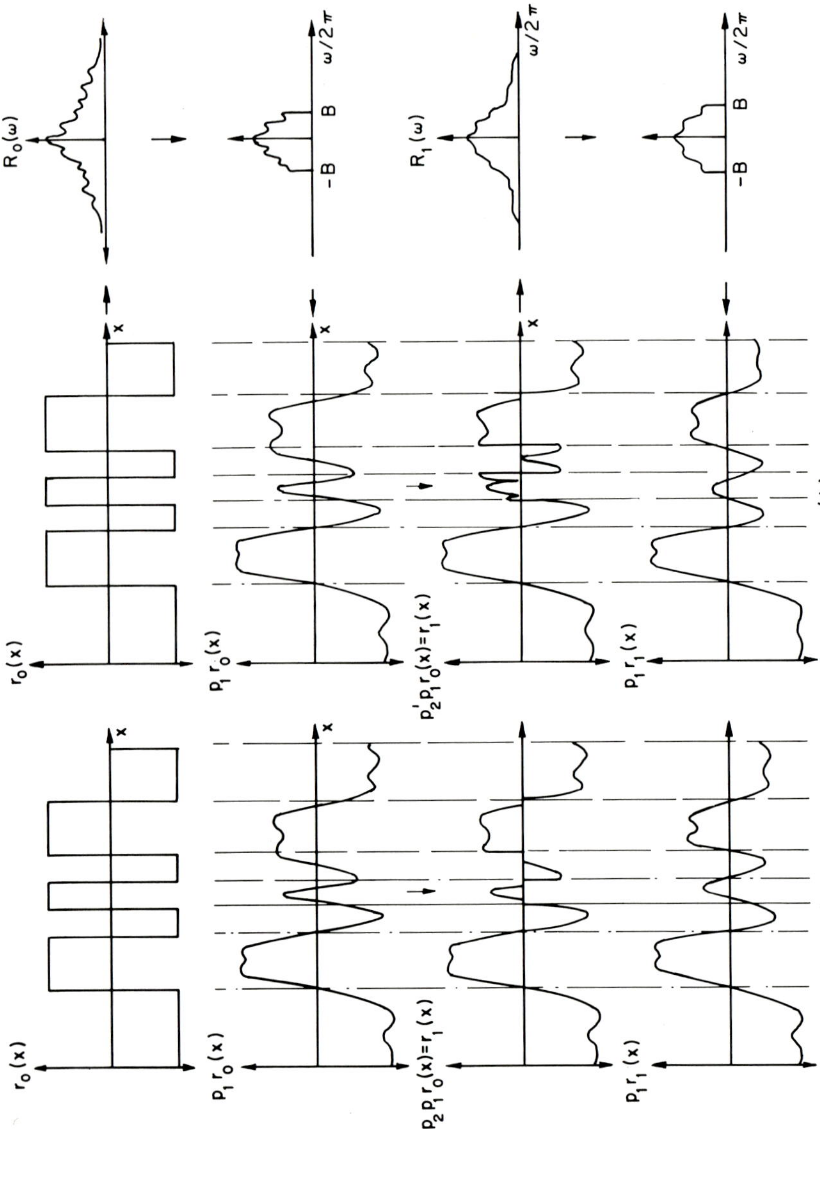

Fig. 12.4-4. Illustration of steps of the algorithm of alternating orthogonal projections, starting with the zeroth iteration $r_0(x)$, computing the first iteration $r_1(x)$, and proceding to compute the next iteration. The two options for the operator P_2 are illustrated in (a) and (b).

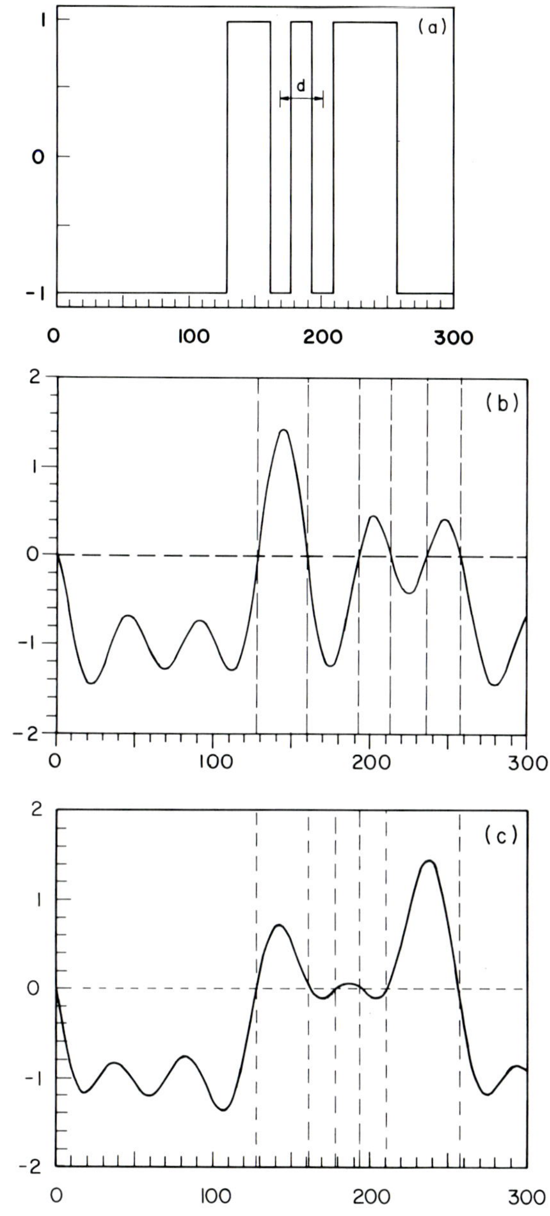

Fig. 12.4-5. Examples of one-dimensional signal synthesis using the algorithm of alternating orthogonal projections. The desired binary signal $g(x)$ is shown in (a). The constructed signal $r(x)$ are shown for (b) $dB = 0.8$ and (c) $dB = 0.6$. (Adapted from Nashold and Saleh [50].)

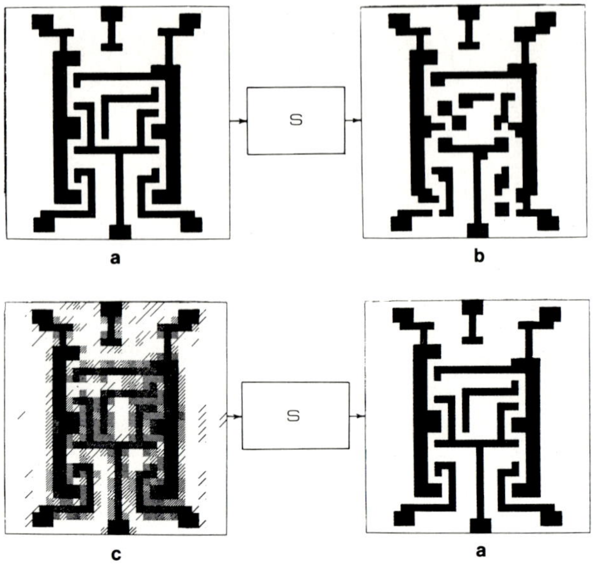

Fig. 12.4-6. Examples of image synthesis using the algorithm of alternating orthogonal projections. (a) Desired image $g(\mathbf{x})$; (b) pattern obtained when $f(\mathbf{x}) = g(\mathbf{x})$, i.e., desired pattern is used as an input mask (note the distortions). (c) Constructed real-valued image $f(\mathbf{x})$ that produces through the system an output pattern identical to the desired pattern in (a). (From Sayegh *et al.* [51].)

One way of avoiding the trivial solution $r(\mathbf{x}) = 0$ is to constrain the solution further by not allowing the absolute value $|r(\mathbf{x})|$ to be smaller than a prescribed level $\Delta > 0$, thus ruling out zero as a solution and guarding against deviations of the hardlimiter from the ideal binary behavior [51]. This constraint can be contained in the operator P_2, with the modification

$$\begin{aligned} P_2' r(\mathbf{x}) &= \phi(|r(\mathbf{x})|), \qquad \mathbf{x} \in \mathscr{D}_+, \\ &= -\phi(|r(\mathbf{x})|), \qquad \mathbf{x} \in \mathscr{D}_-, \end{aligned} \tag{12.4-25}$$

where

$$\begin{aligned} \phi(s) &= s, \qquad s \geqslant \Delta, \\ &= \Delta, \qquad s < \Delta. \end{aligned} \tag{12.4-26}$$

G. Solutions Based on Zero Crossings of One-Dimensional Slices

A simple algorithm for synthesis of a two-dimensional bandlimited function of prescribed zero crossings in an area of finite extent is to dissect the image into parallel slices of one-dimensional bandlimited signals. The function corresponding to each slice is constructed by use of one of the previous algorithms, e.g., Eq. (12.4-4). The slices are then combined to define the two-dimensional image. Since multiplying each of the constructed one-dimensional functions by a constant does not change its bandlimitedness or its zeros, there is an element of arbitrariness in the selection of those constants when the slices are combined to define the image. This arbitrariness may be removed by taking an additional orthogonal slice that cuts across the set of constructed parallel slices (e.g., a set of parallel slices may be cut by an orthogonal slice in a rectilinear pattern, or radial slices may be cut with one circular slice). A bandlimited function that has the zeros of the orthogonal slice is then constructed. The constants associated with the parallel slices are then adjusted to match the functions of the parallel slices with that of the orthogonal slice. This technique is motivated by an identical technique recently suggested for the problem of recovery of bandpass images from their zero crossings [32].

12.5 BINARY IMAGES OF BINARY MASKS

In the previous section, an image $f(\mathbf{x})$ was constructed such that when it is transmitted through a diffraction-limited optical system and detected with an ideal high-contrast detector, it generates a prescribed binary image. The function $f(\mathbf{x})$ was assumed to belong to the set $\mathscr{F}$ of real-valued functions. In this section, we make a more stringent restriction by requiring that $f(\mathbf{x})$ itself be binary. This is appropriate for applications in which the artwork (or mask) must be binary.

The problem can be stated mathematically in the following way: find a function $f(\mathbf{x}) = \{-1, 1\}$ such that

$$
\begin{aligned}
r(\mathbf{x}) = f(\mathbf{x}) * h(\mathbf{x}) &\geqslant 0, \qquad \mathbf{x} \in \mathscr{D}_+, \\
&< 0, \qquad \mathbf{x} \in \mathscr{D}_-,
\end{aligned}
\tag{12.5-1}
$$

where $h(\mathbf{x})$ is the impulse response function of the diffraction-limited system, $*$ represents the operation of convolution, and $\mathscr{D}_+$ and $\mathscr{D}_-$ are the positive and negative regions of the final desired binary image. The found input can be scaled (multiplied by a constant) and biased (added to a constant) to ensure that the input image is nonnegative and that the output threshold is any desired threshold (instead of zero).

Conditions of existence of a solution for this problem are not known. Assuming that a solution exists, we discuss some methods of finding it.

A. Binary Coding of Constructed Real-Valued Images

One approach is to use the results of Section 12.4, in which we found a bandlimited function $r(\mathbf{x})$ that satisfies Eq. (12.5-1). The problem is now to find a binary function $f(\mathbf{x})$ which when bandlimited by the system results in a known real-valued function $r(\mathbf{x})$. This is a well-known problem in image coding.

B. Binary Programming Method

Another approach is to address the problem directly and use techniques of binary programming. The problem is discretized by defining $f_j = f(\mathbf{x}_j)$, $r_j = r(\mathbf{x}_j)$, $h_{i-j} = h(\mathbf{x}_i - \mathbf{x}_j)$. If the image is discretized, then Eq. (12.5-1) becomes: find the binary variables $f_j = \{-1, 1\}$ such that

$$\begin{aligned} r_i = \sum_j f_j h_{i-j} &\geqslant 0, \qquad \mathbf{x}_i \in \mathscr{D}_+, \\ &< 0, \qquad \mathbf{x}_i \in \mathscr{D}_-, \qquad i = 1, 2, \ldots . \end{aligned} \tag{12.5-2}$$

Techniques of binary programming can be used to solve this set of inequalities [52].

Earlier work on this problem [53] was based on searching for an approximate solution by finding the binary sequence $\{f_i\}$ that minimizes a certain measure of distortion between the generated binary pattern $\{r_i\}$ and the desired one. One simple, *ad hoc* (and easy to compute) measure is the deviation of $\{r_i\}$ from zero at points of the boundary C between the regions $\mathscr{D}_+$ and $\mathscr{D}_-$, say

$$\varepsilon = \sum_C \psi(|r_i|) = \min_{f_i}, \tag{12.5-3}$$

where $\psi(\cdot)$ is a convex function, e.g., $\psi(\cdot) = (\cdot)^4$, and $\{r_i\}$ are related to $\{f_i\}$ by Eq. (12.5-2). Ideally the set $\{r_i\}$ should be forced to vanish at points of this contour. Because $\{r_i\}$ is assumed to be a smooth function, its zero crossings are likely to be the only ones in the vicinity.

The criterion does not guarantee that away from the desired boundary C, false zero crossings surrounding small "islands" will not occur. It was found that such a situation indeed arises. One way to circumvent this difficulty is to add a constraint that the set $\{r_i\}$ have the correct sign at a number of properly chosen points away from the contour C. The binary nature of the problem makes such constraints easy to implement. An iterative algorithm for performing the minimization problem is to modify the value of each pixel f_i, one at a time, by selecting the value (1 or -1) that yields a smaller error ε, keeping the values of all other pixels fixed. This is done sequentially for all points. The process is repeated over the entire image for a number of iterations. The algorithm may, of course, encounter a local minimum, but with a reasonable initial guess it was found to be adequate. An example is shown in Fig. 12.5-1.

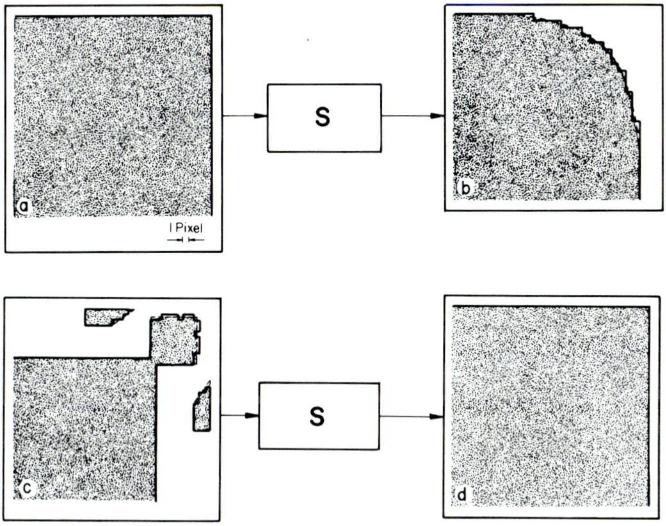

Fig. 12.5-1. Example of synthesis of binary images using binary masks. (a) Desired image $g(\mathbf{x})$; (b) pattern obtained when $f(\mathbf{x}) = g(\mathbf{x})$; (c) constructed binary-valued image $f(\mathbf{x})$ that produces through the system an output pattern identical to the desired pattern in (a). (Adapted from Saleh and Sayegh [53].)

12.6 COMPLEX MASKS FOR COHERENT IMAGING SYSTEMS

This section addresses image synthesis in coherent optical systems. Improving resolution in optical lithography by use of coherent illumination and phase-shifting masks has been demonstrated [54,55]. The idea is that using complex optical fields to coherently generate the desired image offers another degree of freedom (the phase) that can be utilized to generate output images of higher resolution.

A. *Resolution in Coherent Imaging*

Consider the one-dimensional problem of printing two thin lines separated by a distance d through a coherent diffraction-limited imaging system.

The optical system [56] is an ideal bandlimited system of bandwidth $B = \mathrm{NA}/\lambda$ (where λ is the wavelength and NA the numerical aperture) operating on the optical field $f(x)$. The intensity $|r(x)|^2$ of the spatially filtered field $r(\mathbf{x})$ is detected by a high-contrast film of threshold θ, thus generating a binary image of two thin lines.

When $d > 1/B$, the problem is easy; one can come up with numerous input images $f(x)$ that generate the desired output (two thin lines). An example is to choose a real-valued $f(x)$ that has the form of two impulses of amplitude A and separation d:

$$f(x) = A[\delta(x - d/2) + \delta(x + d/2)].$$

The response of the bandlimited system is then

$$r(x) = 2AB\{\mathrm{sinc}[2B(x - d/2)] + \mathrm{sinc}[2B(x + d/2]\}.$$

The detected intensity $|r(x)|^2$ has a double-peaked distribution as long as $d > 0.66/B$. The amplitude A may then be adjusted to make the points at which $|r(x)|^2$ crosses threshold correspond to the desired thickness of the lines. The distance $d = 0.66/B = 0.66\lambda/\mathrm{NA}$ is the two-line Sparrow resolution limit under coherent illumination. (Compare this to the value $0.415\lambda/\mathrm{NA}$ in the incoherent case.)

When $d < 0.66/B$, a double-peaked $|r(x)|^2$ cannot be produced by use of two real-valued impulses separated by a distance d. Several alternatives are possible.

In the first alternative, $f(x)$ is kept real-valued, but distributions other than the two-impulse distribution are allowed. Indeed, it was shown in Sections 12.3 and 12.4 that a solution always exists for arbitrary values of the separation–bandwidth product dB. The solution is not unique. One way to generate a solution is by use of five impulses instead of two [26]. As the product dB decreases, however, the solution becomes unstable, in the sense that slight deviations from the exact computed values for the input will produce large deviations in the output. Moreover, demands on the accuracy of the thresholding device become extremely high.

The second alternative is to use two impulses having a phase difference ϕ:

$$f(x) = A[\delta(x - d/2) + \exp(j\phi)\,\delta(x + d/2)],$$

corresponding to

$$\begin{aligned}|r(x)|^2 = |2AB|^2\{&\mathrm{sinc}^2[2B(x - d/2)] + \mathrm{sinc}^2[2B(x + d/2)]\\ &+2\cos(\phi)\,\mathrm{sinc}[2B(x - d/2)]\,\mathrm{sinc}[2B(x + d/2)]\}.\end{aligned}$$

For a given ϕ, we ask for the minimum separation–bandwidth product dB

for which $|r(x)|^2$ is double-peaked. As Fig. 12.6-1 shows, when $\phi = 180°$, $|r(0)| = 0$, the function is forced to have more than one peak regardless of the product dB. Resolution can obviously be improved in such situations. When $\phi = 90°$, the result corresponds to the case of incoherent imaging (where no interference is allowed). Resolution is clearly inferior in the incoherent case compared to the coherent $\phi = 180°$ case.

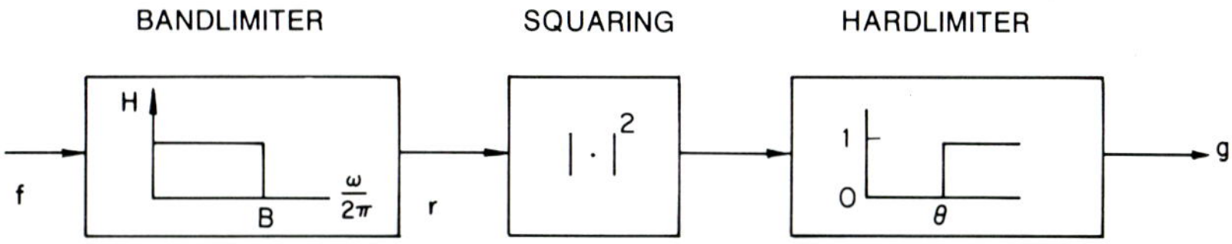

Fig. 12.6-1. Model of a coherent imaging system using high-contrast detection.

A third and more general alternative is to allow the amplitude and phase of $f(\mathbf{x})$ to have arbitrary distributions. This alternative will be examined in the remainder of this section.

B. Problem Formulation

In this synthesis problem, the imaging system S is a cascade of a bandlimited linear shift-invariant system S_1 operating on a complex-valued signal $f(\mathbf{x})$, followed by an operation S_2 which is a combination of absolute-value squaring and a hardlimiter, as illustrated in Fig. 12.6-2. Mathematically,

$$S = S_2 S_1; \tag{12.6-1}$$

the output of the first system is

$$r(\mathbf{x}) = S_1(f(\mathbf{x})) = F^{-1}[H(\boldsymbol{\omega})F(\boldsymbol{\omega})], \tag{12.6-2}$$

where

$$H(\boldsymbol{\omega}) = 0 \qquad \text{for } \boldsymbol{\omega} \notin \mathscr{B} \tag{12.6-3}$$

is the system's transfer function. This output is acted on by the second system to yield

$$\begin{aligned} g = S_2\{r\} &= 1, \qquad |r|^2 \geq \theta, \\ &= 0, \qquad \text{otherwise}, \end{aligned} \tag{12.6-4}$$

where θ is the threshold of the hardlimiter.

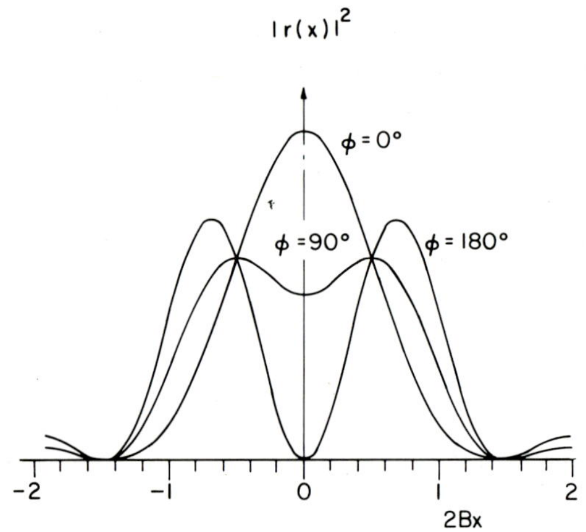

Fig. 12.6-2. Intensity $|r(\mathbf{x})|^2$ at the output of a coherent imaging system when the input is composed of two impulses separated by a sampling distance $1/2B$ and of phase difference zero.

The complex-valued signal f represents the optical field, which may be implemented by use of an amplitude and a phase-shifting mask. The bandlimited system represents the effects of diffraction and aberrations in the optical imaging system. The detector responds to the absolute-value-squared optical field, and the hardlimiter represents a very high contrast printing operation.

The goal is to find a complex-valued function $f(\mathbf{x})$ that produces a prescribed binary image $g(\mathbf{x}) = \{1, 0\}$ through the system S. That image is completely described by the regions $\mathscr{D}_+$ and $\mathscr{D}_-$ within which $g(\mathbf{x})$ has the values 1 and 0, respectively.

The problem is equivalent to finding a complex-valued function $r(\mathbf{x})$, for $\mathbf{x} \in D$, that satisfies the following conditions:

(i) It is bandlimited; its Fourier transform has the support of the system's transfer function

$$R(\boldsymbol{\omega}) = 0, \qquad \boldsymbol{\omega} \in \mathscr{B}. \tag{12.6-5}$$

(ii) Its absolute value satisfies the following inequalities:

$$|r(\mathbf{x})| < \sqrt{\theta}, \qquad \text{for } \mathbf{x} \in \mathscr{D}_-, \tag{12.6-6}$$

$$|r(\mathbf{x})| \geqslant \sqrt{\theta}, \qquad \text{for } \mathbf{x} \in \mathscr{D}_+, \tag{12.6-7}$$

where $\mathscr{D}_+ \cup \mathscr{D}_- = \mathscr{D}$. Such a signal $r(\mathbf{x})$ produces the desired output $g(\mathbf{x})$; because it is bandlimited, it can easily be generated through the bandlimited system. In fact, if the bandlimited system is ideal (i.e., $H(\boldsymbol{\omega}) = 1$, for $\boldsymbol{\omega} \in \mathscr{B}$), $f(\mathbf{x}) = r(\mathbf{x})$.

Another way to state the problem is: find a bandlimited complex-valued function $r(\mathbf{x})$ for $\mathbf{x} \in \mathscr{D}$ whose absolute value has level crossings at points of the contour C that separates the regions $\mathscr{D}_+$ and $\mathscr{D}_-$, which constitute the region $\mathscr{D}$.

Conditions of existence of a solution are not known. Many examples of special cases of prescribed images g for which a solution does exist can be given. However, no example could be found for which a solution can be shown *not* to exist. When a solution exists, it can be found by using techniques similar to those used in Section 12.4. The method of alternating orthogonal projections has been shown to provide such solutions [57].

C. Solution Using Method of Alternating Orthogonal Projections

1. Algorithm

This algorithm aims at finding a complex-valued function $r(\mathbf{x})$ that satisfies the three conditions of Eqs. (12.6-5)-(12.6-7). Those conditions define three subsets in Hilbert space: C_1 is the subset of functions that are bandlimited (to a support $\mathscr{B}$), C_2 and C_3 are the subsets of functions whose absolute values are below or above a prescribed threshold θ in regions $\mathscr{D}_+$ and $\mathscr{D}_-$, respectively.

The projection operators P_1, P_2, and P_3 that correspond to those subsets are defined by:

$$P_1 r(\mathbf{x}) = F^{-1}[H(\omega)R(\omega)], \tag{12.6-8}$$

$$\begin{aligned} P_2 r(\mathbf{x}) &= r(\mathbf{x}), && \mathbf{x} \in \mathscr{D}_+, \\ &= r(\mathbf{x}), && \mathbf{x} \in \mathscr{D}_-, |r(\mathbf{x})| \leqslant \sqrt{\theta}, \\ &= \sqrt{\theta}\exp[j\phi(\mathbf{x})], && x \in \mathscr{D}_-, |r(\mathbf{x})| > \sqrt{\theta}, \end{aligned} \tag{12.6-9}$$

$$\begin{aligned} P_3 r(\mathbf{x}) &= r(\mathbf{x}), && \mathbf{x} \in \mathscr{D}_-, \\ &= r(\mathbf{x}), && \mathbf{x} \in \mathscr{D}_+, |r(\mathbf{x})| \geqslant \sqrt{\theta}, \\ &= \sqrt{\theta}\exp[j\phi(\mathbf{x})], && x \in \mathscr{D}_+, |r(\mathbf{x})| < \sqrt{\theta}, \end{aligned} \tag{12.6-10}$$

where $\phi(\mathbf{x})$ is the phase of $r(\mathbf{x})$. The operator P_1 is simply a bandlimiting operation. The operator P_2 is realized by testing the inequality (12.6-6) at each $\mathbf{x}$ in $\mathscr{D}_-$. If the inequality is not satisfied, the absolute value of $r(\mathbf{x})$ is forced to the threshold value $\sqrt{\theta}$, and its phase is kept the same. Otherwise and elsewhere the function $r(\mathbf{x})$ is kept unchanged. The complement procedure is done for P_3.

The synthesis algorithm starts with an initial guess of $r(\mathbf{x})$ and applies the projection $T = P_1 P_2 P_3$ many times until it converges.

2. *Convergence*

The technique has been applied to a number of one-dimensional cases [57]. In almost all cases it converged to some solution (different solutions were converged to, depending on the initial guess). Convergence is measured by monitoring the error in satisfying condition (i), i.e., monitoring the portion of signal energy outside the desired bandwidth.

Even though this algorithm is motivated by the method of projections on convex sets [43], it actually does not satisfy all of that method's requirements. Convergence is ensured if the sets C_1, C_2, and C_3 are convex sets. In the present case, C_1 and C_2 are indeed convex sets. However, set C_3 (complex functions whose magnitudes are greater than a prescribed value in a prescribed region) is not a convex set. Convergence of the algorithm is therefore not ensured. The fact that convergence occurred in many cases is probably due to "local" convexity when the initial guess is close to a solution. These notions are explored in greater detail by Levi and Stark in Chapter 8.

3. *Performance*

Performance of this synthesis technique was tested by applying it to the one-dimensional binary signal shown in Fig. 12.6-3. The signal is composed of a number of pulses of decreasing widths and separations. Failure of the synthesis technique to correctly generate pulses of a given separation/width d for a given system's bandwidth B is an indication of its limit of performance; it defines the resolution limit for this image. The following observations were reported [57]:

1. When the desired pattern was itself used as a mask (with zero phase), the resolution limit was approximately $1.75/B$.
2. Improvement was obtained by assigning to every other pulse of the pattern a 180° phase shift while keeping the amplitudes unchanged. This is the method used in [54,55]. Resolution was then improved to approximately $1.09/B$.
3. If the iterative algorithm was used to find a better mask but the mask was constrained to be real-valued (this is implemented by using a real-valued initial guess in the algorithm), resolution was brought to about $0.87/B$.

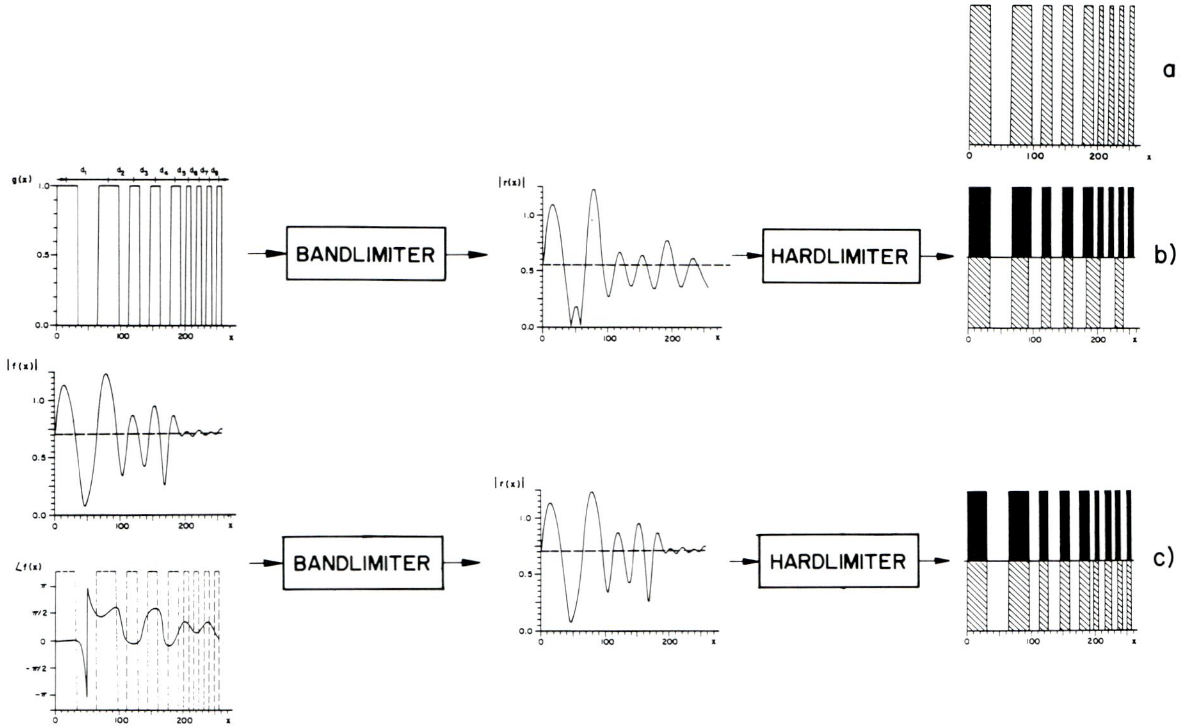

Fig. 12.6-3. Synthesis of a one-dimensional binary image through a coherent imaging system using the method of alternating orthogonal projections. (a) Desired image $g(\mathbf{x})$. (b) Hatched pattern is the pattern obtained when $f(\mathbf{x}) = g(\mathbf{x})$, i.e., desired pattern itself is used as a mask. Blackened pattern is the desired pattern, shown for comparison. (c) Magnitude and phase of constructed complex input $f(x)$ and the corresponding output pattern (hatched). The desired pattern is shown (black) for comparison. (Adapted from Saleh and Nashold [57].)

4. When the iterative algorithm was used in its most general form (allowing the mask to be complex-valued), resolution improved to about $0.44/B$.

12.7 SUMMARY

There are several ways to improve the performance of an imaging system. The most direct way is to improve the image formation system itself by use of larger apertures and higher-quality optics and by better utilization of the available bandwidth. If the purpose of imaging is to extract information present in the object plane (as in a telescope or a microscope), then postprocessing of the information in the imaging plane by use of a computer could substantially enhance the performance of the image recovery process. If the purpose of imaging is to create a prescribed pattern in the image plane to record it on a photodetecting material (as in microphotography, microlithography, and image display devices, including the visual system [58]), then postprocessing is not appropriate. This chapter explored the possibilities of enhancing performance by means of preprocessing, i.e., making changes in the original object (mask). The process of finding the object that produces the desired image is a process of signal synthesis (design, construction, or discovery) instead of restoration (reconstruction, or recovery).

Three image formation systems were examined—a diffraction-limited imaging system (modeled by a bandlimited linear system), a diffraction-limited system employing very high contrast film (modeled by a bandlimited linear system followed by a binary detector, or hardlimiter), and a coherent optical system with high-contrast detection (modeled by a linear bandlimited system operating on complex-valued functions followed by a binary detector operating on the signal absolute value squared). In all systems, we explored the conditions of existence of an input image that generates through the system an arbitrary prescribed output image. When the output image was prescribed only in a region of finite extent, but allowed to have any values elsewhere, it was found that in the one-dimensional case, a solution always exists regardless of the prescribed pattern. In the two-dimensional case, no such assertion could be made. There are, however, two-dimensional patterns whose details within an area of finite extent are finer than the classical resolution limit, for which solutions have been found. A number of algorithms for finding such solutions were reviewed.

Other image formation systems which have not been covered in this chapter include space-variant (nonisoplanatic) systems, systems suffering from geometric distortions, and nonlinear systems (e.g., quadratic systems associated with partially coherent imaging [59–61]). There are some results for the nonlinear problem of compensation of postdetection blur in coherent imaging systems [62]. While compensation of geometric distortions has been practiced in microphotography [1], no theoretical studies of the limitations of the method are available. A great deal of work remains to be done on image synthesis for such systems.

When solving an image synthesis problem, one has the advantage of being able to verify the solution obtained and ascertain that it indeed generates the desired image. This is a luxury not available in the image recovery problem, because the image to be recovered is unknown. In image synthesis uniqueness is not an issue, and the issue of solution existence, although very difficult to prove or disprove in general, can be addressed by a process of search using any of numerous techniques (many of which are rather *ad hoc* in nature).

Image synthesis is a lengthy process in which a custom-made mask is generated for each pattern, for a specific distorting system. In some applications in which the desired pattern is time-varying (e.g., image synthesis to correct distortions resulting from visual impairment, as a visual aid [58]), the process of discovering the correct mask must be automated and performed in real time. The process of computing a mask for a given pattern should then be as simple as passing the pattern through a spatial filter. One would then find the optimum filter, not for a given pattern but for a *class* of patterns. The resultant problem is not unlike finding an optimum filter for postprocessing blurred images.

ACKNOWLEDGMENT

The author acknowledges the support of the National Science Foundation, and fruitful collaboration with J. A. Bucklew, S. I. Sayegh and K. M. Nashold.

REFERENCES

1. G. W. W. Stevens, "Microphotography," Butler and Tanner, London (1968).
2. Eastman Kodak Co., "Techniques of Microphotography," Publications No. P-52, Rochester, New York (1967).
3. M. Born and E. Wolf, "Principles of Optics," 5th ed., Pergamon, Oxford (1975).
4. B. E. A. Saleh and N. Subotic, Pre-inverse filtering of distorted images. Appl. Optics **20**, 3912–3916 (1981).
5. E. Hecht and A. Zajac, "Optics," Addison-Wesley, Reading, MA (1974).
6. G. Toraldo di Francia, Resolving power and information. *J. Opt. Soc. Am.* **45**, 497–501 (1955).

7. C. K. Rushforth and R. W. Harris, Restoration, resolution, and noise. *J. Opt. Soc. Am.* **58**, 539–545 (1968).
8. N. J. Vershad, Resolution, optical-channel capacity and information theory. *J. Opt. Soc. Am.* **59**, 157–163 (1969).
9. B. R. Frieden, On arbitrarily perfect imagery with a finite aperture, *Opt. Acta* **16**, 795–807 (1969).
10. B. R. Frieden, The extrapolating pupil, image synthesis, and some thought applications. *Appl. Opt.* **9**, 2489–2496 (1970).
11. M. Bendinnelli, A. Consortini, L. Ronchi, and B. R. Frieden, Degrees of freedom and eigenfunctions, for the noisy image. *J. Opt. Soc. Am.* **64**, 1498–1502 (1974).
12. R. W. Gerchberg, Super-resolution through error energy reduction. *Opt. Acta* **21**, 709–720 (1974).
13. B. R. Frieden, Image enhancement and restoration. *In* "Picture Processing and Digital Filtering" (T. S. Huang, ed.) New York: Springer-Verlag (1975).
14. B. Saleh, *A priori* information and the degrees of freedom of noisy images. *J. Opt. Soc. Am.* **67**, 71–76 (1977).
15. M. A. Fiddy and T. J. Hall, Nonuniqueness of superresolution techniques applied to sampled data. *J. Opt. Soc. Am.* **71**, 1406–1407 (1981).
16. H. Maitre, Iterative superresolution, some new fast methods. *Optica Acta* **28**, 973–980 (1981).
17. C. K. Rushforth, A. E. Crawford, and Y. Zhou, Least-squares reconstruction of objects with missing high-frequency components. *J. Opt. Soc. Am.* **72**, 204–211 (1982).
18. A. W. Lohmann and D. P. Paris, Super-resolution for nonbirefringent objects. *Appl. Opt.* **3**, 1037–1043 (1964).
19. M. A. Grimm and A. W. Lohmann, Super-resolution image for one-dimensional objects. *J. Opt. Soc. Am.* **6**, 1151–1156 (1966).
20. W. Lukosz, Optical systems with resolving powers exceeding the classical limit. *J. Opt. Soc. Am.* **56**, 1463–1472 (1966).
21. W. Lukosz, Optical systems with resolving powers exceeding the classical limit. II. *J. Opt. Soc. Am.* **57**, 932–941 (1967).
22. G. R. Boyer, Pupil filters for moderate superresolution. *Appl. Opt.* **15**, 3089–3093 (1976).
23. R. Biovin and A. Biovin, Optimized amplitude filtering for superresolution over a restricted field. *Opt. Acta* **27**, 587–610 (1980).
24. J. G. Walker, Optical imaging with resolution exceeding the Rayleigh criterion. *Optic Acta* **30**, 1197–1202 (1983).
25. W. T. Cathey, B. R. Frieden, W. T. Rhodes, and C. K. Rushforth, Image gathering and processing for enhanced resolution. *Opt. Soc. Am.* **A1**, 241–250 (1984).
26. S. I. Sayegh and B. E. A. Saleh, Image design: generation of a prescribed image at the output of band-limited optical systems. *IEEE Trans. Acoustics, Speech and Signal Processing* **33**, 467–470 (1985).
27. B. F. Logan, Jr., Information on the zero crossings of bandpass signals. *Bell Sys. Tech. J.* **56**, 487–510 (1977).
28. B. F. Logan, Jr., Signals designed for recovery after clipping, I. Localization of infinite products. *AT&T Bell Lab. Tech. J.* **63**, 261–285 (1984).
29. B. F. Logan, Jr., Signals designed for recovery after clipping, II. Fourier transform theory of recovery. *AT&T Bell. Lab. Tech. J.* **63**, 287–306 (1984).
30. B. F. Logan, Jr., Signals designed for recovery after clipping, III. Generalizations. *AT&T Bell. Lab. Tech. J.* **63**, 379–399 (1984).
31. H. B. Voelcker and A. P. G. Requicha, Clipping and signal determination: Two algorithms requiring validation. *IEEE Trans. Comm.* **COM-21**, 738–744 (1973).

32. Y. Y. Zeevi and D. Rotem, Image reconstruction from zero crossings. *Faculty of Electrical Engineering, Technion Israel Institute of Technology*, Haifa, Israel, EE Pub. No. 499, June 1984.
33. A. Gavlely, S. Shitz, and Y. Y. Zeevi, Image reconstruction from partial information—Zeros and sine wave crossings. *Dixieme Coloque sur le Traitement der Signal et ses Applications*, Nice, May 1985.
34. S. R. Curtis, S. Shitz, and A. V. Oppenheim, Reconstruction of nonperiodic two-dimensional signals from zero crossings. *IEEE Trans. on Acoustics, Speech, and Signal Processing*, (submitted).
35. J. A. Bucklew and B. E. A. Saleh, Theorem for high-resolution high-contrast image synthesis. *J. Opt. Soc. Am.* **A2**, 1233–1236 (1985).
36. S. I. Sayegh, B. E. A. Saleh, and K. M. Nashold, Image design: generation of a prescribed image through a diffraction-limited system with high-contrast recording. *IEEE Trans. Acoustics, Speech and Signal Processing* **33**, 460–465 (1985).
37. L. Levi, Fitting a bandlimited signal to given points. *IEEE Trans. on Info. Theory*, 372–376 (1965).
38. K. Yao, Applications of reproducing kernel Hilbert spaces—bandlimited signal models. *Information and Control* **11**, 429–444 (1967).
39. R. Martinez-Herrero and M. Arenas, Obtaining any given intensity at discrete close points at the output of band-limited optical systems. *Optica Acta* **31**, 467–470 (1984).
40. J. R. Jackson, On the existence problem of linear programming. *Pac. J. Mathematics* **4**, 29–36 (1954).
41. A. Papoulis, A new algorithm in spectral analysis and bandlimited extrapolation. *IEEE Trans. on Circuits and Systems* **CAS-22**, 735–742 (1975).
42. D. C. Youla, Generalized image restoration by the method of alternating orthogonal projections. *IEEE Trans. Circuits Syst.* **CAS-25**, 695–702 (1978).
43. D. C. Youla and H. Webb, Image restoration by the method of convex projections: Part 1—Theory. *IEEE Trans. Medical Imaging* **MI-1**, 81–94 (1982).
44. H. J. Trussell, The feasible solution in signal restoration. *IEEE Trans. on Acoustic, Speech and Signal Proc.* **ASSP-32** (1984).
45. H. J. Trussell, Convergence criteria for iterative restoration methods. *IEEE Trans. on Acoustic, Speech & Signal Proc.* **ASSP-31** (1983).
46. A. Levi and H. Stark, Image restoration by the method of generalized projections with applications to restoration from magnitude. *J. Opt. Soc. Am.* A**1**, 932–943 (1984).
47. M. I. Sezan and H. Stark, Image restoration by the method of convex projections: Part II—Applications and numerical results. *IEEE Trans. on Medical Imaging* **MI-1**, 95–101 (1982).
48. J. R. Fienup, Phase retrieval algorithms: A comparison. *Appl. Optics* **21**, 2758–2769 (1982).
49. Z. Opial, Weak convergence of the sequence of successive approximations for nonexpansive mappings. *Bulletin of the American Mathematical Society* **73**, 591–597 (1967).
50. K. M. Nashold and B. E. A. Saleh, Image construction through diffraction-limited systems: An iterative approach. *J. Opt. Soc. Am.* **A2**, 635–643 (1985).
51. S. I. Sayegh, Y. Kok, and J. Hong, An algorithm to find two-dimensional signals with specified zero crossings. *IEEE Trans. Acoust., Speech, and Signal Processing* (to be published).
52. T. C. Hu, "Integer Programming and Network Flows," Addison-Wesley, Reading, Mass., 1969, Chapter 14.
53. B. E. A. Saleh and S. I. Sayegh, Reduction of errors of microphotographic reproductions by optimal corrections of original masks. *Optical Engineering* **20**, 781–784 (1981).

54. M. D. Levenson, N. S. Viswanathan, and R. A. Simpson, Improving resolution in photolithography with a phase-shift mask. *IEEE Trans. Electron Devices* **ED-29**, 1828 (1982).
55. D. S. Goodman, M. D. Levenson, H. Santin, and V. Viswanathan, Improved photolithographic resolution with a phase-shifting mask. *In Proceedings, Microcircuit, 1982, the 1982 Grenoble Photolithography Conference.*
56. J. W. Goodman, *Introduction to Fourier Optics*, McGraw-Hill, New York, (1968).
57. B. E. A. Saleh and K. M. Nashold, Image construction: Optimum amplitude and phase masks in photolithography. *Applied Optics* **24**, 1432–1437.
58. B. E. A. Saleh, Optical information processing and the human visual system. *In* "Applications of Optical Fourier Transforms," (H. Stark, ed.), Academic Press, New York (1982).
59. B. E. A. Saleh, Optical bilinear transformations—General properties. *Optica Acta* **26**, 777–799 (1979).
60. B. E. A. Saleh and M. Rabbani, Restoration of bilinearly distorted images: I. FIR Linear Digital Filtering. *J. Opt. Soc. Am.* **73**, 66–70 (1983).
61. B. E. A. Saleh and M. Rabbani, Restoration of bilinearly distorted images: II. Bayesian method. *J. Opt. Soc. Am.* **73**, 71–75 (1983).
62. B. E. A. Saleh and W. C. Goeke, Coherent optical compensation of post-detection spatial distortion. *J. Opt. Soc. Am.* **69**, 1172–1176 (1979).

CHAPTER **13**

The Role of Analyticity in Image Recovery

M. A. Fiddy

Physics Department
Kings College London
Strand, London WC2R 2LS

13.1 INTRODUCTION

The purpose of this chapter is to complement some of those appearing earlier in this volume, especially Chapters 6–8. In many problems of image recovery one finds that it is necessary to invert partial Fourier data to obtain information on the object function of interest. Given the usual constraint that object functions are of compact support, this leads naturally to the Fourier transform of this function having remarkable properties, associated with its analyticity.

Recently it has been shown [1] that the scattering amplitude of a field scattered from a finite domain throughout which the scattering potential is continuous will be an entire function, i.e., analytic throughout the finite complex domain. Moreover, a consequence of this analyticity is that the scattered field necessarily contains both homogeneous and inhomogeneous (evanescent) waves outside the scattering volume.

A brief background to the properties and zero distributions of bandlimited functions is presented followed by a discussion of some of the immediate consequences of this, such as dispersion relations. The relationships between real and imaginary parts and magnitude and phase of bandlimited functions are outlined for one and several dimensions. There is a close link between their zero distributions, the completeness of sets of complex exponentials, and sampling theorems which is also described.

In problems of image recovery from Fourier data, the first major difficulty to overcome is that of limited samples being available and the degree of nonuniqueness of extrapolation of spectral values. Procedures for spectral estimation or extrapolation are the subject of several chapters in this volume and we give an overview, from the viewpoint of the analytic properties of Fourier data, on the nature of the problem and the largely subjective approach to overcoming it.

One of the most instructive aspects of considering analyticity in image recovery problems is the insight that the zero distributions of bandlimited functions provide. The reconstruction of an object from Fourier phase or Fourier magnitude is best viewed in terms of the complex zero structures and the invariance, on their reflection about the real plane, of the available Fourier data. There is a fundamental difference between one- and more than one-dimensional problems of image recovery (or coding) from zero locations or partial Fourier data, which can be clearly seen in this way. Indeed, if a discrete object can be assumed or if a sufficiently "local" area of Fourier data is satisfactory to describe the object, then arguments are put forward to suggest that such data in more than one dimension are uniquely encoded by a single zero distribution encoded by one nonfactorizable analytic function. This being the case, we show why this leads to a unique relationship between Fourier magnitude and Fourier phase.

It is evident from earlier chapters that reconstruction from Fourier magnitude, a problem of wide applicability to inverse problems, is very difficult. Knowledge that a unique phase can be associated with it does not help provide an algorithm to do so. For this reason we describe two methods, based directly on the analytic properties of the Fourier data, for phase retrieval. These are based on the factorization of multivariate polynomials and on zero location. The problems with noise and the size of the data set are discussed.

Finally, we look at the consequences of the analyticity of two-dimensional Fourier data combined with the fact that it may be well represented by a zero distribution described by a single irreducible polynomial of finite degree. One is led to the conclusion that on the real plane the Fourier magnitude will go to zero at isolated points. From this it follows that phase unwrapping, namely imposing continuity on the phase function by removing the usual $2n\pi$ ambiguity, is likely to fail. Image recovery based on methods requiring a continuous phase function are mentioned and the relationship between real point zeros and wave front dislocations is examined.

13.2 MULTIDIMENSIONAL BANDLIMITED FUNCTIONS

Let us define an object distribution $f(\mathbf{t})$ which is assumed to be confined to some finite domain in $\mathbf{t} \equiv t_1, t_2, \ldots, t_N$ in an N-dimensional space. From the theorem of Plancherel and Pólya [2] it follows that a necessary and sufficient condition for this confinement to a finite domain is that $F(\mathbf{x})$, given by

$$F(\mathbf{x}) \equiv F(x_1, x_2, \ldots, x_N) = \int_{-\infty}^{\infty} f(\mathbf{t}) \exp(i\mathbf{t} \cdot \mathbf{x})\, d\mathbf{t} \tag{13.2-1}$$

can be extended to the whole space $\mathbb{C}^N$ of the complex variables $z_j = x_j + iy_j$ $(j = 1, \ldots, N)$, and where $i = \sqrt{-1}$, as an entire function of exponential type or bandlimited function. The entire means that this function is analytic for all finite $\mathbf{z}$, which, in the case of an N-dimensional variable, means that the Cauchy–Riemann equations [3]

$$\frac{\partial \operatorname{Re} F}{\partial x_j} = \frac{\partial \operatorname{Im} F}{\partial y_j}, \qquad \frac{\partial \operatorname{Re} F}{\partial y_j} = \frac{-\partial \operatorname{Im} F}{\partial x_j}, \qquad (j = 1, \ldots, N), \tag{13.2-2}$$

are satisfied for all finite $\mathbf{z}$. Exponential type specifies the characteristics of the growth of such functions. This is expressed in terms of the growth of $F(\mathbf{z})$ in different directions in the space of the variables $y_1, y_2, \ldots, y_N$. If α_j is the angle made by a chosen direction with the y_j axis, then we define

$$h_F(\boldsymbol{\alpha}, \mathbf{x}) = \limsup_{|\mathbf{z}| \to \infty} \frac{\ln|F(x_1 + i|\mathbf{z}| \cos \alpha_1, \ldots, x_N + i|\mathbf{z}| \cos \alpha_N)|}{|\mathbf{z}|},$$

and

$$h_F(\boldsymbol{\alpha}) = \sup h_F(\boldsymbol{\alpha}, \mathbf{x}) \tag{13.2-3}$$

is known as the P-indicator of $F(\mathbf{z})$, where supremum means least upper bound. It is straightforward to show [3, p. 354] that $h_F(\boldsymbol{\alpha})$ is equal to the support function of the smallest convex region outside of which $f(\mathbf{t}) = 0$. Thus the support shape in object space determines the rate of growth of the function $F(\mathbf{z})$ in the complex $\mathbf{z}$ space, and this can be expressed by

$$|F(\mathbf{z})| \leqslant M \exp(|\mathbf{z}| h_F(\boldsymbol{\alpha})), \tag{13.2-4}$$

where M is a constant, i.e., limited by a simple exponential.

It is instructive to consider the one-dimensional (i.e., $N = 1$) case because of some significant differences that occur between this and $N > 1$. It is also useful to relate the one-dimensional case to the higher-dimensional one. This can be readily done by considering projections of $f(\mathbf{t})$ onto, say, one t axis, e.g.,

$$F(x_1) = \int_a^b \left[\int f(\mathbf{t})\, dt_2 \cdots dt_N \right] \exp(it_1 x_1)\, dt_1. \qquad (13.2\text{-}5)$$

One can envisage taking many such one-dimensional projections of $f(\mathbf{t})$ and Fourier transforming them to build up data on $F(\mathbf{x})$. Recovery of $f(\mathbf{t})$ is then possible by inverse Fourier transformation of $F(\mathbf{x})$. Equation (13.2-5) is well known as the central slice theorem or Fourier projection theorem, which underlies image reconstruction by tomography. From Eq. (13.2-5) it follows that $F(z_1)$ is an entire function of exponential type which, for the one-dimensional case, is known as the Paley-Wiener theorem [4]. The rate of growth of the function in the complex plane is such that

$$|F(z_1)| \leq M \exp(|z_1|\sigma) \qquad (13.2\text{-}6)$$

and $\sigma = (b - a)/2$ since the P-indicator $h(\alpha) = [(b - a)/2]|\sin \alpha|$. Thus the maximum value of $h(\alpha)$ lies along the y_1 axis and the minimum rate of growth occurs in the direction of the real axis.

It is well known that one-dimensional functions defined by Eq. (13.2-5) may be described everywhere by their roots or zeros $\mathfrak{z}_j$ to within a linear phase factor [4]. They are the simplest generalization of polynomials and can be written (Weierstrass's factorization theorem)

$$F(z_1) = cz_1^q \exp(i(a + b)z_1) \prod_{j=-\infty}^{\infty} (1 - z_1/\mathfrak{z}_j), \qquad (13.2\text{-}7)$$

where q is the order of a zero at the z_1 origin and c a scaling factor. For the case $a = -b$, $q = 0$, and $c = 1$ the product representation is known as a Hadamard product ([5]; see also Chapter 6) and $F(z_1)$ is entirely encoded by its zero locations $\mathfrak{z}_j$. The way in which zero positions encode information in $f(t_1)$, the object, has been investigated by Fiddy *et al.* [6]. It is interesting to note that asymptotically the zeros are spaced $2\pi/(b - a)$ apart along a line or lines parallel to the real axis at positions given by [7]

$$\mathfrak{z}_j = \frac{2j\pi - i \log\{f(b)/f(a)\}}{b - a}. \qquad (13.2\text{-}8)$$

[Clearly, the points at which $F(z_1)$ equals any constant value could be used in Eq. (13.2-7) [4].]

For $N > 1$ a product representation for $F(\mathbf{z})$ is still available. An extension of the Weierstrass factorization theorem [Eq. (13.2-7)] by Osgood [8] allows one to write a global multivariable factorization of the form

$$F(\mathbf{z}) = \prod_{j=1}^{J} [F_j(\mathbf{z})\, e^{\gamma_j(\mathbf{z})}]^{L_j}, \tag{13.2-9}$$

where the e^{γ_j} are convergence factors and where γ_j is a polynomial and L_j an integer. The $F_j(\mathbf{z})$ are entire functions which are globally nonfactorizable functions in $\mathbb{C}^N$. The significant difference between Eqs. (13.2-9) and (13.2-7) for one dimension is the range $1 \leqslant J \leqslant \infty$, that is, the number of factors in the product representations. In one dimension each factor F_j has the same simple form $(1 - z_1/\mathfrak{z}_j)$ and J is infinite. For $N > 1$, the number of nonfactorizable or prime factors J bears no obvious relationship to $f(\mathbf{t})$ and there exists a much richer variety of algebraic forms for such factors compared with that for $N = 1$. In the special case of a discrete object $f(\mathbf{t})$ represented at a finite number of points, $F(\mathbf{z})$ becomes a multivariate polynomial of finite degree. Under these circumstances it is likely that $J = 1$ and $F(\mathbf{z})$ is completely represented by a single irreducible factor ([9]; see also Chapter 6). In general, any factorizable $F(\mathbf{z})$ is likely to be rendered nonfactorizable very easily by the addition of a small perturbation. For example, the function $z_1^2 + z_2^2$ is reducible but $z_1^2 + z_2^2 + \varepsilon$ is not. Conversely, it is easy to generate examples for $N > 1$ with $J = \infty$. However, as shown by Hayes in Chapter 6, in the set of all multivariate polynomials of finite degree, the subset of factorizable polynomials has measure zero and is not dense in $\mathbb{C}^N$.

It is interesting to study the properties of the nonfactorizable factors $F_j(\mathbf{z})$. In one dimension each factor corresponds to a term depending on the location of a point zero in $\mathbb{C}^1$. In more than one dimension the set of points over which $F(\mathbf{z})$ vanishes is the union of irreducible analytic sets [8, p. 265]. Each analytic set corresponds to a solution to $F(\mathbf{z}) = 0$ on all possible $\mathbf{z}$ reached by analytic continuation from some neighborhood of $z_1 = \mathbf{z}(z_1)$ where $F(\mathbf{z})$ is zero. The Osgood product in Eq. (13.2-9) means that for $N > 1$, $F(\mathbf{z})$ remains determined by its zero locations but the number and nature of these zeros are not clear.

Following an idea of Napier and Bates [10] and using Eq. (13.2-5), one can envisage taking one-dimensional projections of a multidimensional object over sets of arbitrarily small angular increments. Each projection would have a Fourier transform with the characteristics expressed by Eqs. (13.2-7) and (13.2-8). For a sufficiently well-behaved object support one can further envisage the regular asymptotic behavior of the point zeros tracing out zero lines reflecting the support shape as each projection is Fourier transformed. The zeros of functions of N complex variables are $(N - 1)$-dimensional surfaces. Thus in two dimensions the zeros are lines in $\mathbb{C}^2$ (or equivalently two-dimensional surfaces in $\mathbb{R}^4$). Asymptotically, one may therefore expect to have closed zero surfaces, each being an irreducible analytic set and thus represented by a single globally irreducible factor F_j in the Osgood product. For such objects it could follow that J is infinite.

Another example for which J is infinite is the case of a separable Fourier transform $F(\mathbf{z})$. If one can write $F(\mathbf{z})$ as the product of two or more terms, at least one of which is a function of only z_j, for some j, then this factor is expressible in terms of a Hadamard product having an infinity of terms. A simple example of this could be $f(\mathbf{t})$ being a constant in a square support and zero outside. The Fourier transform of this function is of the form

$$F(\mathbf{z}) \equiv F(z_1, z_2) = \frac{\sin \pi z_1}{z_1} \cdot \frac{\sin \pi z_2}{z_2} = \prod_{\substack{j=-\infty \\ j \neq 0}}^{\infty} (1 - z_1/j)(1 - z_2/j). \qquad (13.2\text{-}10)$$

Also, for the special case of $f(\mathbf{t})$ having radial symmetry, one effectively reduces the multidimensional properties to those of a one-dimensional function [11].

13.3 CONSEQUENCES OF ANALYTICITY

The topic of analyticity occurs frequently in physics and offers insight into many problems (e.g. [12]). A bandlimited function is highly constrained in behavior through having to satisfy the Cauchy–Riemann equations. In principle, though not in practice, the analytic properties of $F(\mathbf{z})$ would allow its complete recovery from a small but continuous segment on $\mathbf{z}$. This property has been the stimulus for much of the superresolution work reported in earlier chapters.

One physical manifestation of these analytic properties may be associated with speckle [13], since speckle size is clearly determined by instrument aperture size, which corresponds to the finite $\mathbf{t}$ domain support in this context. Imposing a finite support in the $\mathbf{t}$ domain introduces a fluctuation in amplitude in the $\mathbf{x}$ domain, e.g., a Gibbs' phenomenon, which is due precisely to the necessary presence of zeros in $F(\mathbf{z})$. Reduction of speckle contrast can be achieved by moving zeros away from the real plane. Within the range of validity of the Van Cittert and Zernike theorem, this can be accomplished by apodization of the autocorrelation function of $f(\mathbf{t})$ by the complex degree of coherence [13]. Softening the support constraint by apodization in the $\mathbf{t}$ domain can be seen to be a smoothing in $\mathbf{x}$ which necessarily moves zeros away from the $\mathbf{x}$ plane.

One of the earliest and most important consequences of analyticity was the use of dispersion relations for the recovery of one function from another. For an analytic function in any number of variables one can write Cauchy's integral formula, which relates values of an analytic function in a region to those on the boundary of that region, say D [14], i.e.,

$$F(\mathbf{z}) = \int_D \frac{F(\mathbf{z}')\, d\mathbf{z}'}{\mathbf{z}' - \mathbf{z}}. \tag{13.3-1}$$

By choosing $f(\mathbf{t})$ to be "causal," namely to have its support confined to a region for which the t_j are entirely positive, one can choose D to be the real plane and an infinite semicircular hypersurface from which the contribution to the integral vanishes [15]. Taking real and imaginary parts of the remaining integral on the real plane gives the dispersion relations or Hilbert transform relations. Thus, in one dimension one has

$$F(x_1) = \frac{1}{\pi i} \int_{-\infty}^{\infty} \frac{F(x_1')\, dx_1'}{x_1' - x_1}, \tag{13.3-2}$$

giving

$$\operatorname{Re} F(x_1) = \frac{1}{\pi} \int_{-\infty}^{\infty} \frac{\operatorname{Im} F(x_1')\, dx_1'}{x_1' - x_1}$$

and (13.3-3)

$$\operatorname{Im} F(x_1) = \frac{1}{\pi} \int_{-\infty}^{\infty} \frac{\operatorname{Re} F(x_1')\, dx_1'}{x_1' - x_1}.$$

In the temporal domain causality has a clear meaning: for example, that the polarizability of a medium is zero prior to illumination leads directly to the Kramers–Kronig relations relating the real and imaginary parts of the permittivity in a similar way to Eq. (13.3-3) [14].

It follows that if image recovery is required from $\operatorname{Re} F(\mathbf{x})$ or $\operatorname{Im} F(\mathbf{x})$, this should be easily achieved through Hilbert transformation to obtain $F(\mathbf{x})$. However, in, for example, two dimensions, Cauchy's integral formula leads to dispersion relations analogous to Eq. (13.3-3) of the form [16]

$$\begin{aligned}\operatorname{Re} F(x_1, x_2) = {} & \frac{1}{\pi^2} \iint_{-\infty}^{\infty} \frac{\operatorname{Re} F(x_1', x_2')\, dx_1'\, dx_2'}{(x_1' - x_1)(x_2' - x_2)} \\ & + \frac{1}{\pi^2} \int_{-\infty}^{\infty} \frac{\operatorname{Im} F(x_1', x_2')\, dx_1'}{x_1' - x_1} \\ & + \frac{1}{\pi} \int_{-\infty}^{\infty} \frac{\operatorname{Im} F(x_1', x_2')\, dx_2'}{x_2' - x_2},\end{aligned} \tag{13.3-4}$$

with a similar expression for $\operatorname{Im} F(x_1, x_2)$. These equations constitute a singular integral equation of the form

$$G(x_1, x_2) = \lambda \iint_{-\infty}^{\infty} \frac{G(x_1', x_2')\, dx_1'\, dx_2'}{(x_1' - x_1)(x_2' - x_2)} + H(x_1, x_2), \qquad (13.3\text{-}5)$$

whose solution, obtained by Fourier transformation, is

$$G(x_1, x_2) = \frac{1}{1-\lambda^2\pi^4}\left(H(x_1, x_2) + \lambda \iint_{-\infty}^{\infty} \frac{H(x_1', x_2')\, dx_1'\, dx_2'}{(x_1' - x_1)(x_2' - x_2)}\right), \qquad (13.3\text{-}6)$$

which does not exist for the eigenvalues $\lambda = \pm 1/\pi^2$. Thus Eq. (13.3-4) cannot be solved.

Image recovery from $|F(\mathbf{x})|$ or arg $F(\mathbf{x})$ is also possible under certain circumstances using Hilbert transforms. An expression similar to that in Eq. (13.3-1) can be written for log $F(\mathbf{z})$ which is analytic in those regions where $F(\mathbf{z})$ is also analytic but nonzero. A Hilbert transform pair relates $\log|F(\mathbf{x})|$ and arg $F(\mathbf{x})$ under this condition, allowing one to be calculated from the other. In one dimension the necessary requirement is that $F(z_1)$ has no zeros in D, viz the upper (or lower) half of the complex z_1 plane. A function with this property is referred to as a minimum phase function and one can write

$$\arg F(x_1) = \frac{-1}{\pi} \int_{-\infty}^{\infty} \frac{\log|F(x_1')|\, dx_1'}{x_1' - x_1}, \qquad (13.3\text{-}7)$$

with a similar expression for $\log|F(x_1)|$ in terms of arg $F(x_1)$.

It is not generally known *a priori* that $F(z_1)$ has a zero-free half-plane. This can be ensured, however, by the introduction of a suitable reference function satisfying Rouché's theorem [17]. Such a function could have the form Ae^{icz_1} where $|A| > |F(x_1)|$ and $c > b$ in Eq. (13.2-5), for example. This corresponds to a delta function or reference point located adjacent to the object support (a, b).

In more than one dimension it is not possible to have a zero-free half-space of the kind required to write a dispersion relation between $\log|F(\mathbf{x})|$ and arg $F(\mathbf{x})$, except under very specific circumstances. One can see that each zero surface, described, for example, by the factor $F_j(\mathbf{z}) = z_1 - G_j(z_2, \ldots, z_N)$, is defined for all $z_2, \ldots, z_N$. Since G_j is analytic it must take all values at least once and so $F(\mathbf{z})$ will be zero for some $y_1 > 0$ and $y_1 < 0$. A similar argument holds for each variable. In the case of a separable function such as that given in Eq. (13.2-10), wherever the function of, say, z_2 is zero, the total function $F(\mathbf{z})$ is necessarily zero for all values of z_1, making a truly zero-free half space in $\mathbb{C}^N$ impossible. Also, as for relating

Re $F(\mathbf{x})$ and Im $F(\mathbf{x})$, the dispersion relations constitute a pair of inhomogeneous singular integral equations unless $F(\mathbf{z})$ is separable. Thus in two dimensions, where

$$F(x_1, x_2) = F_1(x_1)F_2(x_2),$$

one can write [16]

$$\arg F(x_1, x_2) = \frac{-1}{\pi}\int_{-\infty}^{\infty} \frac{\log F_1(x_1')\,dx_1'}{x_1' - x_1} - \frac{1}{\pi}\int_{-\infty}^{\infty} \frac{\log F_2(x_2')\,dx_2'}{x_2' - x_2}, \qquad (13.3\text{-}8)$$

with a similar expression for $\log|F(x_1, x_2)|$ in terms of $\arg F(x_1, x_2)$, assuming $F_1(z_1)$ and $F_2(z_2)$ each have zero-free half-planes. This defines the concept of a minimum phase function in two dimensions.

These various difficulties associated with the differences between $N = 1$ and $N > 1$ indicate that for some image reconstruction problems it may be preferable to exploit the analytic properties of one-dimensional strips of a higher-dimensional function.

13.4 IMAGE RECONSTRUCTION FROM LIMITED SPECTRAL DATA

There is a close connection between analytic functions and the completeness of exponentials of the form $\{e^{i\mathfrak{z}_n t_1}\}$.

A sequence of vectors is said to be a basis for a normed vector space A if, for each $\mathbf{a} \in A$, there corresponds a unique sequence of scalars c_1, $c_2, \ldots$ such that

$$\mathbf{a} = \sum_{n=1}^{\infty} c_n \mathbf{a}_n. \qquad (13.4\text{-}1)$$

A sequence of vectors $\{\mathbf{a}_1, \mathbf{a}_2, \ldots\}$ in A is said to be complete if its linear span is dense in A; that is, for each vector $\mathbf{a}$ and each $\varepsilon > 0$ there is a finite combination $c_1\mathbf{a}_1 + \cdots + c_n\mathbf{a}_n$ such that $\|\mathbf{a} - (c_1\mathbf{a}_1 + \cdots + c_n\mathbf{a}_n)\| < \varepsilon$, where we define for any vector $\boldsymbol{f}$ (Parseval's identity)

$$\|\boldsymbol{f}\|^2 = (\boldsymbol{f}, \boldsymbol{f}) \qquad (13.4\text{-}2)$$

and

$$(\boldsymbol{f}, \boldsymbol{f}) = \int_{-\pi}^{\pi} \cdots \int f(\mathbf{t}) f^*(\mathbf{t})\,d\mathbf{t}, \qquad (13.4\text{-}3)$$

the inner product for $L^2(-\pi, \pi)$ and where * denotes complex conjugate.

The system $\{e^{int_1}\}_{-\infty}^{\infty}$ constitutes an orthonomal basis which is complete in $L^2(-\pi, \pi)$ and stable under sufficiently small perturbations of the integers n [18]. This means that if $\{\mathfrak{z}_n\}$ is a sequence of real or complex numbers for which $\{\mathfrak{z}_n - n\}$ is in some sense small, then the system $\{e^{i\mathfrak{z}_n t_1}\}$ will also form a basis in $L^2(-\pi, \pi)$. It follows that every function $f(t_1)$ in $L^2(-\pi, \pi)$ will have a unique nonharmonic Fourier series expansion of the form

$$f(t_1) = \sum_{n=1}^{\infty} c_n \, e^{i\mathfrak{z}_n t_1}. \tag{13.4-4}$$

This is true for the multidimensional case also if we consider the set $\{e^{i\mathfrak{z}_n \cdot \mathbf{t}}\}$. It is well known [18] that for the system $\{e^{i\mathfrak{z}_n t_1}\}$ to be complete in $L_2(-\pi, \pi)$ it is necessary and sufficient that there exist a nontrivial entire function $F(z_1)$, zero for every $z_1 = \mathfrak{z}_n$ and expressible in the form of Eq. (13.2-5), namely

$$F(z_1) = \int_{-\pi}^{\pi} f(t_1) \exp(it_1 z_1) \, dt_1, \tag{13.4-5}$$

with $f(t_1) \in L^2(-\pi, \pi)$. If there are sufficient $\mathfrak{z}_n$'s that $F(z_1)$ vanishes identically whenever $F(\mathfrak{z}_n) = 0$ $(n = 0, \pm 1, \pm 2, \ldots)$ then $\{\mathfrak{z}_n\}$ is a set of uniqueness for $F(z_1)$ and the system of exponentials $\{e^{i\mathfrak{z}_n t_1}\}$ is complete in the t_1 space. In $L^2(-\pi, \pi)$ the set of exponentials is complete whenever $|\mathfrak{z}_n - n| \leq \frac{1}{4}$ $(n = 0, \pm 1, \pm 2, \ldots)$, following a theorem due to Levinson [18]. This is clearly a sufficient rather than necessary condition, and there does not seem to be any simple extension of this to higher dimensions. This desirable regularity in the zero distribution of bandlimited functions is inevitable for large $|z_1|$ [see Eq. (13.2-8)] but considerable freedom is possible in the location of a finite number of zeros [6]. This has been exploited by Bucklew and Saleh (e.g., Chapter 12 and [19]), where a bandlimited function was required to have zero crossings at prescribed locations. The purpose of this was to precisely specify the locations at which an otherwise smoothly varying function would be hardlimited to give a sharp binary image taking the values zero and unity on either side of those points. Any finite number of zeros $\mathfrak{z}_n$ can be manipulated in this way and the required object function producing this distribution can be readily calculated. Also, in more than one dimension one could manipulate a finite number of zeros from a completely factorizable $F(\mathbf{z})$ [e.g., Eq. (13.2-10)] and reconstitute real zero contours of arbitrary shape provided the total order of the zeros thus altered was not increased.

The space of all entire functions of exponential type, say of type π, is frequently referred to as Paley-Wiener space [18], P, and (Parseval's identity)

$$\|F\|^2 = \sum_{n=-\infty}^{\infty} |F(n)|^2 = \|f\|^2. \tag{13.4-6}$$

By taking the Fourier transform of e^{int_1} ($n = 0, \pm 1, \pm 2, \ldots$) the set of functions $\{\sin \pi(z_1 - n)/\pi(z_1 - n)\}_{-\infty}^{\infty}$ also forms an orthonormal basis for P, and here each function $F(z_1)$ in P has a unique expansion of the form

$$F(z_1) = \sum_{n=-\infty}^{\infty} c_n \frac{\sin \pi(z_1 - n)}{\pi(z_1 - n)}, \tag{13.4-7}$$

which is the sampling theorem due to Shannon and requires an infinity of samples of F for a unique representation. Equation (13.4-7) follows because the Fourier transform is an isomorphism between $L^2(-\pi, \pi)$ and P and here [18]

$$(f(t_1), e^{int_1}) = (F(n), \sin \pi(z_1 - n)/\pi(z_1 - n)).$$

Given the allowed small perturbations of the integers n, one can consider restoring a bandlimited function from its unequally spaced samples. The efficiency of such methods will depend on the extent of these perturbations, and the "jitter error" can be reduced by taking the linearly weighted sum of shifted sine functions (e.g. [20] and references therein).

The usual requirement is to retrieve additional data on $F(\mathbf{x})$ from limited samples taken in the real domain at low "spatial frequencies" $\mathbf{x}_n$, $n < M < \infty$. It is most unlikely that all missing samples are zero, which leads to the discrete Fourier transform estimate of the object $f(\mathbf{t})$. Knowing that $F(\mathbf{z})$ is entire in principle allows the possibility of analytic continuation by means of a Taylor series expansion about a point (a_1, a_2) of the form (in two dimensions)

$$F(z_1, z_1) = \sum_{k=0}^{\infty} \frac{1}{k!}\left[(z_1 - a_1)\frac{\partial}{\partial a_1} + (z_2 - a_2)\frac{\partial}{\partial a_2}\right]^k F(a_1, a_2). \tag{13.4-8}$$

Clearly, the partial derivatives of F are determined by any finite segment of $F(\mathbf{x})$ but it is not possible to evaluate these precisely from noisy sampled data. Also, the truncation of this series leads to large errors for large $|\mathbf{z}|$. From this it would appear that direct exploitation of the analyticity of $F(\mathbf{x})$ for higher-resolution reconstruction of $f(\mathbf{t})$ seems unlikely. However, many schemes for the stable extrapolation of Fourier data have been developed, and some of these are described in earlier chapters.

What is clear is that from limited Fourier samples one can construct an infinity of possible functions $F(\mathbf{z})$ compatible with the available samples and certain constraints on $f(\mathbf{t})$ such as its support shape. Lagrange interpolating functions can be designed, having the form

$$L_{\mathbf{n}}(\mathbf{x}) = \frac{G(\mathbf{x})}{(\mathbf{x}-\mathbf{n})G'(\mathbf{n})}, \tag{13.4-9}$$

where $G(\mathbf{x})$ is a factorizable function whose asymptotic zeros can encode the support shape of $f(\mathbf{t})$, if this is known. The $L_{\mathbf{n}}(\mathbf{x})$ are entire functions designed to be zero at all sample locations $\mathbf{m}$ except $\mathbf{m} = \mathbf{n}$, at which it has the value unity. Thus we can choose

$$G(\mathbf{x}) = \prod_{\{\mathbf{m}\}} (1 - \mathbf{x}/\mathbf{m}), \tag{13.4-10}$$

and the interpolated function

$$F(\mathbf{x}) = \sum_{\{\mathbf{n}\}} F(\mathbf{n}) L_{\mathbf{n}}(\mathbf{x}) \tag{13.4-11}$$

is not necessarily equal to the Fourier transform of $f(\mathbf{t})$ but is data consistent and can be made support consistent [21].

In order to select one extrapolated function $F(\mathbf{x})$ from this infinity of possibilities it is usual to maximize or minimize some cost function. For example, the iterative scheme of Gerchberg–Papoulis ([22] and Chapters 7 and 8) determines an estimate $\tilde{f}(\mathbf{t})$ such that $\|f(\mathbf{t}) = \tilde{f}(\mathbf{t})\|^2$ is minimized in a weighted L^2 space. A direct analytical form for this estimate can be written (the generalized moment problem [18]). Given limited Fourier samples $F(\mathbf{n})$ an $f(\mathbf{t}) \equiv \sum_{\{\mathbf{n}\}}^{N} c_{\mathbf{n}} \phi_{\mathbf{n}}(\mathbf{t})$ has this closeness property when the $c_{\mathbf{n}}$ are obtained by solving the system of equations

$$F(\mathbf{m}) = \sum_{\{\mathbf{n}\}}^{N} c_{\mathbf{n}}(\phi_{\mathbf{n}}(\mathbf{t}), \phi_{\mathbf{m}}(\mathbf{t})) \tag{13.4-12}$$

and the $\phi_{\mathbf{n}}(\mathbf{t}) = p(\mathbf{t})e^{i\mathbf{n}\cdot\mathbf{t}}$, where $p(\mathbf{t})$ specifies the object support [23,24] and a weighted L^2 space is considered with norm induced by the inner product

$$(f, g) = \int_{-\pi}^{\pi} \cdots \int f(\mathbf{t}) g^*(\mathbf{t})/p(\mathbf{t})\, d\mathbf{t}. \tag{13.4-13}$$

If higher resolution in the estimate $\tilde{f}(\mathbf{t})$ is desirable, in the sense of spiky features, then a reciprocal polynomial model for the estimate is more likely to be capable of this since real zeros of the polynomial will then result in poles. In one dimension, if we demand that $\tilde{f}(t_1)$ be a real nonnegative function, then by the Fejér–Riesz theorem [21] one can write $\tilde{f}^{-1}(t_1)$ as a product of a minimum phase and maximum phase polynomial and obtain a closed-form solution for this estimate which also corresponds to maximizing the entropy function $-\int_{-\pi}^{\pi} \log f(t_1)\, dt_1$ (see also [23]). Generalization of these ideas to more than one dimension depends on the properties of the zeros of analytic functions in higher dimensions, and as we have seen in Section 13.2 this cannot readily be done.

13.5 IMAGE RECONSTRUCTION FROM ZEROS

We have described how the zeros of a bandlimited function represent it uniquely. In one dimension zero location is straightforward [25], but in higher dimensions one is faced with the problem of locating not a point minimum of the modulus of $F(z_1)$ but sampling this minimum surface in a higher-dimensional space. In two dimensions the zeros are lines in $\mathbb{C}^2$ (or equivalently two-dimensional surfaces in $\mathbb{R}^4$). The irreducible zero sets describe the structure of zero surfaces and can be regarded as units of information for N-dimensional Fourier transforms.

Since the asymptotic behavior of $F(z_1, z_2)$ is determined by the edge of the object support and the factors in the Osgood product [Eq. (13.2-9)] are entire functions, there exist strong conditions on their behavior throughout the whole of $\mathbb{C}^2$. Unlike one-dimensional zeros, the complex zeros in the finite part of $\mathbb{C}^2$ depend on the asymptotic behavior of $F(z_1, z_2)$ [consider Eq. (13.2-10), for example]. Provided the support of the object is convex and the angular increments are sufficiently small, it is possible to generate the asymptotic zeros of $F(x_1, x_2)$ from those of a series of radial slices (as indicated in Section 13.2). Once the asymptotic behavior of all these factors is known, they may be accounted for by a function $G(z_1, z_2)$ of class P (see last section) and the remaining factors by polynomials. Moreover, determining the nonasymptotic zero set of $F(z_1, z_2)$ is made easier by the Weierstrass preparation theorem [26], which states that (in two dimensions) in the neighborhood $U = \{|z_1 - a_1| < r_1, |z_2 - a_2| < r_2\}$ of a point (a_1, a_2) at which an analytic function $F(z_1, z_2)$ vanishes:

$$F(z_1, z_2) = (z_1 - a_1)P(z_1, z_2)G(z_1, z_2), \tag{13.5-1}$$

where $P(z_1, z_2)$ is a polynomial of degree k, namely

$$P(z_1, z_2) = (z_2 - a_2)^k + A_1(z_1)(z_2 - a_2)^{k-1} + \cdots + A_k(z_1), \tag{13.5-2}$$

and $G(z_1, z_2)$ is analytic and nonzero at (a_1, a_2). Polynomial $P(z_1, z_2)$ has the same complex zeros as F in the neighborhood of (a_1, a_2). Locally, the problem is now much simpler because the zeros are those of polynomials rather than entire functions of the exponential type. Clearly if $G(z_1, z_2)$ is rendered relatively unimportant due to attenuation by a window function [e.g., $(\sin(\pi r z_1/\pi z_1)) \cdot (\sin(\pi r z_2/\pi z_2))$] then this sets an effective resolution limit on $f(t_1, t_2)$ (by averaging over boxes of size r by r).

It is well known that almost all polynomials in two or more variables are irreducible [9]. The proof hinges on the fact that the set of reducible polynomials is of measure zero. If $F(\mathbf{z})$ can be written, even locally, as a polynomial in $\mathbf{z}$, then $F(\mathbf{z})$ is extremely unlikely to be reducible.

Consider, for example, the Fourier transform of a featureless circular aperture. This gives a reducible function of the form

$$F(z_1, z_2) = \prod_j^{\infty}(1 - r^2/\alpha_j^2), \tag{13.5-3}$$

where $r^2 = x_1^2 + x_2^2$, and α_j are real constants equal to the radii of zero circles on the real plane. $F(\mathbf{z})$ is an example of a reducible bandlimited function in two dimensions, but the factors are irreducible. The zero set of each factor contains two surfaces and may be rewritten as follows:

$$\begin{aligned} 1 - (x_1^2 + x_2^2)/\alpha_j^2 &= -1/\alpha_j^2(x_1^2 - (\alpha_j^2 - x_2^2)) \\ &= -1/\alpha_j^2(x_1 - \sqrt{(\alpha_j^2 - x_2^2)})(x_1 + \sqrt{(\alpha_j^2 - x_2^2)}). \end{aligned} \tag{13.5-4}$$

If $F(\mathbf{z})$ is divided by one of these factors, the spectrum is no longer bandlimited. Each factor in Eq. (13.5-4) is a branch which is inseparable from its counterpart if $F(\mathbf{z})$ is to remain analytic, and the intersection of the two zero surfaces comprising the single irreducible factor is known as a branch cut. It is possible for two zero surfaces belonging to different irreducible zero sets to intersect. This kind of intersection is known as a pseudobranch since the zero surfaces may be separated without destroying analyticity. It follows that even if zero surfaces intersect the Fourier plane at quite different places, they may be part of a single globally irreducible factor, making their location and subsequent specification difficult.

There has been much interest in the possible recovery of a bandlimited function or polynomial from its zero crossings [27–30] or threshold crossings. It is well known [15] that in one dimension Re $F(z_1)$ and Im $F(z_1)$ are harmonic functions (i.e., satisfy Laplace's equation) and that lines of constant Re $F(z_1)$ or Im $F(z_1)$ are curves in the z_1 plane, the zeros of $F(z_1)$ being the intersections of the zero curves of the real and imaginary parts. At a simple zero these curves intersect at right angles, otherwise at $\pi/2n$ for a zero of order n. If C is a simple closed contour inside and on which $F(z_1)$ is analytic, then Re $F(z_1)$ vanishing at $2k$ distinct points on C implies that $F(z_1)$ has at most k zeros inside C. Thus, if C is the real line and a semicircle of infinite radius in the upper half of the complex plane, then for a causal transform $F(z_1)$ the zero crossings of Re $F(x_1)$ and $|F(x_1)|$ are sufficient to specify $F(z_1)$ uniquely [31]. Knowing the zero crossings of Re $F(x_1)$ is equivalent to knowing $|F(x_1)|$ and one bit of Fourier transform phase [32]. More recently, Curtis *et al.* [28] have shown that in two dimensions if $F(x_1, x_2)$ and $G(x_1, x_2)$ are real nonfactorizable bandlimited functions with the same zero crossings, then $F(x_1, x_2) = cG(x_1, x_2)$, where c is a constant. For F and G polynomials of finite degree (or doubly periodic bandlimited functions) reconstruction from these zero crossings is possible, iteratively using knowledge of the support of $f(\mathbf{t})$ or by solving a set of linear equations for the Fourier series coefficients of the function F, i.e.,

$$\sum_{\{\mathbf{n}\}} f(\mathbf{n})\, e^{i\boldsymbol{z}_m \cdot \mathbf{n}} = 0, \qquad \text{where } \boldsymbol{z}_m \text{ is a zero of } F. \tag{13.5-5}$$

In general it would be the zero crossings of arbitrary complex $F(\mathbf{x})$ one would like to reconstruct from. [In more than one dimension one would not generally expect to be able to directly a relate Re $F(\mathbf{x})$ to Im $F(\mathbf{x})$, as explained in the previous section.] The unique representation of a one-dimensional function by its zero crossings was addressed by Logan [33]. He proved that a bandpass function of bandwidth less than one octave, and which had no zeros in common with its Hilbert transform except real zeros of degree one, is uniquely represented by and can be reconstructed from its zero crossings. These common or "free" zeros, if they occur, can clearly be replaced by a zero at another location without changing the bandpass nature of the function. Zeevi and Rotem have attempted to extend Logan's results to two dimensions by considering one-dimensional strips of a real $F(\mathbf{x})$ (e.g., reconstructing functions along vertical lines and then scaling them according to the function on a horizontal line) [30]. These lines constitute a sampling set on which Logan's conditions are satisfied, and this can be a relatively sparse set considering the continuum of lines one could choose from. This is thus not a strict extension to higher-dimensional functions but appears satisfactory for image reconstruction. Clearly, as the passband [i.e., the support of $f(\mathbf{t})$] moves farther from the $\mathbf{t}$ origin there will be increased sensitivity to estimating zero-crossing positions.

It is interesting to consider the fundamental restrictions that must apply to the real zeros of a bandlimited function or polynomial. While, in one dimension, zeros occurring on the real axis tend to be quite rare in the absence of a carrier (bandpass functions), zero intersections with the real plane in two dimensions are very common. First, it is necessary to define what is meant by the term real zero. The formal definition is as follows.

If $F(\mathbf{z})$ is an entire function, and C_F is its zero set, then if the intersection of C_F with the real plane is a nonempty set and some of its irreducible components disconnect the real plane, these are called the real zeros of $F(\mathbf{z})$. If an intersection does not disconnect the real plane, we call the zero a point zero.

In other words, real zeros are defined as the irreducible principal analytic sets whose intersection with the real plane disconnects it. Each intersection of these zeros with the real plane must be either an infinite or a closed curve. The zero set of $F(\mathbf{z})$ can intersect the real plane either on infinite (or closed) curves, i.e., real zeros, or at isolated points, namely point zeros. A proof of this statement is given below.

As described earlier, the real and imaginary parts of $F(\mathbf{z})$ can be treated as polynomials with real coefficients. The equation $\operatorname{Re} F(x_1, x_2) = 0$ describes a curve in the real plane (given Re F is a polynomial with real coefficients). Usually, the real and imaginary parts go to zero on separate curves in the real plane and intersect only at isolated points. However, there are several cases of degenerate behavior. The first of these is if either the real or imaginary part is zero everywhere on the real plane. In this case, the function $F(\mathbf{x})$ goes to zero at all points where the nonzero imaginary or real part of $F(\mathbf{x})$ is equal to zero; e.g., if the real part is identically zero then the zeros of the function are the zeros of the imaginary part. The second kind of degenerate behavior is when the real and imaginary parts go to zero at the same set of points on the real plane [34].

For the one-dimensional case, the zeros of $F(z_1)$ constitute a denumerable infinity of points on the complex plane and this set of points has measure zero. It can be shown [26] that for two dimensions, the zero set is of measure zero in $\mathbb{R}^4$ but the concept of a zero-free half-space is no longer applicable. The zeros form two-dimensional surfaces in $\mathbb{R}^4$, and for this reason cannot disconnect it. An important consequence of the zeros' inability to disconnect $\mathbb{R}^4$ is that the zero surfaces can extend throughout the whole of $\mathbb{C}^2$ without necessarily intersecting the real plane anywhere.

13.6 IMAGE RECONSTRUCTION FROM POLYNOMIAL REPRESENTATION OF TWO-DIMENSIONAL BANDLIMITED FUNCTIONS

We have made a case in the previous section for why a polynomial model for the bandlimited function $F(\mathbf{z})$ may be appropriate [Eq. 13.5-1]. In many practical applications $F(\mathbf{z})$ is better represented directly by the z-transform of a discrete signal f_{jk}, as described by Hayes in Chapter 6, i.e., in two dimensions. If

$$F(\zeta_1, \zeta_2) = \sum_{j=0}^{M} \sum_{k=0}^{M} f_{jk} \exp(ij\zeta_1 + ik\zeta_2) \tag{13.6-1}$$

we now define

$$F(z_1, z_2) = \sum_{j=0}^{M} \sum_{k=0}^{M} f_{jk} z_1^j z_2^k, \qquad z = \exp(i\zeta). \tag{13.6-2}$$

$F(z_1, z_2)$ is almost certain to be irreducible, although Eisenstein's criterion is one way of ensuring irreducibility [35].

Eisenstein's criterion states that if $F(z_1, z_2) = a_0(z_2) + a_1(z_2)z_1 + \cdots + a_{n-1}(z_2)z_1^{n-1}$ and if there exists an irreducible factor $P(z_2)$ which divides $a_0, a_1, \ldots, a_{n-2}$, but not a_{n-1}, and if $P^2(z_2)$ does not divide a_0, then $F(z_1, z_2)$ is irreducible. Consider as an example the spectrum of a featureless object with square support. Since the support function is separable, the spectrum of the featureless square is also separable and may be expressed as the product of two functions, one dependent on z_1 and one on z_2. The spectrum, which is simply the diffraction pattern of a square aperture, possesses two sets of parallel zero lines on the real plane, one set parallel to the x_1 axis and the other parallel to the x_2 axis.

Provided that the spectrum can be adequately represented by an M-degree polynomial, the object is readily modified to render its spectrum irreducible.

$$\begin{aligned} F(x_1, x_2) = {} & (f_{00} + f_{10}x_1 + \cdots + f_{M0}x_1^M) \\ & + (f_{01} + f_{11}x_1 + \cdots + f_{M1}x_1^M)x_2 \\ & + \cdots \\ & + (f_{0M} + f_{1M}x_1 + \cdots + f_{MM}x_1^M)x_2^M. \end{aligned}$$

Eisenstein's criterion can help construct an irreducible polynomial. In $\mathbb{C}$ the only possible form of the prime is $z_2 + b$, where b is complex. If the region containing the object support is the upper right quadrant described above, then a single nonzero reference point outside the support at $(0, M)$ ensures irreducibility. Thus $F(z_1, z_2)$ is definitely irreducible provided $f_{0k} = 0$ for $k = 0, 1, \ldots, M-1$, $f_{jM} = 0$ for $j = 1, 2, \ldots, M$, $f_{0M} \neq 0$, and $f_{10} \neq 0$. The simplest prime z_2 divides all coefficients except that of the z_1^M-term, and z_2^2 does not divide the z_1^0 coefficient. The coefficient f_{0M} is chosen to be small relative to the rest of the aperture and, if it is sufficiently small, the zero lines of Re $F(x)$ are virtually unaffected by its presence. However, the imaginary part of $F(\mathbf{x})$ now goes to zero only on lines, and the zeros of $F(\mathbf{x})$ correspond to the intersections of the zero lines of Re $F(\mathbf{x})$ and Im $F(\mathbf{x})$. The consequence of imposing irreducibility is to split the original line zeros into isolated point zeros [36]. Increasing the value of f_{0M} distorts the zero lines of Re $F(\mathbf{x})$ and Im $F(\mathbf{x})$, causing them to intersect less frequently. When f_{0M} becomes sufficiently large, no zero crossings at all occur on the real plane. If one is only interested in real objects f_{jk} then a prime factor $P(z_2)$ can be a polynomial with real coefficients irreducible in $\mathbb{R}$. Thus, if a particular $F_1(z_1, z_2)$ is an "Eisenstein" polynomial, so is a sum of weighted shifted replicas of F_1, e.g., of the form $(az_2^2 + bz_2 + c)F_1$ for a z-transform [37a,b]. Moreover, Brames [38] has considered the class of discrete object supports, for which the Eisenstein condition gives a special case and for which irreducibility of the corresponding polynomial is assured. For example, consider two orthogonal axes with a square array of points bounded by the upper left quadrant, i.e., having one corner at the origin. The support geometry obtained on moving this point at the origin into the lower right (includes Eisenstein) quadrant or the upper left quadrant can generate a new support shape whose convex hull (quadrilateral) has no parallel faces provided it does not lie on the axes. The polynomial formed with unit coefficients distributed over this support has been shown to be irreducible, as has the polynomial formed from coefficients distributed over an arbitrary support but one having such a convex hull [38]. A convex hull of a set of points A is the intersection of all convex sets containing A, and A is convex if all points $\mathbf{t}$ on any line segment $\mathbf{t} = \lambda t_{ij} + (1 - \lambda) t_{rs}$, where $0 \leq \lambda \leq 1$, for all t_{ij}, t_{rs} in A are also contained in the set A [39].

There are several consequences of establishing the irreducibility of $F(\mathbf{x})$. The most important concerns the question of the uniqueness of the relationship between Fourier phase and Fourier magnitude, since it is frequently desirable to reconstruct an image from one or the other as described in earlier chapters. It is already clear from Eq. (13.3-7) that in one dimension there is only a unique relationship between these functions in the special case of $F(z_1)$ being known *a priori* to be minimum phase. It was not until the paper of Bruck and Sodin in 1979 [40] that uniqueness appeared to be a more common property in higher dimensions. It is easy to see that $|F(x_1)|^2$ is unchanged by writing $F(x_1) = F(x_1)(1 - x_1/z_0^*)/(1 - x_1/z_0)$ in Eq. (13.2-7) for one or many complex zeros z_j or by "zero flipping." The degree of ambiguity between phase and magnitude was studied in this way many years ago [41,42]. In more than one dimension the degree of nonuniqueness can again be represented by flipping one or many non-self-conjugate factors $F_j(\mathbf{x})$, which of course cannot occur if $F(\mathbf{x})$ is irreducible.

For $F(\mathbf{x})$ irreducible it can be shown that $F(\mathbf{x})$ can be uniquely recovered from $|F(\mathbf{x})|$ or arg $F(\mathbf{x})$, although stable algorithms for achieving this are still the subject of much research effort. The proof of this uniqueness hinges on the uniqueness of the factorization for $F(\mathbf{z})$ in Eq. (13.2-9). One can readily show [43] that one is led to a contradiction in the following manner if uniqueness does not occur. Let $|F(\mathbf{x})|^2 = F(\mathbf{x})F^*(\mathbf{x}) = H(\mathbf{x})H^*(\mathbf{x})$. If C_F denotes the zero set of $F(\mathbf{x})$, then it follows that $C_F \subseteq C_H \cup C_{H^*}$ or $C_F = (C_H \cap C_F) \cup (C_{H^*} \cap C_F)$. Given that C_F is irreducible then $C_F = C_H \cap C_F$ and $C_F = C_{H^*} \cap C_F$, which implies $C_F \subseteq C_H$ and $C_F \subseteq C_{H^*}$. If it is assumed that $C_{F^*} \neq C_{H^*}$ there is a clear contradiction. Similarly, if $F(\mathbf{x})$ and $H(\mathbf{x})$ have the same phase then $F(\mathbf{x})H^*(\mathbf{x}) = F^*(\mathbf{x})H^*(\mathbf{x})$ and $C_F \subseteq C_H \cup C_{F^*}$ or $C_F = (C_F \cap C_H) \cup (C_F \cap C_{F^*})$. If C_F is irreducible then we must have that either $C_F = C_F \cap C_H$ or $C_F = C_F \cap C_{F^*}$. Assuming

$C_F \neq C_{F^*}$, then it follows $C_F \subseteq C_H$. Similarly, one can also show that $C_H = C_H \cap C_F$, thus assuming $C_H \neq C_{H^*}$ one is left with the result that $C_H \subseteq C_F$ and hence $C_F = C_H$. It can be seen from this that reconstruction from Fourier magnitude will be unique only to within entire and zero-free multiplicative factors such as those occurring in the product expansion Eq. (13.2-9). Also, reconstruction from Fourier phase will be unique only to within multiplicative zero phase factors which may not be zero-free but can be neglected on the assumption that $F(\mathbf{x})$ is irreducible. It is interesting to note that since an arbitrary multivariate polynomial of finite degree is generally irreducible [9] the uniqueness between Fourier phase and magnitude can be generally expected. Small changes to the f_{jk} are likely to lead to irreducibility of $F(\mathbf{x})$ since one can expect a neighborhood around each f_{jk} whose elements all lead to irreducible $F(\mathbf{x})$. However, small changes to data values (e.g., noisy data) are likely to lead to no solution since $|F(\mathbf{x})|^2$, a polynomial in its own right, will generally be irreducible (e.g. [44]).

13.7 ANALYTICITY AND PHASE RETRIEVAL

We have briefly described the problems associated with reconstruction from Fourier magnitude, and several chapters in this volume address this question (see also [45–47]). Conditions for the nonuniqueness of the phase retrieval problem have been described and two criteria must be satisfied. First, the zero space of F must be decomposable into the union of several subsets. Second, the subset of $\mathbb{R}^2$ containing the support of the object function must possess a suitable combination of convexity and symmetry. Barakat and Newsam [48] show that both criteria will, in general, be difficult to satisfy so that multiplicity of true solutions to the phase retrieval problem will be very rare in more than one dimension. (We neglect for now the problem of only limited Fourier data being measured [49,50].)

Despite the unlikelihood of ambiguous solutions, examples are known; for example, Huiser and van Toorn [51] constructed a function which was a two-dimensional top-hat function multiplied by a one-dimensional negative exponential. Such a function has a Fourier transform which can be factored and so is reducible. This provides at least one example for which the magnitude does not uniquely determine the phase. Other ambiguous examples can be constructed, in the discrete case, by taking a two-dimensional array and convolving it first with another array $g(m, n)$, and then with $h(m, n) = g^*(-m, -n)$. The two new arrays thus formed will have the same autocorrelation and Fourier intensity but will be different from each other even if they have the same support.

If we assume $F(\mathbf{x})$ is irreducible, then with noise-free magnitude data an obvious procedure to adopt is the factorization of $|F(\mathbf{x})|^2$ into $F(\mathbf{x})$ and $F^*(\mathbf{x})$. Indeed, if $F(\mathbf{x})$ were reducible such a procedure would yield the component factors of $F(\mathbf{x})$ and thus the phase ambiguities. We note that irreducibility of the polynomial $F(\mathbf{x})$ when a z-transform [Eq. (13.6-2)] will almost certainly imply irreducibility of the corresponding Fourier series [Eq. (13.6-1)] [43]. No algorithms exist for general factorization of multivariable polynomials; in fact, none exists even for one-dimensional polynomials, and factorization is generally recognized as a difficult problem. This reflects the fundamental mathematical fact that although there are explicit formulas for generating the roots of quadratics, cubics, and quartics, no formula is possible for quintics and higher-degree polynomials [52]. It is to be expected, then, that there will be limitations inherent in any method for trying to factor a discrete intensity function, which can be expressed as a polynomial of finite degree with arbitrary coefficients in two variables. However, some progress has been made in this direction, and a method employed based on factorization of a one-dimensional polynomial associated with the intensity and factored over a finite field reduced from that of all polynomials. The method developed deals only with real integer coefficient polynomials. This is not necessarily the limitation it might at first seem, since any representation of numbers in a digital computer will have a finite number of bits associated with it. This restricts the set of numbers which are strictly representable in a computer to the rationals because all irrational numbers are implicitly truncated to the accuracy of the machine. Multiplication by a suitable power of 10 will bring the decimal representation up to an integer, albeit a large integer.

The factorization scheme which has been developed [53] is based on a method due to Berlekamp [54] and refined by Knuth [55]. Berlekamp was interested in finding irreducible factors by an integer coefficient polynomial and showed that it is necessary first to reduce all of the coefficients of the original polynomial modulo some prime integer p, and then to factor the reduced polynomial over the Galois field $GF(p)$. [The Galois Field $GF(p)$ is isomorphic to the integers Z_p, which are just the integers $0, 1, \ldots, p$.] If this reduced polynomial factors, then certain constraints apply to the factors of the original polynomial; if the reduced polynomial does not factor over $GF(p)$, then the original polynomial is irreducible over the integers. Berlekamp's procedures, then, form the basis of the new method. However, the two-dimensional polynomial must first be made monic (that is, its leading coefficient must be 1) and then expressed as a one-dimensional polynomial related to the original two-dimensional polynomial. Berlekamp's procedure was applied to this new polynomial and factors modulo p found. Finally, a lifting scheme is needed to produce the factors of the original polynomial.

A z-transform representation can be used and the coefficients in the factorization are then the autocorrelation coefficients, since these are the coefficients of the polynomial representing the intensity FF^*.

One example is:

Autocorrelation coefficients (7×5):

$$\begin{matrix} 1 & 0 & 1 & 1 & 0 & 1 & 0 \\ 0 & 3 & 3 & 2 & 2 & 2 & 0 \\ 1 & 1 & 2 & 9 & 2 & 1 & 1 \\ 0 & 2 & 2 & 2 & 3 & 3 & 0 \\ 0 & 1 & 0 & 1 & 1 & 0 & 1 \end{matrix}$$

[where $(x_2^6 + x_2^4 + x_2^3 + x_2)x_1^4 + (3x_2^5 + 3x_2^4 + 2x_2^3 + 2x_2^2 + 2x_2)x_1^3 + \cdots$ etc., where each row gives a coefficient for each power of x_1 and each row element gives an x_2 coefficient as indicated, thus giving the autocorrelation polynomial, i.e., $F(x_1, x_2)F^*(x_1, x_2)$ or $F(x_1, x_2)F(1/x_1, 1/x_2)$.]

With factor coefficients:

$$\begin{matrix} 1 & 0 & 0 & 1 \\ 0 & 2 & 1 & 0 \\ 0 & 1 & 0 & 1 \end{matrix} \quad \text{and} \quad \begin{matrix} 1 & 0 & 1 & 0 \\ 0 & 1 & 2 & 0 \\ 1 & 0 & 0 & 1. \end{matrix}$$

If the factors themselves are factorable, then the algorithm can be reapplied to each of the factors in turn, and so on. The method thus is capable of producing all the factors from a given intensity or autocorrelation polynomial.

With a given degree n of autocorrelation polynomial and with p as the modulo, the number of computational steps is $O(n^2(\log p)^3 + n^3(\log p)^2)$. The highest coefficient in the autocorrelation polynomial also influences the running time of the program. The conversion of the original nonmonic autocorrelation polynomial to monic form leads to integer overflow problems when the polynomial has large coefficients or is of large degree. Also, the size of the modulo p increases with the size of the coefficients and this leads to slower retrieval of factors.

There is no obvious way of modifying the scheme to take account of noise on the data, and so it remains of more theoretical than practical interest in the solution of the phase problem. [The case of noise on the autocorrelation coefficients such that the intensity still factors (i.e., essentially a noisy object) is considered in [37b].] The two alterations to the scheme which might be considered are "approximate" factorization of some sort and modifying the noisy intensity so that it becomes reducible.

"Approximate" factorization would entail either taking an irreducible polynomial and from it generating the "nearest" reducible polynomial (remembering that it has to be reducible in the very special sense of reducing to FF^*), or using a factorization scheme that produces factors which, when multiplied together, give a polynomial "close" to the original. There is no precise way of defining "nearest" and "close," and this lack of precision makes it very difficult to even begin constructing such a factorization scheme.

Consequently, a method based on one-dimensional zero location was developed, and this can be considered analogous to a factorization of the intensity into the two terms F and F^*. It does not suffer from the problems associated with direct factorization and it is able to cope with noninteger (and even complex) coefficients in the intensity polynomial.

The discrete two-dimensional intensity data can be considered as a collection of one-dimensional strips, one set going in the direction of constant x_1, and an orthogonal set with constant x_2 values. Each one-dimensional strip will be a bandlimited function, with a band limit related to the band limit of the two-dimensional intensity from which it derives. This can be verified by use of the central slice theorem (Section 13.2) and for each one-dimensional intensity strip having $2P$ complex zeros there will be 2^P possible phase functions consistent with it.

Since the intensity distribution is two-dimensional, each one-dimensional strip has an orthogonal strip passing through each sample point. The values of the phase at the point of intersection must be the same for each of the two orthogonal strips, and so this further constraint can be used to determine which of the 2^P combinations is possible. This gives a maximum of 2^{2P} possible combinations to be examined for consistency, although it is explained below how this number can be considerably reduced in practice. Moreover, once the phase of a pixel has been fixed, the phases of all other pixels in that line will also be fixed. The phases of the orthogonal line passing through the given pixel will also be fixed at the same time.

The uniqueness arguments of the previous section lead to the expectation that of all the 2^{2P} possible combinations only one will yield a consistent result for the phase at the point of intersection of two orthogonal lines. Once one horizontal and one vertical line are fixed, in order to fix the remaining vertical lines it is only necessary to examine the phase at the points of intersection of each of the remaining vertical lines with the known horizontal line. This means searching 2^P possibilities only for each vertical line. Thus, in principle, the maximum number of possibilities is 2^{2P} for the first horizontal or vertical line pair, plus 2^P for each of the remaining vertical lines. Once all vertical lines are fixed by the single horizontal line, the problem is solved if the uniqueness arguments are correct, because the phases will be consistent across the whole (x_1, x_2) plane. A threshold has to be set which determines when two values for a given phase are sufficiently close to be considered equal. The value of this threshold depends on the level of noise on the data; if it is set too small then no solutions are found, while if it is too large incorrect solutions will be found.

There are some practical difficulties with this method [56], but zero symmetries and phase symmetries (e.g., for a real object) and the consistency of choice of half-plane for a zero on adjacent lines reduce the number of combinations to be searched. Typical computing times for a 32×32 intensity with 0.1% noise are 20 min on a VAX 11/780 [57,58]. If the program is allowed to generate every possible combination of initial rows and columns then *all* solutions consistent with a given intensity can be computed. This technique then permits an analysis of the degree of ambiguity implicit in the noise intensity data. The requirement of overall analyticity provides a key to dealing with noise. Noise has the effect of disturbing the bandlimited nature of each strip and displacing zeros from occurring in complex conjugate pairs. By imposing these constraints on the zeros of the strips, higher levels of noise can be tolerated.

13.8 ANALYTICITY AND PHASE UNWRAPPING

Image reconstruction has provided much of the motivation for the development of phase unwrapping algorithms. A complex-valued function may be expressed in terms of its magnitude and (principal) phase, the phase being ambiguous to within an additive contribution of $2n\pi$, n an integer. The choice of n does not affect the value of the complex function, but continuity of the phase function is frequently required. Examples include the following:

(a) Homomorphic signal processing, which performs nonlinear operations by using linear filters in the complex cepstral frequency domain of a given signal or image [59]. The cepstrum or the Fourier or z transform of the logarithm of a spectrum is clearly nonunique because of this ambiguity associated with phase values, and a continuous phase curve must be defined.

(b) Phase unwrapping algorithms are needed for inverse scattering algorithms based on the Rytov approximation (see Chapter 10 by Kaveh and Soumekh; see also [60]). The logarithm of the complex scattered field is required in order to generate the complex phase function used to represent the field.

(c) Techniques in speckle interferometry which involve calculating the phase of the Fourier transform of astronomical images need a reliable means for unwrapping the phase values prior to averaging [61]. Averaging is used to remove unwanted variations arising from atmospheric turbulence, but significant errors will occur if only the wrapped or principal phase values are averaged.

(d) A recent application is the interpretation of complex synthetic aperture radar (SAR) imagery. Such data are available, but typically only the image modulus is displayed because the wrapped phase is difficult to interpret. Targets having few scatterers per resolution cell are particularly suited for phase imaging, and it is possible to deduce height information in the limit of small but finite resolution from the degradation of positional information [34].

In one-dimensional problems several methods for phase unwrapping have been proposed based on phase tracking [62], numerical integration of the phase derivative [63], and using a Sturm sequence [64]. We have proposed a method based on the zero locations of the bandlimited function $F(x_1)$ [65]. From the Hadamard product [Eq. (13.2-7)]

$$F(z_1) = \prod_{j=1}^{\infty}(1 - z_1/\mathfrak{z}_j) = \prod_{j=1}^{\infty}(\mathfrak{z}_j - z_1)/\mathfrak{z}_j, \tag{13.8-1}$$

and so

$$\arg F(z_1) = \sum_{j=1}^{\infty}\left\{\arg(\mathfrak{z}_j - z_1) - \sum_{j=1}^{\infty}\arg(\mathfrak{z}_j)\right\}. \tag{13.8-2}$$

Thus for two points (u, v) lying on the real axis,

$$\arg F(v) - \arg F(u) = \sum_{j=1}^{\infty}\left\{\arg(\mathfrak{z}_j - v) - \arg(\mathfrak{z}_j - u)\right\}. \tag{13.8-3}$$

Equation (13.8-3) gives the unwrapped phase differences between two points on the real axis and, provided an initial phase value is known or can be assumed, the unwrapped phase can be found directly. The sum of contributions from all zeros gives the total phase difference but in practice only a finite number will be located [25]. If a zero lies on the real axis there is a choice of contribution to the unwrapped phase of $\pm\pi$. Provided zero location is sufficiently accurate and no zeros lie exactly on the real axis this procedure yields an unambiguous unwrap.

Writing $F(x_1)$ in terms of its real and imaginary parts:

$$F(x_1) = \operatorname{Re} F(x_1) + i \operatorname{Im} F(x_1).$$

The zeros of the real part are solutions of the equation $\operatorname{Re} F(x_1) = 0$. Similarly, the imaginary part goes to zero at points for which $\operatorname{Im} F(x_1) = 0$. The complex-valued function $F(x_1)$ is zero only when both $\operatorname{Re} F(x_1)$ and $\operatorname{Im} F(x_1)$ are zero. It is rare that the zeros of the real part will coincide with any of the zeros of the imaginary part; i.e., for a one-dimensional bandlimited function, nonasymptotic zeros of $F(x_1)$ are hardly ever found on the real axis.

As in one dimension, the zeros of $F(x_1, x_2)$ are points where $\operatorname{Re} F(\mathbf{x})$ and $\operatorname{Im} F(\mathbf{x})$ ($\mathbf{x} = x_1, x_2$) both go to zero. The solution to the equation $\operatorname{Re} F(\mathbf{x}) = 0$ is a set of curves on the real plane [66]. These curves may intersect one another to give multiple zeros of $\operatorname{Re} F(\mathbf{x})$. Likewise, the solution of the equation $\operatorname{Im} F(\mathbf{x}) = 0$ is also a set of curves. In general, the two sets of curves will intersect each other; the points of intersection are zeros in the intensity.

It is apparent that point zeros in $|F(\mathbf{x})|$ are likely to be a common occurrence. To prevent zeros occurring on the real plane these sets of curves must be carefully arranged so that they never intersect one another. Although it is unlikely that the zero lines of $\operatorname{Re} F(\mathbf{x})$ and $\operatorname{Im} F(\mathbf{x})$ will intersect along their whole length, it is equally unlikely that they will be able to avoid doing so at isolated points. In this sense, an analogy can be drawn between real line zeros in two dimensions and point zeros (on the real axis) in one dimension. By way of contrast, point zeros on the real plane have no analog in one dimension—a key distinction between the properties of one- and two-dimensional bandlimited functions.

For random-like objects, the trajectories of the zero lines of the real and imaginary parts on the real plane describe a series of curves wandering about the plane with no apparent structure. These lines are bound to cross at many points in the real plane (see Section 13.5). If $F(\mathbf{x})$ is irreducible but complex on the real plane, then it follows that it can go to zero on the real plane only at points. In the locality of a point zero, $F(\mathbf{x})$ may be written as

$$F(z_1, z_2) = z_1 + W(z_2) \tag{13.8-4}$$

from the Weierstrass preparation theorem [Eq. (13.5-1)]. If only first-order terms are retained and the coordinate origin is suitably rotated and made coincident with the zero, one can write

$$F(z_1, z_2) = \exp(i\alpha)(A_1 z_1 + iA_2 z_2), \tag{13.8-5}$$

where α, A_1, and A_2 are real and

$$\arg F(x_1, x_2) = \alpha + \arctan(A_2 x_2 / A_1 x_1), \tag{13.8-6}$$

from which it can be clearly seen that the phase changes by $\pm 2\pi$ for each closed circuit of this point zero. Thus, the existence of each point zero in the real plane implies an ambiguity in the phase in the neighborhood of that zero and precludes the possibility of obtaining an unambiguous phase unwrap. One can thus expect that for noisy bandlimited functions and specklelike fields, two-dimensional phase unwrapping is likely to fail.

One can reinterpret this property of the phase of two-dimensional bandlimited functions in terms of complex wave fields whose wave front is defined as a surface of constant phase. Where the amplitude of the wave is zero at a point, the phase is indeterminate. In moving around that point the phase changes by 2π. The physical manifestation of this is a wave front dislocation [36]. It is thus possible to find experimentally interference patterns at which a fringe emerges from or ends at zero values in the field's magnitude. This structure is called an edge dislocation by analogy with edge dislocations in crystals. It is a useful analogy since the wave front dislocation has the equivalent of both a Burger's vector and a Burger's circuit [67].

More precise statements of the density of dislocations can be made for random fields. In particular, the dislocation density M in a specklefield with Gaussian statistics is given by [68]

$$M = (k^2/2\pi)(\langle\Theta x_1^2\rangle\langle\Theta x_2^2\rangle)^{1/2},$$

where $k = 2\pi/\lambda$, and $\langle\Theta x_1^2\rangle$, $\langle\Theta x_2^2\rangle$ characterize the beam divergence in x_1 and x_2, respectively.

In other words, M is of the order of the speckle density, $(\Theta/\lambda)^2$, where Θ is the field divergence. This result has recently been confirmed experimentally [68]. Speckle fields, characteristic of coherent systems such as SAR, are therefore expected to possess as many wave front dislocations as there are speckles. If the dislocations could be removed without loss of information, phase unwrapping would then become feasible. By analogy with crystal dislocations it may be possible to remove them by causing them to slip or climb off the real plane or at least away from the region of interest. The use of a large reference point in the original object assists with this [35].

Finally, zeros in the Fourier modulus may prevent iterative algorithms such as Fienup's (see Chapter 7, this volume) from converging. Dislocations in the phase function which are incorrectly positioned at the nth iteration may cause stagnation. Often it is found that the reconstruction contains a stripe artifact originating from harmonics not present in the true object but introduced by the incorrect placement of zeros between sampled points. A method whereby several different iterative solutions are combined to eliminate incorrectly placed zeros can significantly improve the fidelity of the reconstructed image [69]. The result supports the suggestion that dislocations can cause problems for iterative techniques for reconstruction from Fourier magnitude.

Furthermore, methods based on the imposition of phase closure in two dimensions [70] will not always work when the intensity goes to zero at isolated points corresponding to a complex factor. The contour integral of the phase derivative around a closed path is not expected to vanish if the path encloses a single point zero. Therefore, closure of the phase is not always an appropriate choice of constraint.

13.9 SUMMARY AND CONCLUSIONS

A brief description has been given of the properties of bandlimited functions in one and more than one dimension. The factorization of these functions provides considerable insight into the problems of image recovery from partial Fourier data such as reconstruction from Fourier magnitude and phase. To date, these analytic properties have been of more use indirectly than directly. They provide a description of how and why certain procedures succeed (e.g. [71]) and enable questions of uniqueness to be addressed. In the case of image recovery from Fourier magnitude, two methods were described which explicitly exploit the analyticity of the Fourier data. Considerable numerical effort is required to use them, however, and more work is required to overcome problems of noise and limited samples. Ideally one should make use of the expected uniqueness and the known constraints in relationships between the real and imaginary parts of bandlimited functions in a computationally more effective way. Recent work, [72,73] suggests that it may be computationally feasible to determine the location in $\mathbb{R}^4$ of the zero surfaces of a two dimensional bandlimited function. Thus, for image recovery from Fourier magnitude one could distinguish the zero surfaces of F and F^* and then the functions themselves. Similarly, image recovery from a multiply convolved function becomes possible, in principle, if two dimensional bandlimited functions are generally irreducible and the zero surfaces can be reasonably well located and identified with separate factors. The fact that iterative procedures, such as those described in earlier chapters, can be made to work despite there being no uniqueness proof suggests that there is scope for bringing these two directions of approach together.

The physical manifestation of the analytic nature of wave fields was discussed and the high probability of wave front dislocations in two or more dimensions described. The existence of these features in speckle patterns or images necessitates a deeper understanding of the intrinsic characteristics to be expected of entire functions of several complex variables. The results of an investigation of this kind can suggest new procedures for image recovery and coding, new methods for overcoming present difficulties (as exist with phase unwrapping, for example), and stimulate new applications. An example of the latter is the use of phase filters in an optical processor for switching beams between arrays whose geometric distributions are members of a homometric set [74,75], i.e., object distributions having the same Fourier magnitude but different Fourier phases.

ACKNOWLEDGMENTS

The author thanks his colleagues H. V. Deighton, H. M. Powell (née Berenyi), and M. S. Scivier for their contributions to work presented here. He is also grateful to the Mathematics Department at The Catholic University of America for hospitality, to its chairman, C. L. Byrne, for many stimulating discussions, and to ONR grant N00014-85-K-0402 for financial support.

REFERENCES

1. E. Wolf and M. Nieto-Vesperinas (1985). Analyticity of the angular spectrum amplitude of scattered fields and some of its consequences. *J.O.S.A.* **A2**, 886–890.
2. M. Plancherel and G. Pólya (1937, 1938). Fonctions entiéres et intégrales de Fourier multiples. I, II. *Comment. Math. Helv.* **9**, 224–248; **10**, 110–163.
3. B. A. Fuks (1963). Introduction to the theory of analytic functions of several complex variables. *A.M.S. Translations of Mathematical Monographs*, **8**, Providence, Rhode Island.
4. R. P. Boas (1954). "Entire Functions," Academic Press, New York.
5. E. C. Titchmarsh (1925). On the zeros of certain integral functions. *Proc. Lond. Math. Soc.* **25**(2), 283-297.
6. M. A. Fiddy, G. Ross, M. Nieto-Vesperinas, and A. M. J. Huiser (1982). Encoding of information in inverse optical problems. *Opt. Acta* **29**, 23–40.
7. B. J. Hoenders (1975). On the solution of the phase retrieval problem. *J. Math. Phys.* **16**, 1719–1725.
8. W. F. Osgood (1929). "Lehrbuch der Funktionentheorie," **2**, Teubner, Leipzig.
9. M. H. Hayes and J. H. McClellan (1982). Reducible polynomials in more than one variable. *Proc. IEEE* **70**, 197–198.
10. P. J. Napier and R. H. T. Bates (1974). Inferring phase information from modulus information in two dimensional aperture synthesis. *Astron. Astrophys. Suppl.* **15**, 427–432.
11. W. Lawton (1981). Uniqueness results for the phase retrieval problem for radial functions. *J.O.S.A.* **71**, 1519–1522.
12. G. F. Chew (1966). "The analytic *S* matrix," Benjamin, New York.
13. G. Ross and M. A. Fiddy (1978). The speckle effect: A reappraisal. *Opt. Act* **25**, 205–217.
14. H. M. Nussenzveig (1972). "Causality and Dispersion Relations," Academic Press, London.
15. E. C. Titchmarsh (1932). "Theory of Functions," Oxford University Press.
16. M. Nieto-Vesperinas (1980). Dispersion relations in two dimensions: Application to the phase problem. *Optik* **56**, 377–384.
17. R. E. Burge, M. A. Fiddy, A. H. Greenaway, and G. Ross (1976). The phase problem. *Proc. Roy. Soc. Lond.* **A350**, 191–212.
18. R. M. Young (1980). "An Introduction to Nonharmonic Fourier Series," Academic Press, New York.
19. J. A. Bucklew and B. E. A. Saleh (1985). Theorem for high resolution high contrast image synthesis. *J.O.S.A.* **A2**, 1233–1236.
20. E. I. Plotkin, L. M. Roytman, and M. N. S. Swamy (1984). Reconstruction of nonuniformly sampled band-limited signals and jitter error reduction. *Signal Processing* **7**, 151–160.
21. M. A. Fiddy and T. J. Hall (1981). Nonuniqueness of superresolution techniques applied to sampled data. *J.O.S.A. letters* **71**, 1406–1407.
22. A. Papoulis (1981). "Signal Analysis," McGraw-Hill, Kogakusha.
23. C. L. Byrne, R. M. Fitzgerald, M. A. Fiddy, T. J. Hall, and A. M. Darling (1983). Image restoration and resolution enhancement. *J.O.S.A.* **73**, 1481–1487.

24. A. M. Darling, T. J. Hall, and M. A. Fiddy (1983). Stable, non-iterative, object reconstruction from incomplete data using prior knowledge. *J.O.S.A.* **73**, 1466–1469.
25. J. W. Wood, T. J. Hall, and M. A. Fiddy (1983). A comparison study of some computational methods for locating the zeros of entire functions. *Opt. Acta.* **30**, 511–527.
26. R. C. Gunning and H. Rossi (1965). "Analytic functions of several complex variables," Prentice Hall, New Jersey.
27. H. B. Voelcker and A. A. G. Requicha (1973). Clipping and signal determinism: Two algorithms requiring validation. *IEEE* **COM-21**, 738–744.
28. S. R. Curtis, A. V. Oppenheim, and J. S. Lim (1985). Reconstruction of two-dimensional signals from threshold crossing. **ICASSP-85**, 1057–1060.
29. A. A. Requicha (1980). The zeros of entire functions: Theory and engineering applications. *Proc. IEEE* **68**, 308–328.
30. Y. Y. Zeevi and D. Rotem (1986). Image reconstruction from zero crossings. *IEEE ASSP* (in press).
31. R. E. Burge, M. A. Fiddy, T. J. Hall, and G. Ross (1980). Optical phase conjugation for the study of the analytic properties of optical fields. *AIP Conference Proc.* **65**, 31–41.
32. P. L. Van Hove, M. H. Hayes, J. S. Lim, and A. V. Oppenheim (1983). Signal reconstruction from signed Fourier transform magnitude. *IEEE* **ASSP-31**, 1286–1293.
33. B. F. Logan (1977). Information in the zero crossings of bandpass signals. *Bell. Sys. Tech. J.* **56**, 487–510.
34. M. S. Scivier (1985). The interpretation of complex synthetic aperture radar imagery. Ph.D. Thesis, University of London.
35. M. A. Fiddy, B. J. Brames, and J. C. Dainty (1983). Enforcing irreducibility for phase retrieval in two dimensions. *Opt. Lett.* **8**, 96–98.
36. M. S. Scivier and M. A. Fiddy (1985). Phase ambiguities and the zeros of multidimensional band-limited functions. *J.O.S.A.* **A2**, 693–697.
37a. M. Nieto-Vesperinas and J. C. Dainty (1985). A note of Eisenstein's irreducibility criterion for two dimensional sampled objects. *Opt. Comm.* **52**, 94–98.
37b. M. Nieto-Vesperinas and J. C. Dainty (1986). Phase recovery for two-dimensional digital objects by polynomial factorization. *Opt. Comm.* **58**, 83–88.
38. B. J. Brames (1985). Uniqueness and other aspects of the optical phase problem. Ph.D. Thesis, University of Rochester.
39. B. Grünbaum (1967). "Convex Polytopes," Wiley and Sons, London.
40. Y. M. Bruck and L. G. Sodin (1979). On the ambiguity of the image reconstruction problem. *Opt. Comm.* **30**, 304–308.
41. A. Walther (1963). The question of phase retrieval in optics. *Optica Acta* **10**, 41–49.
42. E. Hofstetter (1964). Construction of time-limited functions with specified autocorrelation functions. *IEEE* **IT-10**, 119–126.
43. J. L. C. Sanz and T. S. Huang (1983). Unique reconstruction of a bandlimited multi-dimensional signal from its phase or magnitude. *J.O.S.A.* **73**, 1446–1450.
44. J. L. C. Sanz and T. S. Huang (1985). Polynomial system of equations and its applications to the study of the effect of noise on multidimensional Fourier transform phase retrieval from magnitude, *IEEE* **ASSP-33**, 997–1004.
45. M. A. Fiddy (1983). The phase retrieval problem. *Proc. S.P.I.E.* **413**, 176–181.
46. J. Rosenblatt (1984). Phase Retrieval. *Comm. in Math. Phys.* **95**, 317–343.
47. J. L. C. Sanz (1985). Mathematical considerations for the problem of Fourier transform phase retrieval from magnitude. *SIAM J. Appl. Math.* **45**, 651–664. See also I. Sabba Stefanescu (1985). On the phase retrieval problem in two dimensions, *J. Math. Phys.* **26**, 2141–2160 and R. Barakat and G. Newsam (1985). Algorithms for reconstruction of partially known, band-limited Fourier transform pairs from noisy data. *J.O.S.A.* **A2**, 2027–2039.

48. R. Barakat and G. Newsam (1984). Necessary conditions for a unique solution to two-dimensional phase recovery. *J. Math. Phys.* **25**, 3190–3193.
49. A. M. Darling, H. V. Deighton, and M. A. Fiddy (1983). Phase ambiguities in more than one dimension. *Proc. S.P.I.E.* **413**, 197–201.
50. M. S. Scivier and M. A. Fiddy (1985). Ambiguities in magnitude only reconstruction of band-limited signals. *Opt. Lett.* **10**, 369–371.
51. A. M. J. Huiser and P. van Toorn (1980). Ambiguity of the phase reconstruction problem. *Opt. Lett.* **5**, 499–501.
52. R. B. J. T. Allenby (1983). "Rings, Fields and Groups: An Introduction to Abstract Algebra," Edward Arnold, London.
53. H. M. Berenyi, H. V. Deighton, and M. A. Fiddy (1985). The use of bivariate polynomial factorization algorithms in two-dimensional phase problems. *Opt. Acta* **32**, 689–701.
54. E. R. Berlekamp (1967). Factoring polynomials over finite fields, *Bell. Sys. Tech. J.* **46**, 1853–1859.
55. D. E. Knuth (1981). "The Art of Computer Programming." Vol. 2, Addison-Wesley, New York.
56. H. V. Deighton (1985). Theoretical aspects and computational studies of image reconstruction from limited (power) spectral data. Ph.D. Thesis, University of London.
57. H. V. Deighton, M. S. Scivier, and M. A. Fiddy (1985). Solution of the two dimensional phase-retrieval problem. *Opt. Lett.* **10**, 250–251.
58. H. V. Deighton, M. S. Scivier, H. M. Berenyi, and M. A. Fiddy (1985). Practical phase retrieval based on theoretical models for multi-dimensional band-limited signals. *Proc. S.P.I.E.* **558**, 65–72.
59. A. V. Oppenheim and R. W. Schafer (1975). "Digital Signal Processing," Prentice-Hall, Englewood Cliffs, New Jersey.
60. S. Leeman, M. A. Fiddy, and L. Zapalowski (1985). Born vs. Rytov: Is the debate over? *Proc. S.P.I.E.* **558**; also M. A. Fiddy (1986) Inversion of optical Scattered field data. *J. Phys. D.* **19**, 301–317.
61. B. T. O'Connor and T. S. Huang (1981). Application of phase unwrapping to image restoration. *Comp. Graphics and Image Processing* **15**, 25–44.
62. K. Itoh (1982). Analysis of the phase unwrapping algorithm. *App. Opt.* **21**, 2470.
63. J. M. Tribolet (1977). A new phase unwrapping algorithm. *IEEE* **ASSP-25**, 170–177.
64. R. McGowan and R. Kuc (1982). A direct relation between signal time series and its unwrapped phase. *IEEE* **ASSP-30**, 719–728.
65. M. S. Scivier, T. J. Hall, and M. A. Fiddy (1984). Phase unwrapping using the complex zeros of a bandlimited function and the presence of ambiguities in two dimensions. *Opt. Acta* **31**, 619–623.
66. N. B. Baranova and B. Ya. Zel'dovich (1982). Dislocations of the wavefront surface and zero of the amplitude. *Sov. Phys. JETP* **53**, 925–929.
67. J. F. Nye and M. V. Berry (1974). Dislocations in wave trains. *Proc. R. Soc. Lond.* **A336**, 165–190.
68. N. B. Baranova, A. V. Mamaev, N. F. Pilipetskii, V. V. Shkunov, and B. Ya. Zel'dovich (1983). Wavefront dislocations. *J.O.S.A.* **73**, 525–528 (see also *Sov. Phys. J.E.T.P.* **56**, 983–988 (1982)).
69. J. R. Fienup (1985) (private communication).
70. W. R. Fright and R. H. T. Bates (1982). Fourier phase problems are uniquely solvable in more than one dimension: Computational examples for two dimensions. *Optik* **62**, 219–230.
71. H. Bartelt, A. W. Lohmann, and B. Wirnitzer (1984). Phase and amplitude recovery from bispectra. *Appl. Opt.* **23**, 3121–3129.

72. D. Israelevitz and J. S. Lim (1986). A new algorithm for closed form image reconstruction from Fourier transform magnitude, O.S.A. Topical Meeting on Signal Recovery & Synthesis II, Honolulu, April 2–4, ThB3-1-4.
73. R. G. Lane, W. R. Fright, and R. H. T. Bates, Direct phase retrieval, submitted to *IEEE* ASSP.
74. J. Rosenblatt and P. D. Seymour (1982). The structure of homometric sets. *SIAM J. Alg. Disc. Meth.* **3**, 343–350.
75. H. M. Berenyi and M. A. Fiddy (1986). Application of homometric sets for beam manipulation, *J.O.S.A.* **A3**, 373–375.

Index

G

H

I

K

L

M

N

O

P

R

T

U

V

W

Z